Catenary Optics

Xiangang Luo

Catenary Optics

 Springer

Xiangang Luo
State Key Laboratory of Optical
Technologies on Nano-fabrication
and Micro-engineering
Institute of Optics and Electronics
Chinese Academy of Sciences
Chengdu, Sichuan, China

ISBN 978-981-13-4817-4   ISBN 978-981-13-4818-1   (eBook)
https://doi.org/10.1007/978-981-13-4818-1

Library of Congress Control Number: 2018965891

This Springer imprint is published by the registered company Springer Nature Singapore Pte Ltd.
The registered company address is: 152 Beach Road, #21-01/04 Gateway East, Singapore 189721,
Singapore

Foreword

A catenary is a curve formed by a wire or chain hanging freely under its weight from two points, which was thought as a true mathematical and mechanical form in architecture by Robert Hooke in the 1670s. At first glance, it seems that this curve has little relationship with optics, since it is a concept originated in mechanics and has been widely exploited in architectures and railway electrification systems for centuries. Surprisingly, in some recent demonstrations, the mechanical properties of catenary have been utilized to enable many optical applications such as large solar collectors, geodesic antennas, and membrane lenses for space applications. However, the research area of catenary optics has not formed until the paper entitled *Catenary Optics for Achromatic Generation of Perfect Optical Angular Momentum* published in *Science Advances* in 2015.

Professor Xiangang Luo is an eminent researcher in the areas including engineering optics, micro/nano-optics, plasmonics, and subwavelength electromagnetics. During the last two decades, he has made many important contributions to the fundamental theories, designs, and applications of optical subwavelength structures. Based on the insight into the inner physics and design strategies, he proposed the concept of subwavelength catenary optics, which is the main topic of this book.

In addition to the history and background, this book gives a comprehensive discussion on the catenary functions in the subwavelength scale, i.e., a scale where the characteristic dimension is much smaller than an operating wavelength. Two kinds of catenary functions are involved. First, subwavelength structures defined by a catenary of equal strength (proposed by Davies Gilbert in 1826) are introduced to realize a broadband and continuous geometric phase, forming the basis of a large variety of functional metasurfaces including orbital angular momentum generators and flat lenses. Further, it has been shown that the counter-propagating evanescent waves in coupled subwavelength structures form catenary-shaped optical fields, leading to a giant local field enhancement, a tight focal spot below the diffraction limit, as well as a tunable phase shift and absorption depending on the gap width.

Based on the novel optical properties, it is believed that catenary optics may bring disruptive applications in engineering optics. It is well known that modern optics is based on the laws of reflection, refraction, and diffraction of light waves.

In geometric optics, by applying Fermat's principle and Snell's law, it has been shown that a light ray in a linearly gradient medium follows a catenary curve. Similarly, by controlling the space-variant phase shift generated by the catenary structures, generalized laws of reflection and refraction can be defined, making arbitrary refraction and reflection on a flat surface possible. Meanwhile, by utilizing the high spatial frequency information carried by the catenary optical fields, the imaging resolution could be much higher than the classical diffraction limit. In many additional cases, the catenary optical fields can be described using impedances characterized by a catenary of equal strength. Thus, fast and accurate predictions of optical performances of complex subwavelength structures can be enabled, which is a big breakthrough in the metamaterial and metasurface research field.

This book may serve as a useful and comprehensive reference book for scientists, engineers, and students working on related areas such as subwavelength optics, nanophotonics, and metamaterials. Along with the involvement of intelligences around the world, many new discoveries and revolutionary applications in related areas shall be expected.

Melbourne, Australia   Min Gu
                                                                   RMIT University
                                                         Fellow of the Australian
                                             Academy of Science and the Australian
                                   Academy of Technological Sciences and Engineering
                               Foreign Fellow of the Chinese Academy of Engineering
                                             AIP, OSA, SPIE, InstP, IEEE Fellow

Preface

In 2013, in an accidental experiment involving the interference of light passing from a small aperture, we found that a curved nanoslit in the form of "catenary of equal strength" could impart nearly perfect linear phase shift to a circularly polarized light beam. Because the mathematical similarity, we termed this type of optical structure "catenary of equal phase gradient". Very soon, we realized that the Pancharatnam–Berry phase plays a crucial role during the interaction between light and this special structure. For catenary nanoslits, the properties attracting us most are the significantly improved bandwidth and efficiency compared with previous proposed phase-modulating subwavelength structures. So we started to build different kinds of planar devices with catenary structures, which further confirmed their superior performances. These results were then published in *Science Advances* in 2015 and did attract tremendous interest. In fact, as early as in 2003, I have observed that the hyperbolic catenary function is an intrinsic characteristic of the coupled surface plasmons in layered noble metals. Owing to the so-called catenary optical fields, it is possible to control the high spatial components of light fields, providing an indispensable approach to break the notorious diffraction limit in optical imaging. Subsequently, I paid more attention to this unique form and found that the catenary has multidisciplinary meanings in science, technology, and art in the history (it was thought to be a true mathematical and mechanical form in architecture by Robert Hooke in the 1670s). After searching for the optical performances of a wealth of catenary functions, I expected that the catenary optics would form an important research area in the future.

In this book, I would like to give a systematical discussion of the physics and applications of catenary functions in optics. Besides the catenary of equal phase gradient, I also show that the catenary optical field featured by hyperbolic cosine function is an eigensolution of Maxwell's wave equation in subwavelength structures. Unlike the sine and cosine functions widely utilized in traditional wave optics, the hyperbolic function corresponds to evanescent waves and high spatial frequency components which are often ignored. Consequently, the catenary optical fields are the key to realize sub-diffraction-limited optical imaging and vectorial electromagnetic modulation. For instance, the catenary plasmons in metal–insulator–metal configurations are essential for the plasmonic nanolithography, flat optics, and perfect absorbers. Due to the outstanding optical performances and widespread applications of catenary structures and catenary optical fields, I think it is of great significance to write a comprehensive book on this topic.

There are many people who helped me a lot during the course of writing this book. I deeply thank Prof. Xicheng Lu, Prof. Bingkun Zhou, Prof. Zuyan Xu, Prof. Jishen Li, Prof. Hequan Wu, Prof. Yueguang Lv, and Prof. Guozhen Yang for their help. In the revision of this book, my students have also made important contributions and I would like to express my gratitude to them.

Chengdu, China   Xiangang Luo
2018

Contents

Chapter 1
Introduction

Abstract Catenary optics is a newly emerging branch in optics and nanophotonics, which focuses on the applications of catenary functions in optical and electromagnetic devices. In a more general sense, it may be called catenary electromagnetics to highlight the electromagnetic nature of light. This book is devoted to the physics and applications of catenary optics and catenary electromagnetics. Section 1.1 gives a brief description of the developing history of catenary optics. Section 1.2 describes typical examples which illustrate the universal relation between catenary and optics. Section 1.3 discusses some common misconceptions related to catenary function. Section 1.4 is an overview of this book.

Keywords Catenary electromagnetics · Catenary optical fields · Catenary dispersion · Catenary metasurface

1.1 Concepts and Brief History

The catenary is a curve formed by a flexible wire, rope, or chain hanging freely from two separate points. Under equilibrium, the tension and gravity force would cancel in both the horizontal and vertical directions. As shown in Fig. 1.1, the catenary function can be found in many cases such as the spider webs, glacial valleys, capillary bridges, electricity wires, necklaces, sails, and arches [1, 2].

The English word catenary was derived from its Latin name *catēna*, which means "a connected series of chain." Note that the rope, chain, or wire must be flexible to make sure that there is no shearing force. Also, the rope as well as the gravitational field must be homogeneous to get ideal catenary curves. If these conditions are not strictly met, the shape would deviate from the pure mathematical functions.

Perhaps, the first recorded mathematic consideration of the catenary curve belongs to the Italian painter, scientist, and engineer—Leonardo da Vinci (1452–1519). When he was drawing the *Lady with an ermine* in 1489–1490 (Fig. 1.2), he wondered what is the curve formed by the hanging necklace. Unfortunately, he did not find the answer across his life. Some years later, the Italian physicist Galileo Galilei (1564–1642) wrongly treated the curve of a hanging chain as a parabola. Galileo stated that a

© Springer Nature Singapore Pte Ltd. 2019
X. Luo, *Catenary Optics*, https://doi.org/10.1007/978-981-13-4818-1_1

Fig. 1.1 Catenary curves in natural and artificial structures. **a** Spider webs. **b** Freely hanging transmission line. **c** The Gateway Arch. **d** Glacial valleys. **a–c** Reproduced from Wikipedia [1]. **d** Reproduced from Wikipedia [3]

hanging cord is an approximate parabola, and this approximation improves as the curvature becomes smaller.

It is interesting to note that almost one thousand years before da Vinci, many Chinese drawings have captured the catenary curves formed by either necklace or ropes. Yan Liben (c. 600–673), one celebrated Chinese architect, painter, and politician of the early Tang Dynasty, has drawn the catenary in the *Thirteen Emperors Scroll*. As illustrated in Fig. 1.3 (top left), the rope hanging from the headgear (named "Tianhe Strip") of Emperor Wu of Jin is a catenary. The same rope can be found in the drawing of Bodhisattva discovered in the Mogao cave (top right).

Another famous Chinese drawing featuring catenary curves is the *Court Ladies Wearing Flowered Headdresses* drawn by Zhou Fang in the eighth century. As depicted in the bottom of Fig. 1.3, besides the necklaces, the shape of scarves is also ruled by the catenary. Of course, although these Chinese drawings are much earlier than the *Lady with an Ermine*, there is no evidence that these drawers had thought about the mathematical form of these curves. Even if they did, it was almost impossible for them to get the accurate answer, because the necessary mathematics did not emerge yet.

Fig. 1.2 Lady with an
Ermine. Reproduced from
Wikipedia [4]

The fact that the catenary curve is not a parabola was proven by Joachim Jungius (1587–1657) and Christiaan Huygens (1629–1695). In independent works, the rigorous mathematic equation of catenary, i.e., $y = a\cosh(x/a)$ was derived in 1691 by Gottfried Leibniz (1646–1716), Christiaan Huygens, and Johann Bernoulli (1667–1748) in response to a challenging problem proposed by Jakob Bernoulli (1654–1705, the brother of Johann). It is now currently widely known that the catenary problem can be easily solved using variational methods, as highlighted by the brachistochrone curve, i.e., a curve between two points along which a body can move in a shorter time than for any other curve, proposed by Johann Bernoulli in 1696. A detailed deduction of the shape formed by freely hanging chains can be found in Appendix A of this book.

Note that the catenary curve, i.e., the hyperbolic cosine function, is defined with a new irrational number e ≈ 2.71828, without which it is impossible to get the catenary function. Historically, the first reference to this constant was published in 1618 in the table of an appendix of a work on logarithms by John Napier. However, this did not contain the constant itself, but simply a list of logarithms calculated from the constant. The discovery of the constant itself is credited to Jacob Bernoulli in 1683, who attempted to find the value of the following expression $(1 + 1/n)^n$ when n is increased to be infinitely large. The first recorded use of the constant, represented by the letter

Fig. 1.3 Ancient Chinese drawings featuring catenary curves. Top left: Emperor Wu of Jin drawn by Yan Liben. Top right: Bodhisattva leading the way, discovered in Mogao cave (Circa 875). Bottom: Court ladies wearing flowered headdresses drawn by Zhou Fang (8th century). Reproduced from Wikipedia [5–7]

b, was in correspondence from Gottfried Leibniz to Christiaan Huygens in 1690 and 1691. This is actually when the catenary function is discovered. Consequently, the solution of catenary problem is enabled by the emergence of calculus as well as the Napier constant. Some years later, Leonhard Euler (1707–1783) introduced the letter *e* as the base for natural logarithms in 1731, and thus *e* is also called Euler's number. Catenary of equal strength, i.e., $y = c\ln(|\sec(x/c)|)$, is another important catenary function given by Davies Gilbert in 1826 [8, 9]. Different from the normal catenary, the thickness of the cable (or chain) is thickened according to the extension, and thus its resistance to breaking is constant along its length. Once again, this catenary is associated with the Euler's number, since it is defined with a logarithm and triangular function.

The connection of Euler's number has inspired many imaginations. For instance, the French entomologist Jean Henri Fabre has written in his book *The Life of the Spider* (translated from the French book *Souvenirs Entomologiques*): "With this weird number (Euler's number) are we now stationed within the strictly defined realm of the imagination? Not at all: the catenary appears actually every time that weight and flexibility act in concert. The name is given to the curve formed by a chain suspended by two of its points which are not placed on a vertical line. It is the shape taken by a flexible cord when held at each end and relaxed; it is the line that governs the shape of a sail bellying in the wind; it is the curve of the nanny-goat's milk bag when she returns from filling her trailing udder. And all this answers to the number *e*."

Figure 1.4 illustrates the differences between the normal parabolic function and the two catenary functions. Obviously, when *x* is small, the differences between them are not significant. For large *x*, however, the catenary of equal strength grows the fastest. It should be noted that the asymptotes of catenary of equal strength are perpendicular to the *x*-axis ($x = \pm\pi/2$).

The application of catenary shape to mechanical problem, in particular, the construction of arches, is attributed to the English scientist Robert Hooke (1635–1703). In 1675, Hooke published an encrypted solution as a Latin anagram in an appendix to his *Description of Helioscopes*, where he wrote that he had found "a true mathematical and mechanical form of all manner of Arches for Building." Nevertheless, the anagram, namely "abccc ddeeeee f gg iiiiiiii ll mmmm nnnnn oo p rr sss tttttt uuuuuuuu x," was a secret. In 1705, the unencrypted solution was provided by his executor as "Ut pendet continuum flexile, sic stabit contiguum rigidum inversum," which translates to "As hangs a flexible cable so, inverted, stand the touching pieces of an arch." This means that the catenary shape is an optimal choice for building an arch. Figure 1.5 is a supposed portrait of Hooke, with a catenary chain in his hands.

One of the most famous arches in the world is the Gateway Arch, which was built in 1963–1965 along the west bank of the Mississippi River in St. Louis, Missouri. As shown in Fig. 1.1c, the stainless steel arch is 630 feet (192 m) wide and 630 feet (192 m) high. By measuring in foot, the actual function is $y = -127.7\cosh(x/127.7) + 757.7$. Interesting, the first use of the English word "catenary" recorded in the Oxford English Dictionary is by the Thomas Jefferson, in a letter dated December

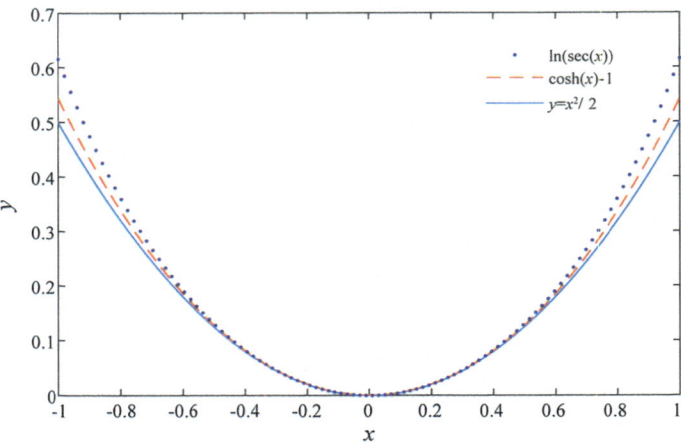

Fig. 1.4 A comparison of the parabolic function, hyperbolic cosine function, and catenary of equal strength. Note that the last one has larger gradient

Fig. 1.5 A memorial portrait of Robert Hooke presented to the Department of Engineering Science at Oxford University in 2009. It shows Hooke in Oxford with important work—a wheel barometer, his microscope and copy of "Micrographia", a pocket watch, optics, spring, and universal joint. He is holding a chain to make a catenary arch. Behind him is a map of the city of London which he helped to survey and rebuild after the Great Fire of 1666. Reproduced from Wikipedia [10]

Fig. 1.6 Top left: Anlan Bridge in Dujiangyan City. Top right: the Luding Bridge. Bottom: the Golden Gate Bridge. Reproduced from Wikipedia [1, 13, 14]

23, 1788 to Thomas Paine, recommending the use of a catenary arch rather than a circular one for a 400-foot span iron bridge that Paine is proposing to build [1]:

"I have lately received from Italy a treatise on the equilibrium of arches, by the Abbé Mascheroni. It appears to be a very scientifical work. I have not yet had time to engage in it; but I find that the conclusions of his demonstrations are, that every part of the catenary is in perfect equilibrium."

As an architect, Jefferson would be pleased to know that the principle of the catenary has formed the core of the largest monument in America in the twentieth century, in honor of Jefferson himself. In fact, the Gateway Arch was designed for the "Jefferson National Expansion Memorial," dedicated to Thomas Jefferson.

Like freely hanging chains, the suspension bridges are essentially thickened cables, and follow catenary curves. Perhaps, the earliest suspension bridge is the Anlan Bridge in Dujiangyan, Chengdu, China [11]. According to the historical records, it was first constructed in about 300 B.C., and then destroyed and rebuilt for many times. As shown in the top left panel in Fig. 1.6, the Anlan bridge spans the 1,000-ft-wide Min River, using pillars that support eight sections of cable to do so. In the year 761, when the bridge was rebuilt, the great Chinese poet Du Fu has written a poem [12]: When cutting bamboo to make a bridge, the construction is the same; not lifting one's robes to ford now, people go back and forth. In cold weather, the white cranes came back to the bridge pillars; At the sun sets, the green dragon could be seen within the water. Here, the bridge pillars are essential to support the catenary suspension bridges, while the inverted images of the catenary look like a green dragon (the reflection image can be seen in the top left of Fig. 1.6).

Besides the Anlan bridge, the Luding bridge is another old suspension bridge located in Sichuan Province, China. As an iron chain structure built during the Qing Dynasty and the reign of Emperor Kangxi (1706), it spans 100 m across the Dadu River (see the top right of Fig. 1.6). It was also known for the calligraphy by Kangxi and the battle during the Long March in 1935. The recorded first iron catenary bridge is the Shenchuan iron bridge built over the Jinsha River (also known as Shenchuan) in circa A.D. 680. This bridge was ruined in A.D. 794 when Nanzhao realigned with the Tang Dynasty in a war with the Tibetan Empire. In contrast to the case in China, the first iron chain suspension bridge in the western countries was built one thousand years later in A.D. 1741 over the Tees near Middleton [9].

Note that if the weight of the roadway per unit length is much larger than the weight of the cable and the wire supporting the bridge, the cable will follow a parabola, like the Golden Gate Bridge shown in the bottom of Fig. 1.6. If the weight of the cable and supporting wires are not negligible, then the analysis and curve shape may become more complex [1].

In above examples, the catenary is related to flexible but solid structures. It should be noted that liquid material such as water could also behave as catenary in some cases such as capillary effects [15–17]. Capillary bridge is a minimized surface of liquid or membrane created between two rigid bodies with an arbitrary shape. For the capillary bridge between two spheres shown in Fig. 1.7, the shape of the surface is catenoid, which is generated by rotating a catenary around its axis. Note that the liquid surface in a cylindrical tube is called meniscus, a word coming from Greek for "crescent". In other words, it was thought as part of a circle. In rigorous mechanical analysis, the shape of meniscus is formed so that the pressure difference across it counters the force of gravity. In general, the shape would not be described by an elementary function, but can be ruled by a differential equation. For thin liquid bridge, the shape becomes a true spherical surface.

The application of catenary shape in optics is a rather new topic. In 1978, the catenary-shaped cylindrical reflector was suggested to concentrate the sun's rays [19]. Compared with common parabolic-shaped mirrors, the catenary reflectors are able to maintain relatively higher performance at much larger incident angles, while the construction cost is greatly reduced, especially for large-size concentrators. Figure 1.8 illustrates a simple catenary reflector formed by hanging flexible and polished sheet steel on two posts. Since light intensity can be greatly increased at the focus, these inexpensive catenary reflectors hold promise as a way of getting more electric power out of smaller photovoltaic panels.

In 1986, Evans and Rosenquist found that the catenary curve is that of light propagated in a linear gradient indexed medium, such as that in road surface mirage [22]. Based on Fermat's principle, they claimed that they found a deeper relation between optics and mechanics. In fact, the variational method is the key to understand both the Fermat's principle and the catenary. Still in 1986, Coleman proposed a catenary theory of accommodation of eye, which well explained the interaction of mechanics and optics in organisms [21, 23]. As illustrated in Fig. 1.9, the lens shape of human eye can be tuned by the suspensory ligament, which acts a flexible string.

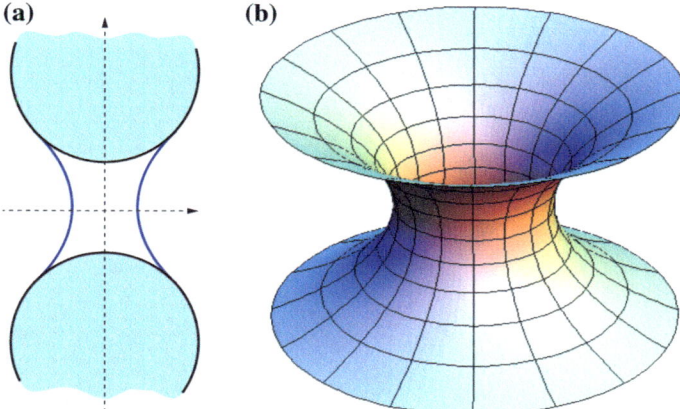

Fig. 1.7 Capillary effects. **a** Capillary bridge between two spherical particles. **b** 3D view of a catenoid. Reproduced from Wikipedia [18]

Fig. 1.8 Photograph of a catenary solar cooker built with steel sheet and wood. Light is focused according to the reflection law. Image courtesy of Tho X. Bui

(a)
Suspensory ligament

lens

(b)

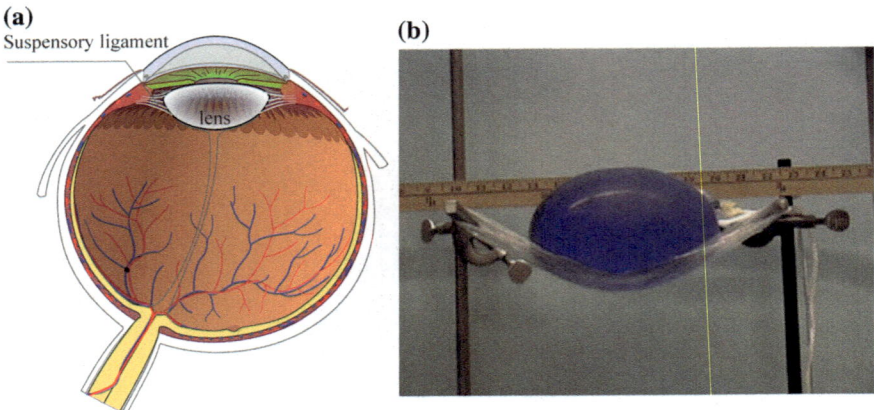

Fig. 1.9 Schematic diagram of the eye and catenary model for accommodation. When the support arms are closer, the shape of the lens becomes steeper. Left: Reproduced from Wikipedia [20]. Right: Reproduced from [21] with permission. Copyright 2014, Springer Science + Business Media New York

Five-hundred-meter Aperture Spherical Radio Telescope (FAST) is a Chinese mega-science project to build the largest single-dish radio telescope. The key to this newly proposed engineering concept is to use the equilibrium shape of the cables under gravity, catenary, to form the parabola. Consequently, the word spherical in the name seems to be not so accurate, since the spherical shape is neither the initial nor the final state [24, 25]. As shown in Fig. 1.10, the main reflector has an opening up to 500 m in diameter. The effective aperture of ~300 m is illuminated by the feed moving on the focus surface halfway from reflector to its center. The telescope is pointed out by moving the focus cabin and simultaneously adjusting the shape of the illuminated area.

Besides FAST, it should be noted that catenary structures are very useful in membrane antennas, membrane lenses, sunshields, as well as solar sails. With some mechanic analysis, it can be shown that by properly designing the shape and materials, very flat membranes can be constructed without wrinkling [26].

The catenary shape also exists in electromagnetic devices with characteristic dimensions comparable to the wavelength. However, different from their optical counterparts where the catenary is described using geometric optics, the electromagnetic catenary must be interpreted using Maxwell's wave equations. In 1973, Cosmahl provided a generalized representation of electric fields in axisymmetric interaction gaps of klystrons and traveling-wave tubes, and found that the real electric fields follow a catenary, but not infinite nor strictly uniform function [27].

For a long time, it was not clear why electric fields amplitude in the gap should have a catenary shape. Recently, it was revealed that the catenary function is a direct result of the evanescent coupling in subwavelength structures. Not only the plasmonic modes in metal–dielectric multilayers and composites (Fig. 1.11) [28–34] but also the

Fig. 1.10 Schematic of the FAST. Reproduced from [25] with permission. Copyright 2006, Science in China Press

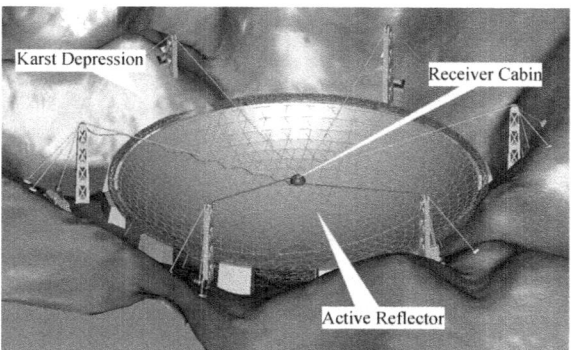

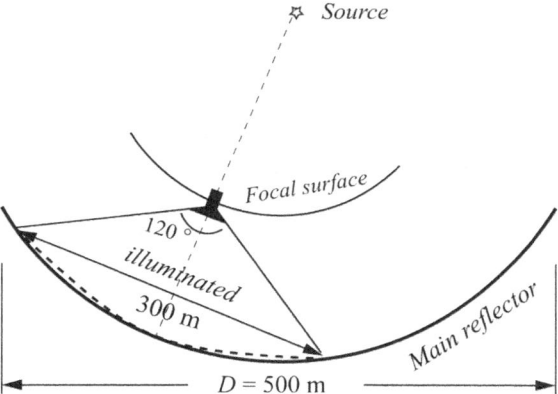

resonant fields in metallic gaps could be described using catenary function (Fig. 1.12) [35, 36]. The introduction of catenary optical fields provides many new perspectives on both the understanding and design of subwavelength structures. On the one hand, the catenary fields are evanescent waves in the horizontal direction, accompanying with much larger propagation constant in the vertical direction. As a result, the equivalent wavelength becomes much smaller than that in vacuum. On the other hand, the catenary optical fields can be arbitrarily tuned via the geometric parameters, providing tremendous freedoms for the modulation of light wave.

Besides the hyperbolic cosine catenary function, recent work shows that the catenary of equal strength also has many optical counterparts when the light–matter interaction at the subwavelength scale is taken into account. For instance, it was shown that a subwavelength aperture with such a shape perforated in a thin metallic film would result in perfect photonic spin–orbit interaction (PSOI), i.e., broadband and efficient conversion of circularly polarized light into optical beams carrying additional momentum or orbital angular momentum (OAM) [38, 39]. As shown in Fig. 1.13, the single catenary aperture may be arranged in different forms to control the phase front of outcoming light.

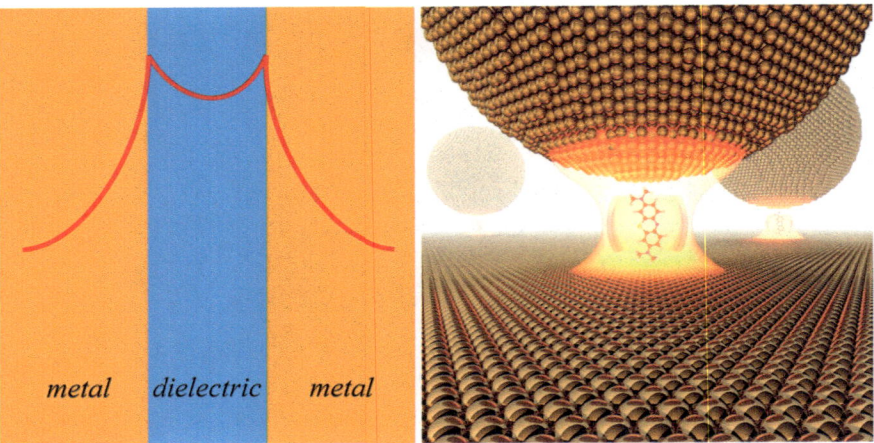

Fig. 1.11 Left: Catenary fields in plasmonic metal–dielectric–metal waveguide. Right: Catenary plasmon in the gap between the metallic sphere and plane. Note this artistic illustration is similar to the capillary bridge, which also follows a catenary shape. Reproduced from [37] with permission. Copyright 2016, Springer Nature

Fig. 1.12 Simulated electric field distribution and the curve fitted using catenary function along the center in a metallic slit. The inset shows the geometric configuration and electric fields in the xz-plane. Reproduced from [35] with permission

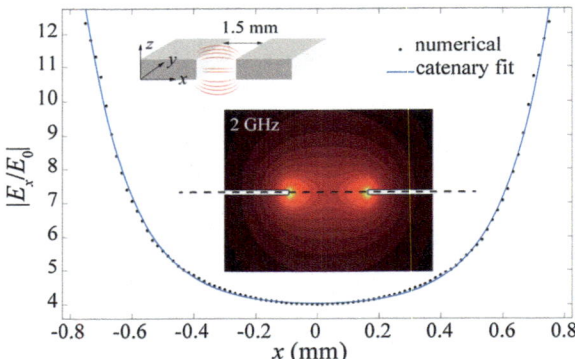

The physical mechanism of the PSOI is attributed to the geometric phase accompanying in polarization conversion, which is also termed as Pancharatnam–Berry phase in honor of S. Pancharatnam and M. V. Berry [40, 41]. In addition to the geometric phase, it was recently found that the electric field of surface plasmon polaritons (SPPs) on a metallic surface approximate to the catenary of equal strength at the resonant wavelength [35]. By analyzing the fundamental spin properties of Maxwell waves, this peculiar property may be attributed to the spin-momentum locking of surface plasmons [42]. Furthermore, we note that the catenary of equal strength has some inherent links to the helix, which has been widely utilized as circular polarizer [43–45]. Figure 1.14 shows a schematic of the transformation from discrete anisotropic elements to continuous helix and catenary structures. Note that in

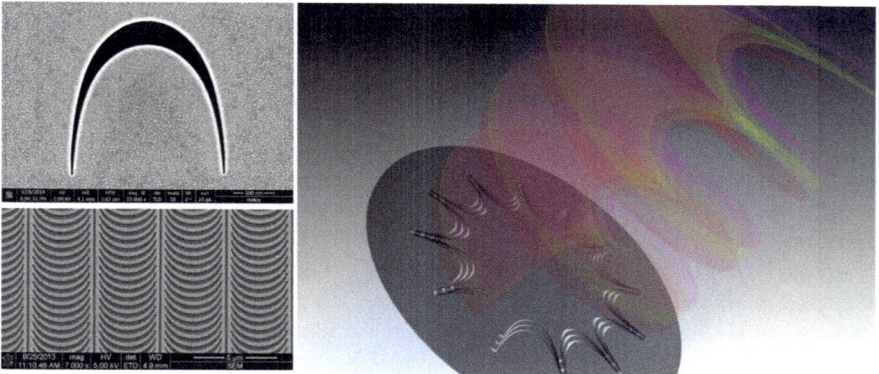

Fig. 1.13 Single catenary and catenary array to control the PSOI. Left shows the scanning electron microscope (SEM) images and the right is an artist drawing of a circular catenary array to generate beams carrying OAM

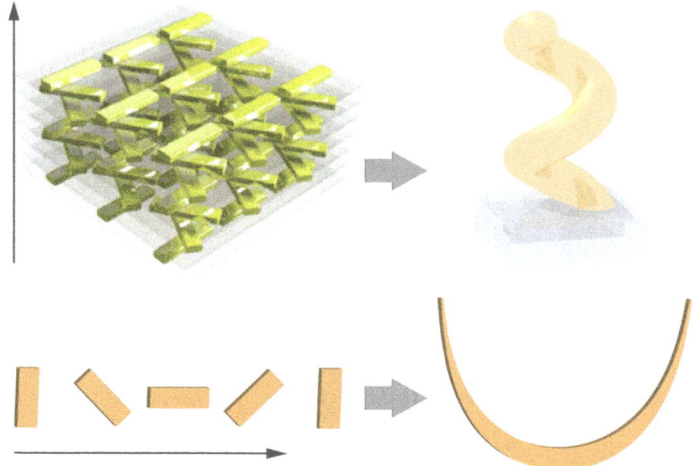

Fig. 1.14 Schematic of the transformation from discrete to continuous structures. The top row shows anisotropic elements with gradient increased rotation angles along the vertical direction, which leads to the 3D helix array. The bottom row shows anisotropic elements with gradient increased rotation angles along the horizontal direction, which lead to the planar catenary of equal strength. Top left: reproduced with permission from Ref. [43]. Copyright 2012, Springer Nature. Top right: reproduced with permission from Ref. [45]. Copyright 2015, John Wiley and Sons

mechanics the helix is well described by Hooke's law, and it has been demonstrated that even DNA double helix still follow this rule.

Based on the above considerations, herein we define the concept of Catenary Optics, which is the subject of this book. The first connotation of Catenary Optics is the catenary optical fields, which deal with either electric or magnetic fields that have intensity distribution, field lines or other properties described by the catenary function [35, 46, 47]. The second connotation is the catenary optical structures, which utilize catenary-shaped subwavelength structures to control light fields [38, 39, 48–50]. The third connotation is the catenary optical dispersion, which relates the dispersion of optical response via catenary function. For example, in the spectral frequency domain, the impedance of a metallic grating is described by the catenary of equal strength, which provides an efficient model for the calculation of similar structures [36, 47]. In the spatial frequency domain, if the dispersion relation follows a catenary rather than hyperbolic function [51, 52], the propagating behavior could be altered to realize many unusual applications.

To provide more backgrounds of Catenary Optics, in the following, we shall describe in detail some classic examples that link the catenary and optics/electromagnetics. Once again, we note that the catenary function is related to many aspects of physics ranging from classic mechanics and special relativity to quantum mechanics and superstring theory [53]. Like the harmonic oscillator, catenary may provide a simple yet powerful approach to understand and predict the core physics in different research areas.

1.2 Catenary Function in Optics and Electromagnetics

1.2.1 The Mirage

Originally described in many myths, a mirage is actually an optical phenomenon in which light rays bend to produce a displaced image of distant objects. In Chinese myths, this is induced by the gas of marine monsters. The word mirage comes to English from the Latin mirare, meaning "to look at, to wonder at" [54]. This is the same root as that for "mirror" and "to admire." Generally speaking, mirages can be categorized as "inferior", "superior", and "Fata Morgana". For exhausted travelers in the desert, an inferior mirage may appear to be a lake of water in some way ahead. An inferior mirage is called "inferior" because the mirage is located under the real object. The real object in an inferior mirage is the (blue) sky or any distant object in that same direction.

In contrast to the inferior mirage, the superior mirage is featured by an image located above the object, which is caused by a temperature inversion. Unlike the desert where temperature is decreasing from ground to high altitude, the temperature of sea surface may be colder as a result of the evaporation of water. Consequently, the light rays would bend toward a reversed direction. In the following discussion, the superior mirage is analytically investigated by considering a linearly varying refractive index along the vertical axis as shown in Figs. 1.15 and 1.16 [22, 55].

According to Snell's law, the refracting angles have a relation of

$$n \cos \alpha = n_0 \cos \alpha_0, \tag{1.2.1}$$

Fig. 1.15 A mirage in the desert. The lake is an inferior mirage that sometimes is seen in deserts. Photographed by in Primm, Nevada on April 4, 2007. Reproduced from Wikipedia [54]

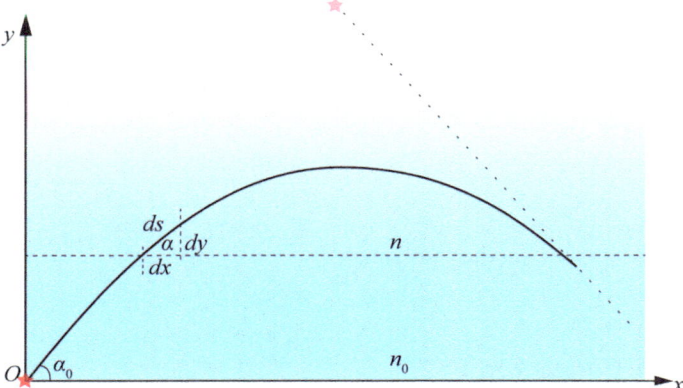

Fig. 1.16 Trajectory of the light beam in medium with linearly gradient refraction index. The angle between the ray at the origin and the x-axis is set to be α_0

where α is the angle between the ray and the horizontal axis, α_0 is the angle at the incidence point. By definition, there is a geometric relation:

$$\left(\frac{dy}{dx}\right)^2 = \left(\frac{1}{\cos \alpha}\right)^2 - 1. \tag{1.2.2}$$

By combining the above two equations, there is

$$\left(\frac{dy}{dx}\right)^2 = \left(\frac{n}{n_0 \cos \alpha_0}\right)^2 - 1. \tag{1.2.3}$$

After differentiating it with x and rearranging the equation, one can obtain

$$\frac{d^2y}{dx^2} = \frac{n}{n_0^2 \cos^2 \alpha_0} \frac{dn}{dy}. \tag{1.2.4}$$

If the refractive index follows a linear relationship with y as $n = n_0 (1 - ky)$, the above equation is written as

$$\frac{d^2y}{dx^2} - \frac{k^2}{\cos^2 \alpha_0} y = -\frac{k}{\cos^2 \alpha_0}. \tag{1.2.5}$$

The general solution is a catenary function:

$$y = C_1 \exp\left(\frac{k}{\cos \alpha_0} x\right) + C_2 \exp\left(-\frac{k}{\cos \alpha_0} x\right) + C_3, \tag{1.2.6}$$

where C_1, C_2, and C_3 are constants. This example illustrates a link between the optics and mechanics. While the ray trajectory is ultimately determined by the Fermat's principle and has the smallest optical path, the catenary-shaped structures in the mechanics also have the smallest potential.

1.2.2 Solar Concentrator

Concentration of light is important for both imaging and the utilization of solar energy. In classic optical designs, this is often achieved using parabolic reflectors and Fresnel Zone plates. Traditional parabolic mirror has two drawbacks: First, the construction of large mirror with low cost is extremely difficult with the state-of-the-art technologies. Second, at different times across the day, the incident angle is different which may introduce significant wavefront error and deterioration of focusing spot. Different from imaging applications, solar concentrators do not need near-diffraction-limited images, and thus one may change the shape to optimize the overall performance. As one natural alternative, the catenary concentrator can be

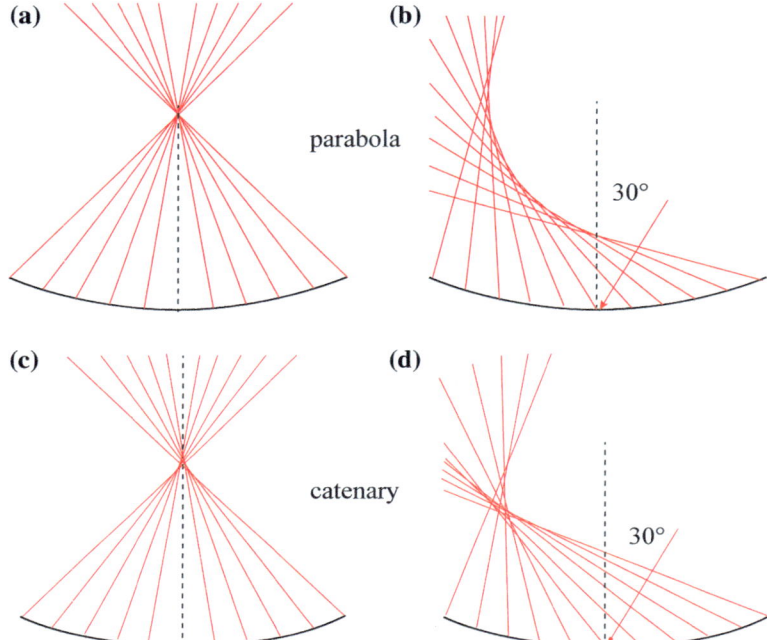

Fig. 1.17 Ray trajectories of parabola and catenary reflectors at normal and off-axis incidence. Adapted from [19] with permission. Copyright 1978, Optical Society of America

realized by simply hanging a sheet between two parallel supporting rods. When the depth of the catenary is small, its optical performance is comparable to the parabolic reflector. However, if we construct bigger catenary with deeper depth, considerable differences would occur.

Besides much easier to construct, the catenary shape also renders much better off-axis performance. As can be seen in Fig. 1.17, the light rays reflected by a common parabola reflector at an incidence angle of 30° produce obvious caustic curves. In contrast, the focus spot is still acceptable for the catenary reflector with the same aperture [19]. This phenomenon may be seen as a compromise between the performances at normal and oblique incidence, i.e., the increase in focusing intensity at large incidence angle is obtained at the expense of that for normal incidence.

With the emergence of generalized laws of reflection and refraction [56], it is interesting to investigate that such wide-angle performance can be realized by flat materials with properly designed phase front.

As illustrated in Fig. 1.18, the phase profile of the lens can be calculated from the optical path lengths [57]. The accumulated phases along the light paths can be expressed as a function of the incident angle α and the position x:

$$AP(x, \alpha) = \frac{2\pi}{\lambda} x \sin \alpha + \Delta \Phi(x, \alpha) + \frac{2\pi}{\lambda} \sqrt{f^2 + (x - x_0(\alpha))^2}, \qquad (1.2.7)$$

Fig. 1.18 Schematic of a perfect flat lens transforming plane waves with any angle of incidence into spherical waves. The accumulated phases along the light paths PP'O and QQ'O are equal. Reproduced from [57] with permission. Copyright 2016, Optical Society of America

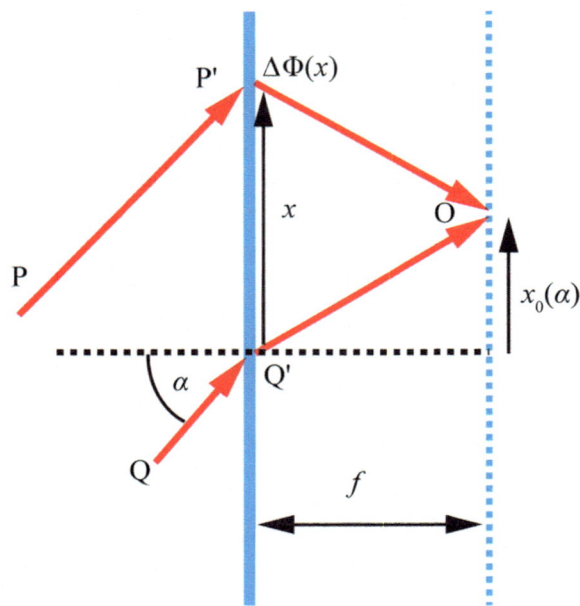

where λ is the wavelength of the incident wave, f is the distance between the lens and the focal plane, and $x_0(\alpha)$ is the position of the focal point as a function of the incident angle. Since the difference of the accumulated phases is zero on any two light paths, the phase change introduced by the metasurface can be expressed for each position and incident angle as

$$\Delta\Phi(x,\alpha) = \Delta\Phi(x_{\text{ref}},\alpha) - \frac{2\pi}{\lambda}$$
$$\left[x\sin\alpha + \sqrt{f^2 + (x - x_0(\alpha))^2} - x_{\text{ref}}\sin\alpha + \sqrt{f^2 + (x_{\text{ref}}(\alpha) - x_0(\alpha))^2} \right], \quad (1.2.8)$$

where $x_{\text{ref}}(\alpha)$ is a reference position on the lens where the phase change is chosen to be zero. Setting $x_{\text{ref}}(\alpha) = 0$, Eq. (1.2.8) is simplified to

$$\Delta\Phi(x,\alpha) = -\frac{2\pi}{\lambda}\left[x\sin\alpha + \sqrt{f^2 + (x - x_0(\alpha))^2} - \sqrt{f^2 + (x_0(\alpha))^2} \right], \quad (1.2.9)$$

which is the incident-angle-dependent phase profile of the ideal flat lens. By substituting $\alpha = 0$ in Eq. (1.2.9) yields the commonly used hyperbolical distribution.

Since the phase profile for either planar or structured surface is almost independent of angle, it seems impossible to maintain good focusing performance for all incident angles with only a single surface. Fortunately, some ingenious methods such as quadratic lens [46, 58, 59] and doublets [60, 61] have been proposed to realize wide-angle focusing and imaging.

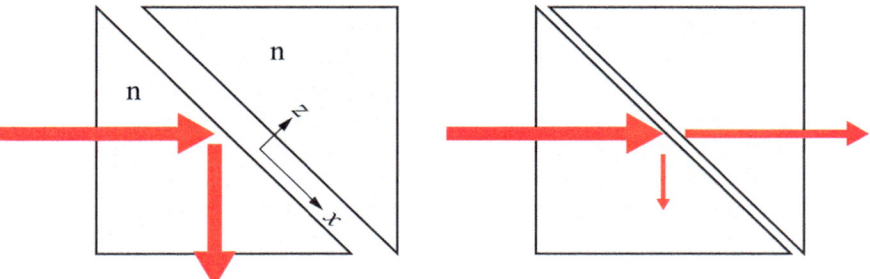

Fig. 1.19 Frustrated total internal reflection. The gap width determines the amount of energy that can be transmitted

1.2.3 Optical and Quantum Tunneling

In optical and electromagnetic waves, the catenary function can be treated as a sum of two counter-propagating evanescent waves. One typical example to see this effect is the frustrated total internal reflection (TIR). As indicated by the Snell's law, when light incidents from an optical dense material to a less dense one, the refraction angle will be larger than the incidence angle. Consequently, there is a critical angle beyond which the refract angle is larger than 90°, and thus all incident light will be totally reflected. In physical optics, it is shown that the TIR is accompanied with evanescent wave decaying exponentially away from the surface. When additional high index material is added close to the first surface, the evanescent waves will bounce back and forward between the two surfaces, which finally let some energy to be transmitted. This effect is called frustrated TIR, which has been utilized in touch screen and sensing applications.

The frustrated TIR is shown in Fig. 1.19. For incidence angle larger than the critical one, the horizontal wave vector k_x is larger than k_0, and thus the vertical wave vector k_z is imaginary. The counter-propagating evanescent waves can be written as

$$U = A \exp(ik_x x - \sqrt{k_x^2 - k_0^2}z) + B \exp(ik_x x + \sqrt{k_x^2 - k_0^2}z), \qquad (1.2.10)$$

where A and B are coefficients. At $x = 0$, the amplitude is

$$U = A \exp(-\sqrt{k_x^2 - k_0^2}z) + B \exp(+\sqrt{k_x^2 - k_0^2}z), \qquad (1.2.11)$$

which is just a modified catenary function.

In dielectric and plasmonic waveguides, the electromagnetic fields in the core are also evanescent waves. When parallel waveguides are placed side by side, the propagating modes will couple with each other and form supermodes. Using transfer matrix theory, these modes can be analytically calculated. Figure 1.20 shows the two supermodes in two coupled identical planar waveguides, one symmetric and

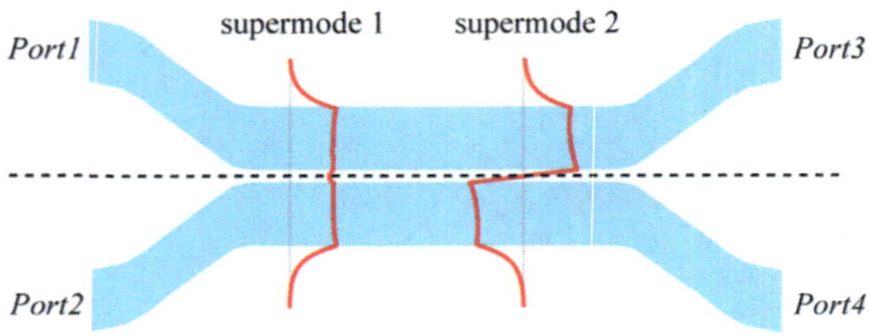

Fig. 1.20 Symmetric and
antisymmetric supermodes
in 3-dB directional coupler.
Adapted from [62] with
permission. Copyright 2010,
Optical Society of America

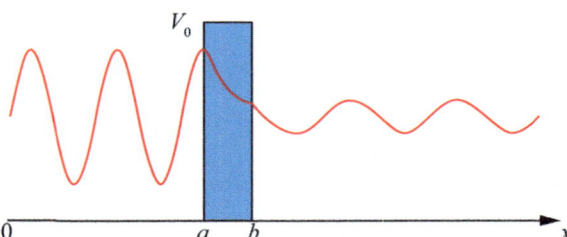

Fig. 1.21 Schematic of the
quantum tunneling. If the
thickness of the barrier
(b–a) is small enough, a
particle trapped in the range
$0 < x < a$ can escape by
tunneling

the other antisymmetric with regard to the mirror plane (the horizontal dashed line).
Since the two modes have different propagation constants, such device may be used
as directional coupler, i.e., the output energy may be switched from one port to
another [62].

The frustrated TIR effect is also known as optical tunneling and shares many
similarities with its quantum counterpart, i.e., the quantum tunneling [63]. As one
of the most bizarre behaviors of quantum physics, quantum tunneling means that
particles with low energy can surpass a barrier with much higher energy level, which
is strictly restricted by classic mechanics. Nevertheless, this effect can be interpreted
if the wave–particle duality and especially the Schrödinger equation are taken into
consideration. In real experiments, the quantum tunneling was first discovered in
atomic phenomena such as alpha decay. As a result of the strong attractive forces
inside the nucleus, only quantum tunneling can explain why the relative low-energy
alpha particle could be emitted.

To have a more clear vision on the quantum tunneling, let us consider the following
problem where a wave packet is trapped behind a barrier and then it escapes to infinity
by tunneling Fig. 1.21. This problem can be viewed as a one-dimensional tunneling
when the potential is infinity for $x \leq 0$. Suppose that a particle of mass m is initially
confined to a segment $0 \leq x \leq a$ in the left of a rectangular potential of height V_0
and width b–a. The wave function is obtained from the solution of the Schrödinger
equation and can be written as [64]

$$\psi_p(x) = \begin{cases} 2\sin(px)/N(p) & \text{for } 0 \leq x \leq a \\ Ae^{\gamma x} + Be^{-\gamma x} & \text{for } a \leq x \leq b \ , \\ Ce^{ipx} + De^{-ipx} & \text{for } b \leq x \end{cases} \qquad (1.2.12)$$

where

$$\gamma = \sqrt{2mV_0 - p^2}. \qquad (1.2.13)$$

The normalization constant $N(p)$ is chosen to satisfy the relation

$$\int_0^\infty \psi_0^*(x)\psi_{p'}(x)\,\mathrm{d}x = \delta(p - p'). \qquad (1.2.14)$$

By imposing the boundary conditions, i.e., the continuity of the logarithmic derivative of the wave function at $x = a$ and $x = b$, the constants A, B, C, and D can be written as

$$A = \frac{i}{N(p)}\left[\sin(pa) + \frac{p}{\gamma}\cos(pa)\right], \qquad (1.2.15)$$

$$B = \frac{i}{N(p)}\left[\sin(pa) - \frac{p}{\gamma}\cos(pa)\right], \qquad (1.2.16)$$

$$C = \frac{1}{2}\left(1 + \frac{\gamma}{ip}\right)Ae^{G/2} + \frac{1}{2}\left(1 - \frac{\gamma}{ip}\right)Be^{-G/2}, \qquad (1.2.17)$$

$$D = \frac{1}{2}\left(1 - \frac{\gamma}{ip}\right)Ae^{G/2} + \frac{1}{2}\left(1 + \frac{\gamma}{ip}\right)Be^{-G/2}, \qquad (1.2.18)$$

where

$$N^2(p) = \frac{1}{2\pi}\left(1 + \frac{\gamma^2}{p^2}\right)\times$$
$$\left\{\left[\sin(pa) + \frac{p}{\gamma}\cos(pa)\right]^2 e^{2(b-a)\gamma} + \left[\sin(pa) - \frac{p}{\gamma}\cos(pa)\right]^2 e^{-2(b-a)\gamma}\right\}. \qquad (1.2.19)$$

From the above equations, it can be concluded that A is very small, and thus the catenary characteristic in the region $0 < x < a$ may be small.

1.2.4 Geodesic Antenna

The catenary sheets have also been exploited as large aperture tunable antennas, as shown in a patent filed in 1965 [65]. As illustrated in Fig. 1.22, when the feed antenna moves along one side of the catenary sheet, the outgoing beams can be dynamically steered from the other side.

This catenary antenna can be interpreted using the geodesic theory. In mathematics, geodesic is relating to the shortest possible line between two points on a sphere or other curved surface. From Fermat's principle, it is known that the light propa-

Fig. 1.22 Schematic of the geodesic antenna based on a cylinder made of a catenary curve. Reproduced from [66] with permission. Copyright 1988, Springer Science Business Media New York

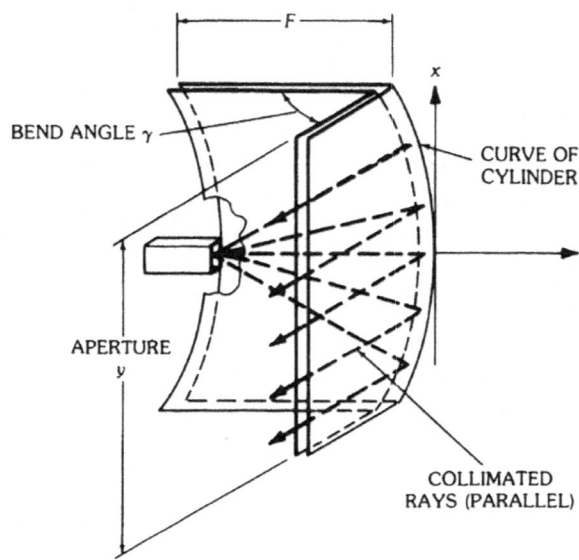

gating on curved surface will follow the geodesic lines. Following the description in Fig. 1.23, the mathematical deduction is as follows:

Assuming CE forms the wavefront of plane wave radiated by a point source at A, the Fermat's principle implies

$$\overline{AE} = \overline{AB} \tag{1.2.20}$$

and

$$(f + y)^2 = f^2 + s^2, \tag{1.2.21}$$

which can be written as

$$s = \sqrt{(f + y)^2 - f^2}. \tag{1.2.22}$$

By differentiating s with x, there is

$$\frac{ds}{dx} = \frac{(f + y)}{\sqrt{(f + y)^2 - f^2}} \frac{dy}{dx}. \tag{1.2.23}$$

By definition, there is

$$\frac{ds}{dx} = \sqrt{1 + \left(\frac{dy}{dx}\right)^2}. \tag{1.2.24}$$

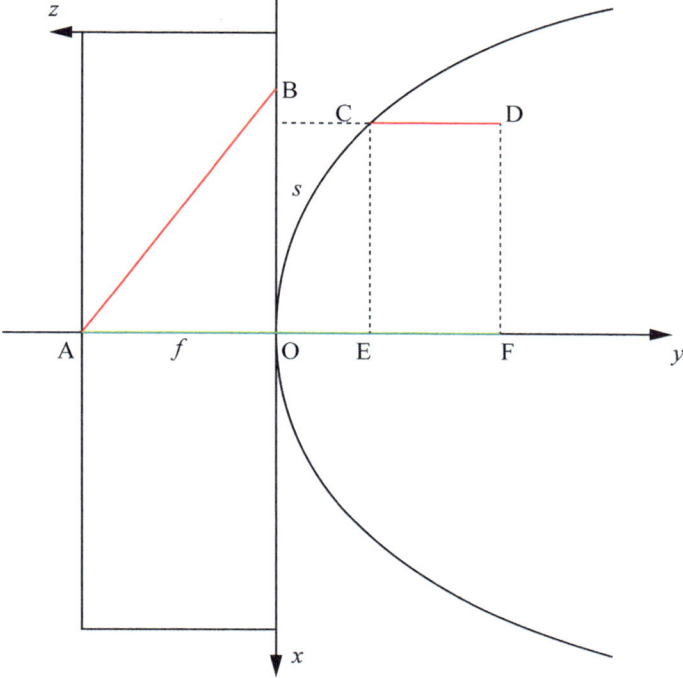

Fig. 1.23 Expanded view of the geometric relations. The xz-plane is tiled from the catenary sheet, xy-plane is the radiating plane. A is the radiating source, B and C are the same points, CD is the radiating ray

By some mathematical manipulation, we have

$$x = f \int \frac{dy}{\sqrt{y^2 + 2fy}}. \tag{1.2.25}$$

By integrating this, the function for the radiating aperture can be obtained as

$$x = f \int \frac{d(y + f)}{\sqrt{(y + f)^2 - f^2}} = f \ln \left| y + f + \sqrt{(y + f)^2 - f^2} \right| \tag{1.2.26}$$

and

$$y = f \cosh\left(\frac{x}{f}\right) - f, \tag{1.2.27}$$

which is just a catenary function.

1.2.5 Wireless Energy Transfer

Evanescent coupling is an efficient method to realize wireless energy transfer [67–69]. The electric field distribution between the two magnetic coils is an ideal catenary. By adding negative index materials in the gap, the evanescent waves may be amplified, thus the coupling strength becomes stronger [67]. For example, Wang et al. proposed a metamaterial with compact size and low loss for the wireless energy transfer. Experiments show that the transfer efficiency at a resonant frequency of ~28 MHz ($\lambda = 10.7$ m) and a distance of 50 cm (~λ/21) can be improved from 17 to 47% by the metamaterial [69].

In 2017, Assawaworrarit et al. proposed theoretically and demonstrated experimentally that a parity–time-symmetric circuit incorporating a nonlinear gain saturation element provides robust wireless power transfer [68]. The results show that the transfer efficiency remains near unity over a distance variation of approximately one meter, without the need for any tuning. Compared with both the fixed-resonant circuits and adaptive-resonant circuits, this new approach has a higher efficiency for a longer distance.

1.2.6 Accelerated Charges in Uniform Electric Fields

It is well known that a moving particle in a uniform electric field takes the form of a parabola. In relativistic physics, the trajectory of particles would transform to catenary. Considering the movement of a charge e in an uniform electric field **E** along the x-direction in the xy-plane as an example [70], the movement equations can be written as

$$\dot{p}_x = e\mathbf{E}, \quad \dot{p}_y = 0. \tag{1.2.28}$$

By integrating with t, there we have

$$p_x = e\mathbf{E}t, \quad p_y = p_0. \tag{1.2.29}$$

The kinetic energy of the charge is

$$\mathcal{E}_{\text{kin}} = \sqrt{m^2 c^4 + c^2 p_0^2 + (ceEt)^2} = \sqrt{\mathcal{E}_0^2 + (ceEt)^2}, \tag{1.2.30}$$

where $\mathcal{E}_0$ is the kinetic energy at $t = 0$. Recalling the velocity equation, we have

$$\frac{\mathrm{d}x}{\mathrm{d}t} = \frac{p_x c^2}{\mathcal{E}_{\text{kin}}} = \frac{c^2 eEt}{\sqrt{\mathcal{E}_0^2 + (ceEt)^2}} \tag{1.2.31}$$

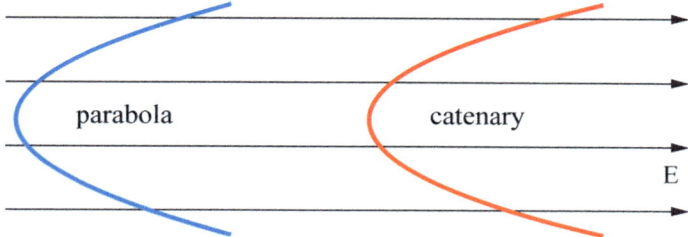

Fig. 1.24 Acceleration of charge in uniform electric field. The left and right are parabola and catenary, respectively

and

$$\frac{dy}{dt} = \frac{p_y c^2}{\mathcal{E}_{kin}} = \frac{p_0 c^2}{\sqrt{\mathcal{E}_0^2 + (ceEt)^2}}. \tag{1.2.32}$$

After some mathematical manipulation, the orbit equation can be written as

$$x = \frac{\mathcal{E}_0}{eE} \cosh \frac{eEy}{p_0 c}, \tag{1.2.33}$$

which is a catenary function. Notably, when the velocity is much smaller than the light velocity, the equation will become

$$x = \frac{eE}{2m v_0^2} y^2 + \text{const}, \tag{1.2.34}$$

which is the well-known solution in classic physics Fig. 1.24.

1.2.7 Coupling Between Atoms and Meta-atoms

In quantum mechanics, the coupling of atoms and molecules is a complicated problem, which can only be rigorously solved for some simple cases. In the molecular orbital approach, the overlapping atomic orbitals are described by wave functions. The 1s atomic orbitals on the two hydrogen atoms interact to form two new molecular orbitals, one produced by taking the sum of the two 1s wave functions, and the other produced by taking their difference [71].

Since the 1s state wave function of a single hydrogen atom is in proportional to $\exp(-r/a_0)$, where r is the distance to the center, a_0 is the Bohr radius, the molecular wave function of H_2 between the two atoms can be written as (Fig. 1.25)

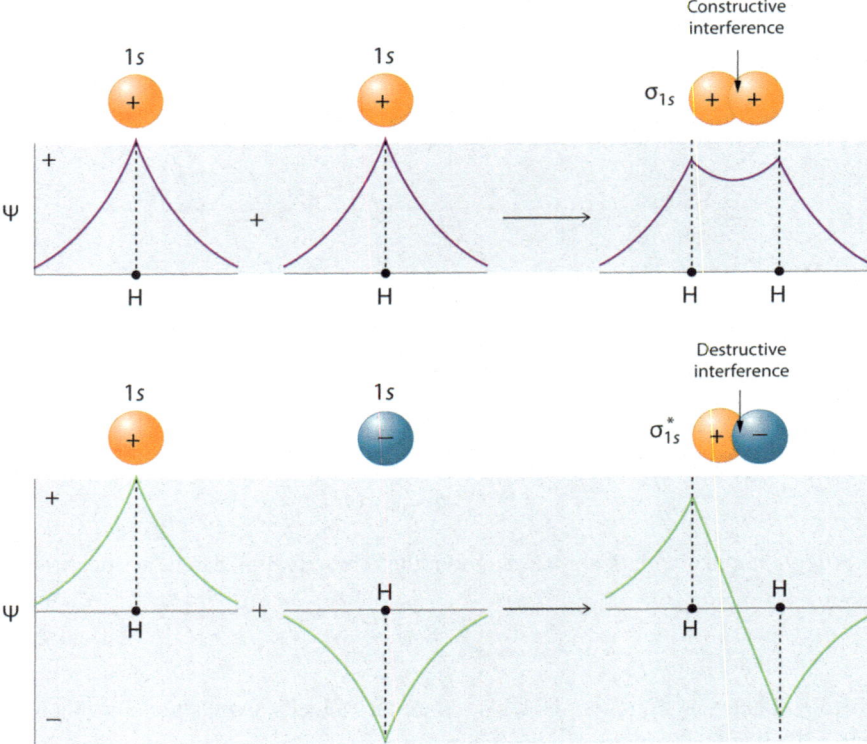

Fig. 1.25 Top, the bonding σ_{1s} molecular orbital for H_2 as the sum of the wave functions (Ψ) of two H 1s atomic orbitals. Bottom, the antibonding σ_{1s}^* molecular orbital for H_2 as the difference of the wave functions of two H 1s atomic orbitals. Reproduced with permission from [71]

$$\sigma_{1s} \sim \exp(x/a_0) + \exp(-x/a_0) = 2\cosh(x/a_0) \qquad (1.2.35)$$

and

$$\sigma_{1s}^* \sim \exp(x/a_0) - \exp(-x/a_0) = 2\sinh(x/a_0). \qquad (1.2.36)$$

The electron density in the σ_{1s} molecular orbital is greatest at the center of two positively charged nuclei, and the resulting electron–nucleus electrostatic attractions reduce repulsions between the nuclei. Thus, the σ_{1s} orbital represents a bonding molecular orbital. Interestingly, the wave function of σ_{1s} is actually a catenary function. In contrast, electrons in the σ_{1s}^* orbital are generally found in the space outside the inter-nuclear region. Because this allows the positively charged nuclei to repel one another, the σ_{1s}^* orbital is an antibonding molecular orbital.

The catenary coupling of Hydrogen atoms is of particular importance because it is so fundamental in materials research. Since this catenary function is a direct result of the coupling of matter wave, it is anticipated that electromagnetic waves would behave similarly. Actually, if we take the atom of materials as scatters for electromagnetic waves, these subwavelength scatters would produce a large amount of evanescent waves. The summation of these catenary waves naturally leads to many catenary electromagnetic fields. It should be noted that when these scatters are much smaller than the wavelength, these catenary fields can be homogenized to obtain an effective response. According to our design experiences, artificial subwavelength structures with dimensions in the range of 0.01λ and λ would form very strong catenary effect.

Note that the bonding and antibonding modes are similar to the symmetric and antisymmetric plasmonic modes in a metallic film and metal–dielectric–metal waveguide [31, 32]. This can be further illustrated in a modified Young's interference experiment. As shown in Fig. 1.26, the widths of the two slits are 20 nm and 40 nm with a separation of $d = 100$ nm [72]. Since the separation is small, the SPPs in the two slits could couple with each other, leading to a symmetric ($\lambda = 1525$ nm) and antisymmetric ($\lambda = 1375$ nm) distribution of magnetic field. Obviously, this Young's interference effect is a further demonstration of the extraordinary Young's interference [35] and related to the directional coupler [62].

Note that the symmetric and antisymmetric modes often have very different propagation lengths. For a simple metal–dielectric–metal plasmonic waveguide with a thin dielectric core, only the symmetric modes (defined by H_y, the symmetry would reverse when defined by electric fields parallel to the boundary) are supported. As another important form in plasmonic applications, metallic dimers and spheres on metallic films also support the symmetric modes. As shown in Fig. 1.27, when the dimensions are only several nanometers or sub-nanometers, the quantum effects such as Landau damping must be considered. Compared with classic bulky damping, the field enhancement factor may decrease, which sets a fundamental limit on the intensity of the catenary optical fields [34].

Figure 1.28 shows the electric field distribution of spherical gold dimer along the cross section at a wavelength of 523 nm. The diameter of the spheres and the gap distance are chosen to be 50 and 30 nm. In order to further check the field distribution within the gap, the electric field distribution along the white curve is retrieved, which is displayed by dotted curve. The catenary function is utilized to fit the electric field profile. It is obvious that the fitting curve can well approach the simulated electric field profile.

Figure 1.29 gives a schematic description of the catenary electromagnetic fields produced by subwavelength atoms. Not only the gap in a single resonator but also the gap between two isolated atoms could be described using capacitances and catenary optical fields. We believe this is an important aspect of metamaterial design that has not been fully exploited.

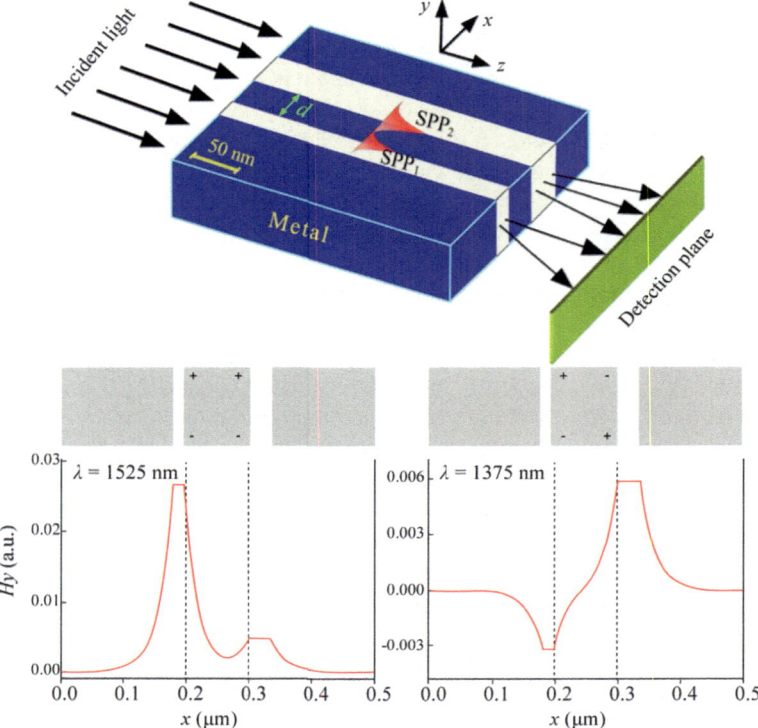

Fig. 1.26 Excited *Hy* profiles in the mid-transverse plane of the slit doublet structure with either symmetric or antisymmetric modes of plasmon hybridization, respectively. Here, the vertical short dash-dot lines mark two lateral boundaries of the slit interval. The upper panels in the two pictures represent surface charge distribution for different hybridization modes. Adapted with permission from [72]. Copyright 2013, IOP Publishing Ltd and Deutsche Physikalische Gesellschaft

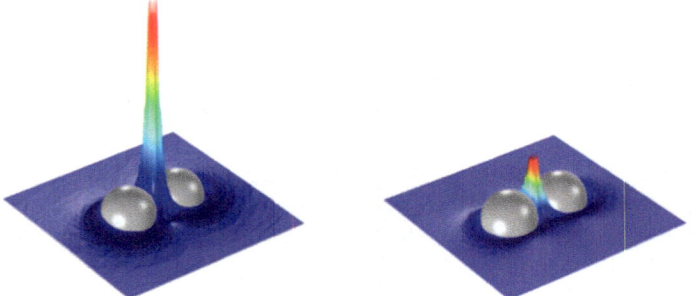

Fig. 1.27 Catenary fields in a plasmonic dimer. The radius and gap of the silver dimer are 2.5 nm and 0.5 nm. While the left considers only bulk damping, the right figure takes both bulk and surface damping into consideration at $\lambda = 663$ nm. Reproduced from [34] with permission. Copyright 2017, American Chemical Society

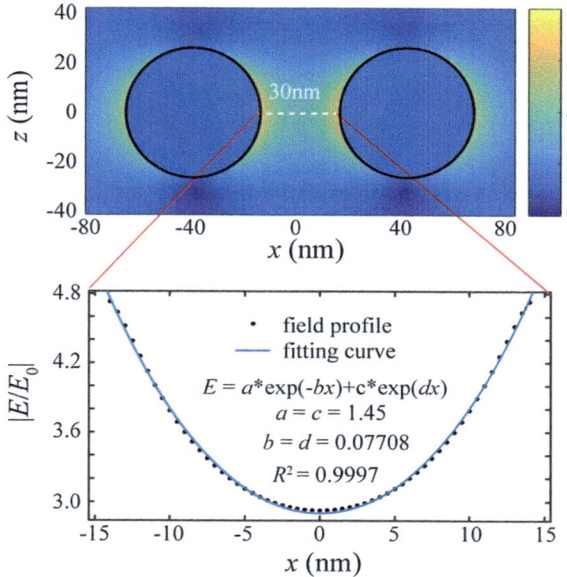

Fig. 1.28 Simulated electric field distribution of the spherical nanoparticle pair along the cross section at a wavelength of 523 nm. The retrieved field profile on the white curve is well fitted by catenary function. Reproduced with permission from [73]

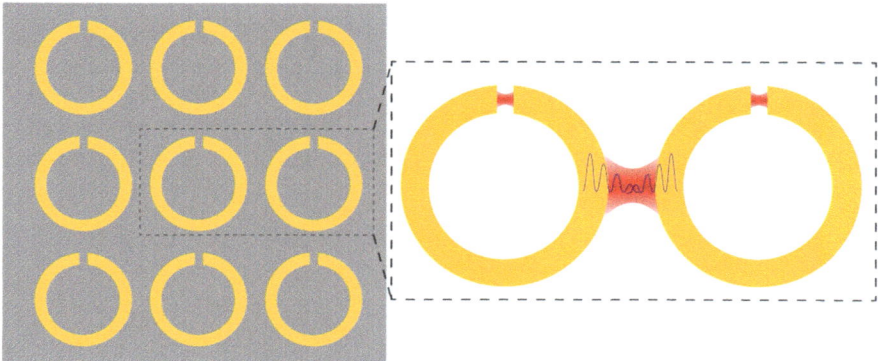

Fig. 1.29 Schematic of the catenary electromagnetic fields in artificial atoms array. The split ring resonator is used as an example to illustrate the catenary coupling in neighboring metallic elements. Note that both the gap in each single resonator and between adjacent resonators support such coupled waves

1.3   Misconceptions and Controversies

It is interesting to note that there are some misconceptions and controversies about the properties of catenary curve. As we noted above, Galileo once wrongly thought that the catenary curve is a parabola. In fact, this misconception is understandable considering that parabola approximates catenary well for small curvature. In the following, we list some other examples.

1.3.1   FAST

As noticed by some observers, the full name of Five-hundred-meter Aperture Spherical Radio Telescope (FAST) seems to be not so accurate [74]. First, the effective aperture with parabolic shape is only 300 m; second, the surface of the telescope is not spherical: Under natural conditions, the cross section of the surface seems to be a catenary, which is tuned by the control cable to be spherical and then parabolic for observing.

1.3.2   Brachistochrone

Brachistochrone is a curve between two points along which a body can move under gravity in a shorter time than for any other curve [75]. The word is coming from Ancient Greek βράχιστος χρόνος (brákhistos khrónos), meaning "shortest time." In 1638, Galileo tried to find the path of the fastest descent from a point to a wall in his *Two New Sciences*. He draws the conclusion that the arc of a circle is faster than any number of its chord:

> "From the preceding it is possible to infer that the quickest path of all [lationem omnium velocissimam], from one point to another, is not the shortest path, namely, a straight line, but the arc of a circle.
>
> ...
>
> Consequently the nearer the inscribed polygon approaches a circle the shorter is the time required for descent from A to C. What has been proven for the quadrant holds true also for smaller arcs; the reasoning is the same."

Many years later, Galileo reviews his own work and noted that the actual solution to this problem is half a cycloid. Galileo studied the cycloid and gave it its name, but the connection between it and his problem had to wait for other scientists including Johann Bernoulli, Huygens, Jakob Bernoulli, Gottfried Leibniz, Isaac Newton, et al.

Interestingly, in 1697, Johann Bernoulli used the Fermat's principle to derive the brachistochrone curve by considering the trajectory of a beam of light in a medium where the speed of light increases following a constant vertical acceleration (that of gravity g). By the conservation of energy, the instantaneous speed of a body v after falling a height y in a uniform gravitational field is given by $v = (2gy)^{1/2}$. Since there is no horizontal force, the speed of motion of the body along an arbitrary curve does not depend on the horizontal displacement. Bernoulli noted that the Snell's law gives a constant of the motion for a beam of light in a medium of variable density:

$$n \sin \theta = n \frac{dx}{ds} = n(v_m),\qquad(1.3.1)$$

where v_m is a constant and θ represents the angle of the trajectory with respect to the vertical axis (normal direction). If we assume that the refractive index is inversely proportional to the velocity, then there is

$$\frac{\sin \theta}{v} = \frac{1}{v}\frac{dx}{ds} = \frac{1}{v_m}.\qquad(1.3.2)$$

The above equations lead to two conclusions:

(1) At the onset, the angle must be zero when the particle speed is zero. Hence, the brachistochrone curve is tangent to the vertical axis at the origin.
(2) The speed reaches a maximum value when the trajectory becomes horizontal and the angle $\theta = 90°$.

Assuming for simplicity that the particle (or the beam) with coordinates (x, y) departs from the point $(0,0)$ and reaches maximum speed after falling a vertical distance D:

$$v_m = \sqrt{2gD}.\qquad(1.3.3)$$

Rearranging terms in the law of refraction and squaring gives

$$v_m^2 dx^2 = v^2 ds^2 = v^2 (dx^2 + dy^2),\qquad(1.3.4)$$

which can be solved for dx in terms of dy:

$$dx = \frac{v\, dy}{\sqrt{v_m^2 - v^2}}.\qquad(1.3.5)$$

Substituting from the expressions for v and v_m above leads to

$$dx = \sqrt{\frac{y}{D - y}}\, dy,\qquad(1.3.6)$$

which is just the differential equation of an inverted cycloid generated by a circle of diameter D (Fig. 1.30).

It was often thought that the curved roof of ancient Chinese buildings resembles the brachistochrone curve. Figure 1.31 shows a newly built classic Chinese building in

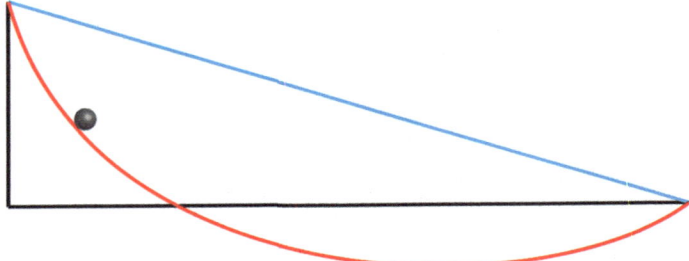

Fig. 1.30 The brachistochrone problem. The ball falling along the brachistochrone curve takes the shortest time among various possible routes

Fig. 1.31 The roof of a classic Chinese building in Chengdu, Sichuan, China

Chengdu, which shows a clearly curved roof. It seems that ancient Chinese engineers know how to make the following rain has largest velocity when leaving the roof. Note that this has been misused in some popular science story like the one entitled *The roof of Style Lei and catenary*.

1.3.3 Glacial Valley

The glacial valley is one of the few landforms which shows great regularity in the cross profile [2]. The form is generally described as a U shape in geomorphology

Fig. 1.32 Schematic of the cross profile of Glacial valley. Reproduced from Wikipedia [3]

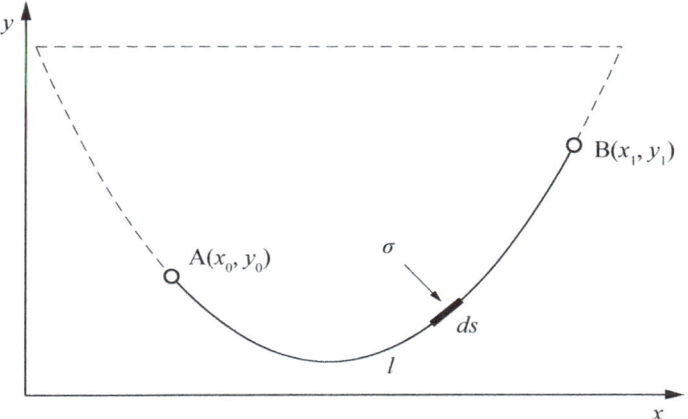

Fig. 1.33 Geometry of the cross profile of a glacial valley. Adapted from [2] with permission. Copyright 2006, John Wiley and Sons

textbooks, as shown in Fig. 1.32. However, the variation principle proves that the ideal or fully developed morphology of the glacial valley should be a catenary.

In the following, we give a simple deduction of the catenary function in glacial valley. A total friction force working at the valley sides from point A to B in Fig. 1.33, within a flowing glacier, can be expressed by integration as

$$f = \int_l \sigma \mu \mathrm{d}s, \tag{1.3.7}$$

where σ is the effective normal pressure and μ is a coefficient of friction. The contact length of a cross profile (perimeter) is given by

$$L = \int_l ds = \int_{x_0}^{x_1} \sqrt{1 + y'^2} dx, \qquad (1.3.8)$$

where $y' = dy/dx$ and the coordinates of the two limiting points. As a first approximation, the normal pressure at the ice–bedrock interface can be taken as approximately proportional to the ice thickness. Thus,

$$\sigma = \eta(y_s - y), \qquad (1.3.9)$$

where η is a constant and y_s is the glacier surface. From the above equations, the total force can be expressed in the form:

$$f = \eta\mu \int_{x_0}^{x_1} (y_s - y)\sqrt{1 + y'^2} dx. \qquad (1.3.10)$$

For a fully developed glacial valley, the friction between ice and rock should be smallest. It is known that a stationary curve which defines the extreme for an integral function

$$J = \int_{x_0}^{x_1} F(x, y, y') dx \qquad (1.3.11)$$

must satisfy the Euler's equation,

$$\frac{d}{dx}\left(\frac{\partial F}{\partial y'}\right) - \frac{\partial F}{\partial y} = 0. \qquad (1.3.12)$$

Now let us introduce a new function with an arbitrary constant λ,

$$F_* = (y_s - y + \lambda)\sqrt{1 + y'}. \qquad (1.3.13)$$

By applying Euler's equation, one can obtain

$$\frac{y_s - y + \lambda}{\sqrt{1 + y'^2}} = c_1. \qquad (1.3.14)$$

This can be separated into variables,

$$\frac{dy}{(y_s - y + \lambda)^2} - c_1^2 = \frac{dx}{c_1}. \qquad (1.3.15)$$

and there is

$$y_s - y + \lambda = -c_1 \cosh z = \frac{-c_1(e^z + e^{-z})}{2}. \qquad (1.3.16)$$

Then, integration about z yield,

$$z = \frac{x}{c_1} + c_2, \qquad (1.3.17)$$

where c_2 is an integral constant. Then, we can obtain the final form as a catenary function,

$$y = y_s + \lambda + c_1 \cosh\left(\frac{x}{c_1} + c_2\right). \qquad (1.3.18)$$

1.3.4 Freely Supported Beam and Cantilever Beam

As we noted in the first paragraph, the curve formed by a chain under gravitational force is catenary only when the chain is flexible and has no shearing force. According to this definition, the curve formed by non-flexible beam will not be catenary. By using rigorous mechanical analysis, the curves could be derived under variant conditions [76].

Since there are many different kinds of beams, we only stress that the cantilever and bow are not catenary. It is listed here to avoid confusion of readers. Indeed, previous researches have shown that catenary may be not more accurate than other curves in many cases [77]. We must bear in mind that the worlds are so complex that catenary is only one particular and important curve.

1.4 Overview of the Book

This book is devoted to the research of catenary function in subwavelength optics and electromagnetics. As shown in Fig. 1.34, most of the building blocks in the newly emerging metasurfaces and materials are composed of subwavelength slits and stripes. Based on the following reasons, they are also the basic elements of catenary optics. First, each scatter would excite evanescent waves and these waves would couple together to form a catenary function. In many cases, the catenary coupling resides in the heart of their exotic optical performances. Second, catenary-shaped continuous structures are ideal candidates for the manipulation of photonic spin–orbit interaction and geometric phase, based on which many novel functionalities such as achromatic orbital angular momentum generation and flat lensing have been demonstrated.

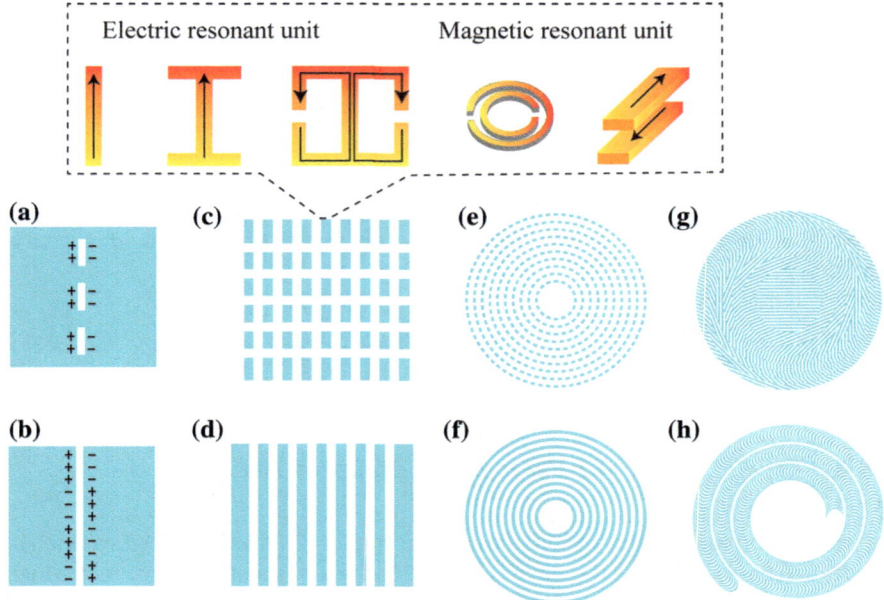

Fig. 1.34 Schematic of the building blocks of metasurfaces. **a** One-dimensional (1D) metallic slits with subwavelength width selectively transmit incident light that is polarized perpendicular to them. **b** A single metallic slit can be taken as a metal–insulator–metal (MIM) waveguide supporting the propagation of SPPs under the illumination of TM polarization. **c** 2D metasurfaces composed of electric or magnetic resonant unit cell arrays. **d** 1D metallic slits or dielectric stripes array, which can be utilized to construct achromatic metasurfaces through dispersion compensation. **e** and **f** Concentric slits or stripes as a platform of spin–orbit interaction for vortex beam generation. **g** and **h** Typical quasi-continuous metasurfaces composed of nanobeams and catenary slits/stripes for high efficiency wavefront manipulation. Adapted from [78] with permission. Copyright 2018. IOP Publishing Ltd

This book is organized as follows:

In the first chapter, the basic concepts and history of catenary and catenary optics are introduced. In particular, the mathematic and physical links between mechanic and optical properties are highlighted.

In Chapter 2, the catenary of equal strength and its applications in spin-controlled beam shaping are discussed. We show that deformed catenary shapes could serve as functional elements in various phase-gradient metasurfaces.

In Chapter 3, catenary structures are used to realize unidirectional excitation of guiding modes in both plasmonic and dielectric waveguides.

In Chapter 4, the catenary optical fields induced by coupling of evanescent waves are discussed in plasmonic metal–dielectric multilayers. Their applications in sub-diffraction-limited imaging and lithography have been presented.

In Chapter 5, the width-dependence coupling of surface plasmons is utilized to realize local phase modulation. Various flat optical devices such as plasmonic beam deflector, flat lens, achromatic flat lens, color filter, and color meta-hologram are introduced.

In Chapter 6, we extend the catenary surface plasmon to catenary electromagnetic fields by leveraging the concept of microscopic meta-surface-wave (M-wave). It is shown that in subwavelength metallic structures, common electromagnetic waves would behave similar as the surface plasmons in the optical regime. Based on these phenomena, the wavefront of microwaves and infrared light could be controlled with greatly increased freedom.

In Chapter 7, the catenary optical fields in lossy subwavelength structures are investigated. We show that with different geometric and materials configurations, the absorption spectrum can be designed almost arbitrarily to realize narrowband, broadband, and multispectral absorption.

In Chapter 8, electromagnetic radiation in the form of thermal emission is discussed. Coupled subwavelength structures could not only enable the on-demand design of emission spectrum but can also change the coherence of thermal emission.

In Chapter 9, we summarize the applications of catenary optics in optical engineering. It is shown that traditional optical laws and theories could be extended and generalized based on catenary optics, which opens a door towards the next-generation engineering optics.

Finally, in the Appendix, the Matlab codes for the catenary structures modeling, catenary dispersion, and transfer matrix calculation, as well as the vectorial diffraction are given to help the readers to reproduce the results presented in this book. Gallery of pictures related to catenary function is also listed.

As the first book devoted to the systematic description of catenary function in optical applications, we must admit that there are still many new exciting things yet to be discovered. So we are just in the starting stage of catenary optics.

References

1. Catenary, http://www.en.wikipedia.org/wiki/Catenary
2. M. Hirano, M. Aniya, A rational explanation of cross-profile morphology for glacial valleys and of glacial valley development. Earth Surf. Process. Landf. **13**, 707–716 (1988)
3. U shaped valley, http://www.en.wikipedia.org/wiki/U_shaped_valley
4. The Lady with an Ermine, https://commons.wikimedia.org/wiki/File:The_Lady_with_an_Ermine.jpg
5. Jin Wu Di, https://commons.wikimedia.org/wiki/File:Jin_Wu_Di.jpg
6. Bodhisattva Leading the Way, https://upload.wikimedia.org/wikipedia/commons/2/28/Anonymous-Bodhisattva_Leading_the_Way.jpg
7. Court Ladies Wearing Flowered Headdresses, https://upload.wikimedia.org/wikipedia/commons/0/06/Zhou_Fang._Court_Ladies_Wearing_Flowered_Headdresses._%2846x180%29_Liaoning_Provincial_Museum%2C_Shenyang.jpg
8. D. Gilbert, On the mathematical theory of suspension bridges, with tables for facilitating their construction. Philos. Trans. R. Soc. Lond. **116**, 202–218 (1826)
9. C.R. Calladine, An amateur's contribution to the design of Telford's Menai Suspension Bridge: a commentary on Gilbert (1826) 'On the mathematical theory of suspension bridges', Philos. Transact. A Math. Phys. Eng. Sci. **373**, 20140346 (2015)

10. "Robert Hooke," https://commons.wikimedia.org/wiki/Robert_Hooke
11. R.K. Temple, *The Genius of China: 3,000 Years of Science, Discovery, and Invention* (Inner Traditions Rochester, VT, 2007)
12. Poem of Du Fu, https://zhuanlan.zhihu.com/p/43666691
13. Anlan Suspension Bridge, https://commons.wikimedia.org/wiki/File:Anlan_Suspension_Bridge-3.jpg
14. Luding Bridge, https://en.wikipedia.org/wiki/Luding_Bridge
15. Q. Chen, X. Yao, L. Xu, Q. Li, Y. Song, L. Jiang, Capillary force restoration of droplet on superhydrophobic ribbed nano-needles arrays. Soft Matter **6**, 2470–2474 (2010)
16. P.A. Kralchevsky, K. Nagayama, Capillary bridges and capillary-bridge forces, in *Particles at Fluid Interfaces and Membranes* (Elsevier, 2001), pp. 469–502
17. Capillary bridges, http://www.en.wikipedia.org/wiki/Capillary_bridges
18. Catenoid, https://en.wikipedia.org/wiki/catenoid
19. G.A. Rottigni, Concentration of the sun's rays using catenary curves. Appl. Opt. **17**, 969–974 (1978)
20. Eye, http://www.en.wikipedia.org/wiki/Eye
21. D.J. Coleman, R.H. Silverman, H. Lloyd, IV.D. Physiology of Accommodation and Role of the Vitreous Body, in *Vitreous: In Health and Disease*, ed. by J. Sebag, (Springer New York, 2014), pp. 495–507
22. J. Evans, M. Rosenquist, "F = ma" optics. Am. J. Phys. **54**, 876–883 (1986)
23. D.J. Coleman, On the hydraulic suspension theory of accommodation. Trans. Am. Ophthalmol. Soc. **84**, 846 (1986)
24. R. Nan, G. Ren, W. Zhu, Y. Lu, Adaptive cable-mesh reflector for the FAST. Acta Astron. Sin. **44**, 13–18 (2003)
25. R. Nan, Five hundred meter aperture spherical radio telescope (FAST). Sci. China, Ser. G **49**, 129–148 (2006)
26. H. Fang, M. Lou, L.-M. Hsia, P. Leug, Catenary systems for membrane structures, in *19th AIAA Applied Aerodynamics Conference*. Fluid Dynamics and Co-Located Conferences (American Institute of Aeronautics and Astronautics, 2001)
27. H.G. Kosmahl, G.M. Branch, Generalized representation of electric fields in interaction gaps of klystrons and traveling-wave tubes. IEEE Trans. Electron Devices **20**, 621–629 (1973)
28. X. Luo and T. Ishihara, Sub 100 nm lithography based on plasmon polariton resonance, in *2003 International Microprocesses and Nanotechnology Conference* (IEEE, 2003), pp. 138–139
29. K. Tanaka, M. Tanaka, Simulations of nanometric optical circuits based on surface plasmon polariton gap waveguide. Appl. Phys. Lett. **82**, 1158–1160 (2003)
30. X. Luo, T. Ishihara, Surface plasmon resonant interference nanolithography technique. Appl. Phys. Lett. **84**, 4780–4782 (2004)
31. X. Luo, T. Ishihara, Subwavelength photolithography based on surface-plasmon polariton resonance. Opt. Express **12**, 3055–3065 (2004)
32. C. Wang, C. Du, X. Luo, Refining the model of light diffraction from a subwavelength slit surrounded by grooves on a metallic film. Phys. Rev. B **74**, 245403 (2006)
33. B. Wood, J.B. Pendry, D.P. Tsai, Directed subwavelength imaging using a layered metal-dielectric system. Phys. Rev. B **74**, 115116 (2006)
34. J. Khurgin, W.-Y. Tsai, D.P. Tsai, G. Sun, Landau damping and limit to field confinement and enhancement in plasmonic dimers. ACS Photonics **4**, 2871–2880 (2017)
35. M. Pu, Y. Guo, X. Li, X. Ma, X. Luo, Revisitation of extraordinary Young's interference: from catenary optical fields to spin-orbit interaction in metasurfaces. ACS Photonics **5**, 3198–3204 (2018)
36. M. Pu, X. Ma, Y. Guo, X. Li, X. Luo, Theory of microscopic meta-surface waves based on catenary optical fields and dispersion. Opt. Express **26**, 19555–19562 (2018)
37. R. Chikkaraddy, B. de Nijs, F. Benz, S.J. Barrow, O.A. Scherman, E. Rosta, A. Demetriadou, P. Fox, O. Hess, J.J. Baumberg, Single-molecule strong coupling at room temperature in plasmonic nanocavities. Nature **535**, 127 (2016)

38. M. Pu, X. Li, X. Ma, Y. Wang, Z. Zhao, C. Wang, C. Hu, P. Gao, C. Huang, H. Ren, X. Li, F. Qin, J. Yang, M. Gu, M. Hong, X. Luo, Catenary optics for achromatic generation of perfect optical angular momentum. Sci. Adv. **1**, e1500396 (2015)

39. X. Luo, M. Pu, X. Li, X. Ma, Broadband spin Hall effect of light in single nanoapertures. Light Sci. Appl. **6**, e16276 (2017)

40. S. Pancharatnam, Generalized theory of interference, and its applications. Part I. Coherent pencils. Proc. Indian Acad. Sci. **44**, 247–262 (1956)

41. M.V. Berry, Quantal phase factors accompanying adiabatic changes. Proc. R. Soc. Lond. Math. Phys. Eng. Sci. **392**, 45–57 (1984)

42. K.Y. Bliokh, D. Smirnova, F. Nori, Quantum spin Hall effect of light. Science **348**, 1448–1451 (2015)

43. Y. Zhao, M.A. Belkin, A. Alù, Twisted optical metamaterials for planarized ultrathin broadband circular polarizers. Nat. Commun. **3**, 870 (2012)

44. J.K. Gansel, M. Thiel, M.S. Rill, M. Decker, K. Bade, V. Saile, G. von Freymann, S. Linden, M. Wegener, Gold helix photonic metamaterial as broadband circular polarizer. Science **325**, 1513–1515 (2009)

45. J. Kaschke, L. Blume, L. Wu, M. Thiel, K. Bade, Z. Yang, M. Wegener, A helical metamaterial for broadband circular polarization conversion. Adv. Opt. Mater. **3**, 1411–1417 (2015)

46. Y. Guo, X. Ma, M. Pu, X. Li, Z. Zhao, X. Luo, High-efficiency and wide-angle beam steering based on catenary optical fields in ultrathin metalens. Adv. Opt. Mater. **6**, 1800592 (2018)

47. M. Zhang, M. Pu, F. Zhang, Y. Guo, Q. He, X. Ma, Y. Huang, X. Li, H. Yu, X. Luo, Plasmonic metasurfaces for switchable photonic spin-orbit interactions based on phase change materials. Adv. Sci. **5**, 1800835 (2018)

48. X. Li, M. Pu, Z. Zhao, X. Ma, J. Jin, Y. Wang, P. Gao, X. Luo, Catenary nanostructures as highly efficient and compact Bessel beam generators. Sci. Rep. **6**, 20524 (2016)

49. Y. Wang, M. Pu, Z. Zhang, X. Li, X. Ma, Z. Zhao, X. Luo, Quasi-continuous metasurface for ultra-broadband and polarization-controlled electromagnetic beam deflection. Sci. Rep. **5**, 17733 (2015)

50. X. Li, M. Pu, Y. Wang, X. Ma, Y. Li, H. Gao, Z. Zhao, P. Gao, C. Wang, X. Luo, Dynamic control of the extraordinary optical scattering in semicontinuous 2D metamaterials. Adv. Opt. Mater. **4**, 659–663 (2016)

51. A. Poddubny, I. Iorsh, P. Belov, Y. Kivshar, Hyperbolic metamaterials. Nat. Photonics **7**, 948–957 (2013)

52. C. Wang, P. Gao, X. Tao, Z. Zhao, M. Pu, P. Chen, X. Luo, Far field observation and theoretical analyses of light directional imaging in metamaterial with stacked metal-dielectric films. Appl. Phys. Lett. **103**, 031911 (2013)

53. D. de Klerk, J. Murugan, and J.-P. Uzan, The catenary revisited: from Newtonian strings to superstrings, arXiv:1103.0788 (2011)

54. Mirage, https://en.wikipedia.org/wiki/Mirage

55. Z. Wang, An explanation of mirage with linearly varying index of refraction. Coll. Phys. **20**, 24–27 (2001)

56. Z. Zhao, M. Pu, Y. Wang, X. Luo, The generalized laws of refraction and reflection. Opto-Electron. Eng. **44**, 129–139 (2017)

57. A. Kalvach, Z. Szabó, Aberration-free flat lens design for a wide range of incident angles. J. Opt. Soc. Am. B **33**, A66–A71 (2016)

58. M. Pu, X. Li, Y. Guo, X. Ma, X. Luo, Nanoapertures with ordered rotations: symmetry transformation and wide-angle flat lensing. Opt. Express **25**, 31471–31477 (2017)

59. W. Liu, Z. Li, H. Cheng, C. Tang, J. Li, S. Zhang, S. Chen, J. Tian, Metasurface enabled wide-angle Fourier lens. Adv. Mater. **30**, 1706368 (2018)

60. A. Arbabi, E. Arbabi, S.M. Kamali, Y. Horie, S. Han, A. Faraon, Miniature optical planar camera based on a wide-angle metasurface doublet corrected for monochromatic aberrations. Nat. Commun. **7**, 13682 (2016)

61. B. Groever, W.T. Chen, F. Capasso, Meta-lens doublet in the visible region. Nano Lett. **17**, 4902–4907 (2017)

62. M. Pu, N. Yao, C. Hu, X. Xin, Z. Zhao, C. Wang, X. Luo, Directional coupler and nonlinear Mach-Zehnder interferometer based on metal-insulator-metal plasmonic waveguide. Opt. Express **18**, 21030–21037 (2010)

63. E. Merzbacher, The early history of quantum tunneling. Phys. Today **55**, 44–50 (2002)

64. M. Razavy, *Quantum Theory of Tunneling*, 2nd ed. (World Scientific, 2014)

65. J.L. McFarland, Catenary geodesic lens antenna, U.S. patent 3,383,691 (1968)

66. J.S. Ajioka, J.L. McFarland, Beam-forming feeds, in *Antenna Handbook* (Chapman & Hall, 1993), Vol. III

67. K.H. Teo, D. Huang, J. Zhang, B. Yerazunis, B. Wang, Wireless energy transfer with negative material, U.S. patent US 20110133566 A1 (2011)

68. S. Assawaworrarit, X. Yu, S. Fan, Robust wireless power transfer using a nonlinear parity–time-symmetric circuit. Nature **546**, 387 (2017)

69. B. Wang, K.H. Teo, T. Nishino, W. Yerazunis, J. Barnwell, J. Zhang, Experiments on wireless power transfer with metamaterials. Appl. Phys. Lett. **98**, 254101 (2011)

70. L.D. Landau, E.M. Lifshitz, *Electrodynamics of Continuous Media* (Pergamon, 1984)

71. B. Averill, P. Eldredge, *Principles of General Chemistry* (2012)

72. B. Zhao, J. Yang, New effects in an ultracompact Young's double nanoslit with plasmon hybridization. New J. Phys. **15**, 073024 (2013)

73. X. Ma, Y. Guo, M. Pu, X. Li, X. Luo, Refined model for plasmon ruler based on catenary shaped optical fields. Plasmonics (2019)

74. Five hundred meter Aperture Spherical Telescope, http://www.en.wikipedia.org/wiki/Five_hundred_meter_Aperture_Spherical_Telescope

75. Brachistochrone Curve, http://www.en.wikipedia.org/wiki/Brachistochrone_curve

76. Beam Deflection Formulae, www.advancepipeliner.com/Resources/Others/Beams/Beam_Deflection_Formulae.pdf

77. S.H. Pepe, Polynomial and catenary curve fits to human dental arches. J. Dent. Res. **54**, 1124 (1975)

78. Y. Guo, M. Pu, X. Li, X. Ma, P. Gao, Y. Wang, X. Luo, Functional metasurfaces based on metallic and dielectric subwavelength slits and stripes array. J. Phys.: Condens. Matter **30**, 144003 (2018)

Chapter 2
Spin-Controlled Beam Shaping with Catenary Subwavelength Structures

Abstract In this chapter, we describe the photonic spin–orbit coupling in catenary-shaped subwavelength structures. First of all, the basic theories of geometric phase, spin-momentum locking are used to introduce the catenary of equal strength, which could generate a phase profile with equal phase gradient. Then, a general mathematical approach is provided to design various continuous structures deformed from the catenary function. These structures are used to realize broadband photonic spin Hall effect, flat lenses, orbital angular momentum, Bessel and Airy beam generators. The intrinsic limitation on the efficiency of a single-layer catenary metasurface is discussed. To realize high-efficiency functional spin-controlled beam shaping, the reflective and all-dielectric configurations have been presented. Finally, it is shown that the coherent control originally utilized in lasers and absorbers could be leveraged to dynamically control the output intensity.

Keywords Spin Hall effect · Geometric phase · Optical vortex · Flat lens · Dielectric metasurface

2.1 Introduction to Spin, Linear and Angular Momentum of Light

In the quantum optical description, the spin and linear/angular momentum are three basic parameters of photons [1]. To understand the spin-controlled beam shaping, it is helpful to make a concise discussion of their definitions.

First of all, the linear momentum of light (p) is associated with the wavevector k of light wave as

$$p = \hbar k, \tag{2.1.1}$$

where $\hbar$ is the reduced Planck's constant. In vacuum, the magnitude of wavevector increases along with frequency, in a way similar to the photon energy.

© Springer Nature Singapore Pte Ltd. 2019
X. Luo, *Catenary Optics*, https://doi.org/10.1007/978-981-13-4818-1_2

Second, the spin-angular momentum of photon corresponds to the circular polarization of light wave in classical optics. The magnitude of the photon spin is quantized to two opposite values:

$$S = \pm\hbar, \tag{2.1.2}$$

where the plus (minus) signs refer to the right-handed (left-handed) circular polarizations (RCP and LCP). In a plane wave, the spin vector is parallel to the wavevector. For a linear polarization, the photons have an equal probability to be either of the spin state.

Besides the spin-angular momentum, light waves with inhomogeneous spatial distribution may possess another kind of angular momentum. For instance, the Laguerre-Gaussian beam characterized by an azimuthal helical phase dependence $\exp(il\varphi)$ has a polarization-independent angular momentum of $l\hbar$. Similarly, the photons in a whispering gallery mode resonator also possess angular momentums depending on the resonant modes [1].

In the following discussion, we will discuss the light–matter interaction from the wave optics and photonics interchangeably. According to some recent studies, James Clerk Maxwell actually revealed a correct relativistic, quantum theory for the light quantum [2]. As a result, almost all of the following theories may be fully described by Maxwell's equations, if proper material dispersion models are adopted.

2.2 Spin-Momentum Locking in Free Space and Guided Waves

In general, the spin, linear, and angular momenta are independent and do not couple with each other. One particular scenario, where the couple is the surface plasmon polariton (SPP), which possesses intrinsic quantum spin Hall effect (QSHE) of light [3]. Since SPP is a transverse magnetic polarized wave, and its two electric components along and perpendicular to the propagation direction have a 90° phase shift, a transverse spin momentum can be defined. Similar to quantum spin Hall effect of the electron, the spin momentum would reverse for oppositely propagating SPPs. It was thus concluded that any interface between free space and a medium supporting surface or guided modes with evanescent tails exhibits counter-propagating opposite-spin edge modes.

In this section, we would like to give the relation of catenary function with the spin-momentum locking. Besides the QSHE of SPP, it is shown that obliquely incident circularly polarized light (CPL) would generate vectorial fields with space-variant polarizations. When this pattern is used as polarization holography, CPL would be diffracted to a direction depending on its handedness. That is to say, this polarization pattern also records one kind of information with spin-momentum locking.

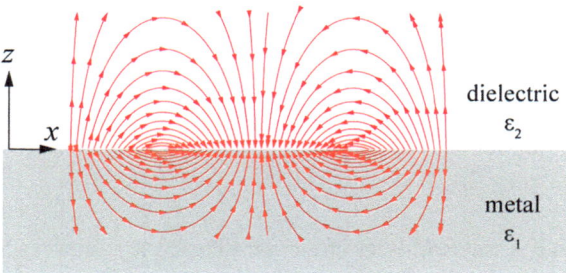

Fig. 2.1 Electric field distribution for a surface plasmon on the metal–dielectric surface

2.2.1 Guided Wave: Surface Plasmon Polaritons

Recent studies showed that guided surface waves such as SPPs also possess universal spin-momentum locking. Using boundary conditions shown in Fig. 2.1, the electric and magnetic fields for TM polarization can be written as [4]

$$H_y(z) = A e^{i\beta x} e^{k_1 z}$$

$$E_x(z) = -iA \frac{1}{\omega \varepsilon_0 \varepsilon_1} k_1 e^{i\beta x} e^{k_1 z}$$

$$E_z(z) = -A \frac{\beta}{\omega \varepsilon_0 \varepsilon_1} e^{i\beta x} e^{k_1 z} \tag{2.2.1}$$

for $z < 0$, and

$$H_y(z) = A e^{i\beta x} e^{-k_2 z}$$

$$E_x(z) = iA \frac{1}{\omega \varepsilon_0 \varepsilon_2} k_2 e^{i\beta x} e^{-k_2 z}$$

$$E_z(z) = -A \frac{\beta}{\omega \varepsilon_0 \varepsilon_2} e^{i\beta x} e^{-k_2 z} \tag{2.2.2}$$

for $z > 0$. Taking the electric fields at $z > 0$ as an example, its x and z components have a phase shift of 90°, which may be reversed if the propagation direction is inverted.

The wavevector relations of the two waves can be written as

$$k_1^2 = \beta^2 - k_0^2 \varepsilon_1,$$

$$k_2^2 = \beta^2 - k_0^2 \varepsilon_2. \tag{2.2.3}$$

Using the imaginary components in Eqs. (2.2.1) and (2.2.2), the differential equations of the electric field lines may be written as

$$\frac{dz_1}{dx} = \frac{\text{Im}(E_z)}{\text{Im}(E_x)} = \frac{\beta}{k_1} \tan \beta x \qquad (2.2.4)$$

and

$$\frac{dz_2}{dx} = \frac{\text{Im}(E_z)}{\text{Im}(E_x)} = -\frac{\beta}{k_2} \tan \beta x. \qquad (2.2.5)$$

By integrating with x, the field lines are found to take the following forms:

$$z_1 = \frac{\beta}{k_1} \int \tan \beta x dx = -\frac{1}{k_1} \ln \cos \beta x, \qquad (2.2.6)$$

$$z_2 = \frac{\beta}{k_2} \int \tan \beta x dx = \frac{1}{k_2} \ln \cos \beta x. \qquad (2.2.7)$$

Equations (2.2.6) and (2.2.7) are often called catenary of equal strength, which was first investigated by Davies Gilbert in 1826 [5].

2.2.2 Free Space: Circularly Polarized Beam at Oblique Incidence

To understand spin-momentum locking in free space, circular polarizations at oblique incidence are investigated. Assuming two light beams propagate in the xz-plane, and the wavenumbers in the x-direction are k_x and $-k_x$, the electric fields can be written as

$$\begin{aligned} E_x &= \exp(ik_x x) + \exp(-ik_x x), \\ E_y &= i \exp(ik_x x) - i \exp(-ik_x x). \end{aligned} \qquad (2.2.8)$$

The real parts can be written as

$$\begin{aligned} \text{Re}(E_x) &= 2\cos(k_x x), \\ \text{Re}(E_y) &= -2\sin(k_x x). \end{aligned} \qquad (2.2.9)$$

The electric field lines are determined by the differential equation

$$\frac{dy}{dx} = \frac{\text{Re}(E_y)}{\text{Re}(E_x)} = -\tan(k_x x), \qquad (2.2.10)$$

which can be integrated to obtain

$$y = \frac{1}{k_x} \ln \cos(k_x x). \qquad (2.2.11)$$

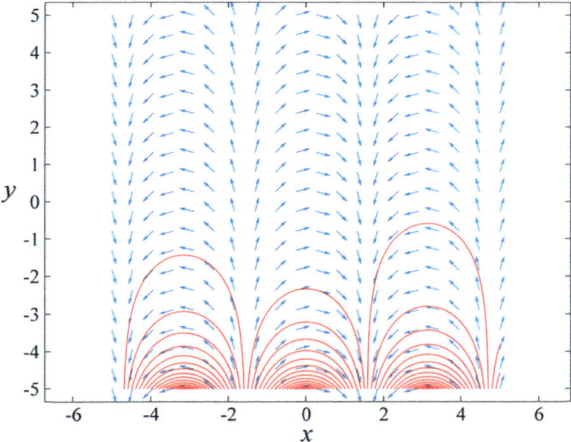

Fig. 2.2 Vectorial electric fields distribution of the interference fields of circular polarization with $k_x = 1$. The red continuous curve is the streamline representation of catenary of equal strength

Figure 2.2 shows the electric fields vectors for $k_x = 1$, which shows a horizontal periodicity of 2π. The electric force lines are obvious catenary with a periodicity of π.

Similarly, the electric vectors of a single oblique circularly polarized beam are

$$E_x = \exp(ik_x x),$$
$$E_y = i \exp(ik_x x). \qquad (2.2.12)$$

At each instantaneous time, the real parts can be written as

$$\mathrm{Re}(E_x) = \cos(k_x x),$$
$$\mathrm{Re}(E_y) = -\sin(k_x x). \qquad (2.2.13)$$

This is just the same as the fields for the interference of two counter-propagating circularly polarized waves. As a result, it can be concluded that this catenary field also corresponds to a spin-momentum locking. For a given polarization distribution, if the incident direction of circular polarization is reversed, the opening direction of the catenary would be flipped because there is

$$\mathrm{Re}(E_x) = \cos(k_x x),$$
$$\mathrm{Re}(E_y) = \sin(k_x x). \qquad (2.2.14)$$

The above knowledge may be used to control the propagation direction by varying the polarization states. By generating such a space-variant polarization state with rotated half-wave plates under circularly polarized incidence, the incident wave can be deflected with a spin-dependent propagation direction [6]. In memory of the scientific contributions of Pancharatnam and Berry, the equivalent phase shift of CPL induced by polarization conversion is also termed geometric phase, topological phase or Pancharatnam–Berry phase [7–9], which has been extensively studied for various applications in recent years [10–15].

2.3 Spin Hall Effect Generated by a Single Catenary Aperture

By taking the geometric phase into consideration, when CPL incidents normally on a catenary-shaped aperture or nano-antennas, the output beam would act as the case of oblique incidence. This polarization-dependent deflection is often called angular spin Hall effect of light (SHEL). To reveal this interesting effect, a single catenary-shaped nano-aperture is investigated with rigorous numerical methods in the following.

Without the loss of any generality, let us first discuss a slim subwavelength aperture perforated on a metallic screen with vanishing thickness. The word "slim" means that the aperture has a length much larger than the width, while the width is much smaller than the operating wavelength. As shown in Fig. 2.3a, such an aperture could be mathematically treated with two parameters, i.e., the inclination angle $\xi(x)$ with respect to the x-axis and the local width w.

Since the early work of Bethe in 1944 [17], it was known that traditional Kirchhoff's diffraction theory does not fulfil the electromagnetic boundary conditions for small apertures even if the thickness of the screen is decreased to zero. In particular, owing to the polarization-dependent transmission [18], the incident fields do not equal to the fields at the output side of the aperture, and the diffraction (or scattering) phenomena may become completely different from that predicted by Kirchhoff's theory. As depicted in Fig. 2.4a, the catenary aperture is obtained by connecting two "catenary of equal strength" curves with a vertical shift of $\Delta \ll \lambda$

$$\begin{cases} y_1 = \frac{\Lambda}{\pi} \ln(|\sec(\pi x/\Lambda)|), \\ y_2 = \frac{\Lambda}{\pi} \ln(|\sec(\pi x/\Lambda)|) + \Delta, \end{cases} \tag{2.3.1}$$

where Λ is the horizontal length of the catenary. Because the value of y is infinite at $x = \pm \Lambda/2$, the curves must be truncated at the two ends with a value of δx thus the span of x is $(-0.5\Lambda + \delta x, 0.5\Lambda - \delta x)$.

Under CPL illumination, the anisotropic transmission would result in a geometric phase for the cross-polarized light, which is twice the inclination angle of the aperture and can be written as $\Phi = 2\sigma\xi(x)$ [9]. Here $\sigma = \pm 1$ denotes the LCP and RCP, respectively. Since the inclination angle of the upright catenary aperture has a form of

$$\xi(x) = \tan^{-1}\left(\frac{dy}{dx}\right) = \frac{\pi}{\Lambda}x, \tag{2.3.2}$$

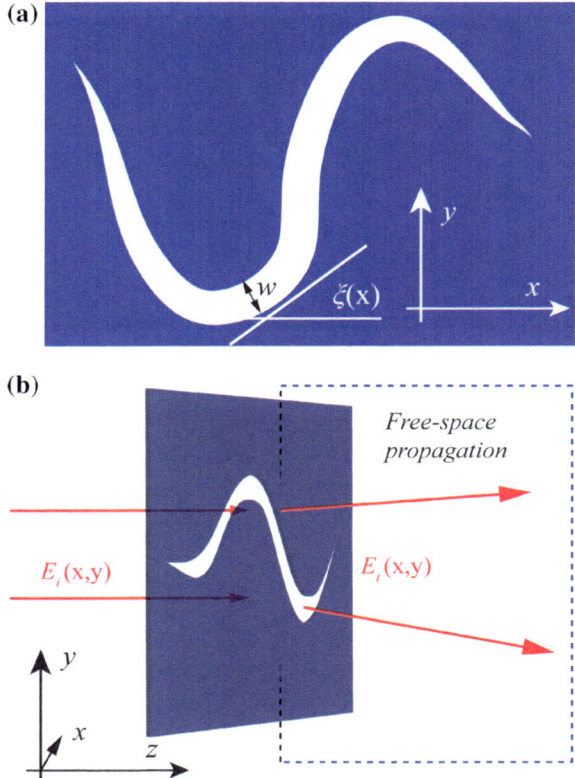

Fig. 2.3 Diffraction of light by an arbitrary aperture on a conducting screen. **a** Geometry of the structure. **b** Sketch of the diffraction problem. Here $E_i(x, y)$ and $E_t(x, y)$ denote the incident and transmitted electric fields. Once $E_t(x, y)$ is obtained, the diffraction fields can be evaluated by free-space propagation methods. Reproduced from [16] with permission. Copyright 2017, the Author(s)

there is a linear phase distribution of $\Phi(x) = 2\sigma\pi x/\Lambda$ at the output boundary of the catenary aperture. Note that the phase is distributed along a curved aperture, thus it can also be written as a function of y

$$\Phi(y) = \begin{cases} -2\sigma \ \arccos\left(e^{-y\pi/\Lambda}\right), x \leq 0, \\ 2\sigma \ \arccos\left(e^{-y\pi/\Lambda}\right), x > 0. \end{cases} \qquad (2.3.3)$$

Owing to the geometric phase, the diffraction of the cross-polarization violates Kirchhoff's theory. Intuitively speaking, a single element of the catenary aperture can act as an artificial bias for the incident CPL, and show dramatic angular SHEL. The theoretical deflection angle of the cross-polarized light (at normal incidence) is $\theta = \sigma \sin^{-1}(\lambda/\Lambda)$, corresponding to an additional horizontal momentum of $\Delta k = 2\sigma\pi/\Lambda$.

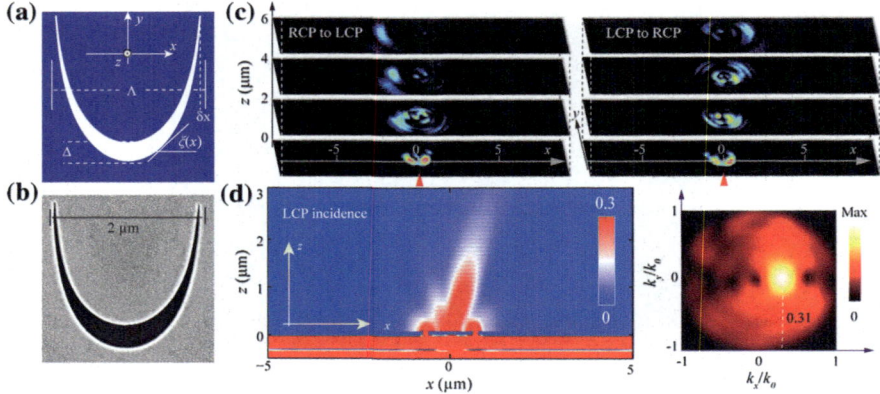

Fig. 2.4 Angular SHEL observed in a single catenary aperture. **a** Schematic of the catenary aperture. **b** Scanning electron microscope (SEM) image of the fabricated catenary aperture with $\Lambda = 2\,\mu m$, $\Delta = 200$ nm and $\delta x = 50$ nm; **c** Measured cross-polarized intensity patterns in the xy-planes for $z = 0, 2, 4, 6\,\mu m$ under RCP (left panel) and LCP (right panel) illumination ($\lambda = 632.8$ nm). The top surface of the sample is set at $z = 0$. **d** Left: cross-polarized electric intensity in the xz-plane ($y = 1\,\mu m$). Right: amplitude of the far-field angular spectrum in the k-space. Reproduced from [16] with permission. Copyright 2017, the Author(s)

In contrast to the cross-polarization, the co-polarized light would not be deflected since it does not acquire any geometric phase.

As shown in Fig. 2.4b, a catenary aperture with $\Lambda = 2\,\mu m$ and $\Delta = 200$ nm was fabricated by focused ion beam (FIB) milling on a 120 nm-thick gold film deposited on a SiO_2 substrate. In the experiment, the incident light was set to propagate perpendicularly to the sample plane. Under LCP and RCP illuminations at $\lambda = 632.8$ nm, the cross-polarized (RCP and LCP) fields transmitted from the aperture were measured using a homemade microscope. Figure 2.4c represents the intensity distribution in the xy-planes, revealing that the CPL has been deflected by the catenary aperture along the x-axis for LCP and RCP, respectively. Figure 2.4d shows the cross-polarized near-field electric fields and far-field scattering of this aperture calculated by a commercial electromagnetic software CST Microwave Studio (MWS). This far-field scattering corresponds well with the experimental results, indicating a deflection angle of ~$\pm 18°$ (the theoretical value is $\pm 18.42°$). In the simulation, the metal film was assumed to be a perfect electric conductor (PEC) since the excited SPP almost does not contribute to the process. It should be noted that the phase retardation originating from the SPP propagation [19] does exist but can be neglected since it is much smaller than the geometric phase for such thin metallic film.

Owing to the geometric nature of the spin–orbit conversion, the handedness scattering could be observed at a broadband frequency range. To demonstrate this, the far-field scattering power is calculated for the single catenary aperture illuminated by LCP light at different wavelengths. As shown in Fig. 2.5a, the cross-polarized scattering is larger than the co-polarization within the whole range. Since the electric

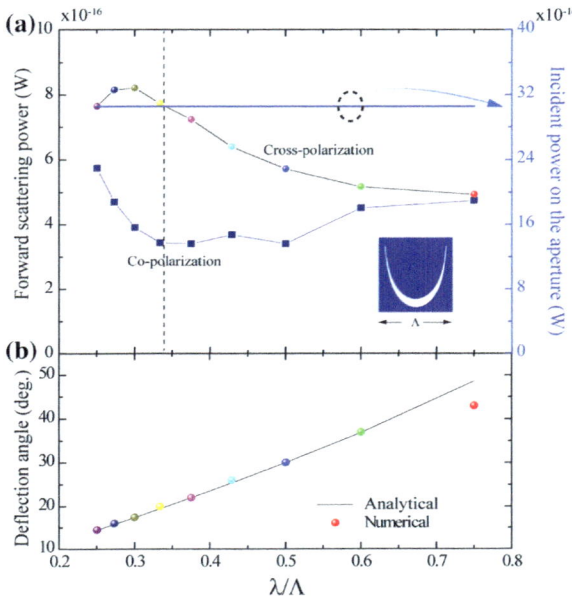

Fig. 2.5 Broadband light scattering by a single catenary aperture. **a** Calculated scattering power in the forward direction for both the co- and cross-polarized components (left-axis), and the total input LCP power integrated on the aperture (right-axis). The scattering power was obtained by integrating the power pattern. **b** Analytical and numerically calculated deflection angles at different wavelengths. Reproduced from [16] with permission. Copyright 2017, the Author(s)

field amplitude of incident plane wave is set as 1 V/m, the power of the light incident on the aperture can be evaluated as 30×10^{-16} W. The conversion efficiency, defined as the ratio of the cross-polarization power to the input power, is below 25%, except for the range where $0.25 \leq \lambda/\Lambda \leq 0.35$. The relatively low efficiency may be attributed to the reflection of the CPL. In addition, the cross-polarized component becomes evanescent and does not contribute to the far field when $\lambda/\Lambda > 1$. In the short wavelength region ($\lambda/\Lambda < 0.25$), the degradation of performance would be attributed to the increased diffraction since the wavelength is comparable to Δ.

Figure 2.5b plots the deflection angles evaluated from CST MWS at different wavelengths, showing good agreement with the analytical results obtained via the relation $\theta = \sigma \sin^{-1}(\lambda/\Lambda)$. The deviation at large angle is mainly because the beam width increases with the deflection angle, which would, in turn, induce stronger scattering by the edge of the aperture.

The spin Hall effect in the catenary aperture is owing to its space-variant anisotropic structures. Although similar effects can be found in other structures, the catenary shows obvious advantages on the operating bandwidth and diffraction efficiency. As shown in Fig. 2.6, there are two kinds of plasmonic chains exhibiting anisotropic transmissivity [20]. On the first hand, although a single coaxial-shaped

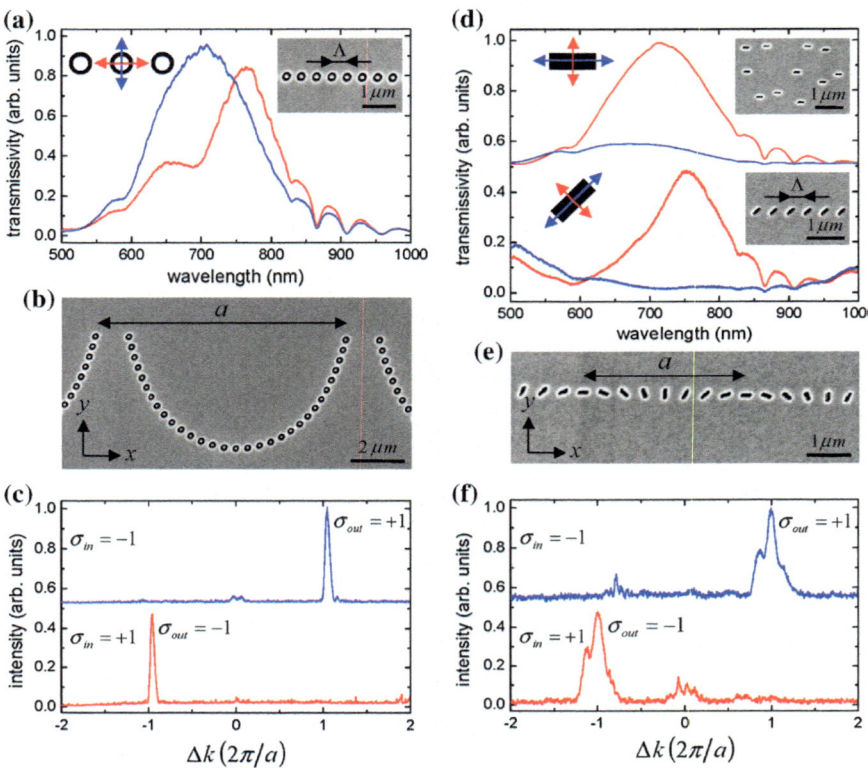

Fig. 2.6 Spin Hall effect in discrete plasmonic chains. **a** Plasmonic chain consisting of coaxial nanoapertures. The blue and red lines/arrows correspond to transversal and longitudinal polarization incidence. The inset shows a SEM image of the chain. **b** SEM image of a catenary-shaped chain whose local orientation angle is varied linearly along the x-axis with a rotation period of $a = 9\ \mu m$. **c** The spin-dependent momentum deviation at a wavelength of 780 nm. **d** Transmission spectra of randomly arranged identically oriented rectangular apertures and of a homogeneous chain with local orientation of 45°. **e** SEM image of a chain in which the nanorods' orientation angle varies linearly along the x-axis with a rotation period of $a = 3.44\ \mu m$. **f** Spin-dependent momentum deviation at a wavelength of 730 nm. Reproduced from [20] with permission. Copyright 2011, American Chemical Society

aperture is intrinsically isotropic, a set of rings would induce slightly different coupling and transmissivity along the two directions parallel and perpendicular to the chain. On the other hand, rectangular apertures, either in the form of random or periodic distribution, show stronger anisotropy in the transmission. Nevertheless, both the two approaches have much smaller bandwidth and efficiency compared with the results shown in Fig. 2.5.

Although the function of a single catenary is relatively simple, a straightforward deformation of the catenary aperture can provide more freedoms for the light manipulation on the nanoscale. As shown in Fig. 2.7a, b, the left half part of the catenary

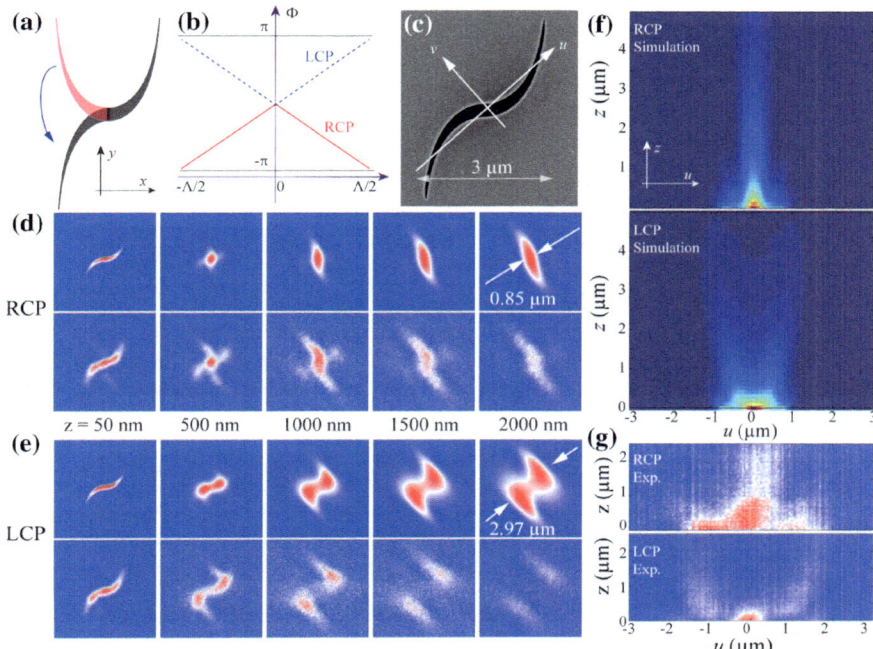

Fig. 2.7 Bessel beam generation by deformed catenary aperture. **a** Schematic of the deformed catenary aperture. **b** Phase distribution of the deformed catenary under LCP (dashed blue) and RCP (solid red) illumination. **c** SEM image of the deformed catenary with $\Lambda = 3\,\mu$m, $\Delta = 300$ nm and $\delta x = 50$ nm. **d, e** Theoretical (top rows) and experimental (bottom rows) cross-polarized intensity for LCP (**d**) and RCP (**e**) incidence. The patterns are recorded at different distances away from the aperture as indicated by the value of z. The dimensions of all panels in (**d**) and (**e**) are $6\,\mu$m $\times$ $6\,\mu$m. **f** Numerically calculated intensity distribution in the uz-plane for RCP and LCP incidence. The angle between the u- and x-directions is 30°. **g** Experimental intensity distribution in the uz-plane. Reproduced from [16] with permission. Copyright 2017, the Author(s)

was flipped upside down, thus the geometric phase can be written as

$$\Phi(x) = \begin{cases} -\sigma\pi x/\Lambda, \ x \in (-0.47\Lambda, 0], \\ \sigma\pi x/\Lambda, \ x \in (0, 0.47\Lambda). \end{cases} \tag{2.3.4}$$

The phase distribution along the y-direction can be obtained as

$$\Phi(y) = \begin{cases} 2\sigma \ \arccos\left(e^{y\pi/\Lambda}\right), \ y \le 0, \\ 2\sigma \ \arccos\left(e^{-y\pi/\Lambda}\right), \ y > 0. \end{cases} \tag{2.3.5}$$

From Eq. (2.3.4), it can be seen that such a deformed catenary aperture may be utilized to create two counter-propagating waves along the x-direction, similar to the case of the diffraction-free Bessel beam, which is made up of plane waves traveling on a cone [21]. In order to test the optical performance of the deformed catenary, a sample with $\Lambda = 3$ μm and $\Delta = 300$ nm was fabricated as shown in Fig. 2.7c. The intensity distribution for cross-polarized component was measured at different distances away from the sample and compared with the results obtained from vectorial diffraction theory [22], where the phase distribution is assumed to follow Eqs. (2.3.4) and (2.3.5). As illustrated in Fig. 2.7d, e, the experimental results agree very well with the theoretical evaluations, except that there is some qualitative mismatch resulting from the approximation in the phase profiles.

Besides phase modulation in the x-direction, the deformed catenary also possesses a phase gradient in the y-direction, as indicated in Eq. (2.3.5). As a result, the deflection direction of the CPL is actually along the diagonal direction (the u-direction) as indicated in Fig. 2.7c and the last column of Fig. 2.7d, e. For LCP illumination, the cross-polarized light fields are gradually separated along with its propagation. At $z = 2$ μm, the separating width becomes 2.97 μm. In contrast, a narrow light spot was obtained for RCP illumination, with a beam width of 0.85 μm at $z = 2$ μm. Figure 2.7f, g plot the intensity profiles in the uz-plane for both the numerical and experimental results. Owing to the low transmission and the noise stemming from the incomplete polarization conversion, the measured intensity became blurred for $z > 2.5$ μm.

To further validate the above theory, a U-shaped aperture is also analyzed both numerically and experimentally. Once again, PEC approximation is used in the numerical simulation. As depicted in the SEM image (Fig. 2.8a), the horizontal length of the U-shaped aperture is 2 μm. For the co-polarized light which does not acquire geometric phase, the diffraction is mainly determined by the traditional Kirchhoff's diffraction theory. Different from the catenary aperture, the U-shaped aperture can generate many diffraction orders for the co-polarization, as shown in both the numerical simulation (the left column of Fig. 2.8b) and experimental measurement (the left column of Fig. 2.8c). For the cross-polarized light, the geometric phase dominates the diffraction behavior. There are many unwanted diffraction orders for the U-shaped aperture (the right column of Fig. 2.8b, c) because the geometric phase is not linearly distributed, in contrast to the catenary aperture. Note that this polarization-dependent effect has also been investigated in similar structures without optimization [23].

2.4 Integration Design of the Catenary Array

The physical mechanism of the geometric phase in catenary arrays is the anisotropic transmission in the local subwavelength structures. By matching the electromagnetic modes at the boundary, Martín-Moreno et al. gave a fitted formula to calculate the anisotropic transmission of a rectangular hole in a perfect conductive screen with zero thickness [24]

$$T = \left(\frac{2\pi a_x}{\lambda} \right)^2 \left(\frac{2\pi a_y}{\lambda} \right)^2 C(\tau), \qquad (2.4.1)$$

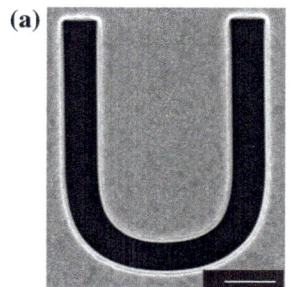

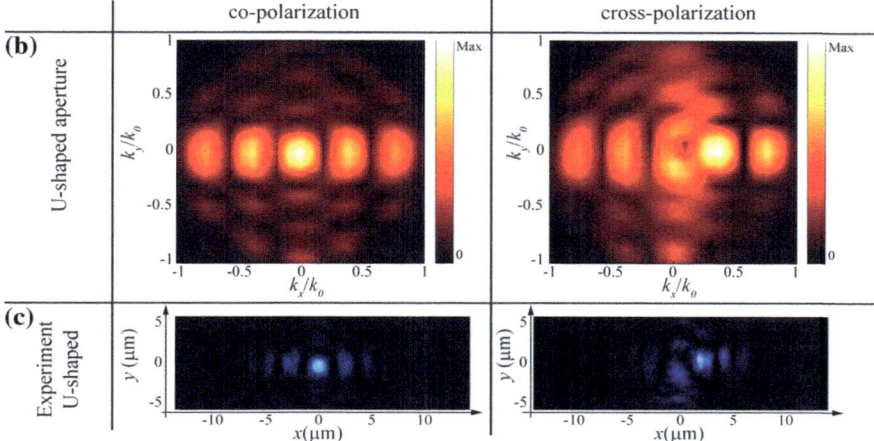

Fig. 2.8 Diffraction patterns of a U-shaped aperture under LCP illumination. **a** SEM image of the U-shaped aperture. **b** Co- and cross-polarized components of the far-field diffraction of the U-shaped aperture with $\Lambda = 2\,\mu$m. **c** Experimental intensity of the U-shaped aperture for the co- and cross-polarized components in the xy-plane at $z = 8\,\mu$m. Reproduced from [16] with permission. Copyright 2017, the Author(s)

where a_x and a_y are the horizontal and vertical lengths of the aperture, λ is the operational wavelength, $\tau = a_x/a_y$ is the aspect ratio. When τ is within the range [1/3, 3], there is $C(\tau) = 0.0132 + 0.2127/\tau + 0.2174/\tau^2$. Clearly, the transmittance ratio of the two polarized stated is $TR = C(\tau)/C(1/\tau)$. For $\tau = 1/3$, there is $TR = 24$, which is large enough to obtain a geometric phase.

For homogenized anisotropic materials, the transmission property for polarized light can be easily obtained by using the Jones matrix formalism. As shown in Fig. 2.9, the Jones matrix for a subwavelength thin metallic grating with a period of p, a width of $w = p/2$, and main axes along the local u-v coordinates takes the form of:

$$J_g = \begin{bmatrix} t_u & 0 \\ 0 & t_v \end{bmatrix}, \tag{2.4.2}$$

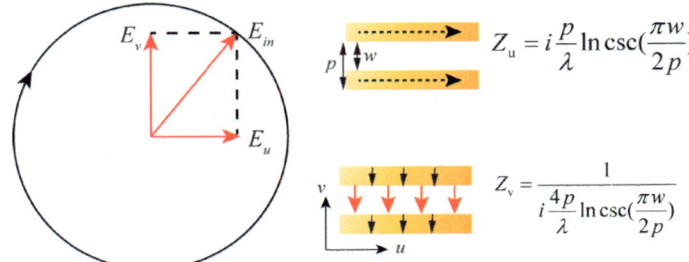

Fig. 2.9 Schematic of the subwavelength metallic grating and the local coordinate. In the u-direction, the grating acts as an inductor, which reflects incident light. In the v-direction, a series circuit is formed which mainly transmit incident light. Both the two sheet impedances could be described by the catenary of equal strength. Reproduced from [6] with permission. Copyright 2015, the Author(s)

where t_u and t_v are the transmission coefficients along the two main axes, which can be calculated using the generalized Fresnel's equations [19] and sheet impedances described by catenary functions [25, 26]. Taking the extreme case $w \ll p$ as an example, there is $t_u = 0$ and $t_v = 1$. Supposing that the u-direction has a rotation angle of ξ with respect to the x-axis, the general Jones matrix in the xy coordinates can be obtained as

$$
J_\xi = \begin{bmatrix} \cos\xi & -\sin\xi \\ \sin\xi & \cos\xi \end{bmatrix} J_g \begin{bmatrix} \cos\xi & \sin\xi \\ -\sin\xi & \cos\xi \end{bmatrix}
$$
$$
= \begin{bmatrix} t_u \cos^2\xi + t_v \sin^2\xi & (t_u - t_v)\sin\xi\,\cos\xi \\ (t_u - t_v)\sin\xi\,\cos\xi & t_u \sin^2\xi + t_v \cos^2\xi \end{bmatrix}. \tag{2.4.3}
$$

Finally, the output fields for CPL input is

$$
\begin{bmatrix} E_x \\ E_y \end{bmatrix} = \frac{J_\xi}{\sqrt{2}} \begin{bmatrix} 1 \\ i\sigma \end{bmatrix} = \frac{1}{2\sqrt{2}} \left((t_u + t_v) \begin{bmatrix} 1 \\ i\sigma \end{bmatrix} + (t_u - t_v)e^{2i\sigma\xi} \begin{bmatrix} 1 \\ -i\sigma \end{bmatrix} \right). \tag{2.4.4}
$$

Therefore, the output fields are composed of two circular polarizations with opposite handedness. The additional phase of the counter-rotating polarization $2\sigma\xi$ is purely geometrical and independent of the working frequency. The conversion efficiency is defined as the cross-polarized transmittance to the overall transmittance

$$
\eta = \frac{|t_u - t_v|^2}{|t_u - t_v|^2 + |t_u + t_v|^2}, \tag{2.4.5}
$$

where $0.5|t_u + t_v|$ and $0.5|t_u - t_v|$ are just the amplitudes of the co-polarized and cross-polarization transmission coefficients for circular polarizations. As indicated by Eq. (2.4.4), the difference in transmission along the u- and v-directions is essential

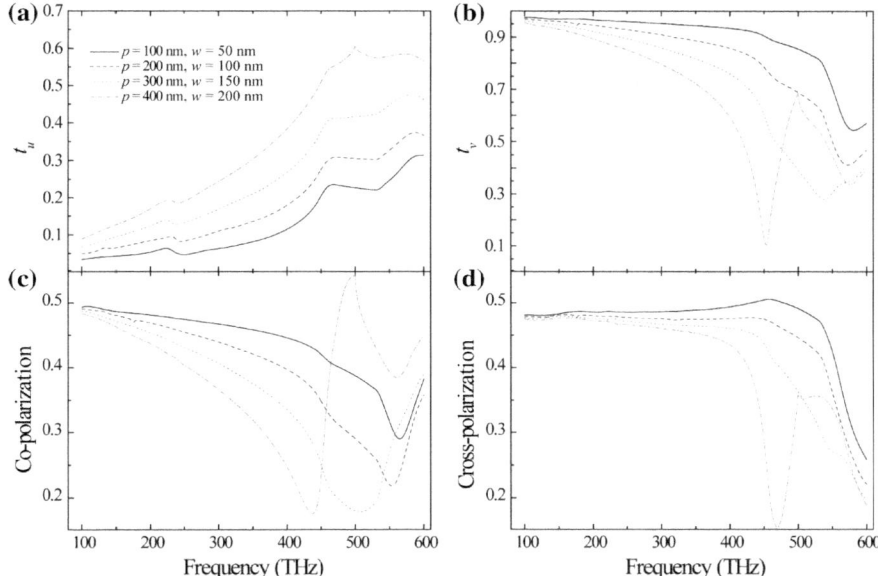

Fig. 2.10 Transmission coefficients of the subwavelength grating. **a** Transmission coefficients of the light polarized along the u-direction. **b** Transmission coefficients of the light polarized along the v-direction. **c** Co-polarized transmission for circular polarization. **d** Cross-polarized transmission for circular polarization. The periods are varying from 100 to 400 nm. The abrupt change for $p = 400$ nm at $f = 500$ THz ($\lambda = 600$ nm) is owing to the diffraction in the substrate, where the dielectric constant is near 1.5 and the effective wavelength in the substrate equals to the period. Reproduced from [6] with permission. Copyright 2015, the Author(s)

to obtain high conversion efficiency. If the anisotropic material is comprised of space-variant half-wave plate, there is $t_u + t_v = 0$ and the co-polarized transmittance would become zero.

Generally speaking, the overall phase shift would deviate from the perfect geometric phase since t_u and t_v may carry additional phase shift. By combining the two kinds of phase, the phase shift can be even independently modulated for two opposite circular polarizations [12, 13, 27, 28]. In the following, full-wave calculations are carried out to obtain the t_u and t_v for the catenary aperture. For simplicity, the catenary is treated as gratings with varying width w and period p. We calculated the transmission coefficients for different period p, while the width is set as $w = p/2$. The co-polarized and cross-polarized transmission coefficients are calculated by $0.5|t_u + t_v|$ and $0.5|t_u - t_v|$. As shown in Fig. 2.10, the transmission coefficients vary with the period at higher frequencies, thus the catenary would also exert amplitude modulation on the incident light. It should be noted that these amplitude fluctuations are much weaker at the low-frequency regime, where the structures are in deep-subwavelength scale and the metals behave more like PEC.

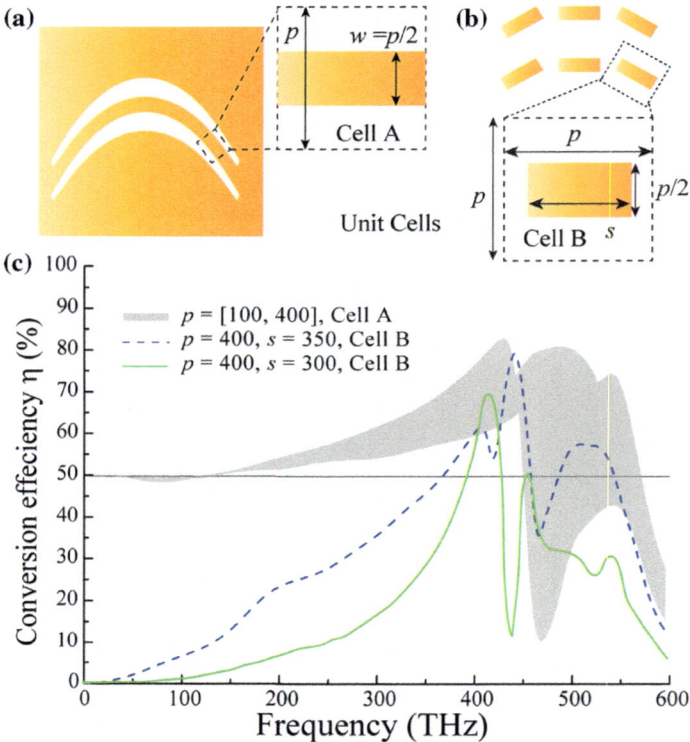

Fig. 2.11 Comparison of the conversion efficiencies and electromagnetic modes of the catenary apertures and discrete nano-antennas. **a, b** Schematic of the unit cells in catenary apertures (**a**) and discrete antennas (**b**). The two cells differ from each other in the length of the metallic bar with respect to the period. **c** Numerically calculated polarization conversion efficiency η for the catenary apertures and nano-antennas. Reproduced from [6] with permission. Copyright 2015, the Author(s)

To compare the performance of discrete and continuous apertures, their conversion efficiencies are calculated as shown in Fig. 2.11. Both the two kinds of structures are modeled with a unit cell comprised of a rectangular metallic antenna: if the length of the antenna is equal to the period, the expanded structure is a continuous grating. Compared with the discrete structures, the continuous gratings have much higher efficiency in almost the entire electromagnetic spectrum. Since the structure is ultrathin compared with the wavelength, the conversion efficiency has a fundamental limit of 50%, i.e., there is 50% energy in the co-polarized light which does not change its polarization state.

In the following, let us discuss a simple scheme for the design of various catenary structures with user-defined phase functions. For the simplicity of description, LCP illumination ($\sigma = -1$) is assumed, i.e., $\Phi = 2\xi$, in the derivation. In general, the design procedure of the catenary apertures consists of two steps. First, write explicitly the space-variant angle $\xi(x, y)$ according to the required phase distributions as given

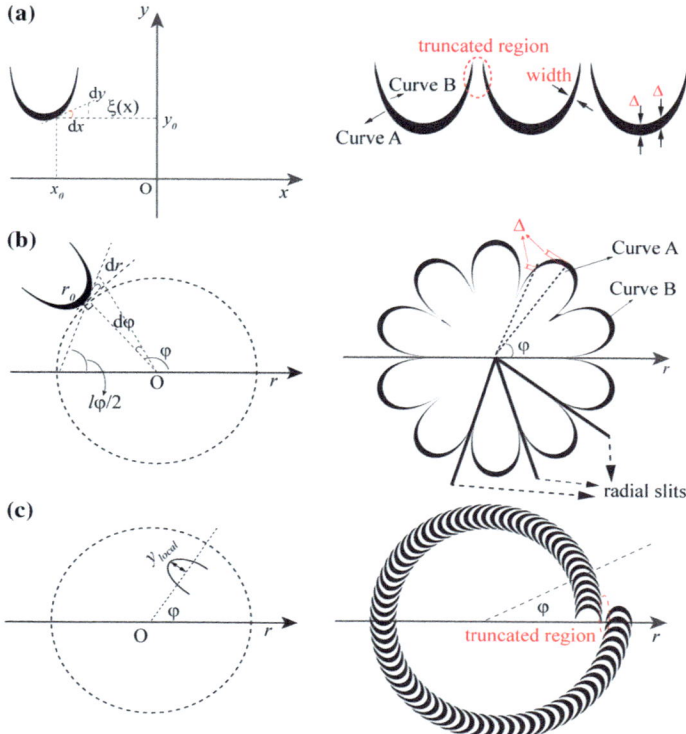

Fig. 2.12 Schematic of the integration procedures to design the catenaries. **a** Linear array in the Cartesian coordinates. Note that the vertical shift of the catenary curve is always Δ, while the width of the catenary aperture vary from place to place. **b** Rotated arrays in the polar coordinates for the generation of the OAM beam. **c** Rotated arrays in the polar coordinates for the generation of Bessel beam and focused OAM beam. Reproduced from [6] with permission. Copyright 2015, the Author(s)

by the geometric relation $\xi(x, y) = \Phi(x, y)/2$; Then, integrate the tangent of the angle, i.e., $\tan \xi(x, y)$, over Cartesian or polar coordinates. For a linear phase distribution along the x-direction, $dy/dx = \tan(\pi x/\Lambda)$ can be directly integrated over the x-axis to obtain the catenary of equal strength. As shown in Fig. 2.12a, a single catenary aperture is obtained by shifting the catenary function with Δ along the y-direction. Once again, because the value of catenary function is infinite for $x = \pm\Lambda/2 + m\Lambda$ (m is an integer), the curves are truncated at the two ends with $\Lambda/20$ in practical design. After the truncation, the four tips of the two catenary curves are connected to obtain the catenary aperture.

For the generation of orbital angular momentum (OAM), the surface structure should be designed so that the phase distribution is $\Phi(r, \varphi) = l\varphi$. As illustrated in the geometric relation in Fig. 2.12b, the tangent angle of the curve at a given position (r, φ) with respect to the azimuthal direction is $\xi = l\varphi/2 - \varphi + \pi/2$. Consequently,

Fig. 2.13 Curves for the generation of perfect OAM. The left and right are obtained with $l = -10$ and 10, respectively. Although they are not the exact catenary curves, the left figure resembles the spider web, which is often regarded as a type of catenary curve

the equation describing the surface profile can be written as

$$\frac{\mathrm{d}r}{r\mathrm{d}\varphi} = -\tan\xi = -\tan\left[\frac{(l-2)\varphi + \pi}{2}\right]. \tag{2.4.6}$$

By some nontrivial mathematical manipulation, the unique curve for generating OAM could be obtained as

$$\begin{aligned} r &= (r_0 + m\Delta)\exp\left\{\frac{2}{2-l}\ln\left(\left|\sec\left[\frac{(l-2)\varphi + \pi}{2}\right]\right|\right)\right\} \\ &= (r_0 + m\Delta)\left(\left|\sec\left[\frac{(l-2)\varphi + \pi}{2}\right]\right|\right)^{\frac{2}{2-l}}, \quad m = 0, 1, 2, 3\dots, \end{aligned} \tag{2.4.7}$$

where $r_0 + m\Delta$ denotes the vertex of a single curve, m is the index of these curves, and Δ is the distance between the adjacent vertexes. In general, the angle "π" in Eqs. (2.4.6) and (2.4.7) can be omitted since it denotes only an azimuthal rotation of the structure. The final aperture arrays are then obtained by connecting the adjacent curves and truncating the curves properly. Obviously, Eq. (2.4.7) has a singularity at $l = 2$ (LCP incidence). In such a specific case, the direct integration of Eq. (2.4.6) leads to a set of concentric rings. Figure 2.13 shows two sets of catenary curves for the generation of perfect orbital angular momenta with $l = -10$ and 10. If the handedness of the incident beam is reversed, the angular momenta would also change sign.

To obtain a Bessel beam carrying OAM, the phase distribution is defined as $\Phi(r, \varphi) = k_r r + l\varphi$, where k_r is the wavenumber along the radial direction. The corresponding catenary is constructed by local approximation assuming that the radial and azimuthal components are decoupled. Along each radial direction with a constant φ, the governing equation becomes

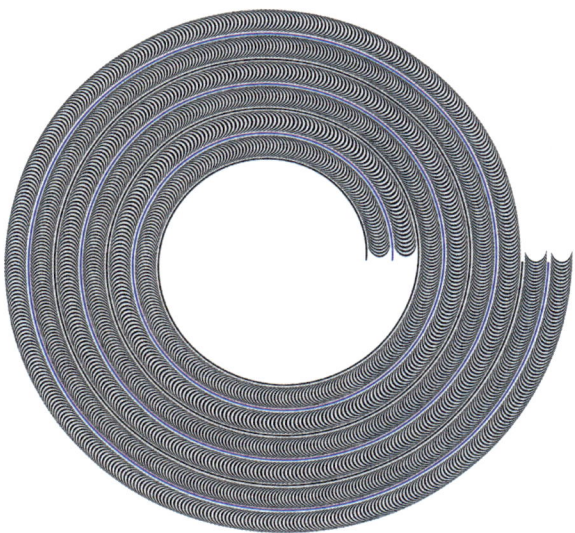

Fig. 2.14 GDSII file for Bessel beam generation ($l = 0$) with a radial period of 2 μm. The diameter of the lens is about 44 μm. The blue and black spirals form the division between adjacent catenaries

$$y_{\text{local}} = \frac{2}{k_r} \ln\left(\left|\sec\left(\frac{k_r r + l\varphi}{2}\right)\right|\right), \tag{2.4.8}$$

where y_{local} is illustrated in Fig. 2.12c. Since the radial wavenumber is a constant, the catenary structures in the Bessel lens have a constant period. Figure 2.14 shows a catenary Bessel lens with a period of 2 μm. Owing to the two offset catenaries with upside opening, the generated OAM is zero. In contrast, when the catenaries are opening downwards (Fig. 2.15), the generated OAM would be ±4. Note that the blue and black curves form the division between adjacent catenaries, which are actual Archimedean spirals described by [29]

$$r = \frac{(l - 2)\varphi + (2m + 1)\pi}{k_r}, \tag{2.4.9}$$

where $m = 1, 2, 3, \ldots$. Similar spirals have been previously used to generate OAM beams [30, 31] but they are suffering from low diffraction efficiency.

In the generation of the focused OAM beam, the radial phase distribution is not linear anymore and no analytic curve can be derived. Instead, numerical integration should be applied to obtain the radial curve at a given azimuthal location

$$y_{\text{local}} = \int \tan\left(\frac{k\sqrt{r^2 + f^2} + l\varphi}{2}\right) dx, \tag{2.4.10}$$

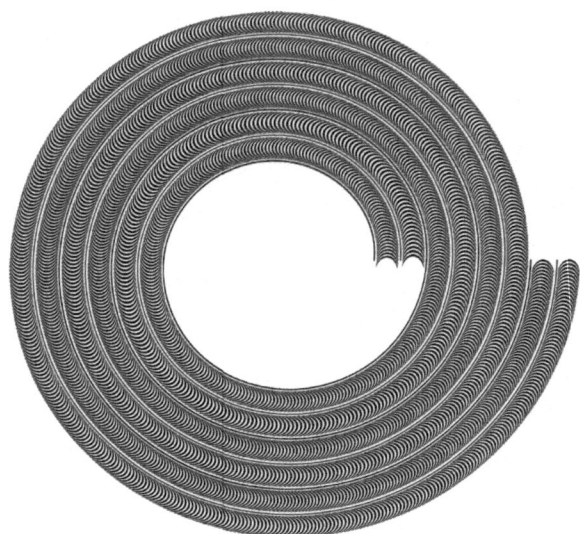

Fig. 2.15 GDSII file for Bessel beam generation ($l = \pm 4$) with a radial period of 2 μm. The diameter of the lens is about 44 μm. The Matlab Code is shown in the Appendix B.5

Fig. 2.16 GDSII file for paraxial focusing with $f = 15$ μm at $\lambda = 632.8$ nm. The horizontal length of the lens is about 21 μm

where k is the free-space wave number, f the focal length, and l the topological charge. Under paraxial approximation, the integration function could be written as

$$y_{\text{local}} = \int \tan\left(\frac{kr^2/(2f) + l\varphi}{2}\right) dx. \tag{2.4.11}$$

Figure 2.16 illustrates the GDSII file of a 1D focus lens obtained using Eq. (2.4.11). Besides the central asymmetrical structures, the shapes in both the left and right sides are deformed catenaries. The periods of these catenaries decrease as they go far away from the center. Similar behaviors can be also found in the 2D catenary lens shown in Fig. 2.17, where the central region is left as a blank.

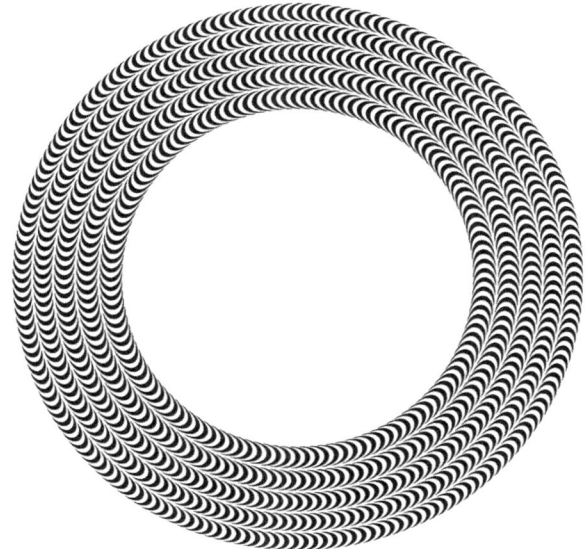

Fig. 2.17 GDSII file for focusing with $|l| = 2, f = 10$ μm and at $\lambda = 632.8$ nm. The diameter of the lens is about 23.5 μm

2.5 Wide-Angle Lenses and Airy Beam Generation

2.5.1 Wide-Angle Flat Lens

As shown in the above deduction, the catenary aperture can be used as a unique building block in optical metasurfaces owing to the continuous phase gradient. As a proof of concept, a one-dimensional (1D) flat lens is demonstrated to focus the incident CPL into a straight line. Under paraxial approximation, the required phase is obtained in a quadratic form

$$\Phi(x) = k\frac{x^2}{2f} = \frac{\pi x^2}{\lambda f}, \tag{2.5.1}$$

where k is the wavenumber in free space, f the focal length. In order to realize such phase distribution using catenary-shaped apertures, the curve shape can be obtained using an integration algorithm

$$y = \int_{x_n + \delta x}^{x_{n+1} - \delta x} \tan\left(\frac{kx^2}{4f}\right) \mathrm{d}x, \tag{2.5.2}$$

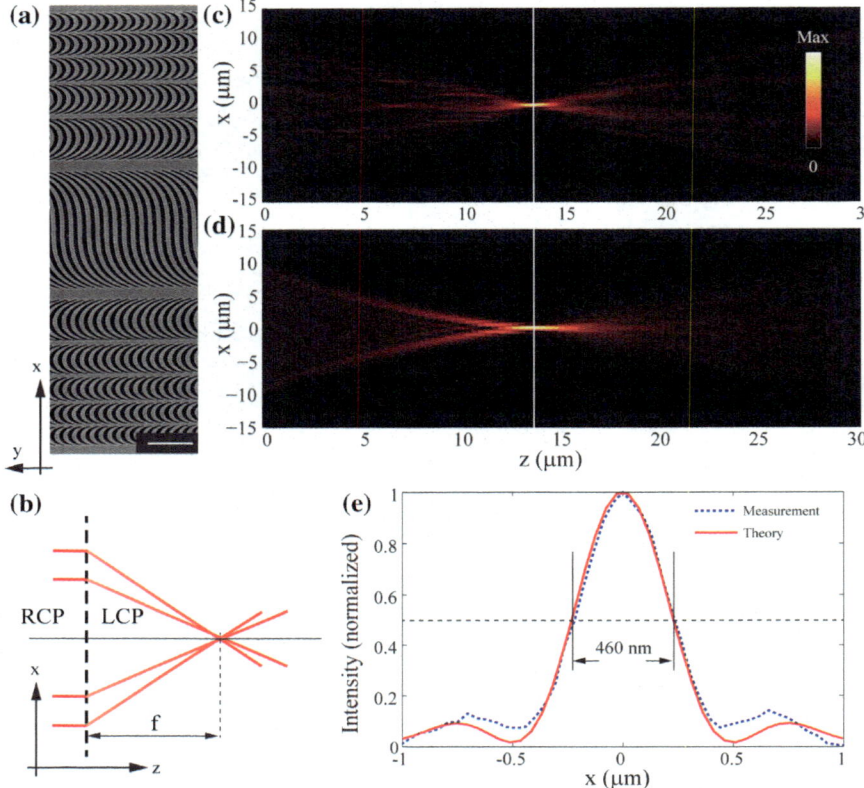

Fig. 2.18 Catenary array as a 1D flat lens. **a** SEM image of the fabricated sample. Scale bar: 2 μm. **b** Schematic of the ray path for the flat lens. **c** Measured and **d** theoretically calculated intensity distribution for the cross-polarized component in the xz-plane. The white line indicates the focal plane at $z = 13.5$ μm. **e** Normalized intensity at the cross section indicated in (**c**) and (**d**). Reproduced from [16] with permission. Copyright 2017, the Author(s)

where x_n and x_{n+1} indicate the left and right sides of each catenary on the x-axis, δx is chosen to be 10 nm. In general, they can be written as

$$x_n = \sqrt{\frac{2f(2n+1)\pi}{k}},\qquad(2.5.3)$$

where n is an integer. In the design, the following parameters are chosen: $\lambda = 632.8$ nm and $f = 15$ μm. The sample was fabricated with a dimension of 20×20 μm². As shown in Fig. 2.18a, the flat lens is composed of many catenary apertures. Since the catenaries are arranged periodically along the y-axis, the phase distribution is independent of y. This is different from the single catenary, where the values of x and y are defined in a closed aperture.

The RCP incident light can be converted to LCP and tightly focused by the catenary array (Fig. 2.18b). Because the diffraction effect of the small aperture and the fact that the quadratic phase is a paraxial approximation, the focal length is actually 13.5 μm, as can be seen in both the measurement (Fig. 2.18c) and theoretical results (Fig. 2.18d). Figure 2.18e shows the normalized intensity at the cross section $z = 13.5$ μm. The measured full width at half maximum (FWHM) is about 460 nm, agreeing well with the theoretical result.

The phase function of the above paraxial 1D lens follows a quadratic form. What will it be if we apply it in the non-paraxial case? As demonstrated recently [32, 33], such phase distribution could make the metasurface function as an ideal optical Fourier analyzer [34]. Assuming that the incident collimated light beam lies in the xz-plane with an arbitrary angle of θ to the normal axis of a 2D quadratic lens, the phase carried by the outgoing light would be

$$\Phi(r) = k_0 \frac{r^2}{2f} + k_0 x \, \sin \theta = \frac{k_0}{2f} \left((x + f \, \sin \theta)^2 + y^2 \right) - \frac{f k_0 \sin^2 \theta}{2}, \qquad (2.5.4)$$

where $k_0 x \sin \theta$ is the gradient phase induced by the oblique incidence. Since the last term in the right hand is independent of r and can be neglected, there is only a transversal shift of $f \sin \theta$ in the x-direction with respect to normal incidence. In this regard, the rotational effect of the oblique incidence light is perfectly converted to the translational symmetry of the focusing beam. As shown in Fig. 2.19a, when the incidence angle changes, the focusing spot just moves horizontally. Compared with the spherical Luneburg and compound lenses which possess wide-angle performances (Fig. 2.19b, c), the flat quadratic lens has an obvious advantage of compactness. It is noteworthy that this symmetry transformation is related to the transformation optics [35], but the quadratic lens is much thinner than that obtained with a direct coordinate transformation [36, 37].

Although the theory of quadratic lens seems straightforward, such lens could not be easily realized with either the refractive or diffractive optics. The main challenge lies in the fact that the radial wavenumber induced by the phase gradient ($k_r = \partial \Phi / \partial r$) is rather larger than that of the conventional case. For instance, there are $k_r = k_0$ and $2k_0$ at $r = f$ and $2f$, respectively. At normal incidence, the light fields passing through a region where $r > f$ will become evanescent and therefore do not contribute to the focal spot. At oblique incidence with angle of θ, the evanescent zone would shift horizontally.

Clearly, the quadratic lens needs a rapid phase variation at its periphery ($k_r = 2k_0$), which in turn requires that the period of the phase-changing unit cell to be $p = \lambda/2N$ (N is the number of discrete levels within $[0, 2\pi]$). If we choose $N = 2$ and $\lambda = 600$ nm, the period should be as small as 150 nm, with the characteristic dimensions of the nanostructures to be much smaller. Besides the fabrication challenge, the deep-subwavelength patterns are less effective in energy efficiency because the electromagnetic resonances corresponding to the phase variation are tightly related with the dimension of these nanostructures [18]. To achieve such phase modulations in the deep-subwavelength scale, catenary-like continuous apertures shown in

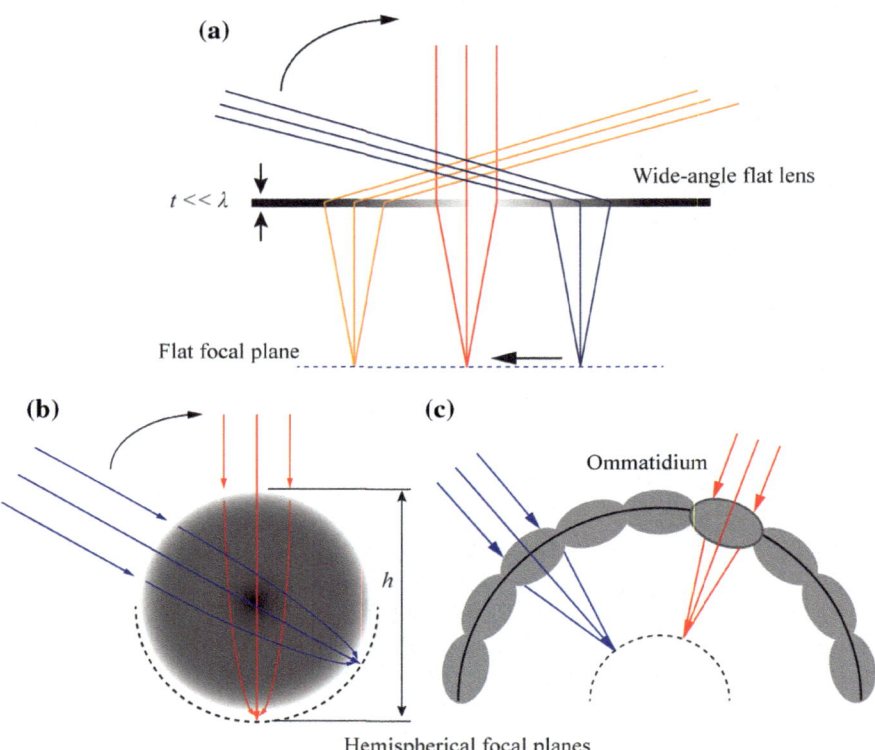

Fig. 2.19 Comparison of different wide-angle lenses. **a** Wide-angle flat lens. The thickness t is much smaller than the wavelength. The red, blue and yellow lines represent the light rays with different incidence angles. **b** Luneburg lens. The height h equals to the diameter. **c** Compound eyes composed of many ommatidia. Reproduced from [32] with permission. Copyright 2017, Optical Society of America

Fig. 2.20 are adopted. Rather than integrating the phase profile, a simple design strategy is provided by rotating long elliptic apertures in a hexagonal lattice [16, 38, 39]. These nanoholes are compactly arrayed so that some of them are connected together, leading to wider bandwidth similar to the catenary apertures.

A sample was fabricated by defining these apertures in a 100 nm-thick gold film using FIB milling (Fig. 2.21a). The radius of the sample is 20 μm and the predefined focal length f is chosen as 10 μm at $\lambda = 532$ nm, which changes to be 8.407 μm at 632.8 nm. To demonstrate the wide-angle performance, four plane waves with different orientation angles with respect to the x-axis are combined and used as the incident light. Finite difference time domain (FDTD) method and vectorial angular spectrum theory are adopted in the numerical simulations. In all the calculations, light is propagating along the $+z$-direction. As can be seen in the simulated results at $\lambda = 632.8$ nm (Fig. 2.21b, c), the light fields for different plane waves are just the same

Fig. 2.20 Flat lenses based on catenary-like apertures. The long and short axes are chosen as 180 and 60 nm. The lattice constant is $p = 150$ nm. Reproduced from [32] with permission. Copyright 2017, Optical Society of America

except that there are some translational shifts corresponding to $\Delta = f \sin \theta$. Owing to the long depth of focus (DOF), the maximal intensity shifts to $z = 7.5$ μm. The values of Δ are -4.4, 5.9, and -8.3 μm for $\theta = -32°$, $45°$, and $-80°$, agreeing well with the theoretical results. Using two He–Ne lasers and a homemade microscope, the focusing behavior of the sample was measured. While one laser beam illuminates at normal incidence, the incidence angle of the other laser beam was tuned at $-32°$ and $-80°$ successively. As shown in Fig. 2.21d–f, the measured results are in good agreement with the numerical ones.

The light fields formed by the quadratic lens do not maintain their width at different positions along the z-direction. Instead, both the intensity and the field distribution vary at each horizontal plane before and after approaching the focal plane, as a result of the interference of the complex electromagnetic fields. Such superoscillatory interference provides a mean to break the diffraction limit in classic optics, although the large side lobes are still needed to be suppressed [40–42]. Another important characteristic of the quadratic lens is that the phase can be simply scaled for different operating wavelengths. That is to say, when the wavelength of light is altered, the focal length changes in a linear way. Recalling the fact that the quadratic lens has a rather long DOF, it is expected that such lens could operate at a constant image distance within a wide spectrum. Such property may help to overcome the chromatic dispersion in traditional flat lens [43, 44]. To demonstrate this, three lasers operating at red (632.8 nm), green (532 nm) and blue (490 nm) wavelengths are used as the light sources, as shown in Fig. 2.22a. The incidence angle for the three lasers is set as: $(\theta_1, \varphi_1) = (0, 0)$, $(\theta_2, \varphi_2) = (32°, 0)$, and $(\theta_3, \varphi_3) = (80°, 270°)$, where $\varphi =$ a tan (y, x) denotes azimuthal angle in the xy-plane. Figure 2.22b, c show the cross sections of the intensity distribution. It is clear that the centroids of the focused beam would shift to larger z when the wavelength is decreased from 632.8 to 532 nm. Nevertheless, the three focal spots can be clearly observed in a common cross plane at $z = 7.5$ μm.

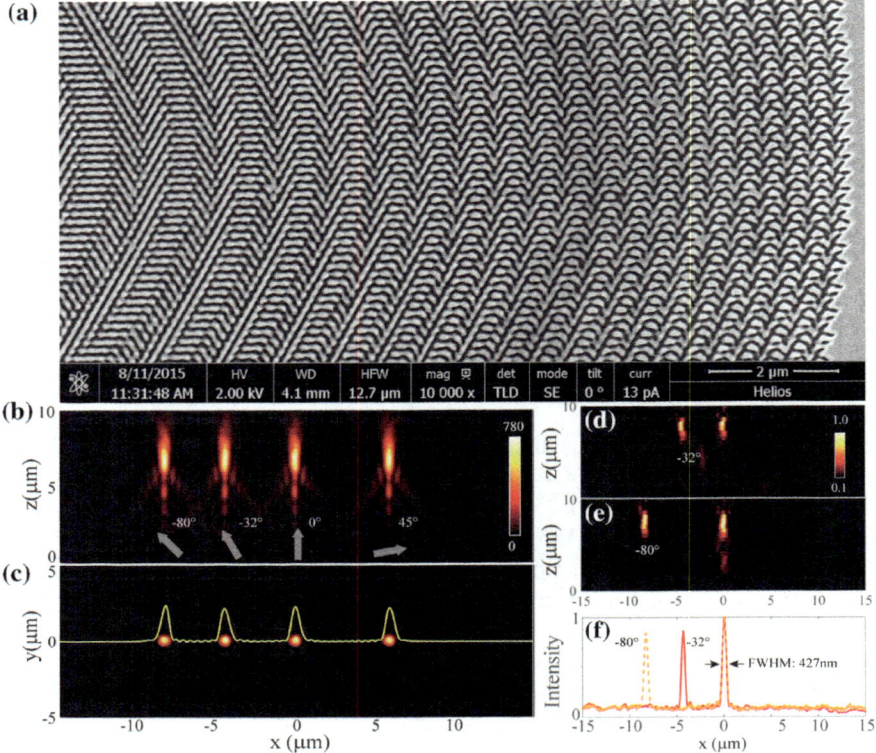

Fig. 2.21 Performance of the wide-angle flat lens based on catenary-shaped nanoapertures. **a** SEM image of one section of the fabricated 2D metalens. Scale bar, 2 μm. **b, c** Simulated light intensity distributed on the xz-plane ($y = 0$) and xy-plane ($z = 7.5$ μm) at a wavelength of 632.8 nm. **d, e** Experimental results of the intensity distribution on the xz-plane ($y = 0$) for $\theta = 0°$, $-32°$ and $\theta = 0°$, $-80°$. **f** Cross-sectional view ($z = 7.5$ μm) of the intensity distribution along the x-direction shown in (**d**) and (**e**). The full width at half maximum (FWHM) is about 427 nm. Reproduced from [32] with permission. Copyright 2017, Optical Society of America

2.5.2 Airy Beam Generation Based on Cubic Phase

As demonstrated in the above section, the catenary structures provide a simple yet powerful approach to realize both linear and quadratic phase manipulation. In principle, it can also be extended to realize cubic, quartic and even higher order phase front. In this section, we show that a cubic phase distribution forms a direct way to produce self-accelerating Airy beams. In quantum mechanics, Berry and Balazs made an important observation in 1979: they demonstrated that the Schrödinger equation describing a free particle may have a nonspreading Airy wave packet solution [45]. The most remarkable feature of this Airy packet is its ability to freely accelerate even in the absence of any external potential. Notably, the concept of Airy beam can

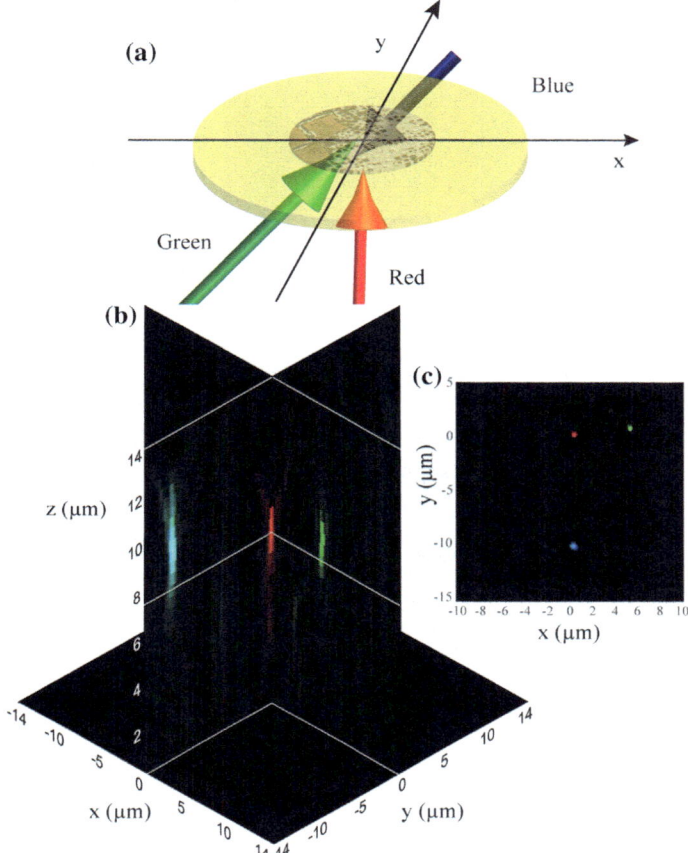

Fig. 2.22 Multiwavelength behavior of the quadratic lens. **a** Schematic of the measuring setup. **b** Cross-sectional view of the intensity distribution in the xy ($z = 0$), xz ($y = 0$) and yz ($x = 0$) planes. **c** Intensity distribution in the focal plane ($z = 7.5\ \mu$m). Reproduced from [32] with permission. Copyright 2017, Optical Society of America

also be extended to optics. It has been demonstrated that such light beam could be generated using a cubic phase [46].

To show the main characteristics of the optical Airy beam, the phase profile

$$\Phi(r) = k_0 \frac{r^3}{2f} \tag{2.5.5}$$

is constructed with catenary structures for $\lambda = 780$ nm and $f = 300\ \mu$m. The catenary curves are written as

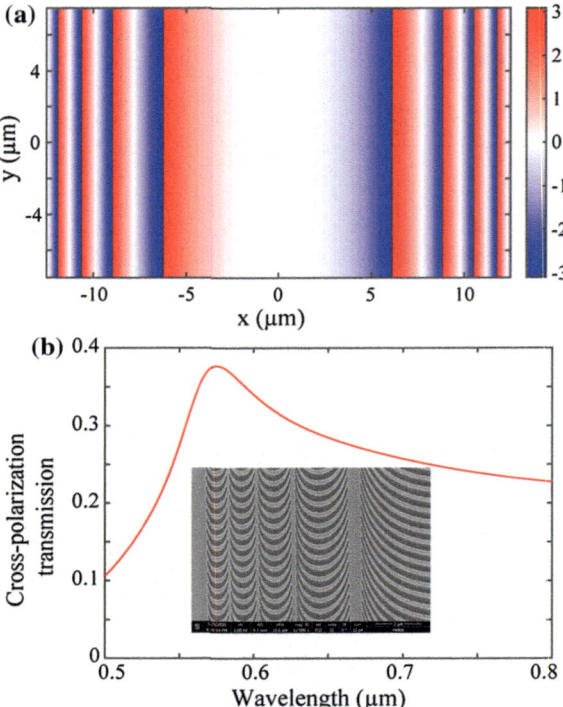

Fig. 2.23 Schematic of phase profile and the catenary structures. **a** Phase distribution. **b** Cross-polarization transmission for metallic grating with a period of 500 nm and a width of 250 nm. Inset is the SEM image of the fabricated sample

$$y = \int\limits_{x_n+\delta x}^{x_{n+1}-\delta x} \tan\left(\frac{kx^3}{4f}\right) dx, \qquad (2.5.6)$$

with

$$x_n = \sqrt[3]{\frac{2f(2n+1)\pi}{k}}. \qquad (2.5.7)$$

As shown in Fig. 2.23, the cubic phase can be well represented by the catenary structures, which show a broadband cross-polarization transmission in the visible range. At the designed wavelength $\lambda = 780$ nm, the conversion efficiency reaches about 0.23.

The performance of the Airy beam is investigated using vectorial diffraction theory and microscopic measurements, in a way similar to the catenary lens. Figure 2.24 shows the theoretical and experimental intensity distribution in the xz- and xy-planes.

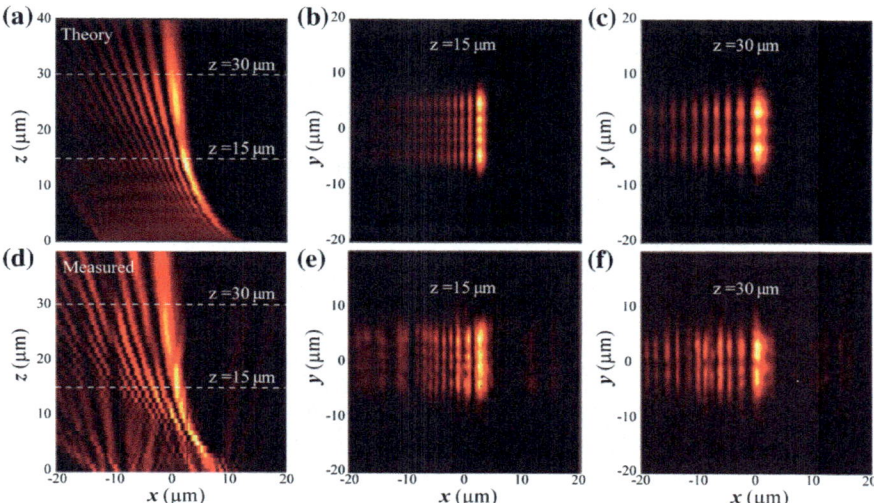

Fig. 2.24 Theoretical and measured intensity distribution in the xz- and xy-planes at a wavelength of 780 nm. **a–c** Theoretical results. **d–f** Experimental results

The deterioration of performance is attributed to the variation of efficiency owing to the fabrication error and nonuniform width.

2.6 Optical Vortex and High-Order Bessel Beam Generation

2.6.1 Achromatic Optical Vortex

As demonstrated in Sects. 2.4 and 2.5, by arranging the catenaries in various forms, it is possible to obtain almost arbitrary phase distributions such as that featured by axicons, lenses and spiral phase plates. In this section, we shall discuss in detail the generation of perfect OAM via catenary apertures. Since the ordinary OAM beam has a spiral phase along the azimuthal direction, the catenary curve in polar coordinates (r and φ) can be derived to represent the corresponding surface topography (see Sect. 2.4 for detail).

$$r = (r_0 + m\Delta)\left(\left|\sec\left[\frac{(l-2)\varphi}{2}\right]\right|\right)^{\frac{2}{2-l}}, \quad m = 0, 1, 2, 3 \ldots, \tag{2.6.1}$$

where r_0 defines the location of the vertex of the innermost curve, Δ is the distance between two vertexes at adjacent curves, and m is the index number of these curves.

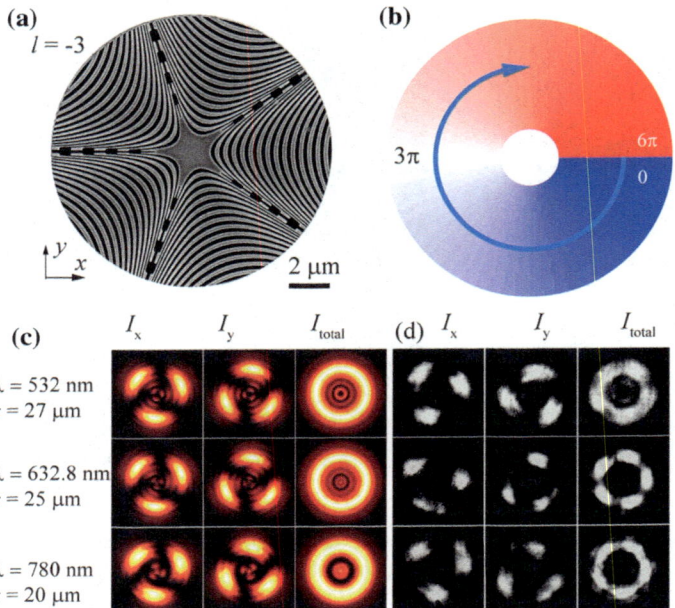

Fig. 2.25 OAM generator based on catenary array. **a** SEM image of the fabricated sample with topological charge of $l = -3$ for LCP illumination. Scale bar: 2 μm. **b** Spiral phase profile. **c** Calculated and **d** measured results of I_x, I_y, and I_{total} for $\lambda = 532$ nm (top row), $\lambda = 632.8$ nm (middle row) and $\lambda = 780$ nm (bottom row). The distances between the recording planes and the sample plane are indicated in each panel. Reproduced from [6] with permission. Copyright 2015, the Author(s)

Following Eq. (2.6.1), one can use the catenaries to produce OAM beams with arbitrary topological charge. As an example, a sample with topological charge of $l = -3$ for LCP illumination ($\sigma = 1$, the topological charge is reversed from l to $-l$ for opposite circular polarization) was fabricated via FIB milling on a 120 nm-thick gold (Au) film, where Δ and r_0 were chosen as 200 nm and 1.5 μm.

Under CPL illumination, the catenary apertures generate simultaneously two kinds of beams with approximately equal intensity, one uniformly phased and the other helically phased with opposite handedness. Based on the interference effect, the topological charge of the OAM can be directly identified, without the use of additional interference beam [47]. In the experiment, three laser sources at $\lambda = 532$, 632.8, and 780 nm were adopted to investigate the broadband performances. The third column of Fig. 2.25d represents the intensity patterns for different wavelengths and polarizations, which were recorded at a few microns away from the sample and in good agreement with the theoretical results obtained by vectorial diffraction theory [22] (Fig. 2.25c). For the x- and y-polarized components, the intensity patterns are manifested by rotating petals encircling the beam centers, where the modulus and sign of l are determined by the number and twisting direction of these petals.

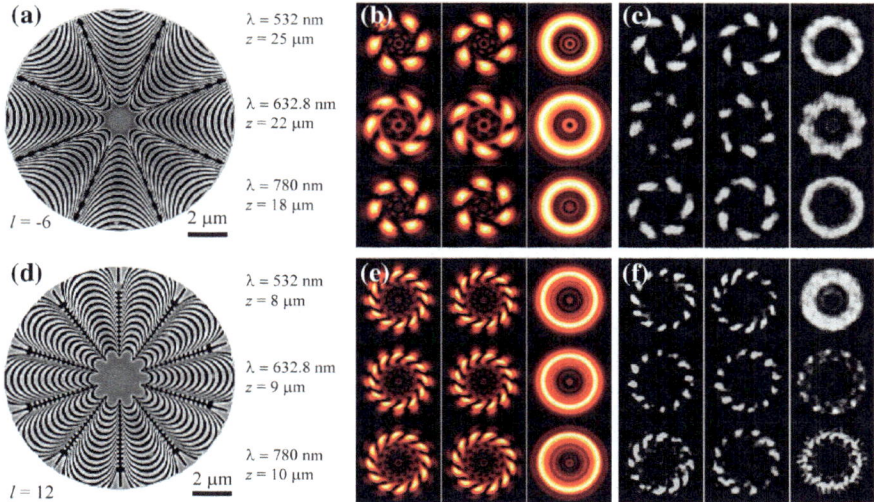

Fig. 2.26 Characterization of OAM generators with topological charges of $l = -6$ and $l = 12$. **a, d** SEM images. **b, e** Simulated and **c, f** experimental field maps for $l = -6$ and $l = 12$. The distances between the recording plane and sample plane are indicated in each panel. Reproduced from [6] with permission. Copyright 2015, the Author(s)

To validate the universal applicability of the design method, additional two OAM generators with topological charges of $l = -6$ and 12 are fabricated and characterized, as shown in Fig. 2.26. As expected, the agreement with theory is good except for some background noise. Subsequently, the conversion efficiency η was measured, which is defined as the ratio of the beam carrying OAM to the overall transmitted power. As depicted in Fig. 2.27, the mean values for $\lambda = 532$, 632.8, and 780 nm are 23.2, 39.8, and 54.4%, exhibiting at least 30-fold enhancement compared with that in circular nano-slits [48]. In principle, the efficiencies can be further increased by utilizing additional reflective layers [49, 50]. As alternatives of the metallic catenaries, high-index dielectric material can also be used to construct all-dielectric catenaries to enhance the energy efficiency [10], although the operating bandwidth may be limited by the resonant nature.

2.6.2 Bessel Beam Carrying Optical Vortex

In 1987, Durnin et al. showed that there exists one kind of non-diffracting beam which has Bessel-type intensity distribution along the radial direction [51]. This so-called Bessel beam is the exact solution of the Helmholtz equation, thus does not violate traditional optical theory. The ideal Bessel beam requires the horizontal dimensions as well as the power carried by the beam are infinite, which cannot be

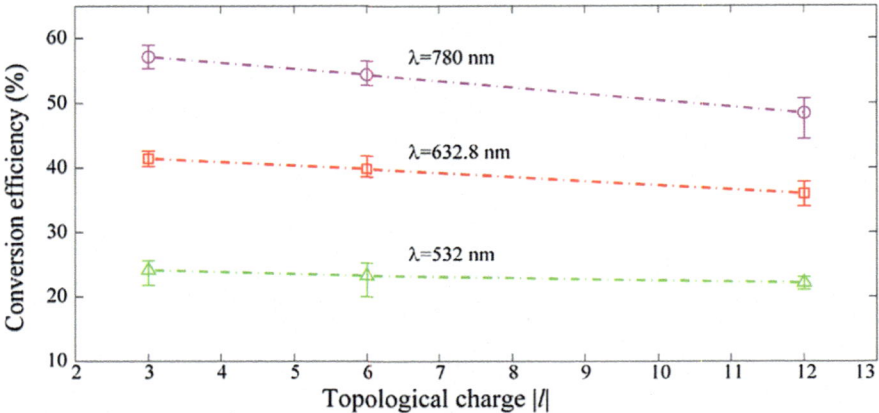

Fig. 2.27 Measured conversion efficiency for the three OAM generators with $l = -3, -6$, and 12. Three laser sources working at $\lambda = 532, 632.8$, and 780 nm were adopted to account the broadband characteristics. Reproduced from [6] with permission. Copyright 2015, the Author(s)

achieved in practical situations. Nevertheless, it was shown that even truncated with a Gaussian function, the Bessel beam can still maintain its unique properties such as diffraction-free and self-healing, which enable a variety of important applications in super-resolution imaging [52], optical micromanipulation and nano-fabrication [53].

Among various Bessel beams, high-order Bessel beams (HOBBs), i.e., Bessel beams carrying OAM have attracted growing interest owing to their ability to transport information encoded in the OAM basis [54]. The diffraction-free property of the Bessel beam and the theoretically infinite freedoms of OAM make HOBBs a promising alternative for high-speed optical and quantum communications systems. Besides, the particular shape of the HOBBs and its ability to retain over an extended propagation distance in a propagation-invariant manner makes it useful in optical manipulation [55].

For Bessel beam, the electromagnetic fields at any cross section perpendicular to the propagation direction can be written as

$$E(r, \varphi) = J_l(k_r r) \exp(il\varphi + ik_z z), \tag{2.6.2}$$

where J_l is the lth-order Bessel function, k_r is the wave vector along the radial direction, φ is the azimuthal angle, and l is the topological charge of the optical vortex. This beam is generated by a phase distribution of $k_r r + l\varphi$ at $z = 0$. Following the procedure described in Sect. 2.4, a catenary array could be constructed to realize such a phase distribution. As illustrated in Fig. 2.28, the designed structures can be treated as a combination of catenary-shaped nano-slits with micrometer/sub-micrometer scale concentric or spiral gratings. These concentric rings and spirals act as zone plates or spiral zone plates in traditional optics.

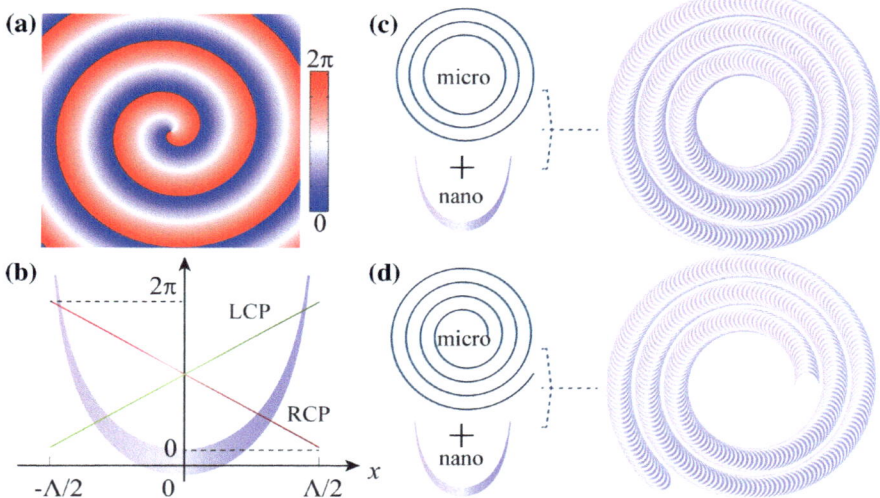

Fig. 2.28 Bessel beam generation process in catenary arrays. **a** Phase pattern of a Bessel beam with $l = 1$. **b** Horizontal phase distributions in $[0, 2\pi]$ under illumination of circular polarizations for a catenary aperture. **c, d** Sketch of the Bessel beam generators for $|l| = 2$ and 3. Reproduced from [29] with permission. Copyright 2016, the Author(s)

Figure 2.28c, d depict the two designs for $|l| = 2$, and 3, respectively. The concentric or spiral gratings can be defined with the trajectory of the catenary ends, which can be written as

$$r = [(l - 2)\varphi + (2m + 1)\pi]/k_r, \quad m = 1, 2, 3, \ldots \quad (2.6.3)$$

Following the analytic design, Bessel beam generators for topological charges $l = 0$, 2, 3 and 4 were fabricated (The GDSII file can be found in Sect. 2.4). All the samples were milled in an Au film on a quartz substrate using FIB. The thickness of the Au film is 120 nm. The period p in the radial direction is 2 μm for all the samples, which determines the radial wave vector of the diffraction beam, by the relation $k_r = 2\pi/p$.

In the experiment, measurements of the diffraction-free transmission properties were performed by using a home-built microscope. A collimated beam from a He–Ne laser at $\lambda = 632.8$ nm was converted into RCP light through the cascaded polarizer and quarter-wave plate, and then illuminated on the samples. The cross-polarization (LCP) component of the transmitted field through the samples was filtered by an additional quarter-wave plate and polarizer. The intensity distribution of the fields was imaged through a 100× objective and a tube lens, then collected by a charge-coupled device (CCD) camera. The intensity distribution was measured in a series of image planes by moving the objective along the z-direction (propagation direction of the Bessel beam) with a step of 1 μm, then mapped the longitudinal cross sections of generated Bessel beams. The output surface of the sample was set as $z = 0$ plane, and

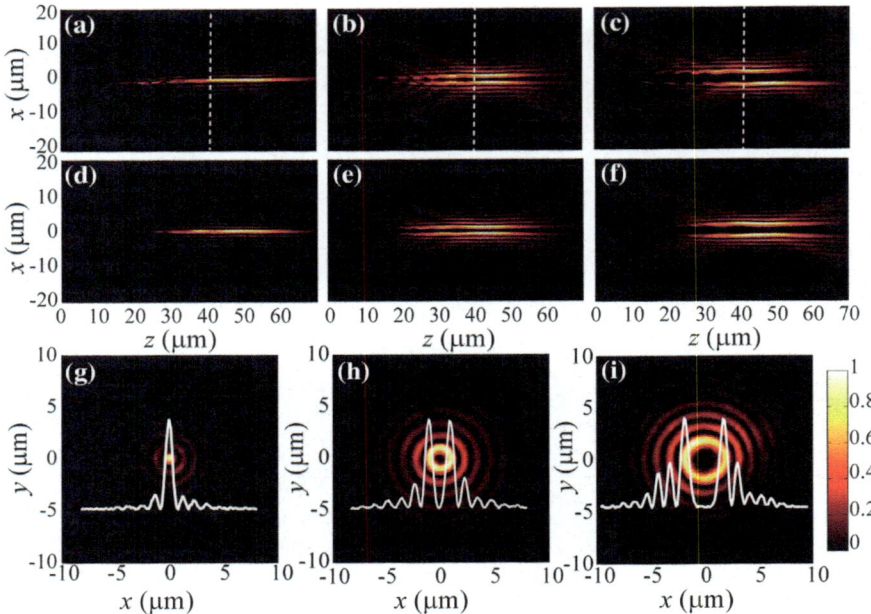

Fig. 2.29 Optical characterization of the Bessel beams generators. **a**–**c** Experimental results of the intensity distribution in the xz-plane (y = 0) for l = 0, 2 and 4. **d**–**f** Calculation results of the intensity distribution in the xz-plane (y = 0) for l = 2, 3 and 4. **g**–**i** Cross-sectional view of the intensity along the dashed line marked in (**a**)–(**c**). Insets show the normalized intensity at y = 0. Reproduced from [29] with permission. Copyright 2016, the Author(s)

the center of beam spot in each xy-plane was set as the original for x and y coordinate. Figure 2.29a–c show the experimental results of the intensity distributions across the center of Bessel beams for l = 0, 2 and 4 in the xz-plane (y = 0).

The theoretical calculations based on vectorial angular spectrum theory are also given as illustrated in Fig. 2.29d–f, and good agreement with experimental data is found. The propagation-invariant manner of the Bessel beams is validated in both experimental and theoretical simulations. In particular, the HOBBs are characterized by hollow centers. The spot size, defined as the radius of the innermost ring of the Bessel beam, increases with the topography charge l for the same radial wavevector and wavelength.

Interferometry is a commonly used method to characterize the specific type of beam carrying OAM. As shown in Fig. 2.30a, a modified experimental setup was adopted to characterize the topography charge of the HOBBs. A linear polarized (x-polarized) light was used to irradiate the samples. After spin–orbit interaction in the samples, the converted RCP and LCP components of the beam were diverged at an angle of 2θ, where θ equals to $\sin^{-1}(\lambda/p)$. After diverging, the RCP and LCP beam carries OAM with $\pm l$ respectively. The two beams were then collimated and focused by two lenses. The y-polarized components of the interference patterns around the

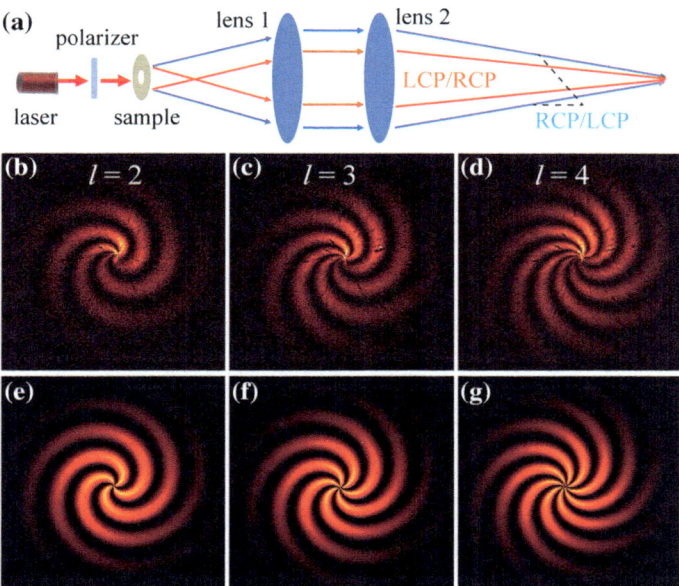

Fig. 2.30 Interference characterization of the OAMs of HOBBs. **a** Experimental setup. **b–d** Experimental interference patterns for $l = 2$, 3 and 4. **e–g** Calculated interference patterns for $l = 2$, 3 and 4. Reproduced from [29] with permission. Copyright 2016, the Author(s)

focal spot were recorded by a large area CCD. The interference patterns of $l = 2$, 3 and 4 from experiment are illustrated in Fig. 2.30b–d, which elegantly agree with the calculation results shown in Fig. 2.30e–g. The petal-like intensity patterns originate from the interference between the beam carrying OAM with opposite signs [56]. The number of the petals in the intensity patterns is twice of $|l|$.

As the geometric phase distribution is determined by the structural design, the proposed structure inherently works in a broadband frequency range. The propagation property of light with $l = 3$ was measured at a variety of wavelengths in the experiment. Figure 2.31a–c show the experimental results of the intensity distributions across the center of beams at wavelength $\lambda = 808$, 780, and 532 nm. Theoretically, the bandwidth of the proposed structures is not limited except the wavelength is too long that k_r is larger than the wavenumber in free space. The theoretical propagation properties of the HOBBs at the corresponding wavelengths were also calculated, as shown in Fig. 2.31d–f. The results from the theoretical model and experimental measurement match well. Figure 2.31g illustrates the wavelength dependent spot size, defined as the peak-to-peak distance along the x-direction. The spot size of the HOBBs is approximately linearly dependent on the operating wavelength. The small discrepancy between the experimental data and a theoretical model is owing to the fabrication and measurement errors.

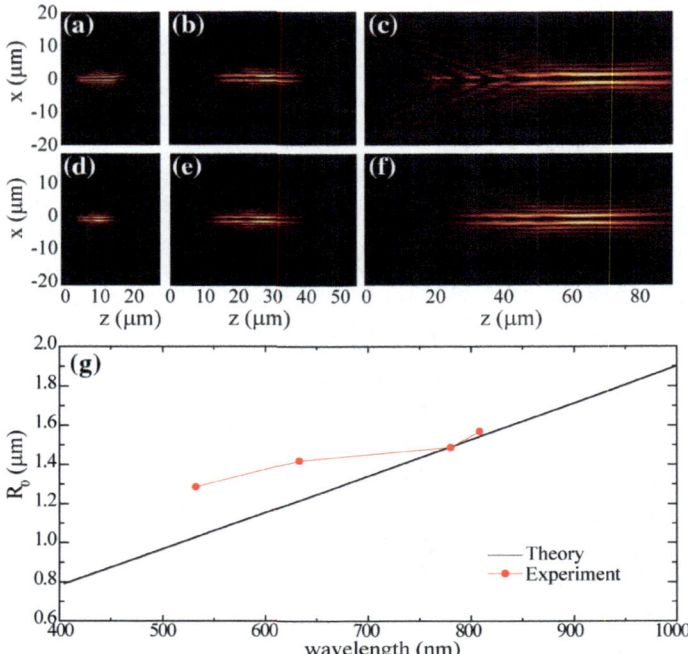

Fig. 2.31 Broadband characterization of the Bessel beam generators. **a–c** Experimental intensity distributions in the *xz*-plane at wavelengths of λ = 808, 780 and 532 for *l* = 3. **d–e** The corresponding theoretical results. **g** The radii of the Bessel beams for various wavelengths with *l* = 3. Reproduced from [29] with permission. Copyright 2016, the Author(s)

The Bessel-type radial phase distribution could enable the diffraction-free transfer of OAM states, while the focused OAM beam has potential application in sub-diffraction imaging and micromanipulation. As a result, it is beneficial to compare the intensity distribution of the Bessel and ordinary lens. Following the design procedure in Sect. 2.4, a catenary lens has been designed and fabricated with phase distributions of

$$\Phi(r, \varphi) = \sigma\left(k\sqrt{r^2 + f^2} + l\varphi\right). \tag{2.6.4}$$

The design parameters are $l = 2$, and $f = 40\,\mu m$. The inner and outer radii, defined as r_1 and r_2 are 10.6 and 20.8 μm. Figure 2.32a–c show the SEM images of the Bessel axicon ($l = 3$) and OAM lens. The transmitted intensity patterns at λ = 632.8 nm are measured under RCP illumination. As shown in Fig. 2.32d, the intensity distribution along the propagation direction reveals the main two characteristics of the HOBB, i.e., the diffraction-free property and the intensity singularity. The intensity map for the planar OAM lens, however, is more constrictive along the propagation direction (Fig. 2.32e). For the two kinds of devices, the measured center-to-center distances

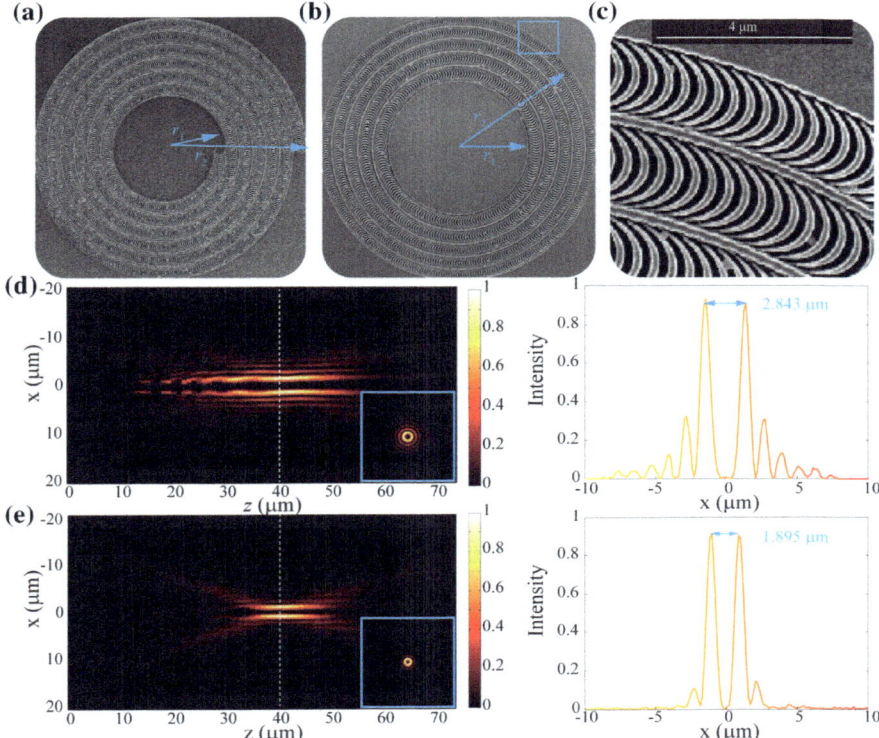

Fig. 2.32 Catenary arrays for the generation of HOBB and focused OAM beam. **a, b** SEM images of the catenary arrays for HOBB and focused OAM. **c** Scaled view of the rectangular region shown in (**b**). **d, e** Measured cross-polarized intensities of the HOBB and focused OAM at a wavelength of 632.8 nm. The intensities at the cross section of $z = 40\ \mu$m are plotted in the insets of (**d**) and (**e**). Reproduced from [6] with permission. Copyright 2015, the Author(s)

of the doughnut-shaped intensity patterns at $z = 40\ \mu$m are 2.843 and 1.895 μm, which are a bit larger than the theoretically calculated 2.54 and 1.56 μm.

2.7 Catenary Devices with Maximized Efficiency

The polarization conversion efficiency of geometric phase metasurface is often not high enough for many applications. Although a 24.7% efficiency has been demonstrated in the microwave region with a thickness of $\lambda/1000$ [57, 58], the theoretically predicted upper limit 25% provides an intrinsic obstacle for its further improvement.

To overcome the efficiency barrier, the thickness of the devices must be increased to produce a 180° phase shift between the two main axes. As demonstrated by the following results, this can be achieved in either reflection or transmission mode, with either metallic or dielectric materials.

2.7.1 Metal–Dielectric Composites

As shown in Fig. 2.33, a metallic ground plane and a dielectric substrate are added below the catenary metasurface. When the thickness of the dielectric spacer is adjusted properly, the metasurface, dielectric spacer, and the metallic ground plane would form a space-variant waveplate. For half-wave plate, circular polarization would be reversed after reflection [60, 61], with a space-variant geometric phase. When illuminated with a linearly polarized wave, two circular components which undergo different deflection should be considered. The anomalous reflection angle θ_r can be calculated by

$$\theta_r = \arcsin\left(\frac{\lambda}{2\pi}\frac{d\Phi}{dx}\right) = \arcsin\left(\frac{\sigma\lambda}{\Lambda}\right). \tag{2.7.1}$$

Through the transfer matrix method, the whole reflection electric field is written as

$$\begin{aligned}
\begin{bmatrix} E_x \\ E_y \end{bmatrix} &= \frac{1}{2}\exp\left(i\frac{2\pi x}{\Lambda}\right)\begin{bmatrix} 1 \\ -j \end{bmatrix} + \frac{1}{2}\exp\left(-i\frac{2\pi x}{\Lambda}\right)\begin{bmatrix} 1 \\ j \end{bmatrix} \\
&= \begin{bmatrix} \cos\left(\frac{2\pi x}{\Lambda}\right) \\ \sin\left(\frac{2\pi x}{\Lambda}\right) \end{bmatrix}.
\end{aligned} \tag{2.7.2}$$

As inspired by the concept of virtual shaping [49], one could make an object of arbitrary shape and material properties appear exactly the same like another object. To verify the reduction of monostatic radar scattering section (RCS), full model simulations with a lateral dimension of 314×314 mm^2 are carried out using CST MWS. The period of the catenary metallic strips along the y-direction is set as $s = 10$ mm, less than the incident wavelength λ to prevent the diffraction in yoz plane. The period of the catenary metallic strips along x-direction is $\Lambda = 62.8$ mm, corresponding to a deflection angle of $30°$ at $f = 9.55$ GHz. The width of catenary metallic strips is $\Delta = 5$ mm to obtain a maximal polarization conversion efficiency. Since the loss of copper is tiny in the microwave band, it is simplified as PEC in the simulations.

Figure 2.34a shows the monostatic RCS of metallic reflection plane and the catenary metasurface in X band (8–12 GHz). The RCS of the proposed metasurface is lower than that of a metallic plate with the same size in the whole X band. Especially, the maximum of the RCS reduction exceeds 20 dB at frequencies of 8 and 9.4 GHz for TE and TM polarizations, respectively. To get more physical insight into the RCS reduction, the RCS of proposed metasurface under the illumination of RCP and LCP at 9.4 GHz is investigated. It can be found that the catenary metasurface forces the reflected beam of opposite handedness to propagate in opposite directions. The

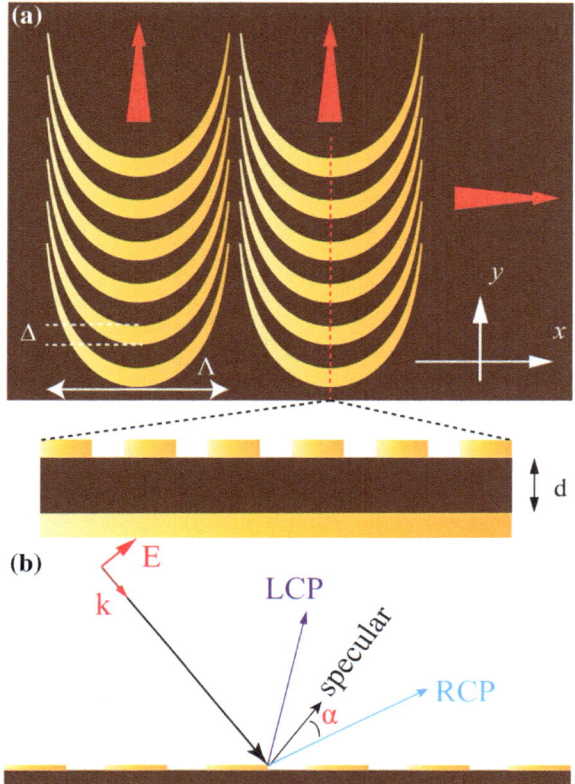

Fig. 2.33 Operation principle of the reflective catenary deflector. **a** Geometry of the catenary array. Six rows and two columns are shown. **b** Schematic of the polarization-dependent reflection. Under linear incidence, RCP and LCP are reflected to opposite directions with respect to the specular direction. Reproduced from [59] with permission. Copyright 2017, Institute of Optics and Electronics, Chinese Academy of Sciences

simulated deflection angle at 9.4 GHz is $\pm 30.5°$ with respect to the normal direction, agreeing with the theoretically calculated value of $30.7°$ from Eq. (2.7.1). Note that although the geometric phase originated from SOI is frequency-independent, the wavelength and polarization conversion efficiency vary at these frequencies, therefore the scattering angular spectra presented in Fig. 2.34c changes with the frequency and the abnormal deflection angle decreases with the incident frequency.

The reflected electromagnetic fields in the xz-plane for two frequencies (8 and 9.4 GHz) are depicted in Fig. 2.35. As predicted by Eq. (2.7.2), the whole reflection electric fields are formed by the interference between the field $\sin(2\pi x/\Lambda)$ and $\cos(2\pi x/\Lambda)$. The constant phase fronts demonstrate the presence of anomalous refraction are $36.5°$ and $30.5°$ at 8 and 9.4 GHz, consistent with the theoretically calculated $36.8°$ and $30.7°$. At the two frequencies, the catenary metasurface can

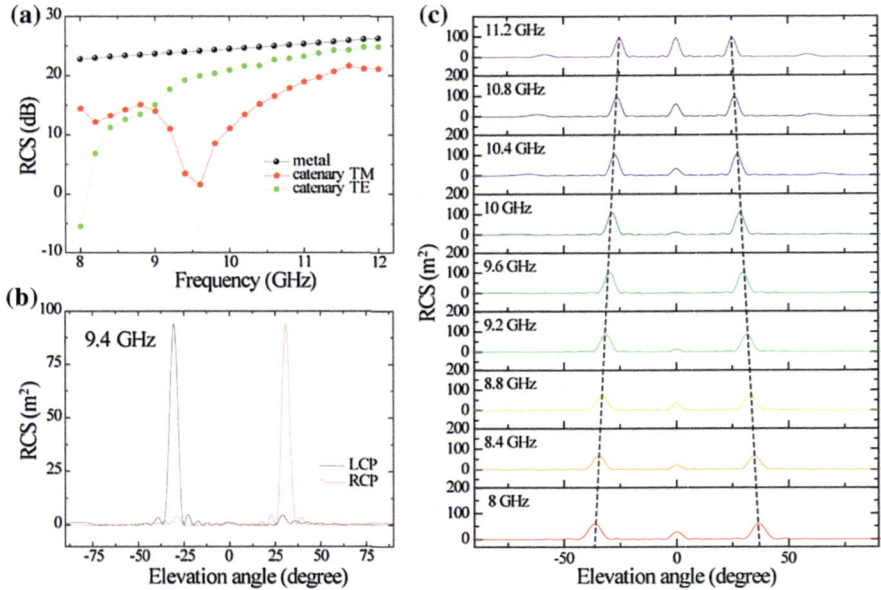

Fig. 2.34 Simulation of the catenary structure. **a** RCS of the catenary array compared with PEC for TE and TM polarizations. **b** RCS in the incidence plane as a function of azimuthal angle at 9.4 GHz. **c** 2D RCS for frequencies ranging from 8 to 11.2 GHz. The two dashed lines indicate the dispersion of the diffraction angle. Reproduced from [59] with permission. Copyright 2017, Institute of Optics and Electronics, Chinese Academy of Sciences

be taken as a simple circular beam splitting structure supporting background-free circular components in two distinct diffraction directions.

In order to prove the numerical results, a sample was fabricated with printing circuit board technique as shown in Fig. 2.36b. Reflection measurements were taken in a microwave anechoic chamber with a network analyzer. Two standard linearly polarized horn antennas (the electric field is parallel to the x-axis) are used as transmitter and receiver, respectively. The result shown in Fig. 2.36c displays reflection reduction of 20 dB from 9.7 to 11 GHz. Considering the imperfection in the fabricating and measuring process, this is in reasonable agreement with the simulated results in Fig. 2.34a.

2.7.2 All-Metallic Catenary Meta-Mirror

Similar to the infrared all-metallic devices [62], properly designed structured metals can also be used to scatter electromagnetic waves to predefined direction. In the following, it is demonstrated that a metallic plate engraved with catenary-shaped

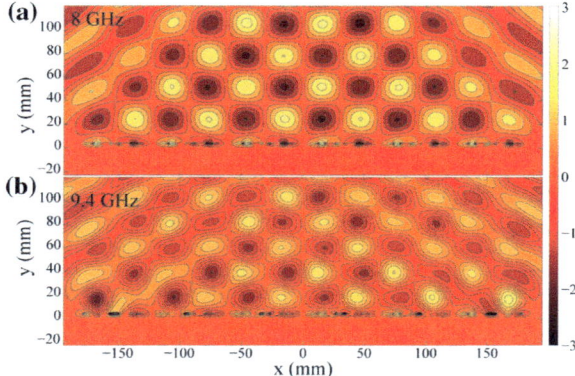

Fig. 2.35 Calculated electric fields (E_z) distributions in the xz-plane. **a** $f = 8$ GHz. **b** $f = 9.4$ GHz. Adapted from [59] with permission. Copyright 2017, Institute of Optics and Electronics, Chinese Academy of Sciences

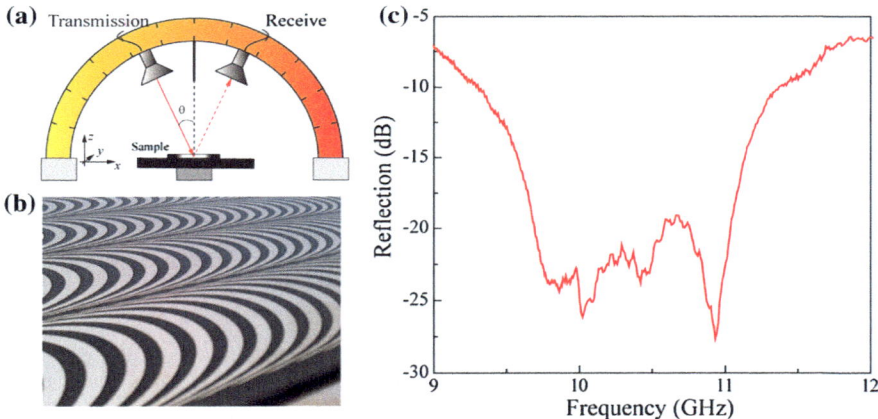

Fig. 2.36 Characterization of the reflective catenary. **a** Schematic of the measurement setup. **b** Photograph of the sample. **c** Measured reflection in dB. Reproduced from [59] with permission. Copyright 2017, Institute of Optics and Electronics, Chinese Academy of Sciences

grooves could generate high-efficient geometric phase, thus deflecting the incident waves.

As illustrated in Fig. 2.37a, a microwave detection system was used to measure the reflection amplitude as functions of the scattering angle and operating frequency. A pair of wideband horn antennas connected to a vector network analyzer are utilized as transmitter and receiver. The incident and reflected angles can be easily changed by rotating the guide rails. The fabricated catenary sample is shown in Fig. 2.37b. Owing to the finite size of the horn antennas, the incident angle is fixed to 5° and the angle of the receiving antenna changing from −70° to 70° with a 5° step at

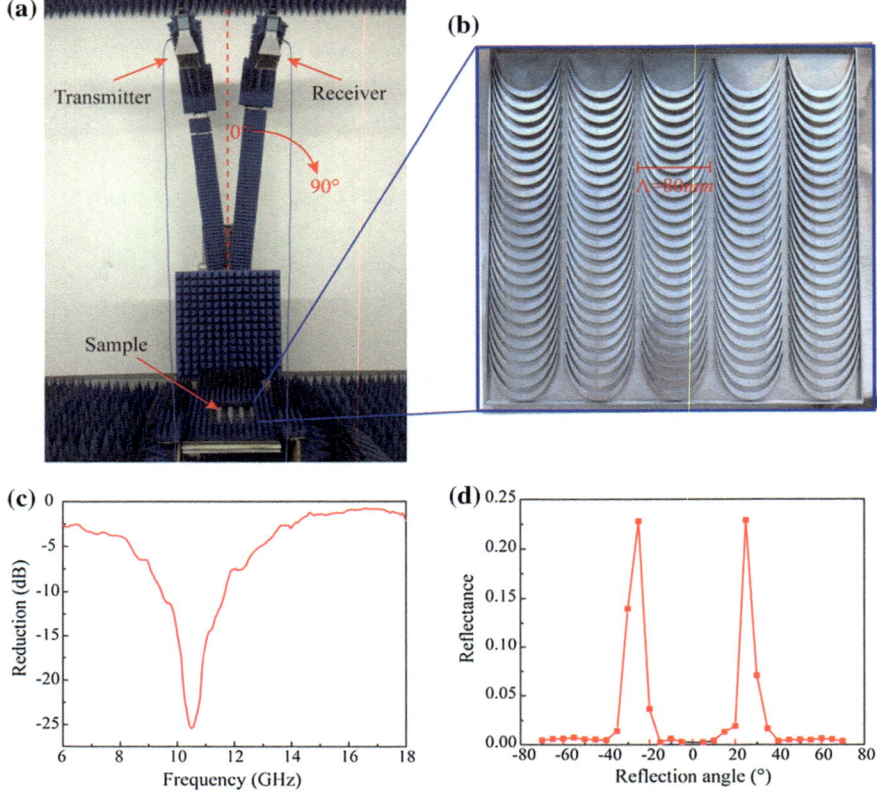

Fig. 2.37 a The schematic of experimental setup. **b** Fabricated all-metallic catenary metasurface. **c** Measured reflection spectrum at the frequencies ranging from 6 to 18 GHz. **d** Measured reflectance in wide angles from −70° to 70° at the frequency of 10.5 GHz

the same distance. The reflected field intensity is normalized to be a fraction of the incident field intensity, which is obtained by measuring the reflected field from an ordinary metallic plate at a small incident angle (~5°), with the same distances between two antennas and the sample. First, the receiving angle is set at 5° to measure the specular reflection in the broad frequency range from 6 to 18 GHz. As shown in Fig. 2.37c, the reflection is dramatically reduced at the working frequency of 10.5 GHz. Subsequently, the reflectance over wide angles from −70° to 70° was measured to check the anomalous deflection performance, as shown in Fig. 2.37d. It can be seen that the incoming linearly polarized wave is deflected in two opposite sides with equal angle of 26° and equal efficiencies.

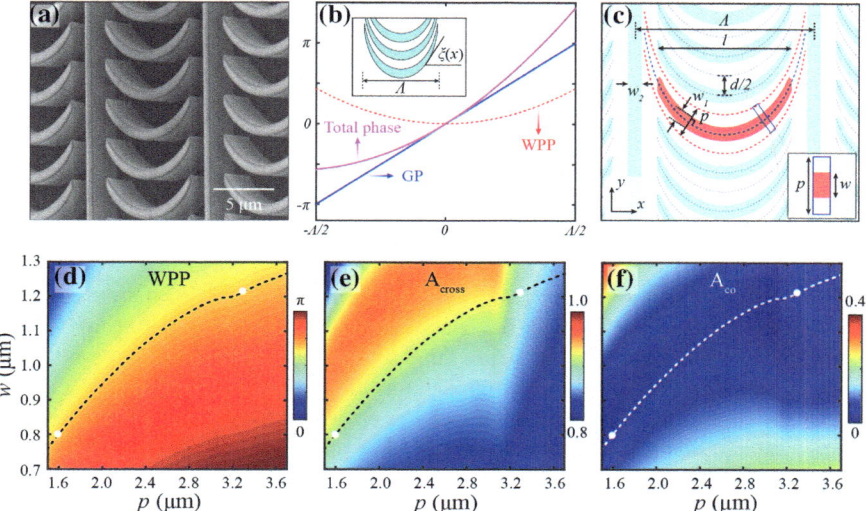

Fig. 2.38 **a** SEM image of the fabricated OCG. **b** The phase distributions of nonoptimal catenary structure along *x-direction* with RCP illumination. The structure is formed by two catenaries with a shift along *y*-direction. **c** The schematic illustration of the OCG. **d–f** Simulated WPP, cross-polarized amplitude (A_{cross}), and co-polarized amplitude (A_{co}) distributions as a function of the width and period of the ordinary grating at a wavelength of 10.6 μm

2.7.3 All-Dielectric Catenary Devices

Dielectric materials with low loss provide another good choice to build efficient metasurfaces, as shown in many recent experimental demonstrations [10, 11, 13]. However, pixel-patterned dielectric metasurfaces usually suffer from discrete and inaccurate phase distribution. In order to construct continuous dielectric catenaries, previous method of building a metal catenary structure by shifting along the vertical direction (Fig. 2.38b) is invalid as it would introduce unwanted phase shifts. On the one hand, dielectric elements usually have a large aspect ratio in order to realize high polarization conversion efficiency. The large aspect ratio makes the waveguide propagation phase (WPP) of the dielectric grating sensitive to its width and period. On the other hand, the previous catenary structure is featured by a nonuniform width and period, thus unwanted WPP gradient is introduced into the geometric phase (GP) along the *x*-direction.

In order to suppress the WPP gradient, the width of the catenary grating should be adjusted according to the optical database constructed by ordinary gratings. Figure 2.38c shows the schematic illustration of the optimal catenary grating (OCG). The dashed curves are an array of "catenary of equal strength" with an interval of $d/2$ along the *y*-direction. The local part of the catenary structure can be regarded as an ordinary grating with a period of p and a width of w. The period of p is defined as the distance between two intersections formed by the normal line of one catenary and its two adjacent catenaries. The top and bottom boundaries of a catenary structure are formed by equidistant width of $w_1/2$ in two directions of the normal line. As a proof of concept, the design is aimed at mid-infrared band (8–14 μm) with a central wavelength of 10.6 μm. Both materials of substrate and structure are silicon (Si). The height of the catenary is optimized as 4.7 μm. RCP light is chosen as incidence. Figure 2.38d–f display simulated WPP, cross-polarized amplitude and co-polarized amplitude distributions of normal grating structure, respectively, as a function of its width and period at the wavelength of 10.6 μm. As predicted, the WPP and amplitudes have obvious changes with varied width and period, thus it is necessary to employ special design to suppress the WPP gradient and nonuniform amplitude.

The equiphase line is chosen as the basis of constructing catenary grating in order to suppress the WPP gradient. There are three points that need to be considered in the choice of equiphase line, including high and uniform cross-polarized amplitude, low co-polarized amplitude, and simple fabrication process. Considering the above three aspects, the equiphase line is plotted in the form of dashed curve. The coordinates of the starting point are (1.6, 0.8 μm), and the maximum period is limited to be $d = 3.3$ μm. High and uniform cross-polarized amplitude (~0.91) and low co-polarized amplitudes (<0.04) are supported along the equiphase line. Since the period of p on both sides of the catenary is almost equal to zero, the catenary is truncated ($l = 0.75\Lambda$) in order to guarantee the feasibility of fabrication. Vertical structures are added in truncated regions to increase the phase continuity. The number (m) of vertical structures is determined by an integral part of $3\Lambda/8d$, and its equivalent period is equal to $\Lambda/4\,m$. According to the numerical calculations, the widths of the catenary structure and vertical structure (w_1 and w_2) can be determined.

According to the design method mentioned above, five OCGs (D1–D5) are designed, fabricated and characterized, corresponding to diffractive angles of −30°, −45°, −60°, −75°, and −90° at normal incidence. The OCG is angular insensitive and has high efficiencies even at extremely large oblique incidence. For an incident angle of θ_1, the diffractive angle of an OCG can be calculated by $\theta = \sin^{-1}[\sin(\theta_1) - \lambda/\Lambda]$. Figure 2.39 shows the simulated efficiencies of five devices at different incident angles. Within the entire simulated wavelength region, the efficiencies of other orders (except for −1st and 0th orders) are lower than 3%. The energy of transmitted light is concentrated into −1st and 0th orders, which can be further proved by measured efficiencies and diffraction patterns shown in the form of spots and insets in Fig. 2.39. The measured relative efficiencies are higher than 90% and have an average value at around 92.7%. The maximal and minimal measured absolute efficiencies are 71.8 and 67.1%, corresponding to D2 under 30° incidence and D3 under 60° incidence, respectively. In principle, the absolute efficiency can be further improved by employing materials with a low refractive index as the substrate, although at the cost of greater procedure complexity. The gradient of the refractive index can introduce a Fabry–Pérot cavity effect along the propagating direction, leading to a higher efficiency. Similar to the metallic catenary lens and OAM generators, the dielectric catenaries can be applied in the flat catenary lens for Fourier analysis and imaging, which is one of our current work.

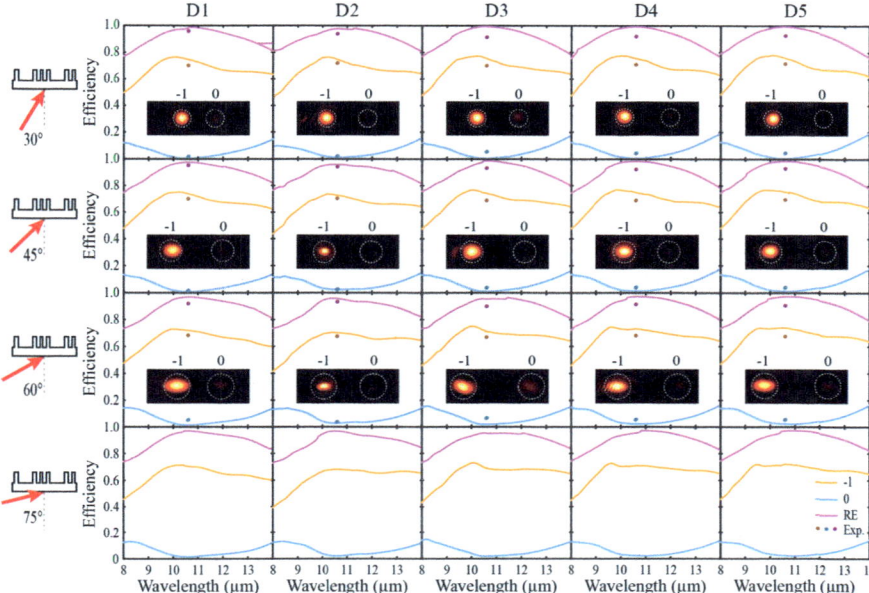

Fig. 2.39 The simulated absolute efficiencies of −1st and 0th orders and relative efficiency (RE) at oblique incidence as a function of the wavelength. The spots indicate measured efficiencies at 10.6 µm wavelength. The insets show measured diffractive patterns

2.8   Coherent Control of the Diffraction Efficiency of Catenary Metasurface

Interference is one of the most basic characteristics of waves. In general, all light—matter interactions in subwavelength structures are associated with multiple interferences. It was recently demonstrated that coherent control with two or more input channels could be used to dynamically change the light–matter interaction on the metasurface [63–68]. Electromagnetic wave could be completely absorbed by an ultrathin metasurface in an unexpected broad frequency range with coherent illumination, which broke the classic bandwidth-thickness limit [67, 69]. Meanwhile, the polarization states could also be coherently controlled as demonstrated in previous work, and the efficiency could reach nearly 100% with frequency-independent property [70, 71]. The coherent control method provides a viable alternative to signal processing based on nonlinear materials. Since nonlinear process is not needed anymore, the signal intensity, as well as the thermal problems, can be dramatically reduced [72, 73].

In this section, we show that catenary arrays perforated in a thin metallic screen could be combined with coherent control method to surpass the intrinsic efficiency

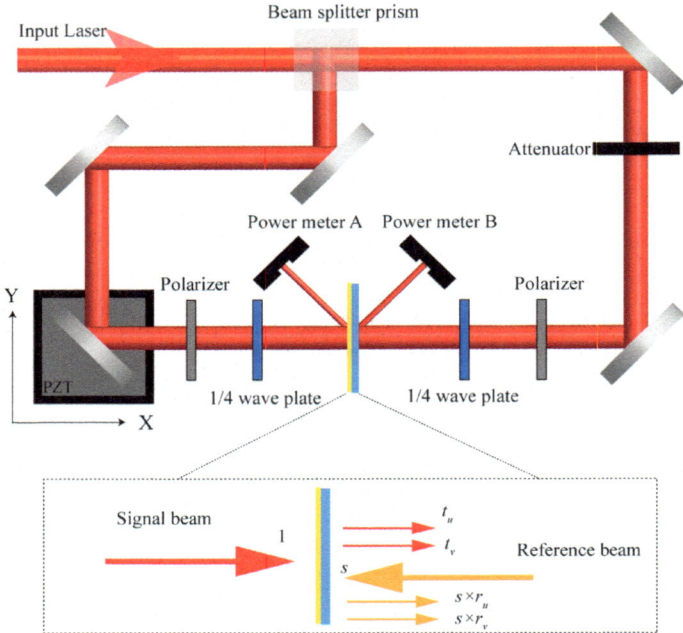

Fig. 2.40 Experimental configuration of the coherent control. Laser beams from a diode-pumped solid-state laser (532 nm) or a He–Ne laser (632.8 nm) enter a beam splitter. The two split beams are directed normally onto opposite sides of the sample. A phase delay in one of the beam paths controls the relative phase difference between the two beams, by moving a piezo-actuated nano-positioning stage with 0.8 nm resolution. An additional attenuator ensures that the transmitted beams have equal intensities, compensating for imbalances in the beam splitters and the inherent asymmetry of the sample. The two abnormal output beams, which are the superposition of abnormal transmitted beam and abnormal reflected beam, are collected by two power meters synchronously. The inset is a schematic of various components. Reproduced from [74] with permission. Copyright 2016, Wiley-VCH Verlag GmbH & Co. KGaA, Weinheim

limit of the metasurface in transmission mode and realize dynamic control over the generalized Snell's law [74].

As illustrated in Fig. 2.40, two collimated counter-propagating laser beams (the signal and control beams) are directed onto opposite surfaces of the sample. Before illuminated on the sample, the two beams are converted to be RCP and LCP through the cascaded polarizer and a quarter-wave plate. Assuming that the amplitudes of the two beams are 1 and s, and the reflection and transmission amplitudes are equal, the electric fields in the right side of the sample can be written as

$$\begin{bmatrix} E_x \\ E_y \end{bmatrix} = \begin{bmatrix} E_{tx} \\ E_{ty} \end{bmatrix} + s \begin{bmatrix} E_{rx} \\ E_{ry} \end{bmatrix}$$

$$= \{(t_u + t_v) + s(r_u + r_v)\}\begin{bmatrix} 1 \\ i\sigma \end{bmatrix}$$

$$+ \{(t_u - t_v) + s(r_u - r_v)\}e^{2i\sigma\xi}\begin{bmatrix} 1 \\ -i\sigma \end{bmatrix}, \tag{2.8.1}$$

where r and t denote the complex reflection and transmission coefficients, u and v represent the local coordinates of the anisotropic elements, ξ is the orientation angle of u with respect to the x-axis. The output intensity of the cross-polarized component can be written as

$$I_{\text{cross}} = |(t_u - t_v) + s(r_u - r_v)|^2. \tag{2.8.2}$$

Obviously, I_{cross} is dependent on s, thus can be dynamically tuned by the control beam. The modulation depth, defined as the ratio of the maximum to the minimum intensity, is then

$$M = \frac{\max\big(|(t_u - t_v) + s(r_u - r_v)|^2\big)}{\min\big(|(t_u - t_v) + s(r_u - r_v)|^2\big)}. \tag{2.8.3}$$

For ideal anisotropic structure, the transmission and reflection coefficients can be approximated as $t_u = 0$, $t_v = 1$, $r_u = -1$, and $r_v = 0$. In this case, the modulation depth can approach infinite since there is $I_{\text{cross}} = 0$ for $s = -1$.

An ultrathin metallic metasurface comprised of catenary slits array is used to generate linear phase distribution with a period of $P_x = 2$ μm and $P_y = 2\Delta = 400$ nm along the x-direction and y-direction respectively. As shown in Fig. 2.41, a sample was fabricated by FIB milling in a 120-nm-thick gold layer deposited on a quartz substrate.

The anomalous refraction and reflection intensities were experimentally measured with respect to the phase difference between the two coherent light beams by controlling the piezo-actuated nano-positioning stage. In order to test the modulation ability and the efficiency enhancement of this method, the refracted and reflected power of the single signal beam and control beam were also recorded. The total refraction (T_{total}) is defined as the summation of the refraction of the signal beam and the reflection of the control beam. Similarly, the total reflection (R_{total}) is defined as the summation of the reflection of the signal beam and the refraction of the control beam.

A diode-pumped solid-state laser at $\lambda = 532$ nm was used as the laser source of coherent control. For single beam illumination, the values of both T_{total} and R_{total} nearly keep constant as the stage moving. For coherent illumination, a sinusoidal function type of intensity with respect to the movement was obtained through the phase difference control, as illustrated in Fig. 2.42a. The position for the constructive interference corresponds to the peak of the sinusoidal function. Similarly, the destructive interference occurs at the position of the valley of the sinusoidal function. The average output power of coherent anomalous refracted beams and the anomalous

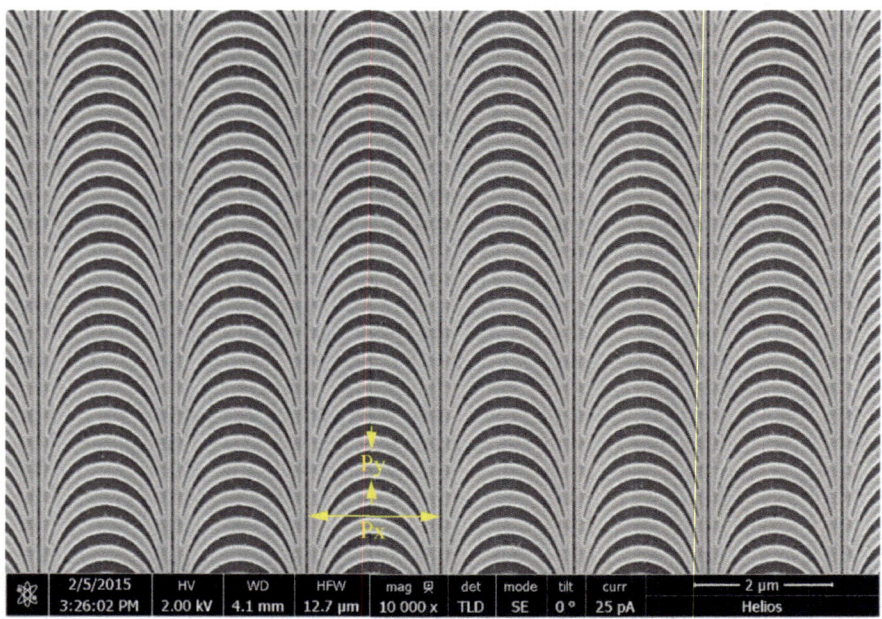

Fig. 2.41 SEM image of the linear catenary array fabricated by FIB

reflected beams of signal light at the peaks are enhanced by 78 and 62% compared
with that with no coherent illumination (T_{total} and R_{total}). The modulation of intensity
is dynamic with the movement of the stage, and the average period of the intensity
modulation from the experiment is 531.4 nm, matching well with the theoretical
value 532 nm (wavelength of the incident light). The inset in Fig. 2.42a plots the
anomalous refracted light spots at the marked peak and valley by a CCD camera.
The experimental modulation depth of the sample is about 8:1. The finite value of
modulation depth is mainly attributed to the asymmetry of the sample due to the
existence of the substrate. Consequently, by employing a substrate-free design, the
performance may be greatly improved.

To demonstrate the broadband performance of the coherent control method, a
He–Ne laser with λ = 632.8 nm was also adopted in the control experiment. Similar
results were obtained as shown in Fig. 2.42b. Compared with that with no coherent
illumination, the average output power of the coherent anomalous refracted beams
and the anomalous reflected beams are enhanced by 62 and 88% at the peaks, respectively.

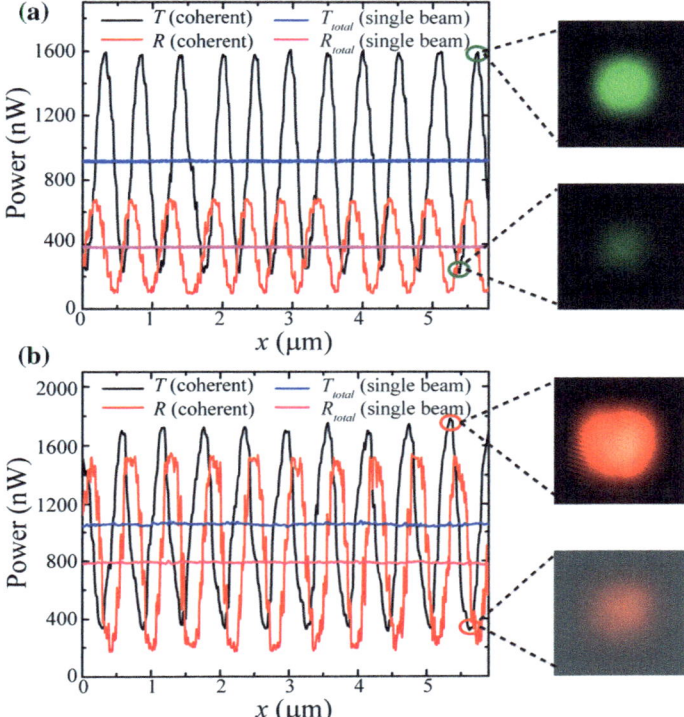

Fig. 2.42 Dynamically tuned abnormal transmission and reflection. **a, b** Measured results of the abnormal transmission and reflection at **a** $\lambda = 532$ nm and **b** $\lambda = 632$ nm. The refraction spots in the inset were captured at 20 cm away from the output plane of the sample. The modulation depth in refraction at $\lambda = 532$ and 632 nm is 8:1 and 5.8:1, respectively. Reproduced from [74] with permission. Copyright 2016, Wiley-VCH Verlag GmbH & Co. KGaA, Weinheim

References

1. B.E.A. Saleh, M.C. Teich, *Fundamentals of Photonics*, 2nd edn. (Wiley, Hoboken, 2007)
2. M.G. Raymer, B.J. Smith, The Maxwell wave function of the photon, *The Nature of Light: What Is a Photon?* (SPIE, Bellingham, 2005), pp. 1–5
3. K.Y. Bliokh, D. Smirnova, F. Nori, Quantum spin Hall effect of light. Science **348**, 1448–1451 (2015)
4. S.A. Maier, *Plasmonics: Fundamentals and Applications* (Springer Science & Business Media, Berlin, 2007)
5. D. Gilbert, On the mathematical theory of suspension bridges, with tables for facilitating their construction. Philos. Trans. R. Soc. Lond. **116**, 202–218 (1826)
6. M. Pu, X. Li, X. Ma, Y. Wang, Z. Zhao, C. Wang, C. Hu, P. Gao, C. Huang, H. Ren, X. Li, F. Qin, J. Yang, M. Gu, M. Hong, X. Luo, Catenary optics for achromatic generation of perfect optical angular momentum. Sci. Adv. **1**, e1500396 (2015)
7. M.V. Berry, The adiabatic phase and Pancharatnam's phase for polarized light. J. Mod. Opt. **34**, 1401–1407 (1987)

8. Z. Bomzon, V. Kleiner, E. Hasman, Pancharatnam-Berry phase in space-variant polarization-state manipulations with subwavelength gratings. Opt. Lett. **26**, 1424–1426 (2001)

9. R. Bhandari, Polarization of light and topological phases. Phys. Rep. **281**, 1–64 (1997)

10. D. Lin, P. Fan, E. Hasman, M.L. Brongersma, Dielectric gradient metasurface optical elements. Science **345**, 298–302 (2014)

11. M. Khorasaninejad, W.T. Chen, R.C. Devlin, J. Oh, A.Y. Zhu, F. Capasso, Metalenses at visible wavelengths: diffraction-limited focusing and subwavelength resolution imaging. Science **352**, 1190–1194 (2016)

12. R.C. Devlin, A. Ambrosio, N.A. Rubin, J.P.B. Mueller, F. Capasso, Arbitrary spin-to-orbital angular momentum conversion of light. Science **358**, 896–901 (2017)

13. F. Zhang, M. Pu, X. Li, P. Gao, X. Ma, J. Luo, H. Yu, X. Luo, All-dielectric metasurfaces for simultaneous giant circular asymmetric transmission and wavefront shaping based on asymmetric photonic spin–orbit interactions. Adv. Funct. Mater. **27**, 1704295 (2017)

14. S. Chen, Z. Li, Y. Zhang, H. Cheng, J. Tian, Phase manipulation of electromagnetic waves with metasurfaces and its applications in nanophotonics. Adv. Opt. Mater. **6**, 1800104 (2018)

15. X. Luo, Subwavelength optical engineering with metasurface waves. Adv. Opt. Mater. **6**, 1701201 (2018)

16. X. Luo, M. Pu, X. Li, X. Ma, Broadband spin Hall effect of light in single nanoapertures. Light Sci. Appl. **6**, e16276 (2017)

17. H.A. Bethe, Theory of diffraction by small holes. Phys. Rev. **66**, 163–182 (1944)

18. A.Y. Nikitin, D. Zueco, F.J. Garcia-Vidal, L. Martin-Moreno, Electromagnetic wave transmission through a small hole in a perfect electric conductor of finite thickness. Phys. Rev. B **78**, 165429 (2008)

19. X. Luo, Principles of electromagnetic waves in metasurfaces. Sci. China-Phys. Mech. Astron. **58**, 594201 (2015)

20. N. Shitrit, I. Bretner, Y. Gorodetski, V. Kleiner, E. Hasman, Optical spin Hall effects in plasmonic chains. Nano Lett. **11**, 2038–2042 (2011)

21. D. McGloin, K. Dholakia, Bessel beams: diffraction in a new light. Contemp. Phys. **46**, 15–28 (2005)

22. A. Ciattoni, B. Crosignani, P. Di Porto, Vectorial free-space optical propagation: a simple approach for generating all-order nonparaxial corrections. Opt. Commun. **177**, 9–13 (2000)

23. A.V. Krasavin, A.S. Schwanecke, N.I. Zheludev, M. Reichelt, T. Stroucken, S.W. Koch, E.M. Wright, Polarization conversion and "focusing" of light propagating through a small chiral hole in a metallic screen. Appl. Phys. Lett. **86**, 201105 (2005)

24. F.J. García-Vidal, E. Moreno, J.A. Porto, L. Martín-Moreno, Transmission of light through a single rectangular hole. Phys. Rev. Lett. **95**, 103901 (2005)

25. M. Pu, X. Ma, Y. Guo, X. Li, X. Luo, Theory of microscopic meta-surface waves based on catenary optical fields and dispersion. Opt. Express **26**, 19555–19562 (2018)

26. L.B. Whitbourn, R.C. Compton, Equivalent-circuit formulas for metal grid reflectors at a dielectric boundary. Appl. Opt. **24**, 217–220 (1985)

27. Y. Guo, M. Pu, Z. Zhao, Y. Wang, J. Jin, P. Gao, X. Li, X. Ma, X. Luo, Merging geometric phase and plasmon retardation phase in continuously shaped metasurfaces for arbitrary orbital angular momentum generation. ACS Photonics **3**, 2022–2029 (2016)

28. F. Zhang, M. Pu, J. Luo, H. Yu, X. Luo, Symmetry breaking of photonic spin-orbit interactions in metasurfaces. Opto-Electron. Eng. **44**, 319–325 (2017)

29. X. Li, M. Pu, Z. Zhao, X. Ma, J. Jin, Y. Wang, P. Gao, X. Luo, Catenary nanostructures as highly efficient and compact Bessel beam generators. Sci. Rep. **6**, 20524 (2016)

30. N.R. Heckenberg, R. McDuff, C.P. Smith, A.G. White, Generation of optical phase singularities by computer-generated holograms. Opt. Lett. **17**, 221–223 (1992)

31. G. Rui, W. Chen, D.C. Abeysinghe, R.L. Nelson, Q. Zhan, Beaming circularly polarized photons from quantum dots coupled with plasmonic spiral antenna. Opt. Express **20**, 19297–19304 (2012)

32. M. Pu, X. Li, Y. Guo, X. Ma, X. Luo, Nanoapertures with ordered rotations: symmetry transformation and wide-angle flat lensing. Opt. Express **25**, 31471–31477 (2017)

33. W. Liu, Z. Li, H. Cheng, C. Tang, J. Li, S. Zhang, S. Chen, J. Tian, Metasurface enabled wide-angle Fourier lens. Adv. Mater. **30**, 1706368 (2018)
34. J. Kedmi, A.A. Friesem, Optimal holographic Fourier-transform lens. Appl. Opt. **23**, 4015–4019 (1984)
35. J.B. Pendry, D. Schurig, D.R. Smith, Controlling electromagnetic fields. Science **312**, 1780–1782 (2006)
36. H. Ma, T. Cui, Three-dimensional broadband and broad-angle transformation-optics lens. Nat. Commun. **1**, 124 (2010)
37. N. Kundtz, D.R. Smith, Extreme-angle broadband metamaterial lens. Nat. Mater. **9**, 129–132 (2010)
38. L. Marrucci, C. Manzo, D. Paparo, Optical spin-to-orbital angular momentum conversion in inhomogeneous anisotropic media. Phys. Rev. Lett. **96**, 163905 (2006)
39. N. Shitrit, I. Yulevich, E. Maguid, D. Ozeri, D. Veksler, V. Kleiner, E. Hasman, Spin-optical metamaterial route to spin-controlled photonics. Science **340**, 724–726 (2013)
40. E.T.F. Rogers, J. Lindberg, T. Roy, S. Savo, J.E. Chad, M.R. Dennis, N.I. Zheludev, A super-oscillatory lens optical microscope for subwavelength imaging. Nat. Mater. **11**, 432–435 (2012)
41. D. Tang, C. Wang, Z. Zhao, Y. Wang, M. Pu, X. Li, P. Gao, X. Luo, Ultrabroadband super-oscillatory lens composed by plasmonic metasurfaces for subdiffraction light focusing. Laser Photonics Rev. **9**, 713–719 (2015)
42. Z. Li, T. Zhang, Y. Wang, W. Kong, J. Zhang, Y. Huang, C. Wang, X. Li, M. Pu, X. Luo, Achromatic broadband super-resolution imaging by super-oscillatory metasurface. Laser Photonics Rev. **12**, 1800064 (2018)
43. F. Aieta, M.A. Kats, P. Genevet, F. Capasso, Multiwavelength achromatic metasurfaces by dispersive phase compensation. Science **347**, 1342–1345 (2015)
44. F. Qin, K. Huang, J. Wu, J. Jiao, X. Luo, C. Qiu, M. Hong, Shaping a subwavelength needle with ultra-long focal length by focusing azimuthally polarized light. Sci. Rep. **5**, 09977 (2015)
45. M.V. Berry, N.L. Balazs, Nonspreading wave packets. Am. J. Phys. **47**, 264 (1979)
46. G.A. Siviloglou, J. Broky, A. Dogariu, D.N. Christodoulides, Observation of accelerating airy beams. Phys. Rev. Lett. **99**, 213901 (2007)
47. S. Franke-Arnold, L. Allen, M. Padgett, Advances in optical angular momentum. Laser Photonics Rev. **2**, 299–313 (2008)
48. E. Brasselet, G. Gervinskas, G. Seniutinas, S. Juodkazis, Topological shaping of light by closed-path nanoslits. Phys. Rev. Lett. **111**, 193901 (2013)
49. M. Pu, Z. Zhao, Y. Wang, X. Li, X. Ma, C. Hu, C. Wang, C. Huang, X. Luo, Spatially and spectrally engineered spin-orbit interaction for achromatic virtual shaping. Sci. Rep. **5**, 9822 (2015)
50. G. Zheng, H. Mühlenbernd, M. Kenney, G. Li, S. Zhang, Metasurface holograms reaching 80% efficiency. Nat. Nanotechnol. **10**, 308–312 (2015)
51. J. Durnin, Exact solutions for nondiffracting beams. I. The scalar theory. J. Opt. Soc. Am. A **4**, 651–654 (1987)
52. H. Gao, M. Pu, X. Li, X. Ma, Z. Zhao, Y. Guo, X. Luo, Super-resolution imaging with a Bessel lens realized by a geometric metasurface. Opt. Express **25**, 13933–13943 (2017)
53. A. Dudley, M.P.J. Lavery, M.J. Padgett, A. Forbes, Unraveling Bessel beams. Opt. Photonics News **22**, 24–29 (2013)
54. J. Wang, J.-Y. Yang, I.M. Fazal, N. Ahmed, Y. Yan, H. Huang, Y. Ren, Y. Yue, S. Dolinar, M. Tur, A.E. Willner, Terabit free-space data transmission employing orbital angular momentum multiplexing. Nat. Photonics **6**, 488–496 (2012)
55. K. Dholakia, P. Reece, M. Gu, Optical micromanipulation. Chem. Soc. Rev. **37**, 42–55 (2008)
56. J. Jin, J. Luo, X. Zhang, H. Gao, X. Li, M. Pu, P. Gao, Z. Zhao, X. Luo, Generation and detection of orbital angular momentum via metasurface. Sci. Rep. **6**, 24286 (2016)
57. X. Ding, F. Monticone, K. Zhang, L. Zhang, D. Gao, S.N. Burokur, A. de Lustrac, Q. Wu, C.-W. Qiu, A. Alù, Ultrathin Pancharatnam-Berry metasurface with maximal cross-polarization efficiency. Adv. Mater. **27**, 1195–1200 (2015)

58. Y. Wang, M. Pu, Z. Zhang, X. Li, X. Ma, Z. Zhao, X. Luo, Quasi-continuous metasurface for ultra-broadband and polarization-controlled electromagnetic beam deflection. Sci. Rep. **5**, 17733 (2015)
59. X. Tan, Anomalous scattering-induced circular dichroism in continuously shaped metasurface. Opto-Electron. Eng. **44**, 87–91 (2017)
60. Y. Guo, Y. Wang, M. Pu, Z. Zhao, X. Wu, X. Ma, C. Wang, L. Yan, X. Luo, Dispersion management of anisotropic metamirror for super-octave bandwidth polarization conversion. Sci. Rep. **5**, 8434 (2015)
61. M. Pu, P. Chen, Y. Wang, Z. Zhao, C. Huang, C. Wang, X. Ma, X. Luo, Anisotropic meta-mirror for achromatic electromagnetic polarization manipulation. Appl. Phys. Lett. **102**, 131906 (2013)
62. X. Xie, X. Li, M. Pu, X. Ma, K. Liu, Y. Guo, X. Luo, Plasmonic metasurfaces for simultaneous thermal infrared invisibility and holographic illusion. Adv. Funct. Mater. **28**, 1706673 (2018)
63. D.G. Baranov, A.E. Krasnok, T. Shegai, A. Alù, Y.D. Chong, Coherent perfect absorbers: linear control of light with light. Nat. Rev. Mater. **2**, 17064 (2017)
64. Y.D. Chong, L. Ge, H. Cao, A.D. Stone, Coherent perfect absorbers: time-reversed lasers. Phys. Rev. Lett. **105**, 053901 (2010)
65. W. Wan, Y. Chong, L. Ge, H. Noh, A.D. Stone, H. Cao, Time-reversed lasing and interferometric control of absorption. Science **331**, 889–892 (2011)
66. C. Yan, M. Pu, J. Luo, Y. Huang, X. Li, X. Ma, X. Luo, Coherent perfect absorption of electromagnetic wave in subwavelength structures. Opt. Laser Technol. **101**, 499–506 (2018)
67. M. Pu, Q. Feng, M. Wang, C. Hu, C. Huang, X. Ma, Z. Zhao, C. Wang, X. Luo, Ultrathin broadband nearly perfect absorber with symmetrical coherent illumination. Opt. Express **20**, 2246–2254 (2012)
68. M. Pu, Q. Feng, C. Hu, X. Luo, Perfect absorption of light by coherently induced plasmon hybridization in ultrathin metamaterial film. Plasmonics **7**, 733–738 (2012)
69. S. Li, J. Luo, S. Anwar, S. Li, W. Lu, Z.H. Hang, Y. Lai, B. Hou, M. Shen, C. Wang, Broadband perfect absorption of ultrathin conductive films with coherent illumination: superabsorption of microwave radiation. Phys. Rev. B **91**, 220301(R) (2015)
70. Y. Wang, M. Pu, C. Hu, Z. Zhao, C. Wang, X. Luo, Dynamic manipulation of polarization states using anisotropic meta-surface. Opt. Commun. **319**, 14–16 (2014)
71. M. Crescimanno, N.J. Dawson, J.H. Andrews, Coherent perfect rotation. Phys. Rev. A **86**, 031807 (2012)
72. J. Zhang, K.F. MacDonald, N.I. Zheludev, Controlling light-with-light without nonlinearity. Light Sci. Appl. **1**, e18 (2012)
73. M. Papaioannou, E. Plum, J. Valente, E.T.F. Rogers, N.I. Zheludev, Two-dimensional control of light with light on metasurfaces. Light Sci. Appl. **5**, e16070 (2016)
74. X. Li, M. Pu, Y. Wang, X. Ma, Y. Li, H. Gao, Z. Zhao, P. Gao, C. Wang, X. Luo, Dynamic control of the extraordinary optical scattering in semicontinuous 2D metamaterials. Adv. Opt. Mater. **4**, 659–663 (2016)

Chapter 3
Catenary Structures for Spin-Dependent Coupling of Waveguide Modes

Abstract The geometric phase in catenary structures can not only be exploited to control the propagation direction in free space, but also be used to convert freely propagating beams to guiding waves. In this chapter, we first describe how catenary apertures could realize efficient and polarization-dependent excitation of surface plasmon polaritons (SPPs). Compared with discrete structures with the same size, catenary structures have led to much higher conversion efficiency and extinction ratio between the left and right channels. Besides plasmonic optical circuits, silicon photonics is of great importance in future engineering optics, which has enabled new types of nanophotonic applications, including optical interconnects for data communications and ultra-fast optical communications systems. Especially, the high index contrast and well-established complementary metal–oxide–semiconductor (CMOS) compatible processing make silicon-on-insulator (SOI) as an attractive platform for optical information processing. Section 3.2 presents a combination of catenary subwavelength structures with SOI waveguide, which provides much more freedom for the spin-dependent light coupling. In Sect. 3.3, it is shown that a single-curved SOI waveguide could also couple circular polarized beam to different channels when the curvature is comparable to the wavelength.

Keywords Unidirectional excitation · Catenary waveguide · Integrated optics

3.1 Catenary Apertures for Unidirectional Excitation of SPP

In this section, the geometric phase taking place in catenary-shaped aperture is utilized to realize high-efficient unidirectional excitation of SPP with high extinction ratio, which has a wide range of applications in SPP source, on-chip signal processing, and spin-controlled photonics. Owing to the continuous phase modulation provided by the catenary aperture, the horizontal size of the SPP launcher is smaller than the wavelength. Simultaneously, the catenary aperture exhibits a distinctive directionality much higher than that of traditional structures.

3.1.1   Discrete Spin-Controlled Unidirectional Coupler

It has been demonstrated that when a circularly polarized light (CPL) illuminates on a single slit obliquely, surface plasmons will be excited at only one side. This effect can be understood by considering the scattering of two orthogonal linear polarizations: As shown in Fig. 3.1a, for transverse electric (TE) polarization, the z-components of SPPs at the two sides of the slit has a phase shift of 90°; For transverse magnetic (TM) polarization, however, the two components have the same phase when excited. Since CPL is composed of two orthogonal linear polarizations, the coherent addition of these components would induce unidirectional excitation of SPPs [1]. These phenomena can be explained using the magnetic and electric dipoles, as illustrated in Fig. 3.1b, c. Alternatively, vectorial near-field interference is also a powerful mean to describe these results [2], which can be further extended to more complicated cases [3].

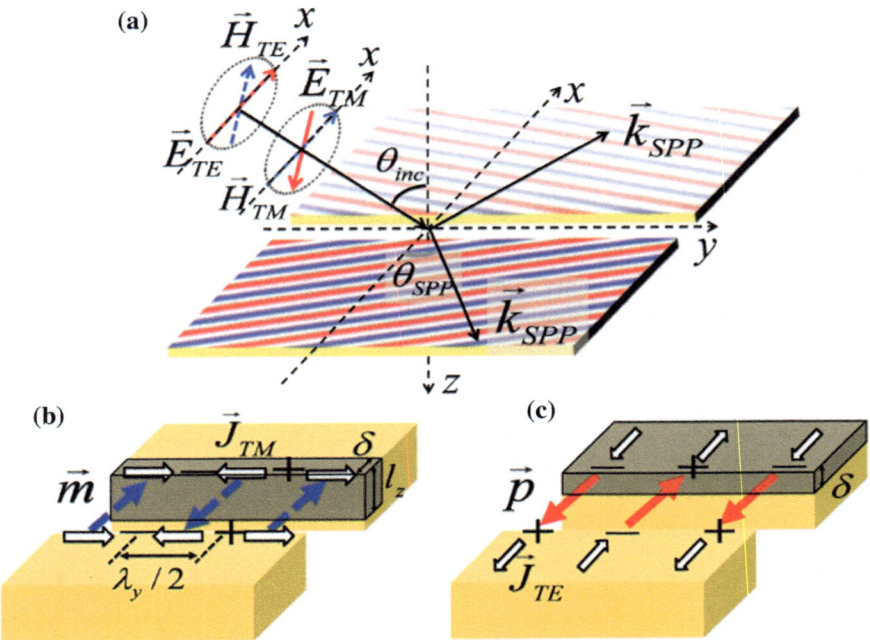

Fig. 3.1 a The configuration of the experiment. The SPP field distributions are illustrated on the metal surface. The incident wave vector of the plane wave is always laid on the yz-plane so it does not break symmetry with respect to the slit structure. **b, c** The surface currents, accumulated charges, and induced dipole moments are illustrated for **b** TM wave and **c** TE wave incidence. Reproduced from [1] with permission. Copyright 2012, The American Physical Society

Under normal incidence, SPP can only be excited by TM polarization, i.e., the magnetic field is parallel to the slit. Since TM wave is composed of two circularly polarized components, both left-handed circular polarization (LCP) and right-handed circular polarization (RCP) would excite two identical beams toward the opposite directions. This effect, together with the above experiment under oblique incidence, can be also explained by considering the symmetry of light–matter interaction. It is the oblique incidence which breaks the symmetry of spin–orbit interaction.

Besides oblique incidence, asymmetric surface structures could also induce unidirectional interference, as demonstrated by Lin et al. in 2013 [4]. One key to understand this effect is the fact that the z-components of electric fields have opposite phase along the two directions perpendicular to the long axis. As shown in Fig. 3.2, two parallel columns of apertures are spaced at a distance S apart and the rectangular aperture has a width of W and a length of L and $W \ll L < \lambda_{SPP}$. The apertures of one column are spaced a distance D apart, with $D < \lambda_{SPP}$. The two columns are offset by half a period ($D/2$) along their axes to reduce near-field coupling and scattering of the SPPs by neighboring apertures.

Any circular polarization can be decomposed into two linear polarizations, with one (E_1) perpendicular to the long axis of the left apertures and the other (E_2) parallel to the long axis of the right-side apertures. For circular polarizations, E_1 and E_2 have a relative phase shift of $\pi/2$. More generally, they can have a relative phase δ that is preserved by the coupling process. If the propagation loss of the SPPs over the small distance S is neglected, interference of the SPP waves launched by each column

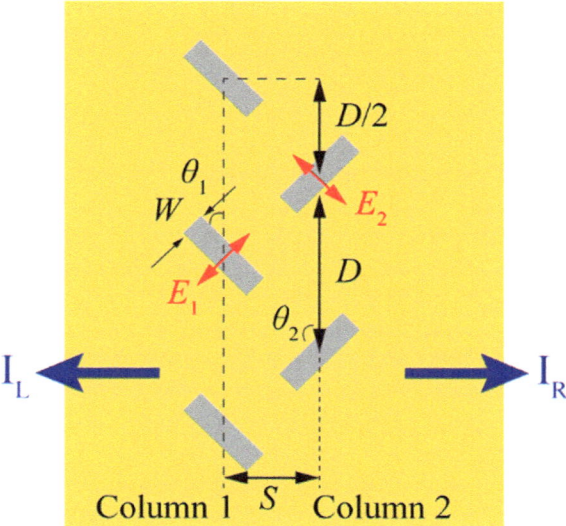

Fig. 3.2 Closely spaced subwavelength apertures as polarization-selective SPP couplers. Two columns of apertures (1 and 2) are positioned in parallel with spacing S. The columns couple to the field components

results in time-averaged SPP field intensities propagating to the right (I_R) and the left (I_L) as

$$I_R \propto \left[\left(C_1 E_1^2 + C_2 E_2^2 \right) + 2 E_1 E_2 \cos(k_{SPP} S + \delta) \right],$$
$$I_L \propto \left[\left(C_1 E_1^2 + C_2 E_2^2 \right) + 2 E_1 E_2 \cos(k_{SPP} S - \delta) \right], \qquad (3.1.1)$$

where the intensities are calculated in the dielectric, k_{SPP} is the wavenumber of SPP. By choosing $\theta_1 = 45°$ and $\theta_2 = 45° + 90° = 135°$, there is $|E|^2 = |E_1|^2 + |E_2|^2$ and $C_1 = C_2 = C$. If the lateral spacing S of the columns with respect to the SPP wavelength $S = \pi/(2k_{SPP})$, Eq. (3.1.1) becomes

$$I_R \propto C \left[\left(E_1^2 + E_2^2 \right) - 2 E_1 E_2 \sin \delta \right],$$
$$I_L \propto C \left[\left(E_1^2 + E_2^2 \right) + 2 E_1 E_2 \sin \delta \right]. \qquad (3.1.2)$$

Note that, the combined intensity of the coupled SPP fields is proportional to the intensity of the incident light but is independent of δ and of the relative magnitude of the coupled field components. The total power that is converted to SPPs can thus be made independent of the polarization of the incident light. Moreover, the polarization dependence in Eq. (3.1.2) can be used to adjust the fraction of the total converted power that is propagating in the SPP waves launched either toward the left or toward the right side of the column pair by changing the polarization state of the incident light.

For circular polarizations, there are $E_1 = E_2$ and $\delta = \pm\pi/2$. Consequently, Eq. (3.1.2) could be rewritten as

$$I_R \propto 2C E^2 [1 \mp 1],$$
$$I_L \propto 2C E^2 [1 \pm 1], \qquad (3.1.3)$$

which implies a perfect enhancement and extinction of the intensity at the two directions.

In the experiment, 10 column pairs were fabricated with $S = 150$ nm, $D = 300$ nm, $W = 50$ nm, and $L = 200$ nm in a 150-nm-thick gold film using focused ion (FIB) beam milling. The devices were back-illuminated with linearly polarized 633-nm laser, and the SPPs were measured using scanning near-field optical microscopy (SNOM). The field distributions are in good agreement with the theoretical model. In particular, CPL of opposite handedness could generate counter-propagating SPP beams, whereas linearly polarized light launches SPP beams of equal intensity toward the two sides of the coupler.

Although unidirectional excitation only requires a relative phase shift of π between two surface scatters, some researches also try to introduce the phase-gradient subwavelength structures to this application. The rationale is clear: by increasing the diffraction angle of a common beam deflector to be larger than 90° [5], unidirectional excitation of guided modes can be realized [6]. Similarly, when geometric phases in space-varying rotating elements are taken into account [7–9], the excitation direction may be switched by using circular polarization state.

Figure 3.3a shows a unidirectional SPP coupler consisted of an array of elongated apertures with a constant gradient of orientation angle along x-direction [10]. In electromagnetic theory, each aperture may be considered as an electric dipole. Under CPL incidence, the transmitted light is featured by space-variant linear polarizations with an additional phase shift equal to the rotation angle. In the far-field, the linear polarizations can be further decomposed into two circular polarizations. Finally, a geometric phase twice the rotation angle is obtained for the cross-polarized component.

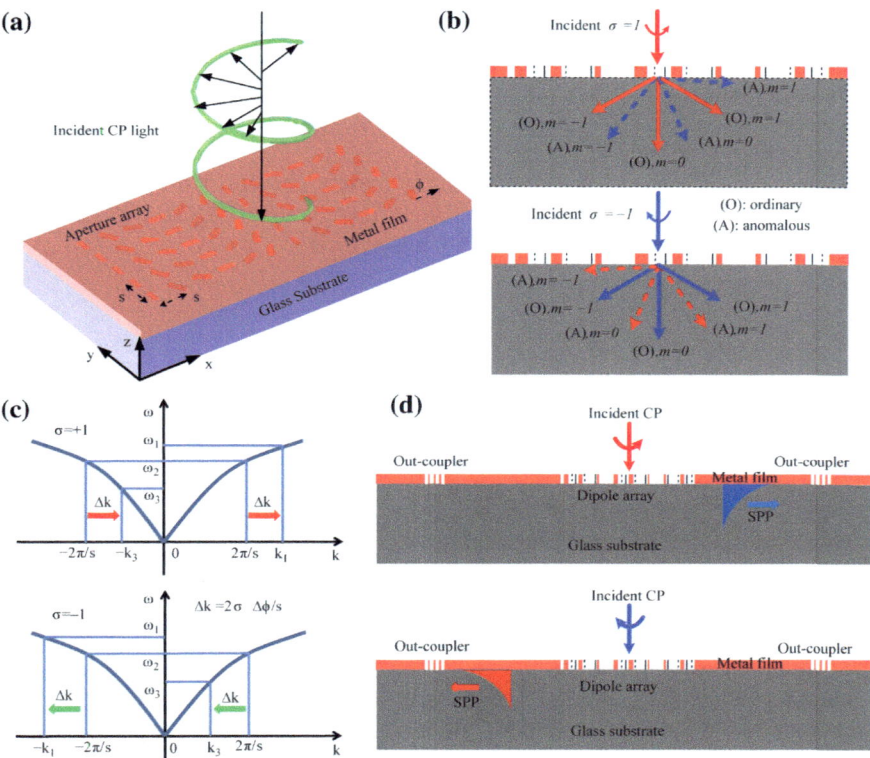

Fig. 3.3 **a** Schematic of a unidirectional SPPs coupler consisting of an array of rectangular apertures with spatially varying orientations on a metal film. **b** Ordinary (O) and anomalous (A) refraction and diffraction for the two circular polarization states. The anomalous diffraction orders are asymmetric about the normal direction. **c** Dispersion curves of SPPs and the momentum matching condition for ordinary and anomalous diffraction orders. **d** Only one of the anomalous diffraction orders can be matched to the SPP dispersion relation to launch unidirectional SPP due to the asymmetry in the anomalous diffraction orders. Reproduced from [10] with permission. Copyright 2013, CIOMP

In the original paper, Huang et al. directly applied the far-field concepts to the waveguides. By taking into account the contribution from the phase gradient, the angles of the ordinary and anomalous refracted and diffracted beams for a normal incident beam are given as

$$n_t \sin \theta_t = m \frac{\lambda_0}{s} \text{ (ordinary)}, \tag{3.1.4}$$

$$n_t \sin \theta_t = m \frac{\lambda_0}{s} + 2\sigma \frac{\lambda_0}{2\pi} \frac{\Delta\varphi}{s} = \frac{\lambda_0}{s} \left(m + 2\sigma \frac{\Delta\varphi}{2\pi} \right) \text{(anomalous)}, \tag{3.1.5}$$

where n_t and θ_t are the refractive indices and angles for the transmitted beams, m is the diffraction order, s is the lattice constant, and σ represents the helicity of the incident beam, which takes the values of ± 1. Equations (3.1.4) and (3.1.5) describe the ordinary and anomalous diffraction, respectively. Note that, the sign of the phase change depends on the polarization state of the incident beam for anomalous refraction and diffraction.

The geometric gradient phase offers the necessary phase matching condition for a beam at normal incidence to excite SPPs, as illustrated by Fig. 3.3c. The in-plane wave vector of light for generating SPPs can be calculated by

$$k_{SPP} = \frac{2m\pi}{s} + \sigma \frac{2\Delta\varphi}{s}. \tag{3.1.6}$$

For ordinary refraction, both of the two first-order diffracted beams ($m = \pm 1$) couple to SPPs at the same optical frequency ω_2, but with opposite propagation directions. For anomalous refraction with an incident beam of right-handed helicity, the phase matching condition is shifted to a higher frequency ω_1 for SPPs propagating along $+x$-direction, and to a lower frequency ω_3 for propagating along $-x$-direction.

Considering the fact that SPPs are linearly polarized rather than circular polarization, the above theory may be not accurate. In fact, it has been demonstrated the SPPs generated by a rotating aperture has a phase shift equal to, rather than twice the rotating angle. This effect seems to be in contrary to the above theory. Here, we show that the two explanations can be unified, and it seems that the unidirectional excitation phenomenon could be further improved. As shown in Fig. 3.4, along with the rotation of rectangular apertures, the generated SPPs under CPL illumination would obtain a phase shift directly proportional to the rotational angle, similar to the analysis made by Lin et al. [4]. However, the directions of these SPPs are also rotated correspondingly. If we only consider the components propagating along the right directions, the last four elements would have an additional phase shift of π (180°). Consequently, a rotation of 180° will induce a geometric phase of 360°, as predicted by the theory given by Huang et al. [10]. Note that, the SPPs components propagating along the right direction have an amplitude depending on the rotation angle, which can be approximated as a cosine function. Consequently, the middle element actually does not contribute to the unidirectional excitation.

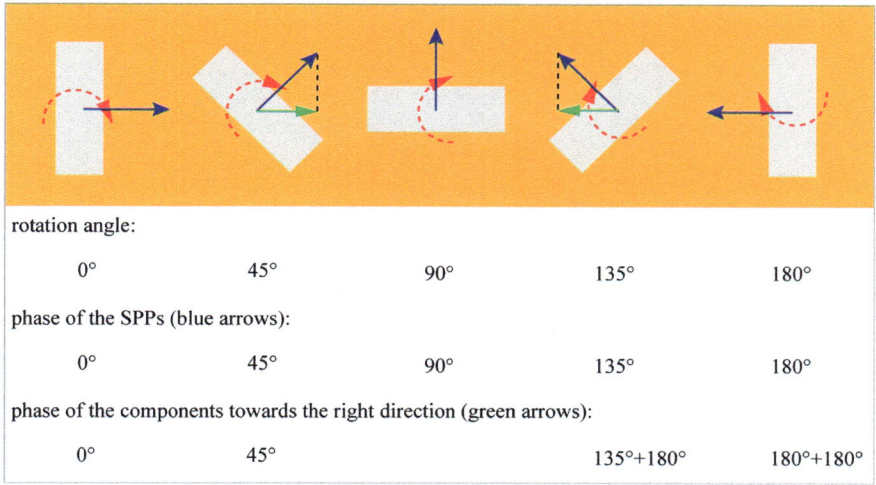

rotation angle:				
0°	45°	90°	135°	180°
phase of the SPPs (blue arrows):				
0°	45°	90°	135°	180°
phase of the components towards the right direction (green arrows):				
0°	45°		135°+180°	180°+180°

Fig. 3.4 Schematic of the phase of SPPs generated by rotated rectangular apertures

3.1.2 Catenary Unidirectional Coupler

Since the geometric phase only provides an approximate theory, the unidirectional performance should be optimized with rigorous electromagnetic software. In this section, we show that a continuous catenary aperture could produce higher directionality regarding the extinction ratio between the two directions. For simplicity, the geometric phase theory is used to design the catenary.

As shown in Fig. 3.5a, a catenary aperture can be formed by rotating a rectangular aperture. Under normally illumination of CPL, the phase of SPP excited by a single aperture is approximately $2\sigma\xi$. The z-component of the electric field of the SPP propagating along x-axis is given by

$$E_z \propto \pm e^{i2\sigma\xi}, \qquad (3.1.7)$$

where the upper (+) sign corresponds to an SPP wave propagating in the positive (right) x-direction, the lower (−) sign to an SPP wave propagating in the negative (left) x-direction. In order to realize directional launching effect, the generated SPP wave must interfere constructively along one direction. Different from free space, the wavevector of SPP must be considered. Note that, the apertures may scatter the induced SPP many times, thus the theoretical analysis is only an approximation.

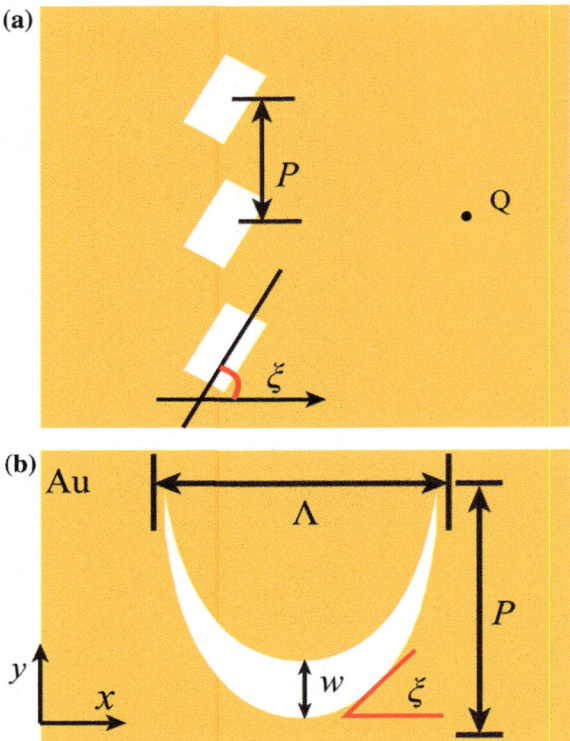

Fig. 3.5 **a** A rectangular nanohole with the rotation angle of ξ. **b** Geometric parameters of a single catenary aperture. The vertical shift of the catenary curve is $w = 126$ nm. The periodicity along the y-axis is $P = 550$ nm

The superiority of the catenary is that it can produce continuous and linear phase shift covering $[0, 2\pi]$ using a single subwavelength block under the illumination of CPL. According to the relationship between the rotation angle of geometric meta-surface and the phase shift of excited SPPs, the catenary aperture with carefully designed parameters can be used to modulate the direction of SPP propagation at the metal surface. And, the horizontal size of the launcher can be squeezed into sub-wavelength scale. The schematic of the unidirectional SPP launcher featured by the subwavelength catenary perforated in the metal film is depicted in Fig. 3.5b. Under CPL illumination, the SPP is excited and propagated along one side of the catenary apertures. The catenary on the gold film is closed by two same catenary curves with a certain vertical shift w. The catenary curve can be written as follows:

$$y = -\frac{\Lambda}{\pi} \ln\left|\cos\left(\frac{\pi x}{\Lambda}\right)\right|, \tag{3.1.8}$$

where Λ is a constant, which determines the horizontal length of a single catenary. Under normal illumination of CPL, the catenary aperture would introduce a continuous phase shift which related to the inclination angle of the catenary curve to the excited SPP light. The relationship between the phase of excited SPP and the inclination angle $\xi(x)$ of the catenary aperture is $\varphi = \sigma\xi(x)$. And meanwhile, each point of the catenary fulfills the condition of construction interference.

In order to control the SPP at visible wavelength, we optimized the parameters of the catenary aperture. The geometric parameters of a single catenary were chosen as $w = 126$ nm, $\Lambda = 600$ nm, and a thickness of 120 nm. FDTD method was employed to perform the directional launching effect of the catenary aperture. In the numerical simulation, catenary apertures with periodic boundary condition along y-direction and perfectly matched layer boundary condition along x- and z-directions were modeled, and the periodicity along y-direction was $P = 550$ nm, as shown in Fig. 3.5c. A circularly polarized plane wave was normally incident on the catenary apertures along $+z$-axis. At the wavelength of 618 nm, the intensity distributions of z-component electric field for the two circularly polarized incident light at $z = 10$ nm are shown in Fig. 3.6a, c, respectively. And the corresponding field intensity distribution at the $y = 0$ plane are displayed in Fig. 3.6b, d.

Clearly, the catenary apertures coupled the left CPL into the SPP propagating along the $-x$-direction and for the right circularly polarized incident light, the SPP propagated along the $+x$-direction. The field intensity of the SPP on the two sides of the catenary presents a striking contrast. The extinction ratio defined by the ratio of SPP power flowing in the two directions of the x-axis was employed to characterize the directionality of the catenary apertures. The SPP power was obtained by integrating the Poynting's vector on a vertical plane positioned 5 μm away to the catenary apertures.

To experimentally verify the ability of directional SPP excitation, the catenary apertures perforated in a 120 nm-thick gold film deposited on the glass substrate were fabricated using FIB milling. The scanning electron microscopy images of the sample are shown in Fig. 3.7a. The dimensions of the catenary array along the x- and y-directions are 0.552 and 27.5 μm, respectively. Two groove arrays with periodicity of 800 nm, width of 400 nm and depth of 50 nm were symmetrically fabricated on the two sides of the catenary apertures to decouple the excited SPP light. In order to efficiently decouple the excited SPP, the distance between the catenary apertures and gratings is optimized as 5 μm. A home-made microscope was used to characterize the performances of the fabricated catenary aperture, as shown in Fig. 3.7b. A continuously tunable laser (SuperK EXTREME) was chosen as the source of the test system. The monochromatic light was converted into the CPL by using the combination of a linear polarizer and a quarter wave plate. Then, the CPL was focused on the fabricated sample from the substrate side through a 50× objective lens. The CPL focusing beam with a radius of 2 μm was incidented onto the center of the catenary apertures by adjusting the position of the sample and the distance between the sample and objective lens. The electric field of the excited directional SPP can be coupled out by the groove array. The field intensity images collected by the 100× objective lens were recorded by a CCD camera.

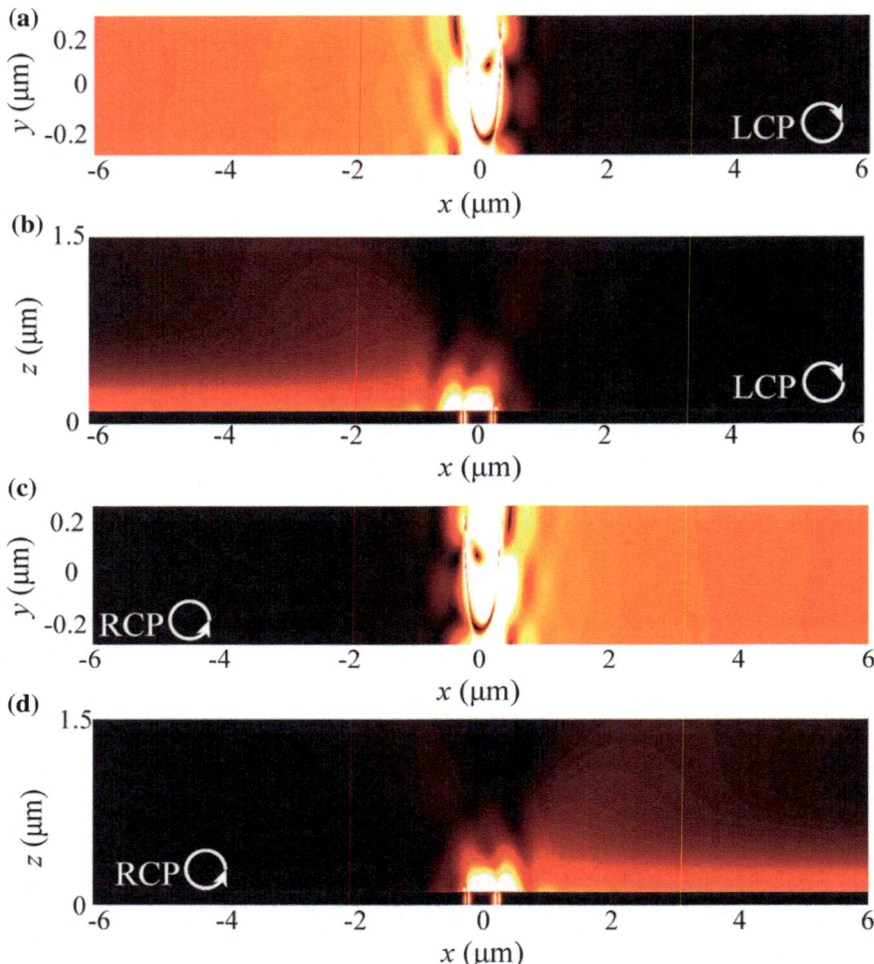

Fig. 3.6 a The z-component of the electric field intensity distribution in the xy-plane 10 nm above the apertures. The incident light is LCP. **b** The z-component of the electric field intensity distribution in the xz-plane positioned at the center of the y-axis. **c** The z-component of the electric field intensity distribution in the xy-plane 10 nm above the apertures. The incident light is RCP. **d** The z-component of the electric field intensity distribution in the xz-plane, $y = 0$

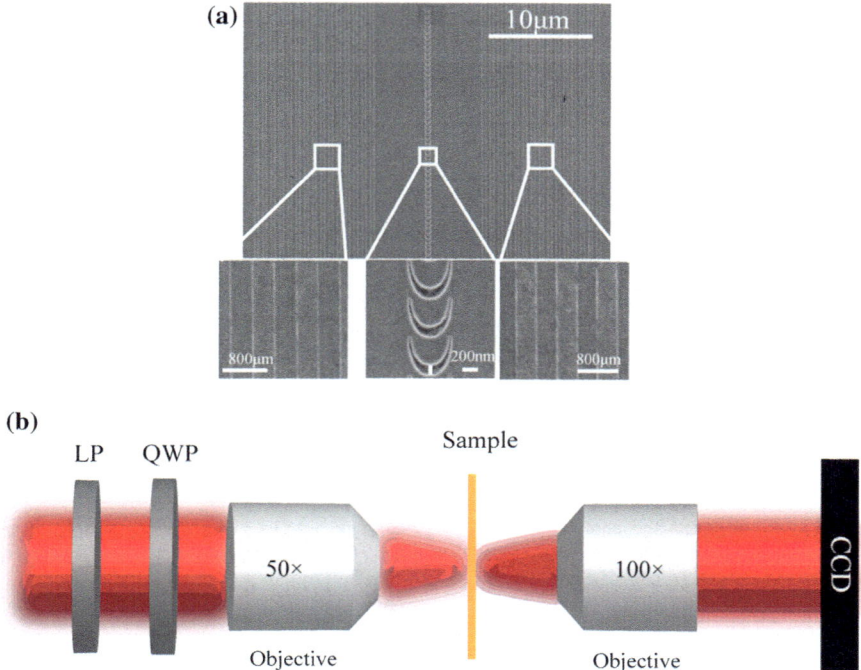

Fig. 3.7 **a** The schematic of the SEM image of the catenary apertures array. The insert image is the magnification of the local structures. **b** The experimental setup

Figure 3.8 depicts the captured field intensity images. When the focused left-circularly polarized beam normally incidents on the sample from the substrate side, the excited SPP propagating along the $-x$-direction was decoupled by the left side groove array, as sketched in Fig. 3.8a. As can be seen from the diagram, the light at the left side groove array is much brighter than the right side (there is nearly no light observed on the right side). Then, we convert the handedness of the incident light into right circular polarization, the SPP light was scattered by the groove array sitting on the right side of the catenary apertures, as shown in Fig. 3.8b. And when the catenary apertures were illuminated by the linear polarized light, the SPP propagating along the $-x$- and $+x$-directions, so we can observe that the light scattered from the two sides of the grooves arrays, as shown in Fig. 3.8c. Therefore, by adjusting the handedness of the incident light, the catenary apertures can realize the directional control of the SPP. To further ascertain the extinction ratio of the fabricated device, under the left CPL illumination, the ratio between the light intensity decoupled by the left groove array and the intensity decoupled by the right groove array were evaluated. The calculated extinction ratio at different wavelength is illustrated in Fig. 3.8d.

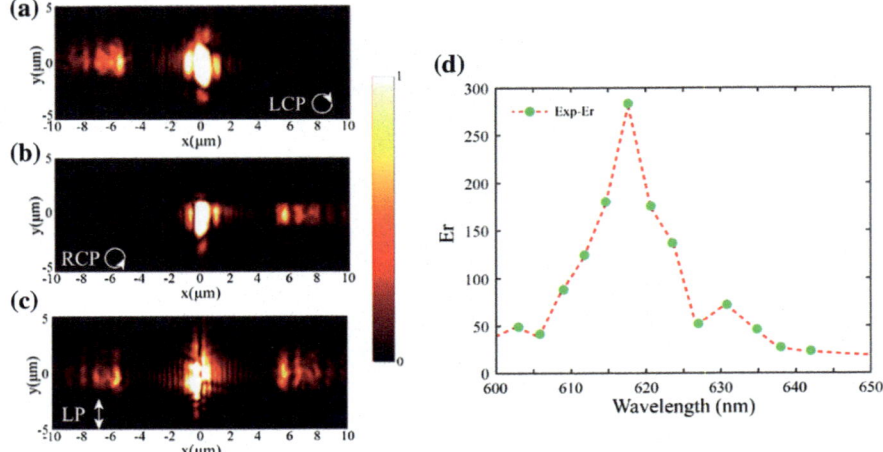

Fig. 3.8 Experimental results. **a** The light intensity distribution captured by the CCD when the catenary apertures were illuminated by the LCP at the wavelength of 618 nm. The central spot is the light transmitted directly from the catenary apertures. The left side light spot is the decoupled SPP scattered by the grooves. **b** The light intensity distribution captured by the CCD under the RCP illumination. **c** The light intensity distribution captured by the CCD under the linear polarized illumination. **d** The measured extinction ratio at different wavelengths

The measured extinction ratio at the wavelength of 618 nm is 283, which is a bit lower than the numerical results owing to the fabrication and measurement errors. Nevertheless, this extinction ratio is much larger than the discrete structure with the same size, as a result of the continuous structure. As we noted in the introduction, the theory for the unidirectional excitation is still not perfect owing to the polarization mismatch between the incident and guided waves. As a result, we expect that further improvement of the device is possible by simultaneously modifying the shape and size of the catenary apertures.

3.2 Spin-Controlled Router for SOI Waveguide

As two research hot spots, plasmonic and SOI waveguides have their own merits and demerits regarding the compactness and propagation length. On the one hand, plasmonics provides a possibility to realize sub-diffraction-limited localization of optical fields, which makes the control of vectorial properties of light relatively easy. For instance, by perforating nanoslits in a metallic film, the phase profile of incoming light can be readily adjusted by using either the propagating phase or the geometric phase [5, 8]. The unidirectional coupling can also be easily achieved, as demonstrated in the previous section.

On the other hand, silicon-based waveguides are building blocks for silicon photonics, which have enabled a large amount of nanophotonic applications, including optical interconnects for data communications and ultra-fast optical communications systems. Especially, the high index contrast and well established CMOS-compatible processing make SOI as an attractive platform for optical information processing. However, one of the major problems to be solved is the interface coupling between the high confined waveguide and the free space due to the mismatched transversal wavevector. Grating couplers, both one dimensional [11] and two-dimensional ones [12, 13], are efficient means to couple light between them, which can generate a transversal wavevector $k_x = \pm 2\, m\pi/\Lambda$, where Λ is the period of grating and m is an arbitrary integer. Generally, the in-coupling of light is equally divided into two opposite directions along the waveguide, which means half of the optical flow is wasted. In many cases, it is more expected to direct the nanoscale optical flow to an expected direction. Although it is possible to realize such a function by adding Bragg reflectors at one side of the grating, such a configuration prohibits the switching of excitation direction.

Recently, various hybrid plasmonic-silicon optical waveguide platforms have been developed, as they can inherit the advantages of high mode confinement of plasmonic antennas and low loss of dielectric waveguide [14, 15]. For example, a Yagi-Uda antenna was deposited on SOI waveguide to function as a directional coupler [16]. Alternatively, one can exploit the coherent interference between double-element antennas [17] or multipolar interference of single metallic antenna [18] for unidirectional in-coupling. Phase-gradient metasurfaces have been integrated on the optical waveguide to control the propagation of guide mode [19]. Nevertheless, most of the reported directional waveguide couplers only operate for specific linear polarization and cannot fulfill the bidirectional routing and sorting function.

Inspired by the broadband spin Hall effect in single catenary nanoaperture as shown in the previous section and in Ref. [9], here a spin-controlled bidirectional router and splitter are realized by integrating single catenary nanoantenna on standard single-mode SOI waveguide [20, 21]. The spin–orbit interaction of single catenary is leveraged to directionally scatter light from a focused free-space beam into SOI waveguide. FDTD simulations demonstrate that the extinction ratio of directional coupling is higher than 15 dB in a broadband from 1450 to 1600 nm, with a peak value of 25 dB around 1550 nm. By reversing the spin of incidence, opposite light flow is excited in the integrated waveguide. These results reveal a novel roadmap toward future nanophotonic integrated circuits.

The proposed router is formed by an optical catenary placed on top of an SOI slab waveguide, as shown in Fig. 3.9. The silicon waveguide has a height of 200 nm and a width of 500 nm, which exhibits a single-mode character with high transmittance. Along the catenary profile, the inclination angle between the curve tangent and the x-axis, $\xi(x)$, varies from $-\pi/2$ to $\pi/2$ between the left and right endpoints, yielding continuous geometric phase as a result of the geometric relation $\Phi(x) = 2\sigma\xi(x)$, where $\sigma = \pm 1$ denotes the left- and right-handed circular polarizations. With the additional phase shift, incident beam can be converted to guiding waves to one particular direction depending on the spin state.

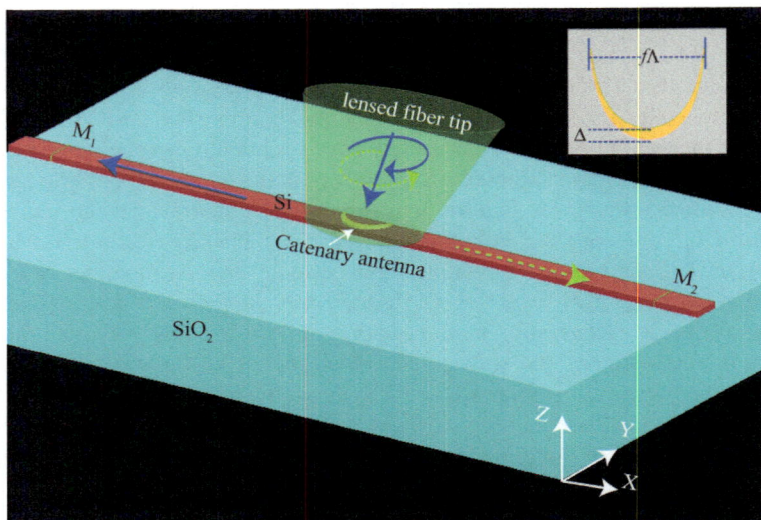

Fig. 3.9 Schematic of the designed spin-controlled router based on an optical catenary located on an SOI waveguide. Reproduced from [20] with permission. Copyright 2018, The Japan Society of Applied Physics

In order to validate the design principle above, the transmission characteristics are numerically simulated via FDTD methods. As shown in Fig. 3.9, two power monitors (M_1 and M_2) are located at a distance of 10 μm to record the power through the waveguide. A circularly polarized focused light is normally projected on the optical centenary with a thickness of 90 nm, $\Lambda = 900$ nm, $\Delta = 100$ nm. The optical properties of the silicon and gold are taken from the data of Palik [22] and Johnson and Christy [23]. The substrate material is SiO_2 whose refractive index is set as 1.44. Moreover, the absorbing boundary, i.e., perfectly matched layer (PML) is added in all directions to avoid the boundary reflection.

Figure 3.10 shows the power that the catenary couples into the different directions of the waveguide under the illumination of RCP. The blue (T_1) and red (T_2) curves indicate the power through monitor M_1 and M_2. Obviously, most of the power scattered from the optical catenary is coupled to the left side of the waveguide while the power coupled to the right side nearly can be neglected. The performance can be further measured through the extinction ratio, defined as $10 \log\big(\max(T_1, T_2) \big/ \min(T_1, T_2)\big)$. Since the directional scattering originates from linear gradient phase of optical catenary rather than the constructive interference between different antennas, the extinction ratio can keep larger than 15 dB in a broadband (1450–1600 nm) with a peak value of 25 dB around 1550 nm. Compared with previous design that utilizes multiple discrete antennas, the proposed design based on single optical catenary not only possesses more compact volume but exhibits broader bandwidth and higher extinction ratio, which is mainly attributed to the linear and

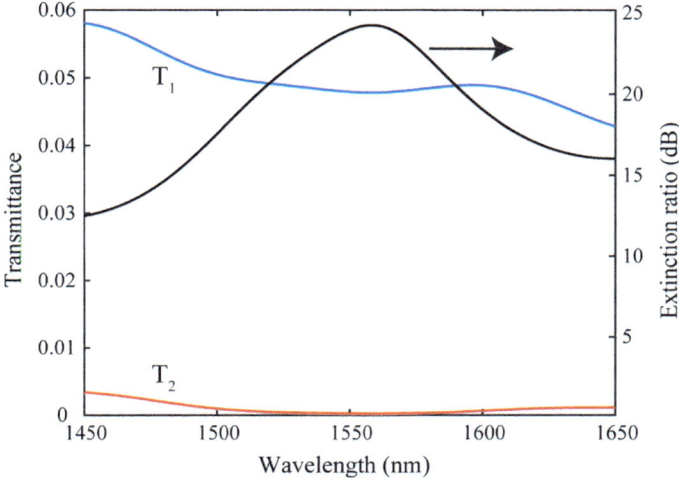

Fig. 3.10 Retrieved transmittance from two power monitors and the calculated extinction ratio between them with respect to the wavelength. Reproduced from [20] with permission. Copyright 2018, The Japan Society of Applied Physics

continuous gradient phase. Owing to the fact that the geometric gradient phase is polarization-dependent, when the handedness of the incidence is reversed, opposite coupling will be achieved. The electric field distributions are shown in Fig. 3.11 clearly shows how the scattered light is coupled into the left side of the waveguide for RCP incidence and right side for LCP incidence.

Although the proposed design is optimized at 1550 nm, it should be mentioned that the proposal can also be scaled to other band. Figure 3.12 shows the extinction ratio as a function of wavelength under different catenary lengths. It can be seen that the peak wavelength has a redshift with the catenary length Λ due to the decreased phase gradient. Therefore, the proposed prototypes can be extended to other communication band (e.g., 980 and 1310 nm) by adjusting the length of the optical catenary. Note that when the length changes, the magnitude of peak also varies, which can be optimized separately by considering the influence of other parameters such as vertical width, thickness of the film as well as the constituting materials.

Despite the proposed gold antenna can perform well as tunable directional coupler and polarization router, the gold is not CMOS compatible. In order to check the proposed design can also be suitable for CMOS-compatible platform, we replace the gold catenary by silicon catenary and germanium catenary without changing the geometries and investigate the directional coupling capability in SOI waveguide again. The retrieved transmission and extinction ratio from two power monitors are shown in Fig. 3.13. We can see the proposed directional coupler can still operate well, no matter what the catenary antenna is metallic or dielectric, which is also an extra advantage of spin–orbit interaction and geometric phase gradient compared to that based on constructive interferences [16–18]. Just like its metallic counterpart,

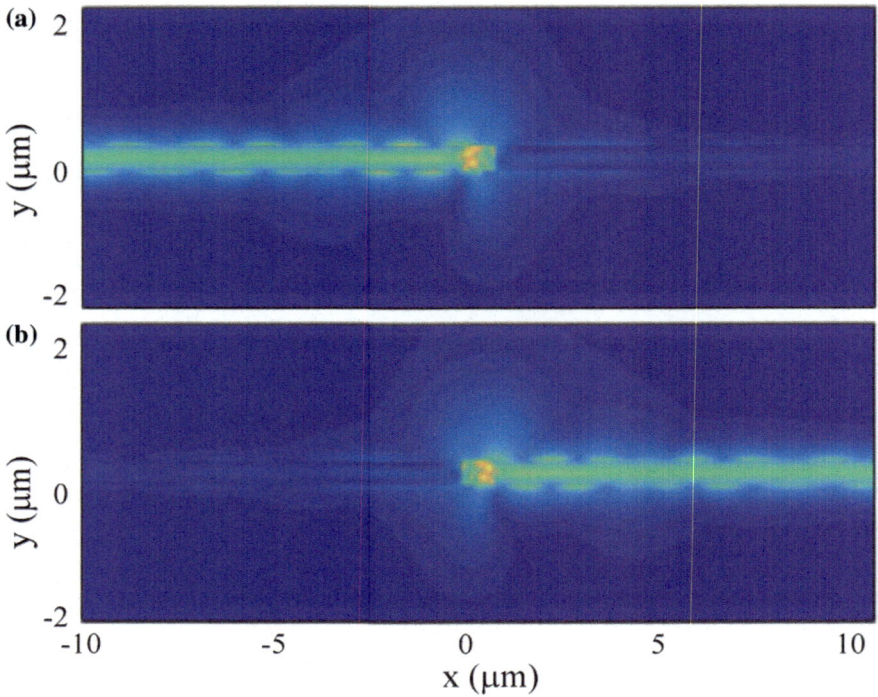

Fig. 3.11 Electric-field intensity in the *xy*-plane at the wavelength of 1550 nm for RCP incidence (top panel) and LCP incidence (bottom panel). Reproduced from [20] with permission. Copyright 2018, The Japan Society of Applied Physics

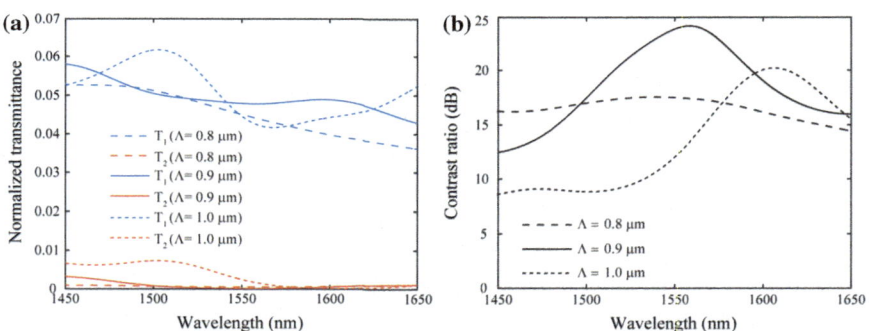

Fig. 3.12 Influence of the horizontal length on the performance. **a** Retrieved transmittance from two power monitors and **b** contrast ratio as a function of wavelength for different catenary length. Reproduced from [20] with permission. Copyright 2018, The Japan Society of Applied Physics

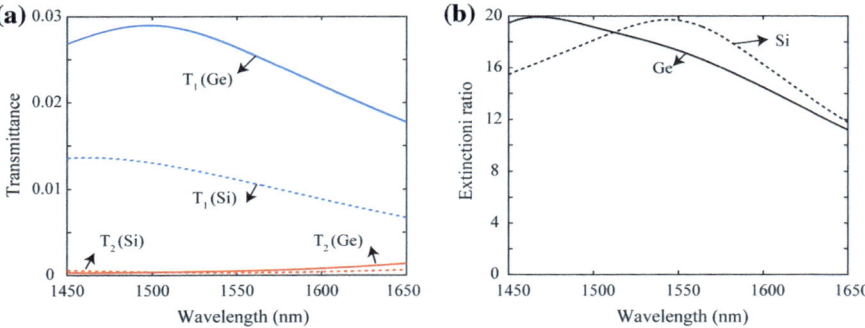

Fig. 3.13 Retrieved **a** transmittance and **b** extinction ratios with respect to the wavelength from two power monitors for the CMOS-compatible case (the gold catenary is replaced with a silicon catenary and a germanium catenary without changing the geometries). Reproduced from [20] with permission. Copyright 2018, The Japan Society of Applied Physics

the dielectric catenary antenna can also be taken as a succession of spatially variable anisotropic elements where spin–orbit interaction occurs. Nevertheless, the phase difference between the two axes of anisotropic elements is mainly attributed to the different propagation constant [24], rather than the circuit-like resonances in metallic antennas [25, 26]. Although the transmittance is lower than that of the metallic case since the geometries are not optimized for an ideal half-wave plate, it can be further improved by increasing the dielectric catenary thickness.

3.3 Catenary-Shaped Waveguides

As illustrated in the previous two sections, nanostructure on the waveguides may be used to realize spin-controlled excitation of waveguide modes. In this section, we show a different approach to realize such applications. Instead of adding sub-wavelength structures on the straight waveguides, the waveguide itself is bended to generate direct spin–orbit coupling.

As shown in Fig. 3.14a, a single-mode SOI waveguide is patterned into a catenary shape. When a CPL is normally incident onto the nanostructured catenaries, it is primarily scattered into waves of the same handedness as that of the incidence without phase change, and waves of the opposite handedness with an abrupt phase change. Such phase change results from the spin conversion that accumulates a Pancharat-nam–Berry phase from the polarization path on the Poincaré sphere [27], which is geometric in nature and dispersionless. The relationship between the phase change Φ and the orientation φ of elementary slice of the waveguide (Fig. 3.14b) can be simply expressed as $\Phi = 2\sigma\varphi$, where $\sigma = \pm 1$ denotes the left- and right-handed circular polarizations (LCP and RCP).

(a)

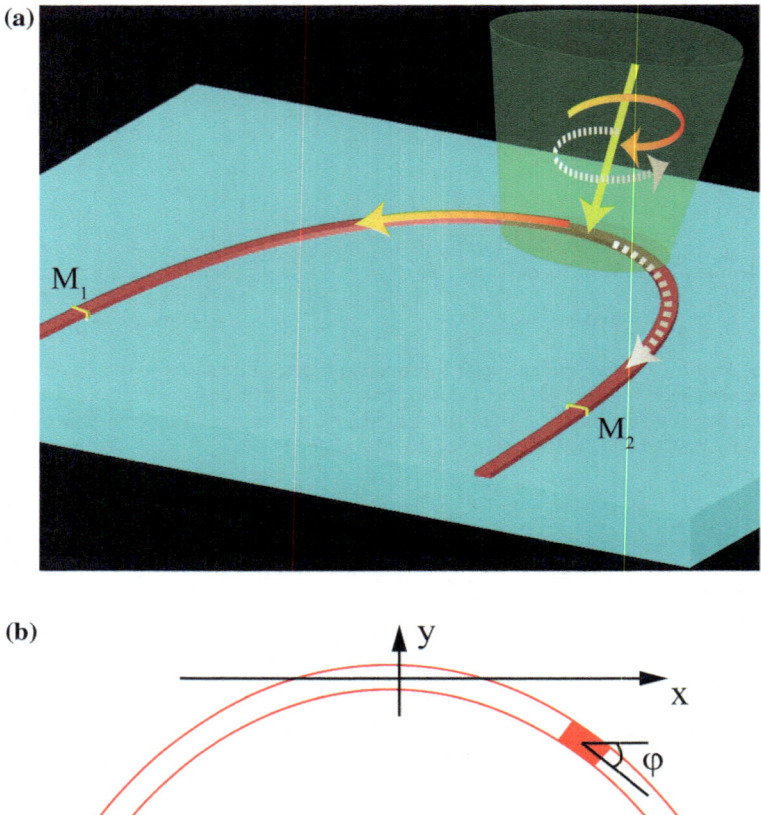

(b)

Fig. 3.14 **a** Artistic drawing of the designed directional coupling and spin routing in catenary-shaped SOI waveguide. **b** Profile of a catenary-shaped waveguide with an elementary slice of the waveguide is emphasized in red

In order to ensure the steady single-mode propagation, the width of catenary-shaped SOI waveguide is fixed as 400 nm. FDTD simulation software is utilized to carry full wave simulation. The optical property of silicon is extracted from the experimental data of Palik. Two total-field–scattered-field (TFSF) sources with orthogonal polarization and ±90° phase difference are used to simulate CPL with different handedness. Within the illumination of TFSF, the field is the superimposition of the incidence field and the scattering field while outside illumination of TFSF only the scattering field remains. The absorbing boundary, i.e., perfectly matched layer (PML) is adopted in all directions to avoid the boundary reflection. Two power monitors (M_1 and M_2) are utilized to record the power through the waveguide.

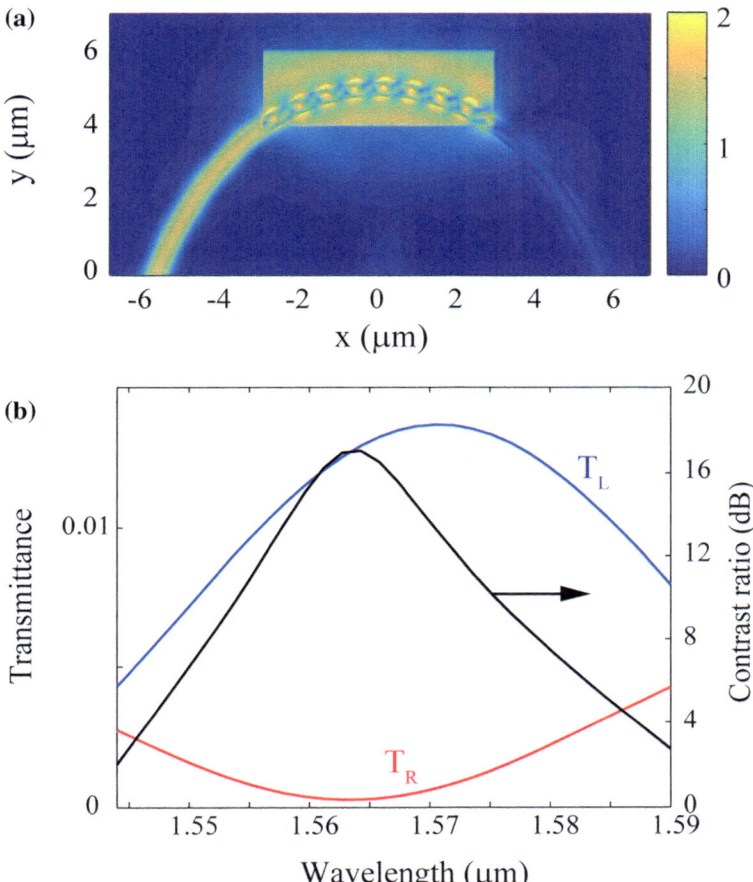

Fig. 3.15 **a** Electrical field magnitude distribution in the *xy*-plane through the waveguide for RCP incidence with a wavelength of 1565 nm. **b** Transmittance of coupled light toward either side of the waveguide (red and blue curves, respectively, represent the transmittance along left and right side of the waveguide) and the contrast ratio between them as a function of wavelength

Figure 3.15a displays the simulated electric field distributions in the *xy* cross section through the waveguide for normal RCP illumination at the wavelength of 1565 nm. The in-coupling light mainly propagates toward the left side of the catenary SOI waveguide, while the optical flow toward the opposite direction nearly can be ignored. In order to quantify the unidirectional in-coupling property, the transmittances from the preestablished power monitors are retrieved and plotted in Fig. 3.15b. The transmittances toward the left and right sides of the catenary waveguide are denoted as T_L and T_R. The contrast ratio between them (black curve) is defined as $10 \times \log_{10}(T_L/T_R)$ with a maximum of 17 dB around 1565 nm. The unidirectional in-coupling performance of the catenary-shaped SOI waveguide is comparable with

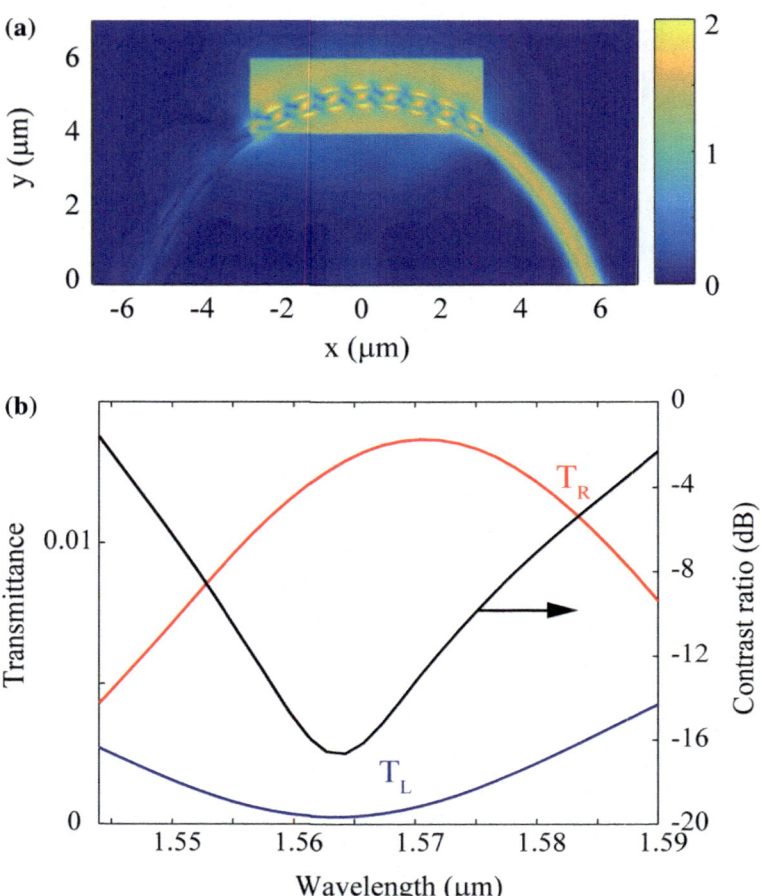

Fig. 3.16 a Electrical field magnitude distribution in the *xy*-plane through the waveguide for LCP incidence with a wavelength of 1565 nm. **b** Transmittance of coupled light toward either side of the waveguide and the contrast ratio between them as a function of wavelength

that reported in the previous literature [18, 21, 28]. Since there is no additional metallic or dielectric subwavelength structure, the proposed strategy greatly reduces the fabrication difficulty.

In order to verify the proposed catenary-shaped SOI waveguide can behave as a spin router/switcher, the in-coupling property under the illumination of LCP is also investigated. As expected, reversed in-coupling behavior occurs due to the polarization-dependent spin–orbit interaction. In other words, most in-coupling optical flow propagates toward to the right side of the catenary-shaped waveguide, as indicated by the field distribution in Fig. 3.16.

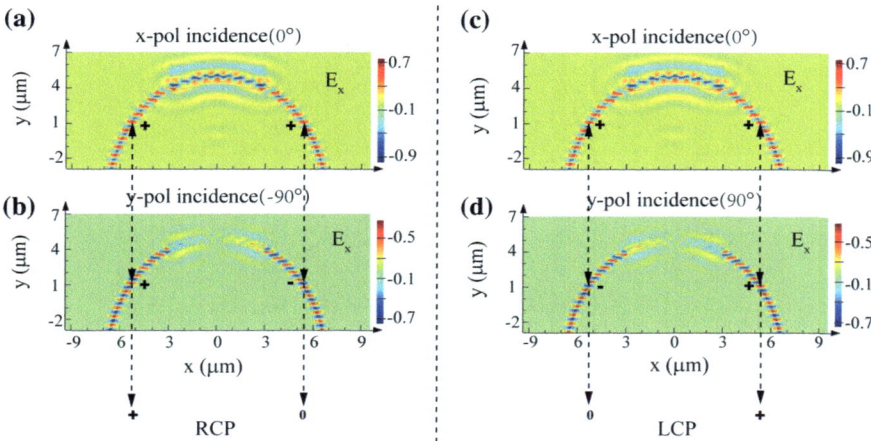

Fig. 3.17 Field distribution of E_x for **a** x polarization with the initial phase of $0°$ and **b** y polarization with the initial phase of $-90°$. Field distribution of E_x for **c** x polarization with initial phase of $0°$ and **d** y polarization with the initial phase of $-90°$

Figure 3.17a illustrates the field distribution of E_x for x polarization with the initial phase of $0°$, from which we can see the field is symmetric about the center of the waveguide. Figure 3.17b displays the field distribution of E_x for y polarization with the initial phase of $-90°$. It can be seen that the field distributions are antisymmetrical about the center of the waveguide. For RCP incidence, the field distribution can be taken as a superimposition of that shown in Fig. 3.17a, b. Consequently, the field distribution at left side of the waveguide is interferential enhancement due to in-phase between them while the field at the right side is interferential suppression due to out-of-phase between them. Figure 3.17c, d shows the field distribution of E_x for x polarization with the initial phase of $0°$ and y polarization with the initial phase of $90°$, which can be taken as an LCP incidence. In this case, the field is interferential suppression at the left side of the waveguide while interferential enhancement at the right side of the waveguide.

According to the scalability of Maxwell's equations, by changing the length of catenary waveguide Λ, the wavelength for unidirectional in-coupling can be adjusted without decreasing the contrast ratio. As shown in Fig. 3.18, when the length of catenary waveguide changes to 14.5 and 15.5 μm, the wavelength of optimized unidirectional in-coupling shit to 1550 and 1573 nm with the max contrast ratio being about 17 dB. The peak wavelength of as a function of the length of catenary waveguide Λ is shown in Fig. 3.18c.

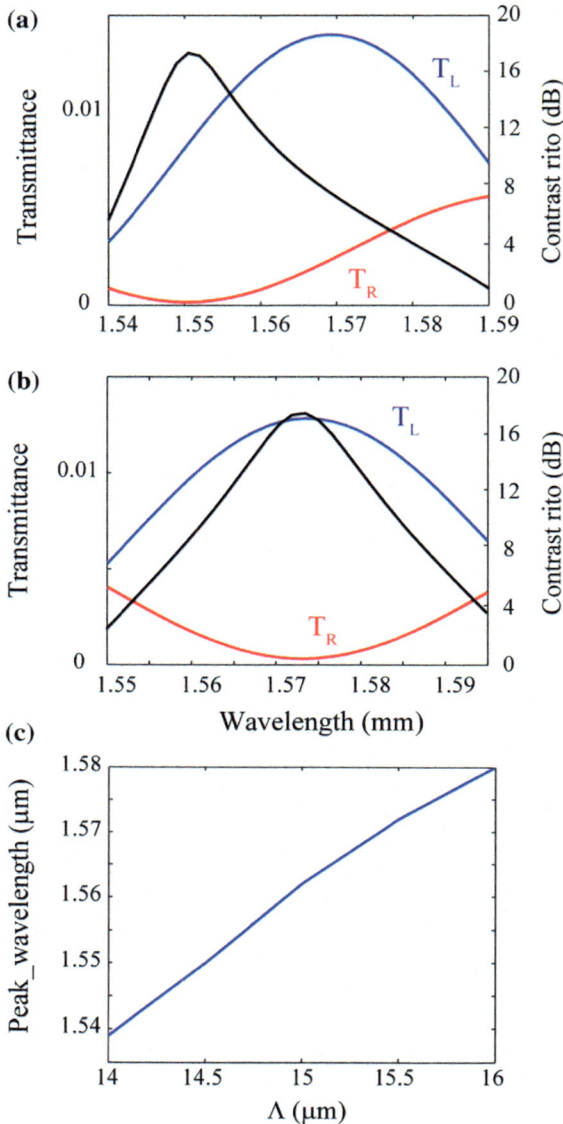

Fig. 3.18 Transmittance of coupled light toward either side of the waveguide (blue and red curves) and the contrast ratio between them (black curve) as a function of wavelength for **a** $\Lambda = 14.5\,\mu$m, **b** $\Lambda = 15.5\,\mu$m. **c** The peak wavelength as a function of the length of catenary-shaped waveguide

References

1. S.-Y. Lee, I.-M. Lee, J. Park, S. Oh, W. Lee, K.-Y. Kim, B. Lee, Role of magnetic induction currents in nanoslit excitation of surface plasmon polaritons. Phys. Rev. Lett. **108**, 213907 (2012)
2. F.J. Rodríguez-Fortuño, G. Marino, P. Ginzburg, D. O'Connor, A. Martínez, G.A. Wurtz, A.V. Zayats, Near-field interference for the unidirectional excitation of electromagnetic guided modes. Science **340**, 328–330 (2013)
3. M. Pu, Y. Guo, X. Li, X. Ma, X. Luo, Revisitation of extraordinary Young's interference: from catenary optical fields to spin-orbit interaction in metasurfaces. ACS Photonics **5**, 3198–3204 (2018)
4. J. Lin, J.P.B. Mueller, Q. Wang, G. Yuan, N. Antoniou, X.-C. Yuan, F. Capasso, Polarization-controlled tunable directional coupling of surface plasmon polaritons. Science **340**, 331–334 (2013)
5. T. Xu, C. Wang, C. Du, X. Luo, Plasmonic beam deflector. Opt. Express **16**, 4753–4759 (2008)
6. T. Xu, Y. Zhao, D. Gan, C. Wang, C. Du, X. Luo, Directional excitation of surface plasmons with subwavelength slits. Appl. Phys. Lett. **92**, 101501 (2008)
7. X. Chen, L. Huang, H. Mühlenbernd, G. Li, B. Bai, Q. Tan, G. Jin, C.W. Qiu, S. Zhang, T. Zentgraf, Dual-polarity plasmonic metalens for visible light. Nat. Commun. **3**, 1198 (2012)
8. M. Pu, X. Li, X. Ma, Y. Wang, Z. Zhao, C. Wang, C. Hu, P. Gao, C. Huang, H. Ren, X. Li, F. Qin, J. Yang, M. Gu, M. Hong, X. Luo, Catenary optics for achromatic generation of perfect optical angular momentum. Sci. Adv. **1**, e1500396 (2015)
9. X. Luo, M. Pu, X. Li, X. Ma, Broadband spin Hall effect of light in single nanoapertures. Light Sci. Appl. **6**, e16276 (2017)
10. L. Huang, X. Chen, B. Bai, Q. Tan, G. Jin, T. Zentgraf, S. Zhang, Helicity dependent directional surface plasmon polariton excitation using a metasurface with interfacial phase discontinuity. Light Sci. Appl. **2**, e70 (2013)
11. D. Taillaert, W. Bogaerts, P. Bienstman, T.F. Krauss, P. Van Daele, I. Moerman, S. Verstuyft, K. De Mesel, R. Baets, An out-of-plane grating coupler for efficient butt-coupling between compact planar waveguides and single-mode fibers. IEEE J. Quantum Electron. **38**, 949–955 (2002)
12. R. Halir, P. Cheben, S. Janz, D.-X. Xu, Í. Molina-Fernández, J.G. Wangüemert-Pérez, Waveguide grating coupler with subwavelength microstructures. Opt. Lett. **34**, 1408–1410 (2009)
13. D. Benedikovic, P. Cheben, J.H. Schmid, D.-X. Xu, B. Lamontagne, S. Wang, J. Lapointe, R. Halir, A. Ortega-Moñux, S. Janz, M. Dado, Subwavelength index engineered surface grating coupler with sub-decibel efficiency for 220-nm silicon-on-insulator waveguides. Opt. Express **23**, 22628–22635 (2015)
14. Z. Ge, L. Zhang, G. Wang, W. Zhang, M. Liu, S. Li, L. Wang, Q. Sun, W. Ren, J. Si, W. Zhao, On-chip router elements based on silicon hybrid plasmonic waveguide. IEEE Photonics Technol. Lett. **29**, 952–955 (2017)
15. S. Wang, T. Liu, Four-port polarization and topological charge controlled directional plasmonic coupler. IEEE Photonics Technol. Lett. **28**, 2391–2394 (2016)
16. F. Bernal Arango, A. Kwadrin, A.F. Koenderink, Plasmonic antennas hybridized with dielectric waveguides. ACS Nano **6**, 10156–10167 (2012)
17. T.P.H. Sidiropoulos, M.P. Nielsen, T.R. Roschuk, A.V. Zayats, S.A. Maier, R.F. Oulton, Compact optical antenna coupler for silicon photonics characterized by third-harmonic generation. ACS Photonics **1**, 912–916 (2014)
18. D. Vercruysse, P. Neutens, L. Lagae, N. Verellen, P. Van Dorpe, Single asymmetric plasmonic antenna as a directional coupler to a dielectric waveguide. ACS Photonics **4**, 1398–1402 (2017)
19. Z. Li, M.-H. Kim, C. Wang, Z. Han, S. Shrestha, A.C. Overvig, M. Lu, A. Stein, A.M. Agarwal, M. Lončar, N. Yu, Controlling propagation and coupling of waveguide modes using phase-gradient metasurfaces. Nat. Nano **12**, 675–683 (2017)

20. Y. Guo, M. Pu, X. Li, X. Ma, X. Luo, Ultra-broadband spin-controlled directional router based on single optical catenary integrated on silicon waveguide. Appl. Phys. Express **11**, 092202 (2018)
21. Y. Guo, M. Pu, X. Li, X. Ma, S. Song, Z. Zhao, X. Luo, Chip-integrated geometric metasurface as a novel platform for directional coupling and polarization sorting by spin-orbit interaction. IEEE J. Sel. Top. Quantum Electron. **24**, 1–7 (2018)
22. E.D. Palik, *Handbook of Optical Constants of Solids* (Academic Press, Cambridge, 1985)
23. P.B. Johnson, R.W. Christy, Optical constants of the noble metals. Phys. Rev. B **6**, 4370–4379 (1972)
24. F. Zhang, M. Pu, X. Li, P. Gao, X. Ma, J. Luo, H. Yu, X. Luo, All-dielectric metasurfaces for simultaneous giant circular asymmetric transmission and wavefront shaping based on asymmetric photonic spin–orbit interactions. Adv. Funct. Mater. **27**, 1704295 (2017)
25. M. Pu, P. Chen, Y. Wang, Z. Zhao, C. Huang, C. Wang, X. Ma, X. Luo, Anisotropic metamirror for achromatic electromagnetic polarization manipulation. Appl. Phys. Lett. **102**, 131906 (2013)
26. Y. Guo, Y. Wang, M. Pu, Z. Zhao, X. Wu, X. Ma, C. Wang, L. Yan, X. Luo, Dispersion management of anisotropic metamirror for super-octave bandwidth polarization conversion. Sci. Rep. **5**, 8434 (2015)
27. M.V. Berry, Quantal phase factors accompanying adiabatic changes. Proc. R. Soc. Lond. Math. Phys. Eng. Sci. **392**, 45–57 (1984)
28. F.J. Rodríguez-Fortuño, I. Barber-Sanz, D. Puerto, A. Griol, A. Martínez, Resolving light handedness with an on-chip silicon microdisk. ACS Photonics **1**, 762–767 (2014)

Chapter 4
Catenary Plasmons for Sub-diffraction-Limited Imaging and Nanolithography

Abstract According to its definition, the interference of two counter-propagating evanescent waves would form a typical hyperbolic cosine catenary optical field. At a given wavelength, the attenuation would be much larger for higher order evanescent waves, leading to a deeper catenary curve. As a result, the shape of catenary optical fields is deeply related to the horizontal wavevector of evanescent wave. In this chapter, we first discuss the catenary optical fields induced by coupled surface plasmons (catenary plasmons) in metal–dielectric multilayers. Then their applications in sub-diffraction-limited imaging and lithography are systematically investigated.

Keywords Catenary plasmon · Coupled plasmon · Plasmonic lithography · Diffraction limit

4.1 Catenary Optical Fields in Plasmonic Waveguides

In the simplest case, the coupling of optical waves can be illustrated in planar multilayer structure. By applying plane wave expansion and transfer matrix analysis, the transmission and reflection properties, as well as the eigenmodes could be easily obtained [1].

4.1.1 Transfer Matrix Analysis of Metal–Dielectric Multilayer

Considering the transverse magnetic (TM) polarized light in layered 1D structures with nonmagnetic materials ($\mu = 1$) shown in Fig. 4.1, the continuity of the tangential electric and magnetic fields implies [2, 3]:

$$
\begin{aligned}
H_y^j = H_y^{j+1} &\Rightarrow a_j e^{ik_j d_j} + b_j e^{-ik_j d_j} = a_{j+1} + b_{j+1} \\
E_z^j = E_z^{j+1} &\Rightarrow \frac{k_{xj}}{\varepsilon_j}\left(a_j e^{ik_j d_j} - b_j e^{-ik_j d_j}\right) = \frac{k_{j+1}}{\varepsilon_{j+1}}\left(a_{j+1} - b_{j+1}\right),
\end{aligned}
\tag{4.1.1}
$$

© Springer Nature Singapore Pte Ltd. 2019
X. Luo, *Catenary Optics*, https://doi.org/10.1007/978-981-13-4818-1_4

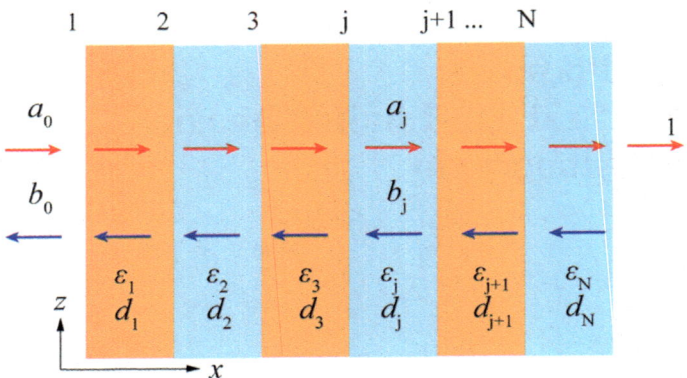

Fig. 4.1 Schematic of a planar multilayer. The red arrows depict the wave component propagating along the positive x-direction, while the blue lines represent the negative direction

where a and b stand for the coefficients for counter-propagating waves in each layer, and k is the wavenumber along the propagating direction, j is the layer index. Equation (4.1.1) can also be written as a matrix form as

$$\begin{bmatrix} a_j \\ b_j \end{bmatrix} = T_{j,j+1} \begin{bmatrix} a_{j+1} \\ b_{j+1} \end{bmatrix}, \tag{4.1.2}$$

where $T_{j,j+1}$ is the transfer matrix between the jth and $j+1$th layer

$$T_{j,j+1} = \frac{1}{2}\begin{bmatrix} (1+K)e^{ik_jd_j} & (1+K)e^{ik_jd_j} \\ (1-K)e^{-ik_jd_j} & (1+K)e^{-ik_jx_j} \end{bmatrix}. \tag{4.1.3}$$

Specially, $T_{0,1}$ should be written as

$$T_{0,1} = \frac{1}{2}\begin{bmatrix} 1+K & 1+K \\ 1-K & 1+K \end{bmatrix}. \tag{4.1.4}$$

For TM polarization, we have $K = k_{j+1}\varepsilon_j/(k_j\varepsilon_{j+1})$. The above equations are also applicable for transverse electric (TE) polarization, if the K-factor is changed as $K = k_{j+1}/k_j$.

For cascaded N layers, the coefficients at the left side can be calculated by directly multiplying these transfer matrixes together:

$$\begin{bmatrix} a_0 \\ b_0 \end{bmatrix} = T_{0,1} \times T_{1,2} \times T_{2,3} \times \cdots \times T_{N-1,N} \times T_{N,N+1} \begin{bmatrix} 1 \\ 0 \end{bmatrix}. \tag{4.1.5}$$

The transmission and reflection coefficients are then $t \equiv 1/a_0$, $r \equiv b_0/a_0$. For reflective device with a reflective layer, the $N+1$ layer should be considered with its real material parameters such as conductivity. In this case, the transmission coefficient is virtually 0, thus the absorbance is $1 - |r|^2$.

Note that the above formula is suitable not only for normal incidence but also for oblique incidence. When the horizontal wavenumber, i.e., the propagation constant becomes larger than that in the background material, the incident wave is evanescent wave. At some specific cases, both the transmission and reflection coefficients may become much larger than unity. When the two coefficients become infinite, it is said that a guided mode occurs (when there is materials loss, the propagation constant is a complex value). As will be illustrated in the following discussion, many of the interesting effects are actually stemming from the waveguiding effect. In addition, the above theory can also be extended to curved multilayers such as cylindrical and spherical multilayers [4, 5].

4.1.2 Catenary Optical Fields as Plasmonic Eigenmodes

Coupled surface plasmons in multilayered structures are typical catenary optical fields. For simplicity of discussion, here we consider a three-layered structure with either an insulator–metal–insulator (IMI) or metal–insulator–metal (MIM) configuration (Fig. 4.2). The materials for the two cladding layers are set to be the same, thus the modes would possess good symmetry.

For TM polarization, the electric fields can be written as [7]:

$$H_y = A e^{i\beta x} e^{-k_2 z}$$

$$E_x = i A \frac{1}{\omega \varepsilon_0 \varepsilon_2} k_2 e^{i\beta x} e^{-k_2 z}$$

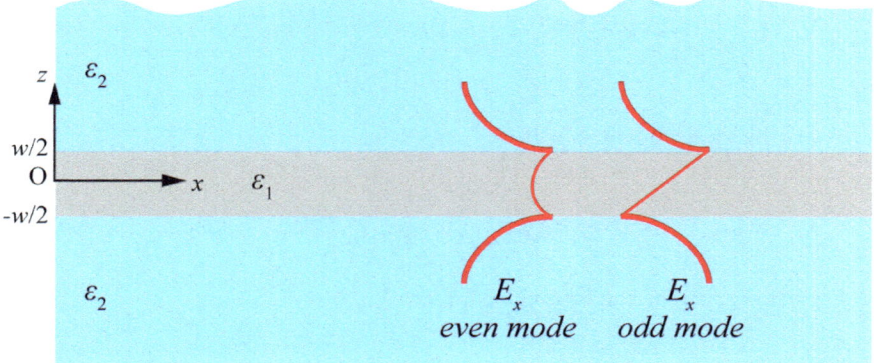

Fig. 4.2 Geometry and field distributions in a symmetric three-layered structure. Reproduced with permission from [6]

$$E_z = -A \frac{\beta}{\omega \varepsilon_0 \varepsilon_2} e^{i\beta x} e^{-k_2 z} \tag{4.1.6}$$

for the region $z < -w/2$, and

$$H_y = B e^{i\beta x} e^{k_2 z}$$
$$E_x = -i B \frac{1}{\omega \varepsilon_0 \varepsilon_2} k_2 e^{i\beta x} e^{k_2 z}$$
$$E_z = -B \frac{\beta}{\omega \varepsilon_0 \varepsilon_2} e^{i\beta x} e^{k_2 z} \tag{4.1.7}$$

for region $z > -w/2$. In the central region, the fields are a summation or interference of two counter-propagating evanescent waves:

$$H_y = C e^{i\beta x} e^{k_1 z} + D e^{i\beta x} e^{-k_1 z}$$
$$E_x = -i C \frac{1}{\omega \varepsilon_0 \varepsilon_1} k_1 e^{i\beta x} e^{k_1 z} + i D \frac{1}{\omega \varepsilon_0 \varepsilon_1} k_1 e^{i\beta x} e^{-k_1 z} . \tag{4.1.8}$$
$$E_z = C \frac{\beta}{\omega \varepsilon_0 \varepsilon_1} e^{i\beta x} e^{k_1 z} + D \frac{\beta}{\omega \varepsilon_0 \varepsilon_1} e^{i\beta x} e^{-k_1 z}$$

By applying the boundary conditions, one can obtain

$$A e^{-k_2 w/2} = C e^{k_1 w/2} + D e^{-k_1 w/2}$$
$$\frac{A}{\varepsilon_2} k_2 e^{-k_2 w/2} = -\frac{C}{\varepsilon_1} k_1 e^{k_1 w/2} + \frac{D}{\varepsilon_1} k_1 e^{-k_1 w/2} \tag{4.1.9}$$

at $z = w/2$ and

$$B e^{-k_2 w/2} = C e^{-k_1 w/2} + D e^{k_1 w/2}$$
$$-\frac{B}{\varepsilon_2} k_2 e^{-k_2 w/2} = -\frac{C}{\varepsilon_1} k_1 e^{-k_1 w/2} + \frac{D}{\varepsilon_1} k_1 e^{k_1 w/2} \tag{4.1.10}$$

at $z = -w/2$.

By solving these equations with

$$k_1^2 = \beta^2 - k_0^2 \varepsilon_1$$
$$k_2^2 = \beta^2 - k_0^2 \varepsilon_2 , \tag{4.1.11}$$

the dispersion relations can be obtained:

$$\tanh \sqrt{\beta^2 - k_0^2 \varepsilon_1} \, w/2 = -\frac{\varepsilon_1 \sqrt{\beta^2 - k_0^2 \varepsilon_2}}{\varepsilon_2 \sqrt{\beta^2 - k_0^2 \varepsilon_1}}$$

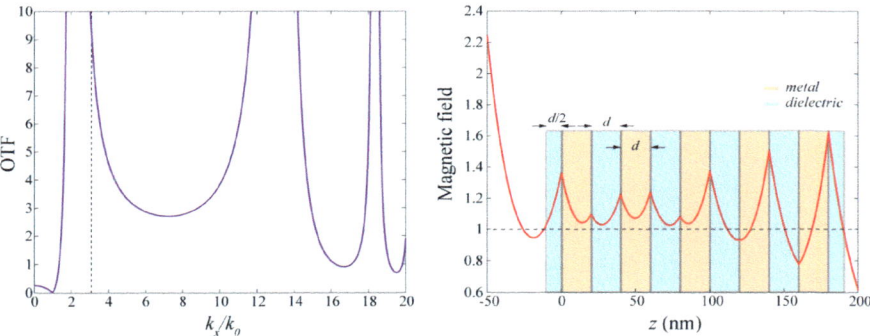

Fig. 4.3 OTF and electric field distribution in metal–dielectric multilayer at a wavelength of $\lambda = 365$ nm. The thickness of the outer two dielectric layers are $d/2$ ($d = 20$ nm). The permittivities for the dielectric and metal are 1 and -1, respectively. Note that the electric field is unit at the outmost interface, owing to the definition in the transfer matrix calculation

$$\tanh\sqrt{\beta^2 - k_0^2\varepsilon_1}\,w/2 = -\frac{\varepsilon_2\sqrt{\beta^2 - k_0^2\varepsilon_1}}{\varepsilon_1\sqrt{\beta^2 - k_0^2\varepsilon_2}} \tag{4.1.12}$$

for odd and even modes, respectively. Note that here the symmetry is defined by the symmetry of transverse electric field E_x. Regarding the magnetic field (H_y), the symmetry would reverse.

Now let us consider the fields' distribution rather than the dispersion equation of surface plasmon polariton (SPP). Since these structures are symmetric, there are $C = -D$ for the symmetric (even) mode, and $C = D$ for the antisymmetric (odd) mode. Consequently, E_x is in proportional to $\exp(k_1z) + \exp(-k_1z)$, i.e., a catenary function. For the antisymmetric mode, although E_x is in proportional to $\exp(k_1z) - \exp(-k_1z)$, H_y and E_z are characterized by the catenary function. More generally, if the three layers are not symmetric, the magnitudes of C and D are not equal, the field curves would be asymmetric catenaries. In the following discussion, we shall use the terminologies catenary optical fields and catenary plasmons interchangeably.

Both the MIM and IMI waveguides are basic structures in plasmonic waveguides. When more layers are added, the waveguides modes may become much more complex. To illustrate this, the optical transfer function (OTF) and electric fields corresponding to $k_x = 3k_0$ are calculated for a multilayer composed of five metal layers and six dielectric layers. As shown in Fig. 4.3, the electric field in each layer is actually an asymmetric catenary. As the evanescent wave propagates from the left to the right side, the signature of catenary become much weaker in the last 5 layers, where only forward (backward) propagating wave is kept in the dielectric (metallic) layer.

If the thickness is much smaller than the wavelength, the attenuation of the evanescent waves is not significant, and the signature of catenary may become weaker.

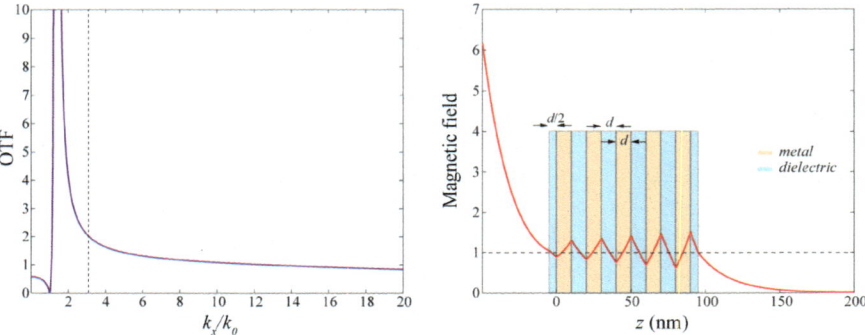

Fig. 4.4 OTF and electric field distribution in metal–dielectric multilayer at $\lambda = 365$ nm. The thickness d is reduced to be 10 nm. The permittivities for the dielectric and metal are 2 and -2, respectively

Figure 4.4 illustrates the OTF and electric fields distribution for a thinner multilayer. Obviously, the incident evanescent wave decays in the dielectric layers and grows exponentially in the metallic layers. As predicted by Pendry in 2000 [8], this effect shows that a thin metallic film can act as a negative index material, which could amplify the incident evanescent waves.

In the deep-subwavelength scale, the evanescent wave is attenuated and increased periodically in the dielectric and metallic layers. As shown in the right panel of Fig. 4.4, the electric field amplitude tends to oscillate above and below an average line, thus the fields can be regarded as homogeneous. In this case, a multilayer of metal–dielectric slab could be treated as effective anisotropic materials [9]. When the electric fields are parallel to the interface, the continuity of tangent electric fields requires [10]:

$$E_1 = E_2 = \langle E_{\parallel} \rangle, \tag{4.1.13}$$

where E_1 and E_2 are the electric fields in the two materials, $\langle\ \rangle$ is the average operator. The averaged displacement current is

$$\langle D_{\parallel} \rangle = \frac{D_1 d_1 + D_2 d_2}{d_1 + d_2} = \varepsilon_0 \frac{\varepsilon_1 d_1 + \varepsilon_2 d_2}{d_1 + d_2} \langle E_{\parallel} \rangle. \tag{4.1.14}$$

D_1 and D_2 are the displacement currents. Consequently, the averaged permittivity can be written as

$$\varepsilon_{\parallel} = \frac{\varepsilon_1 d_1 + \varepsilon_2 d_2}{d_1 + d_2} = \frac{\varepsilon_1 + \varepsilon_2 \eta}{1 + \eta}, \tag{4.1.15}$$

where $\eta = d_2/d_1$ is the thickness ratio of the two materials. In the case shown in Fig. 4.4, there are $\eta = 1$, $\varepsilon_1 = 2$ and $\varepsilon_2 = -2$. As a result, there is $\varepsilon_{\parallel} = 0$.

On the other hand, when the electric field is perpendicular to the interface, there are

$$D_1 = D_2 = \langle D_\perp \rangle, \tag{4.1.16}$$

and

$$\langle E_\perp \rangle = \frac{E_1 d_1 + E_2 d_2}{d_1 + d_2} = \frac{1}{\varepsilon_0} \frac{\frac{d_1}{\varepsilon_1} + \frac{d_2}{\varepsilon_2}}{d_1 + d_2} \langle D_\perp \rangle. \tag{4.1.17}$$

The effective permittivity is

$$\frac{1}{\varepsilon_\perp} = \frac{\frac{1}{\varepsilon_1} + \frac{\eta}{\varepsilon_2}}{1 + \eta}. \tag{4.1.18}$$

Thus there is $\varepsilon_\perp = \infty$. Since the dispersion equation is

$$\frac{k_x^2}{\varepsilon_\parallel} + \frac{k_z^2}{\varepsilon_\perp} = k_0^2, \tag{4.1.19}$$

the dispersion curve would overlap with $k_x = 0$. In this case, all wave components, regardless its horizontal wavenumber k_z, would propagate in a straight line along the x-direction.

4.1.3 Catenary Plasmons in Superlens

John Pendry has shown that the coupled surface plasmon in a single metallic film may be used to realize perfect optical imaging with resolution beyond the classic diffraction limit [8]. To understand the role of catenary plasmons in this process, the OTF and electric fields distributions are analyzed in this section.

Figure 4.5 plots the OTFs of a thin metallic film for three thicknesses. At a thickness of 40 nm, there is only a single transmission peak located at $3.5k_0$. However, when the thickness decreases, the OTF separates to two branches, one shifts to the right side and the other shifts to the left side, corresponding to the symmetric and antisymmetric modes, respectively. Notably, the propagation constant of the symmetric mode increases with the reduction of thickness, thus a higher resolution is achievable [11]. This can be understood by investigating the modes distribution shown in the inset of Fig. 4.5, for the symmetric mode, most energy is concentrated in the metallic region, thus the propagation constant is very large; for the antisymmetric mode, most energy is concentrated in the dielectric host, thus the propagation constant is close to that in the dielectrics.

Another way to understand the super-resolution imaging via superlens can be found in the extraordinary Young's interference (EYI) originally observed since

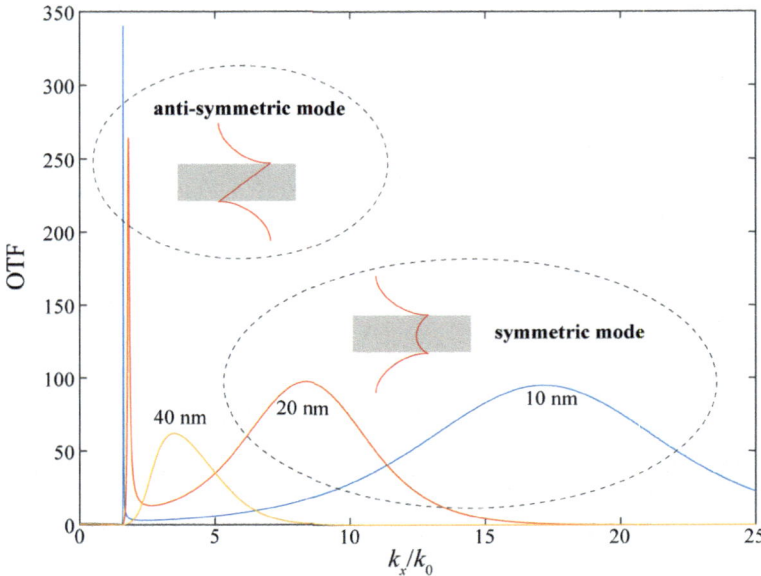

Fig. 4.5 OTF of a thin metallic film with different thickness embedded in a dielectric host. The permittivities of metal and dielectric are 2.4 and $-2.4012 + 0.2488i$ at a wavelength of 365 nm. These peaks are not infinite because the propagation constant is a complex value but the horizontal axis only represent their real parts. The inset shows the tangential electric fields distribution for the two modes

2004 [6, 12]. As shown in Fig. 4.6, two 10 nm wide nanoslits are used to excite the EYI on a 20-nm-thick silver film. Under x-polarized incidence, antisymmetric E_z (corresponding to symmetric E_x) is generated with a hyperbolic sine shaped amplitude profile. When the wavelength increases from $\lambda = 365$–380 nm, the interference period anomalously decreases from 100 nm to about 25 nm, which implies that one may increase the optical resolution without reducing the wavelength. In particular, the wavelength of the catenary SPP at the surface is about 45 nm, which is 1/8.4 times the vacuum wavelength (380 nm). It should be noted that Young's interference is a milestone in classic wave optics. The Young's interference of single electrons was even though as the most beautiful physical experiment [13]. As a collective excitation of photons and electrons, catenary SPPs may lie at the heat of modern nanophotonics.

Figure 4.7 shows the OTF at the wavelengths of $\lambda = 365$ and 380 nm. Obviously, there is a transmission peak at $8.4k_0$ for $\lambda = 380$ nm, which means that the excitation efficiency is at its maximum. In contrast, the highest peak is located at about $5.5k_0$ for $\lambda = 365$ nm, corresponding to an effective wavelength of 66 nm. It should be noted that the near-field interference cannot be observed in numerical simulations if the mesh size is not small enough (the minimal mesh is set to be $\lambda/100$ in Fig. 4.6).

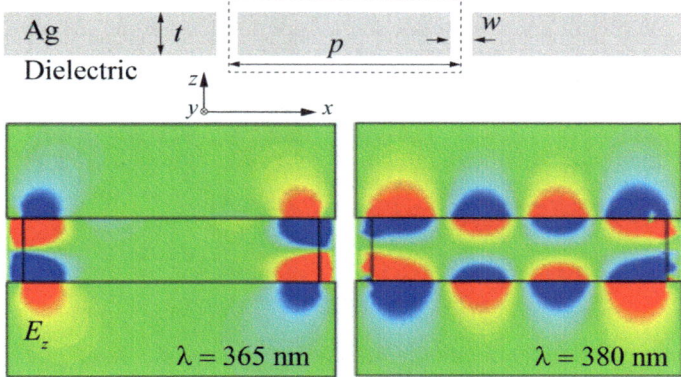

Fig. 4.6 The top panel shows the basic configuration of Young's interference experiment, while the dashed rectangle indicates the simulation region. The bottom figure shows the instantaneous E_z at 365 and 380 nm. The values of p, w, and t are set as 100, 10, and 20 nm. The refractive index of the dielectric host is 1.7. Reproduced from [6] with permission. Copyright 2018, American Chemical Society

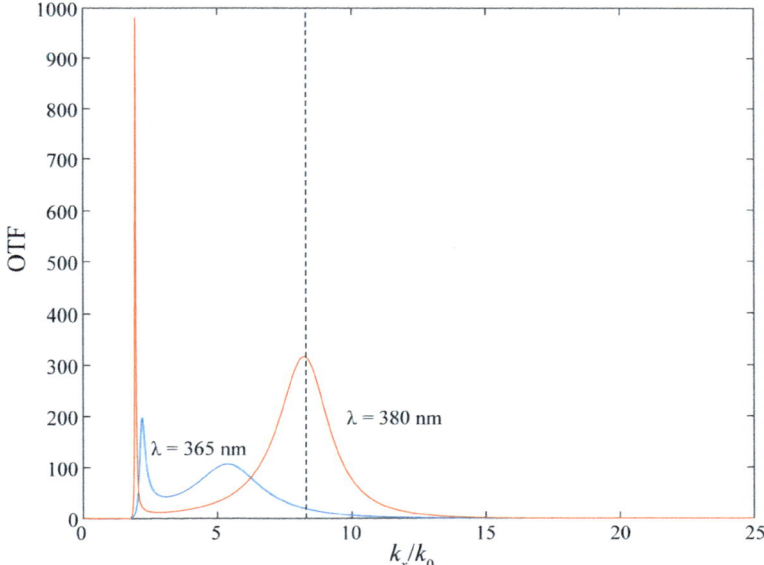

Fig. 4.7 OTF of a 20-nm-thick silver film embedded in a dielectric host ($n = 1.7$). A peak occurs at $k_x = 8.4k_0$ for $\lambda = 380$ nm

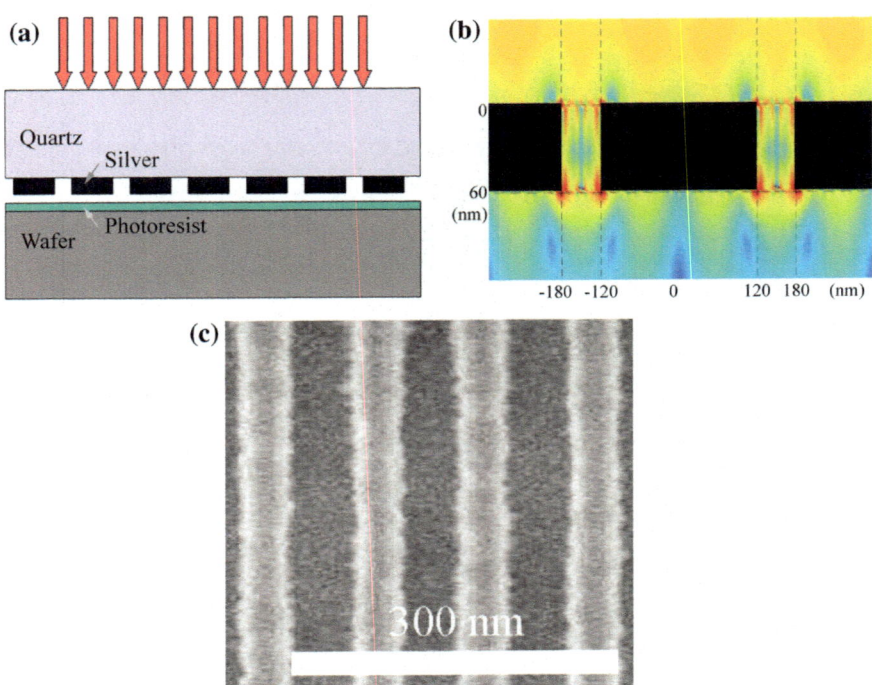

Fig. 4.8 Interference lithography based on EYI effect. **a** Schematic of the experimental configuration. **b** Simulated intensity distribution for the electric fields. **c** SEM image of the interference pattern recorded in the photoresist. Reproduced from [12] with permission. Copyright 2004, American Institute of Physics

The EYI effect can be also interpreted as an imaging process: The interference pattern is first excited in the front surface and then imaged into the bottom interface. In fact, the light–matter interaction in this simple configuration is not so simple, thus it is difficult to completely separate the interference and imaging process. In 2004, the EYI phenomena were experimentally demonstrated by placing the photoresist layer below the silver film (60 nm thick), as shown in Fig. 4.8 [12]. With an illumination wavelength of 436 nm, the near-field interference pattern recorded in the photoresist has a periodicity (~100 nm) different from that of the opening features on the mask ($p = 300$ nm). The effective wavelength of SPP is about 200 nm, corresponding to a propagation constant of $2.18k_0$. Note that this is much smaller than the value predicted by previous calculation because higher order evanescent waves tend to decay more quickly, thus much more delicate experiments should be devolved [14, 15].

In 2005, a plasmonic imaging experiment based on superlens was reported [16, 17], where 35-nm-thick silver film is taken as a superlens, as shown in Fig. 4.9. The experimental setup is just the same as that shown in Fig. 4.8. Because the impedance

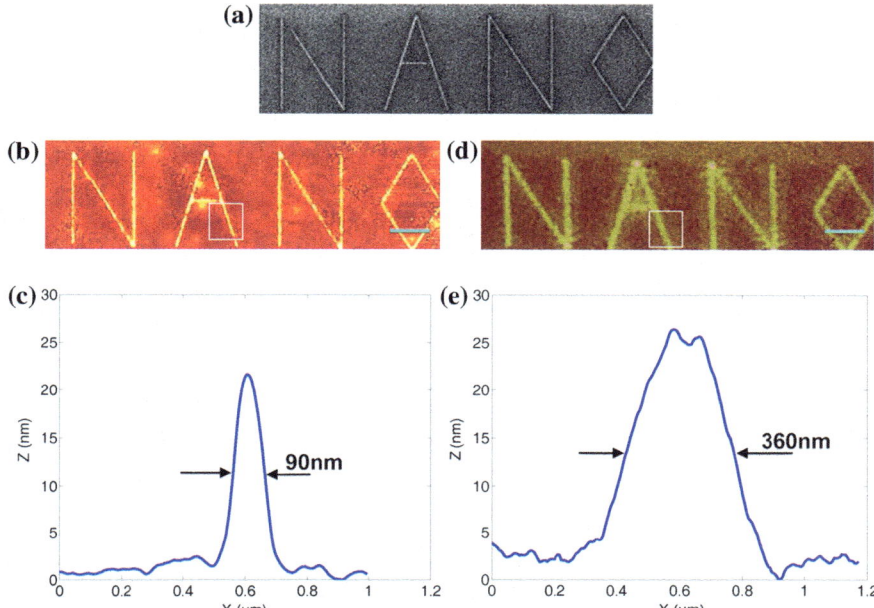

Fig. 4.9 The imaging of a two-dimensional arbitrary object "NANO". **a** SEM image of the object "NANO" fabricated on Cr film. **b** Superlensing image (scale bar 2 μm) and **c** its cross-section profile. **d** Control imaging result of the same object (scale bar 2 μm) and **e** its average line cross section. Line width of ~90 nm is achieved with superlensing while the control imaging produced diffraction-limited image with line width ~360 nm. Reprinted from [17] with permission. Copyright 2005, IOP Publishing Ltd and Deutsche Physikalische Gesellschaft

match condition is met at the surface of the silver, the excitable surface plasmon band of k_x is significantly broadened, resulting in the superlensing effect. 60-nm half-pitch dense grating patterns and a "NANO" characters pattern with 90 nm line width were presented. In the meantime, Blaikie's group reported similar plasmonic imaging nanolithography experiment [18]. Nevertheless, the recording photoresist patterns are quite weak and drown in the background roughness of the developed photoresist. To improve the resolution and performance of superlens lithography, subsequent investigations mainly concentrated on the aspects of geometrical parameters optimization, reducing films roughness and damping loss inside. For example, by resorting to smooth Ag superlens, Chaturvedi et al. observed a 30 nm half-pitch resolution [19].

Notably, the superlens effect can be extended to any frequency band if proper negative permittivity can be found. In 2006, near-field microscopy through a 440-nm-thick SiC superlens membrane was demonstrated at infrared band [20]. The object plane is covered by an Au film patterned with holes of different diameters and separations. Scattering type scanning near-field optical microscopy (s-SNOM) was utilized to record both the amplitude and the phase of the optical field distribution. At

$\lambda \sim 11 \, \mu$m where superlensing is expected, the infrared amplitude and phase images resolve the 1200 and 860-nm holes. Even the smaller 540-nm holes exhibit sufficient optical contrast to allow for the detection of $\lambda/20$-sized objects 880 nm away from the near-field probe.

4.2 Sub-diffraction-Limited Nanolithography with Planar Lens

4.2.1 Reflective Superlens

One of the most important applications of plasmonic coupling is nanolithography and nano-fabrication [21]. Although superlens could resolve the subwavelength features of near-field objects, the obtained photoresist patterns are usually characterized with shallow profiles, low contrast, and great aberrations compared with masks. The decaying feature of evanescent waves is believed to be the dominating reason for this problem. In 2005, SPP-assisted near field photolithography was demonstrated by introducing a Titanium (Ti) shield slab below the photoresist [22]. It was shown that the evanescent waves amplification in reflection manner by appropriate metal shield help to enhance the imaging performance in the near field. Subsequently, a metal-cladding superlens structure was proposed for localizing SPPs and projecting deep-subwavelength patterns [23]. In this section, we present an experimental demonstration of plasmonic slab imaging lithography with evanescent wave amplification in reflection mode [24, 25]. Owing to the catenary-shaped optical fields formed by the summation of counter-propagating waves, two-dimensional nanopatterns with critical dimensions (about 30–50 nm) far below the wavelength are well imaged in lithography resist.

Figure 4.10 presents the schematic of the subwavelength imaging lithography structure with a reflective plasmonic slab. Photoresist layer is sandwiched between the chromium mask and the silver slab. The illuminating light comes from an i-line mercury lamp exposure system with a central wavelength of 365 nm. The thickness of the Cr mask and the photoresist (Pr) is 50 nm. The thickness of the silver plasmonic slab is 100 nm, which is much larger than the skin depth of light in silver. The Cr mask is milled with a transparent pattern of characters "OPEN" with 36 nm line width. The total length in the y-direction of each character is about 500 nm. The width of the character "O" is about 420 nm and the other characters are about 320 nm.

If we choose the interface between the Cr mask and the photoresist as $z = 0$, the OTF in the photoresist (the total thickness is d) can be simplified as

$$T(k_x) = \exp(ik_z z) + r(k_x) \exp(ik_z 2d) \exp(-ik_z z), \qquad (4.2.1)$$

where k_x is the transversal wavenumber, d is the thickness of the photoresist, $r(k_x)$ is the reflection coefficient of the reflective slab, and $k_z = \left(\varepsilon k_0^2 - k_x^2\right)^{1/2}$ is the

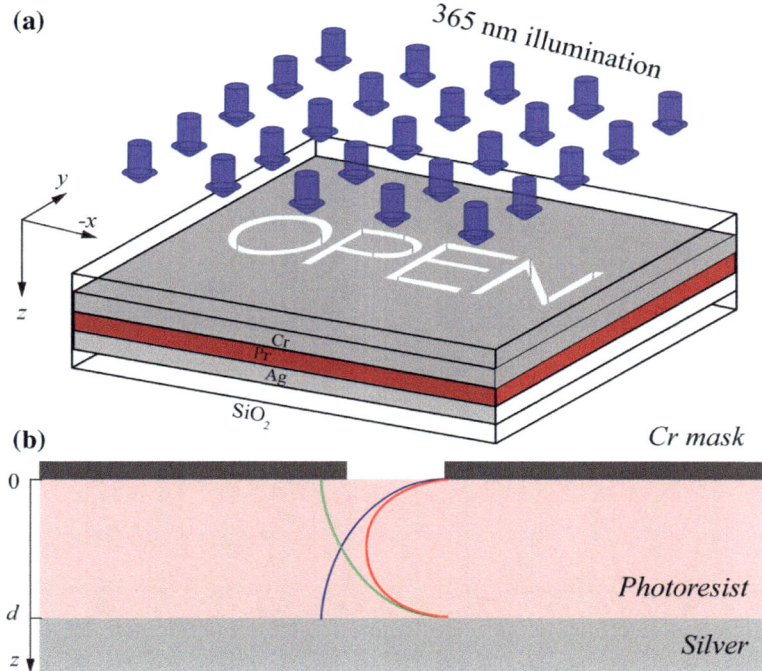

Fig. 4.10 **a** Schematic of imaging lithography structure with a reflective plasmonic slab. **b** Formation of the catenary optical fields by the summation of counter-propagating evanescent waves. Adapted from [24] with permission. Copyright 2013, Optical Society of America

wavenumber along the z-direction. For evanescent wave component, k_z is imaginary and Eq. (4.2.1) is converted to

$$T(k_x) = \exp\left(-\sqrt{k_x^2 - \varepsilon k_0^2}z\right) + r(k_x)\exp\left(-2\sqrt{k_x^2 - \varepsilon k_0^2}d\right)\exp\left(\sqrt{k_x^2 - \varepsilon k_0^2}z\right).$$

(4.2.2)

Obviously, this is an asymmetric catenary function which has two maximal values at mask-photoresist and photoresist-reflector interfaces. As can be seen in the following results, this catenary function is helpful to obtain a longer depth of focus for nanolithography. In special configuration, there is $r(k_x)\exp\left[-2\left(k_x^2 - \varepsilon k_0^2\right)^{1/2}d\right] = 1$, thus Eq. (4.2.2) transforms to an ideal catenary function.

The numerical simulation of the imaging performance is performed by the finite element method with Comsol Mulitphysics 4.0. Figure 4.11a, b illustrates the calculated electric field intensity distribution at 30 nm below the Cr–photoresist (Pr) interface for linear polarization in the x and y-direction, respectively. Owing to the polarization-dependent transmission of subwavelength aperture, only the lines

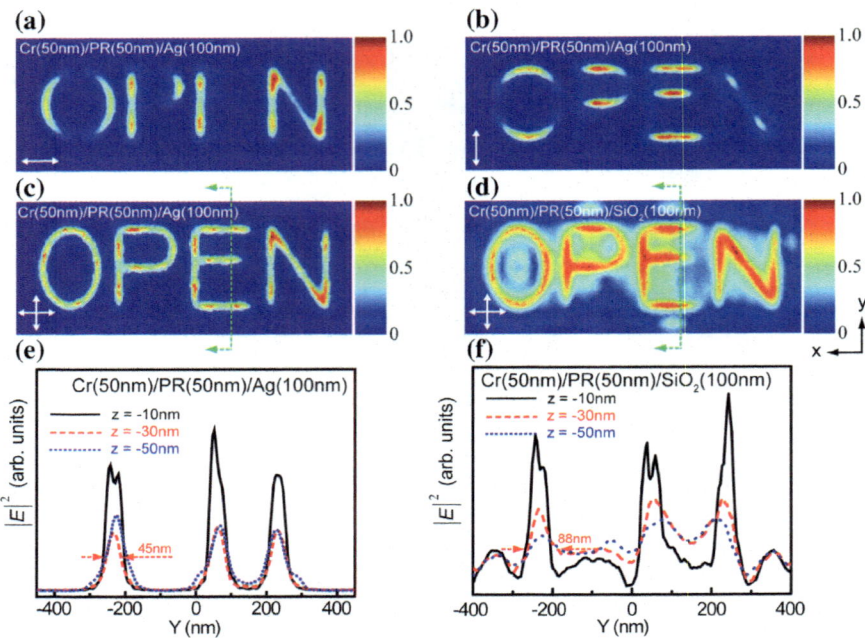

Fig. 4.11 **a–c** Calculated electric field intensity distributions at 30 nm below the Cr–Pr interface illuminated by different polarizations with a reflective silver slab. The white arrows show the polarization direction. **d** Control simulation result where the reflective silver slab is replaced by fused silica. **e, f** The cross-section profiles at 10, 30, and 50 nm below the Cr–Pr interface. Reproduced from [24] with permission. Copyright 2013, Optical Society of America

perpendicular to the polarization direction are imaged. The sum of the electric field intensity is plotted in Fig. 4.11c to represent light illumination with natural polarization in the experiments. Furthermore, the calculated results without the silver slab structure are shown in Fig. 4.11d, showing great distortions compared with both the original image and the reflective superlens.

Figure 4.11e, f plots the light distributions at variant z position. The narrow width of the line (36 nm) leads to a strong electromagnetic coupling with SP excited at the interface of Ag and photoresist. The image distribution in the photoresist layer with a reflective slab demonstrated a line width of about 45 nm at any thickness of photoresist. The intensity contrast of images at 10 and 30 nm below the Cr–Pr reach about 0.98 and 0.91. Even at the bottom of the Pr (50 nm thickness), the intensity contrast is larger than 0.8. For the structure without reflector, the intensity contrasts at 10, 30, and 50 nm below the Cr/Pr interface are 0.66, 0.3, and 0.1, showing a rapid decrease of imaging property. For positive photoresist, the required intensity contrast usually should be larger than 0.3. Therefore, the effective imaging depth for Fig. 4.11f is less than 30 nm. In contrast, the reflective plasmonic slab structure gives an imaging depth of 50 nm and even larger. This effect could be simply attributed

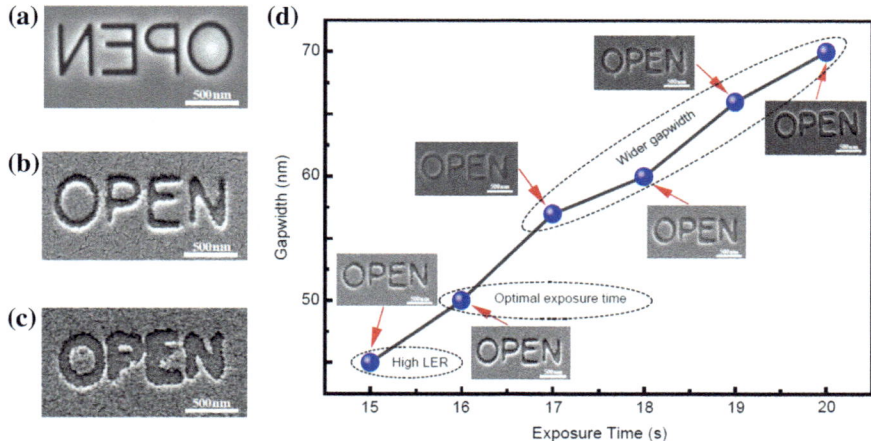

Fig. 4.12 **a** SEM image of the mask pattern "OPEN" characters with a line width of about 36 and 500 nm height. **b** SEM image of resist pattern with reflective plasmonic slab and exposure time of 16 s. **c** Control experiment result without the reflective slab but with the same photoresist thickness, exposure dose and development time as (**b**). **d** The line width of "OPEN" resist pattern for variant exposure time. Reproduced from [24] with permission. Copyright 2013, Optical Society of America

to the catenary fields formed by the superposition of the evanescent waves produced by the masks and the reflected light.

To perform the experiment, masks with nanopatterns were first fabricated on a 50-nm-thick Cr layer by focused ion beam (FIB) milling (Fig. 4.12a). The plasmonic slab was fabricated by depositing 110-nm-thick Ag film on a fused silica substrate with magnetron sputtering. Upon it was spun with about 50-nm-thick positive photoresist made by 3000 rpm and diluted AR-P 3170 to record the near-field images. After 2 min of prebake of photoresist at 100 °C hotplate, the substrate with reflective slab was physically contacted with the mask with the help of an air pressure (~0.3 MPa), and then flood-exposed under an i-line (365 nm) mercury lamp illuminating system. The postexposure substrate was developed by diluted AR 300-35.

Compared with the mask pattern, the recorded resist pattern in Fig. 4.12b shows good fidelity and about 50 nm line width by applying the reflective plasmonic slab and appropriate exposure time of 16 s. With the same photoresist thickness, mask, exposure dose, and development time, the control experiment result without reflector in Fig. 4.12c gives a greatly deteriorated "OPEN" resist images with line width larger than 100 nm. In the experiment, it is also found that the reflective plasmonic slab configuration contributes to increase the process latitude for variant exposure time, as shown in Fig. 4.12d. Although the line width of resist pattern can be further narrowed to 45 nm with shorter exposure time (15 s), the line edge roughness (LER) of this pattern is apparent. On the other hand, increase of exposure time would deliver wider line width of resist pattern but not destruct the imaging fidelity.

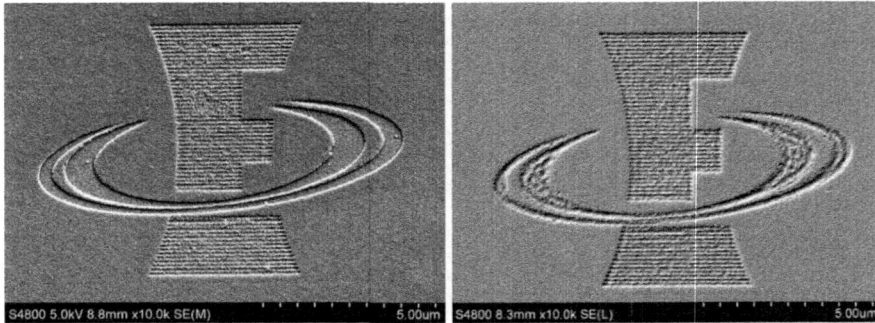

Fig. 4.13 SEM images for dense lines array pattern by reflective plasmonic slab lithography with dense lines half-pitch of 55 nm. The right is a superlens lithography without reflective layer

Figure 4.13 shows another comparison of the reflective lithography and contact lithography. It is clear that the image fidelity has been greatly improved via the reflective layer as well as the catenary optical fields therein. In principle, by using photoresist with smaller molecule and higher resolution, the lithography performance could be further improved.

4.2.2 Plasmonic Imaging of Dense Lines

Although reflection plasmonic lithography could improve the imaging resolution to about 50 nm, it would be more competitive if pattern resolution with high aspect profile and fidelity is extended to 32 nm and beyond. In this section, a plasmonic cavity lens is experimentally demonstrated to enhance aspect profile and resolution of the plasmonic lithography. The profile depth of 32-nm half-pitch resist patterns is enhanced to about 23 nm. The resist patterns are then transferred to the 80 nm deep bottom layer resist pattern using hard mask technology and etching process [14].

Figure 4.14a shows the cross-sectional view of the fabricated plasmonic cavity lens composed of 21 nm thickness top Ag, 30 nm Pr and 54 nm bottom Ag. To control surface roughness of Ag films, a 2 nm germanium seed layer is deposited on a silicon substrate and the AFM measured surface roughness root mean square (rms) of the bottom Ag reflective layer is 0.6 nm. Benefiting from the ultrasmooth photoresist layer (rms ~ 0.4 nm), the surface roughness rms of the top Ag layer is only 0.5 nm. Figure 4.14b, c shows the top view and cross-sectional SEM images of the obtained 32-nm half-pitch resist pattern, where the pattern lines are clearly resolved and the sloped profile approximately extends to the bottom Ag film. The line edge roughness (LER) rms of lines pattern is about 8.38 nm (3σ). The LER of the photoresist pattern could be further reduced by utilizing a high γ-value photoresist or reducing surface roughness of top Ag film. The measured profile depth in the plasmonic cavity lens lithography is about 23 nm, which illustrates a significant improvement of aspect

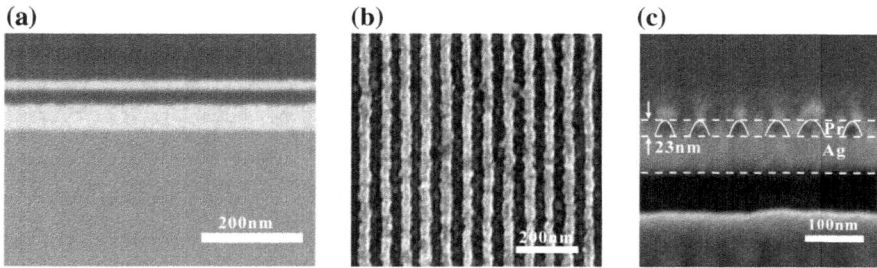

Fig. 4.14 **a** SEM cross-sectional view of the fabricated plasmonic cavity lens. **b** SEM image of 32-nm half-pitch dense lines resist pattern. **c** Cross-sectional view of pattern in (**b**). Reproduced from [14] with permission. Copyright 2015, AIP Publishing LLC

profile compared with the 5 nm profile depth in the near-field lithography and 9 nm in the superlens control experiment.

The physical mechanism of profile depth improvement could be obtained by analyzing the field distribution inside the plasmonic cavity lens. As shown in Fig. 4.15a, the cavity lens is composed of a photoresist layer sandwiched between a bottom Ag reflective layer and a top Ag superlens layer. The red curves indicate the E-field distribution of upper Ag–Pr and bottom Pr–Ag interfaces associated with the symmetric mode. Clearly, the E_x component inside the resist region of the plasmonic cavity lens is greatly larger than the E_z component, which shows a $\pi/2$ phase difference with respect to E_x and makes negative contribution in superlens imaging [26]. From another point of view, because E_z is mainly localized at the edges of object, its width may be greatly expanded compared with E_x. Consequently, E_x is preferred for imaging lithography.

Figure 4.15b, c shows the dependence of simulated intensity contrast of the superlens and the plasmonic cavity lens on air spacer thickness and photoresist depth position. The mask pattern is assumed to be 32-nm half-pitch dense lines. For an air spacer thickness below 5 nm, a low-intensity contrast could be observed and attributed to the damped excitation of surface plasmons with extremely thin air layer sandwiched between chromium and silver films. It is surprising and counterintuitive that high-intensity contrast imaging does not appear as the mask pattern is physically contacted with superlens and plasmonic cavity lens. This phenomenon is welcome in the lithography, considering the issues of physical degradation and challenges to close proximity. For an air spacer thickness ranging from 5 to 10 nm, the superlens scheme shows an intensity contrast $V > 0.4$ (Fig. 4.15b) at a limited photoresist depth range, while plasmonic cavity lens helps to achieve $V > 0.4$ in the whole photoresist depth direction (Fig. 4.15c). For air spacer thickness larger than 10 nm, the rapid damping of evanescent waves results in intensity contrast decreasing in both of the superlens and plasmonic cavity lens.

According to the simulated intensity contrast of the plasmonic cavity lens, theoretical profile depth enhancement would occur with 5–10 nm air spacer. In the exposure and development process, the profile depth is not exactly determined by contrast cri-

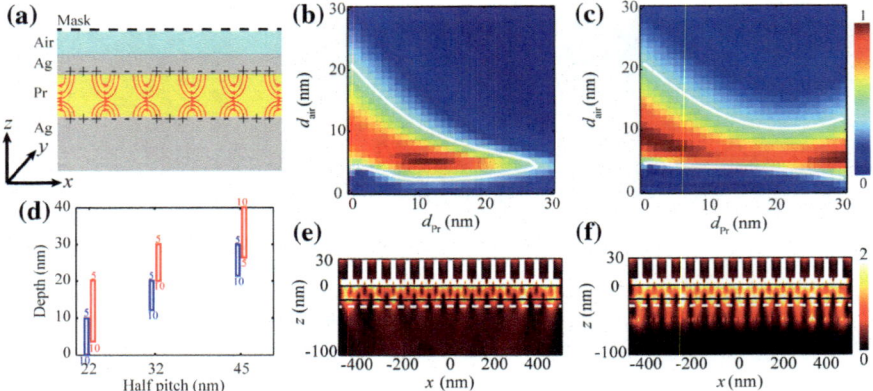

Fig. 4.15 **a** Schematic of plasmonic cavity lens. **b, c** Simulation of intensity contrast (32-nm half-pitch) versus photoresist depth and air spacer with **b** superlens and **c** plasmonic cavity lens. The white solid curves are the intensity contrast contour of 0.4. **d** Profile depth for 22, 32 and 45-nm half-pitch. The blue and red blank bars represent profile depth of the superlens and plasmonic cavity lens. The data below and above the blank bars indicate the air spacer. The photoresist thickness of 20, 30 and 40 nm are selected for 22, 32 and 45-nm half-pitch patterns, respectively. **e, f** Simulated *E*-field intensity distributions of 32-nm half-pitch grating in superlens (**e**) and plasmonic cavity lens (**f**). Reproduced from [14] with permission. Copyright 2015, AIP Publishing LLC

terion of 0.4 on basis of the simulated *E*-field intensity distribution. Figure 4.15d illustrates the dependence of the calculated exposure depth for 22, 32, and 45-nm half-pitch patterns on the specified air spacer thickness of 5–10 nm. The calculations are made by finite element method simulations incorporated with the near-field exposure model. The blue blank bars show an obtainable profile depth range of the superlens lithography. Due to decaying characteristics, the profile depth is decreasing with reducing feature size. The red blank bars indicate that plasmonic cavity lens could significantly improve the profile depth.

The surface roughness of the silver films also affects the *E*-field intensity distribution. To show this point, Fig. 4.15e, f plot the numerically simulated *E*-field intensity distribution of superlens and plasmonic cavity lens, where AFM measured surface roughness rms of Ag superlens layer (0.5 nm), Pr layer (0.4 nm), and Ag reflective layer (0.6 nm) are considered. In Fig. 4.15e, the *E*-field intensity contrast with the superlens decreases in the photoresist depth direction. In contrast, Fig. 4.15f shows that the *E*-field intensity in the plasmonic cavity lens is clearly resolved with high-intensity contrast even at the bottom of the photoresist layer.

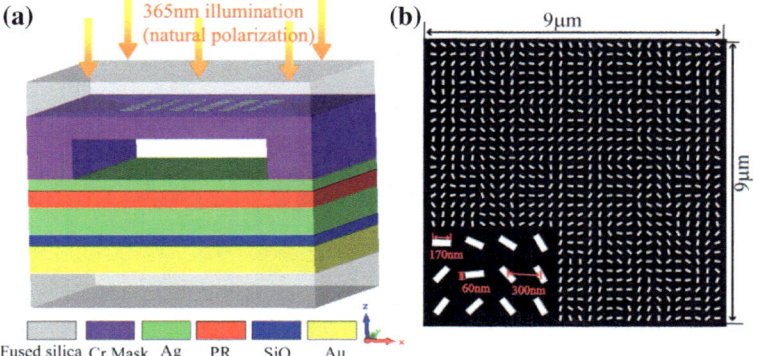

Fig. 4.16 **a** Schematic of plasmonic cavity lens lithography system in separated mode. **b** Top view of the designed Cr mask patterns. Reproduced from [27] with permission. Copyright 2017, WILEY-VCH Verlag GmbH & Co. KGaA, Weinheim

4.2.3 Plasmonic Imaging of Complex Patterns

Based on the catenary plasmons in the metal–dielectric multilayer, the plasmonic cavity lens is also suitable for the fabrication of complex arbitrary nanopatterns to realize functional devices such as flat lenses and holograms [27]. As an example, an Au plasmonic hologram with nanoaperture size about 95×175 nm^2 was fabricated, and the reconstructed holographic image is in good agreement with the design.

The schematic configuration of plasmonic cavity lens lithography in separated mode (it means that the mask and the lens are separated) is shown in Fig. 4.16a. The geometric parameters of the cavity lens are optimized as 20-nm-Ag/30-nm-Pr/50-nm-Ag and 25-nm gap distance. A plane wave at $\lambda = 365$ nm normally illuminates on the Cr mask. The cavity lens is physically contacted with Cr spacer in experiment. In addition, a 20-nm-thick SiO$_2$ film is placed below the cavity lens for pattern transfer, and a 50-nm-thick Au film on the fused silica substrate is adopted for plasmonic holography. Figure 4.16b is the top view of the designed Cr mask patterns with a nanoaperture size of 60×170 nm^2, a period of 300 nm and a pattern area of $9 \times 9\ \mu$m^2. Under a working wavelength of 365 nm, the relative permittivity of materials used for simulation is $\varepsilon_{Cr} = -8.55 + 8.96i$, $\varepsilon_{air} = 1$, $\varepsilon_{Ag} = -2.17 + 0.36i$ and $\varepsilon_{Pr} = 2.59$.

Compared with the near field and superlens lithography, the cavity lens could greatly improve the imaging performance. In order to demonstrate the whole imaging effect, the light intensity distributions in the arrayed nanoapertures were simulated. Figure 4.17a illustrates the distinguishable and uniform imaging patterns in the middle of Pr layer, which closely resembles the used mask patterns. Figure 4.17b gives the cross-sectional light field distribution in the xz-plane. Clearly, the light field in Pr layer shows strong intensity and uniformity inside Pr layer, which would effectively increase the exposure depth. Figure 4.17c further proves this point, showing

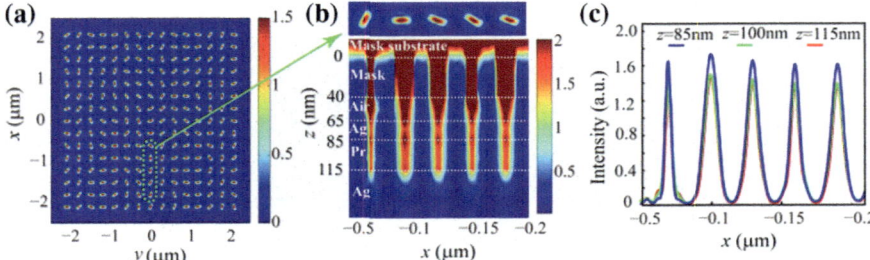

Fig. 4.17 **a** Calculated light field distribution in the middle of Pr layer for natural polarized light. **b** The top graph is the local magnification of the green dashed line region, and the bottom graph is the cross-sectional light intensity distribution in the xz-plane. **c** The light intensity distribution is plotted at different z in Pr layer. Reproduced from [27] with permission. Copyright 2017, WILEY-VCH Verlag GmbH & Co. KGaA, Weinheim

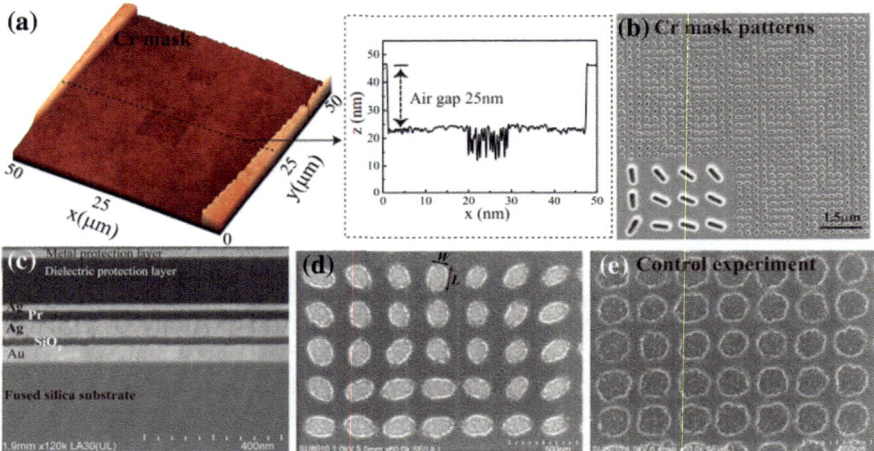

Fig. 4.18 **a** The 3D AFM image of the Cr mask with Cr spacer, and the right graph shows the 25 nm air separation. **b** The SEM image of mask patterns. **c** The SEM cross-sectional image of the sample. **d** The SEM image of Pr patterns generated from the cavity lens lithography. **e** The SEM image of Pr patterns generated from control experiment. Reproduced from [27] with permission. Copyright 2017, WILEY-VCH Verlag GmbH & Co. KGaA, Weinheim

nearly equal intensity and similar profiles in Pr layer for $z = 85$, 100, and 115 nm, respectively.

The above lithography design was performed in experiment to demonstrate the fabrication of metasurface holograms. To keep a certain distance between mask patterns and cavity lens, a 25-nm-thick groove was fabricated in Cr mask, as shown in Fig. 4.18a. Mask patterns displayed in Fig. 4.18b were prepared by FIB milling inside the groove, with a nanoaperture size of $(60 \pm 5) \times (170 \pm 5)$ nm^2, a period of

300 nm, and an area size of 9×9 μm^2. Figure 4.18c illustrates the cross-sectional of the cavity lens.

Figure 4.18d displays the recorded Pr patterns with distinguishable anisotropy. Therefore, the experimental results suggest that the plasmonic cavity lens lithography could break the diffraction limit and be used to fabricate anisotropic nanoapertures. In contrast, the control experiment, namely the near-field lithography, generated almost circular patterns without any anisotropy, regardless of the optimization of lithographic processes.

It is worth mentioning that the nanoaperture size of about 160×238 nm^2 is much larger than that of mask nanoaperture size about 60×170 nm^2, and the calculated increasing factors are about 2.7 and 1.4 times for width and length. Obviously, the size increase in the short axis direction is much larger than that in the long axis direction. This is owing to the optical proximity effect (OPE), which is the diffraction effect of light generated from mask patterns and generates deviation between mask patterns and Pr patterns. The OPE effect is more obvious with smaller size. During the design of mask patterns, the OPE effect is corrected by increasing the aspect ratio of the nanoaperture.

To obtain the Au plasmonic holograms, further etching transfer was accomplished by ion beam etch (IBE) dry etching and HF solution wet etching. The schematic of etching procedure was shown in Fig. 4.19a. After the etching of 50-nm-Ag/20-nm-SiO_2/50-nm-Au films, the corresponding SEM images are shown in Fig. 4.19b–d. The Au plasmonic holograms were successfully obtained with a nanoaperture size of $(95 \pm 6) \times (175 \pm 6)$ nm^2. Figure 4.19e, f present the AFM image of Au holograms with a depth of about 53 nm, indicating that the 50-nm-thick Au film has been completely etched.

The varying ratios of length/width in Fig. 4.19b–d deserve special attention. Owing to the 25-nm air separation and the OPE effect, the ratio for simulated Pr patterns is about 2.1 (209 nm/99 nm) despite the ratio for mask patterns is about 2.8 (170 nm/60 nm). The ratio for experimental Pr patterns is about 1.5 (238 nm/160 nm), and the errors may be brought from many factors, such as lithographic process, fabrication deviation, the film roughness, etc. Fortunately, owing to the shadowing effect during the IBE etching process, the ratio of Au patterns is increased to about 1.8 (175 nm/95 nm). During the IBE etching, the ion beams impinge the sample on a rotated platform with 30° oblique angle. As a result, the etching rate in the middle region of antenna pattern would be faster than the shadow region because the ion beams are partially screened by the Pr wall. By further optimizing the lithography processes, the final patterns may become much better with higher efficiency and phase modulating accuracy. In a more general sense, the full lithography and etching processes may be numerically modeled to give the optimal design of masks.

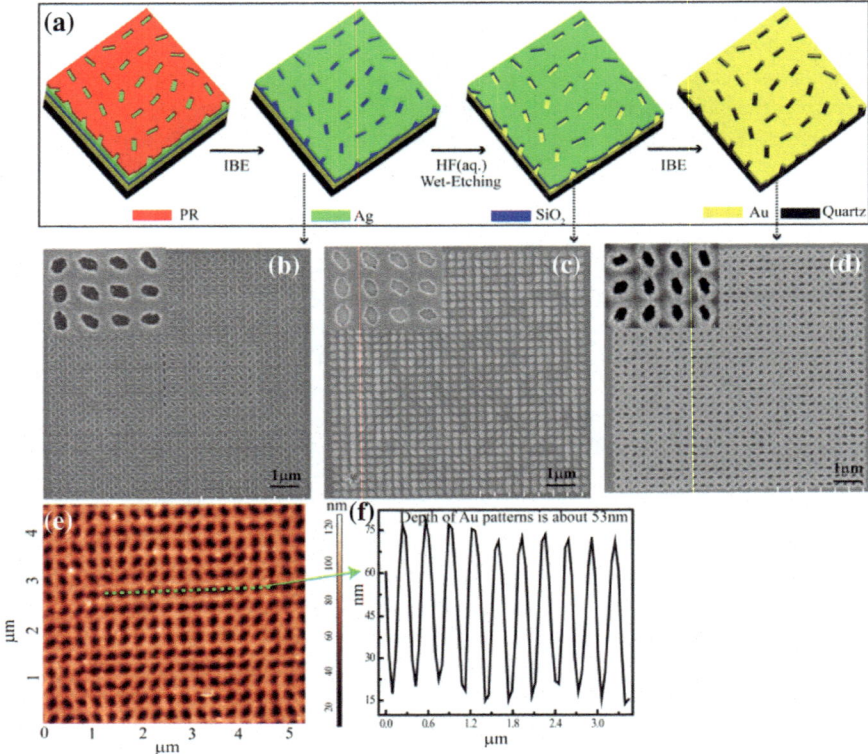

Fig. 4.19 **a** The schematic of pattern transfer by IBE dry etching and HF wet etching. **b–d** The SEM images of the corresponding etching patterns in (**a**). **e** The 2D AFM image of Au patterns in (**d**). **f** The plot of Au pattern's depth in (**e**). Reproduced from [27] with permission. Copyright 2017, WILEY-VCH Verlag GmbH & Co. KGaA, Weinheim

4.3 Demagnifying Imaging Based on Curved Hyperlens

4.3.1 Numerical Simulation

As demonstrated in both numerical and experimental results [9, 28], light may trans-mit in metal–dielectric multilayers without diffraction as a result of the coupling of surface plasmons. By bending these multilayers, light would transmit along the radial direction, which could be used to magnify deep-subwavelength objects to the far-field, or demagnify objects on the mask to the photoresist [4, 29]. Since the mag-nifying hyperlens has been intensively investigated, here we focus our discussion in demagnifying lithography. Similar to previous reflective superlens and cavity lens, the plasmonic reflection is also considered to increase the imaging performance.

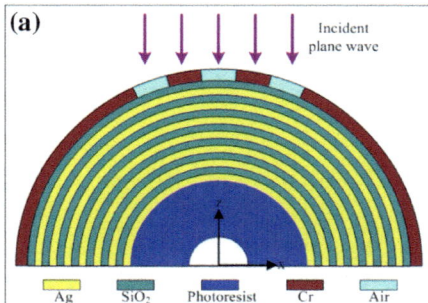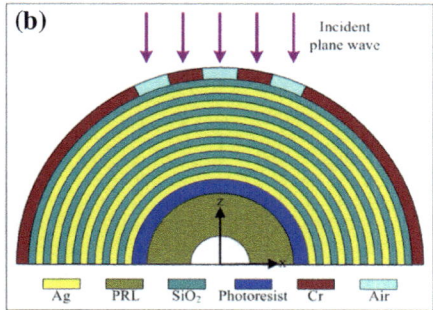

Fig. 4.20 Schematic drawing of hyperlens without (**a**) and with (**b**) a PRL. The photoresist layer is 20 nm thick, and the last silver PRL is about 200 nm thick in the simulation. Reproduced from [4] with permission. Copyright 2013, Springer Science + Business Media New York

For demagnifying imaging and nanolithography, it is necessary to see the dependence of polarization states on the resolution of hyperlens. Figure 4.20a, b shows the schematic drawing of hyperlens structures without and with a plasmonic reflective layer (PRL). The hyperlens is composed of seven pairs of Ag/SiO_2 layers in cylindrical form with 20-nm layer thickness. At an incident TM(p) polarized plane wave at $\lambda = 365$ nm, the dielectric constants are $\varepsilon_{Ag} = -2.4012 + 0.2488i$, $\varepsilon_{SiO_2} = 2.13$, and $\varepsilon_{Pr} = 2.56$. The inner and outer radius of the hyperlens is 320 and 600 nm, respectively. Nano-objects represented by transparent slits at chromium layer are positioned closely at the outer radius.

Figure 4.21 presents numerical simulations of light fields distribution in the hyperlens without a PRL. To get a full understanding of the polarization dependence of the imaging behavior, the total electric intensity $|E|^2$, radial electric intensity $|E_r|^2$, tangential electric intensity $|E_\theta|^2$, magnetic intensity $|H_y|^2$, and radial Poynting vector S_r are given under TM polarized illumination.

In the simulations, the slit width of the object is 100 nm and center-to-center space is 250 nm. As mentioned before, the image should give light distribution with center-to-center space down to about 125 nm. However, only $|H_y|^2$ and S_r distributions at the imaging region show distinguished images. For $|E|^2$ distribution that determines the exposure and pattern quality of photoresist in lithography, the two slits could not be resolved. Compared with $|E_r|^2$, $|E_\theta|^2$ shows a better resolving ability. But the ratio between $|E_r|^2$ and $|E_\theta|^2$ at the imaging region is close to 0.5 which leads to blurring images. On the other hand, S_r turns into a negative value at the output surface of the hyperlens, indicating that the light cannot be coupled to the image region effectively. So, it is clear that the simple reciprocity does not hold for hyperlens applied in microscope and lithography processes. For magnification hyperlens, the image shows weak dependence of light polarization. However, in the demagnification mode, hyperlens usually exhibits greatly degraded resolution and poor imaging quality evaluated with electric intensity. This would bring a serious problem in high-resolution nanolithography. In fact, the perfect imaging performance based on the

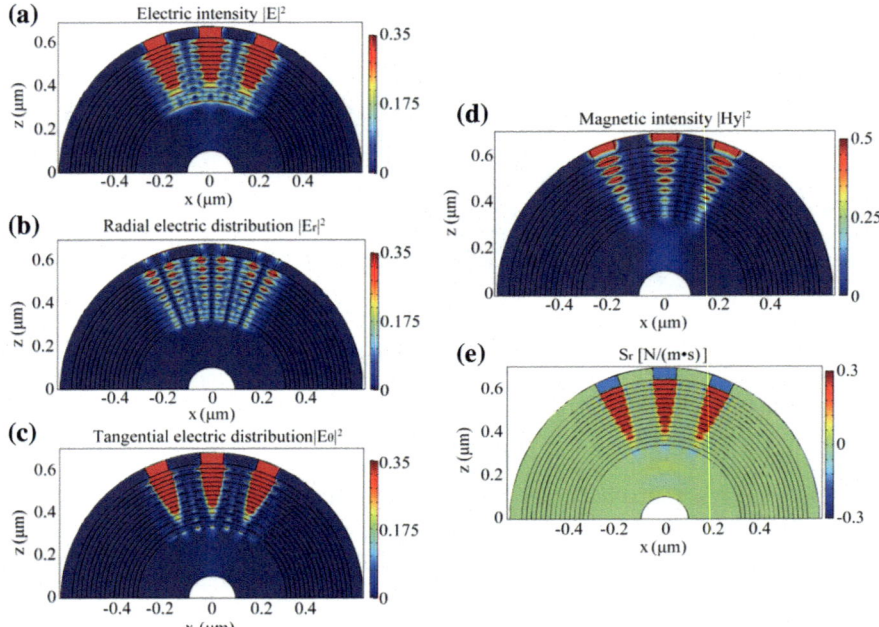

Fig. 4.21 Demagnification imaging performance of the hyperlens without a PRL. **a** Total electric field intensity $|E|^2$. **b** Radial electric intensity $|E_r|^2$. **c** Tangential electric intensity $|E_\theta|^2$. **d** Magnetic field intensity $|H_y|^2$. **e** Radial Poynting vectors S_r. Reproduced from [4] with permission. Copyright 2013, Springer Science + Business Media New York

exact reciprocity theorem of electromagnetic field requires that the stacked dielectric and metal layers have opposite sign of both permeability and permittivity, which is hard to realize in practice.

By introducing a PRL to a cylindrical hyperlens, catenary optical fields could be established in the photoresist layer. Consequently, the imaging performance may be significantly improved even in the measurement of $|E|^2$, as shown in Fig. 4.22a. The improved resolution is achieved by high spatial frequency components in the transmitted light, which can excite SPPs at the boundary of photoresist and plasmonic metal. The interference between the SPPs and incident light creates a catenary optical field inside the photoresist, which improves the image confinement in the lateral direction. From Fig. 4.22b, c, the ratio of $|E_r|^2/|E_\theta|^2$ at the imaging plane is greatly reduced. Furthermore, the value of S_r keeps positive in all regions, which implies that light is coupled well in the image region.

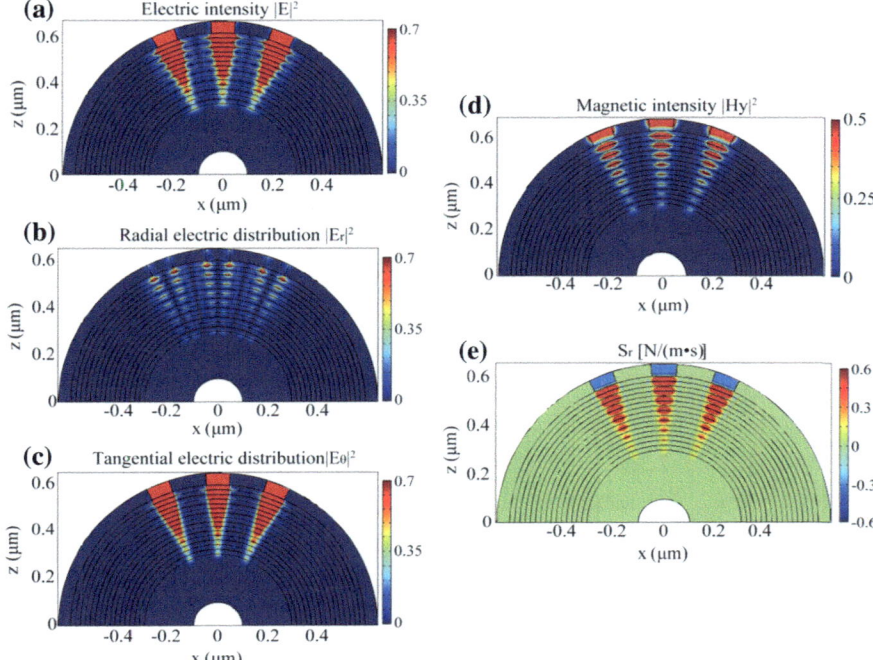

Fig. 4.22 Simulations of demagnification imaging for the hyperlens with a PRL. **a** Electric field intensity $|E|^2$. **b** Radial electric intensity $|E_r|^2$. **c** Tangential electric intensity $|E_\theta|^2$. **d** Magnetic field intensity $|H_y|^2$. **e** Poynting vectors S_r. The photoresist layer is 20 nm thick, and the last silver PRL is about 200 nm thick. Reproduced from [4] with permission. Copyright 2013, Springer Science + Business Media New York

4.3.2 Experimental Demonstration

In this section, the experimental demonstration of a preliminary hyperlens for sub-diffraction demagnification imaging lithography is presented. A plasmonic reflector layer was combined with hyperlens to obtain high contrast and good fidelity images by engineering electric field components in the photoresist area. The resist patterns with 55-nm line width and a demagnification factor of about 1.8 were experimentally demonstrated at a wavelength of 365 nm [30].

Figure 4.23a shows the schematic of the designed hyperlens consisting of mask pattern, cylindrical hyperlens with Ag and SiO$_2$ films, photoresist layer, and a Ag reflector upon it. The process of the experiment is as follows: First, a V shape groove with 90-nm width and 200-nm depth was milled on quartz substrate by FIB, and etched in buffered oxide etch (BOE) solution to get a nearly half cylindrical (it is actually a catenary as the case of glacial valley [31]) groove with a radius of 500 nm, a width of 800 nm and a thickness of 200 nm. For the convenience of fabricating

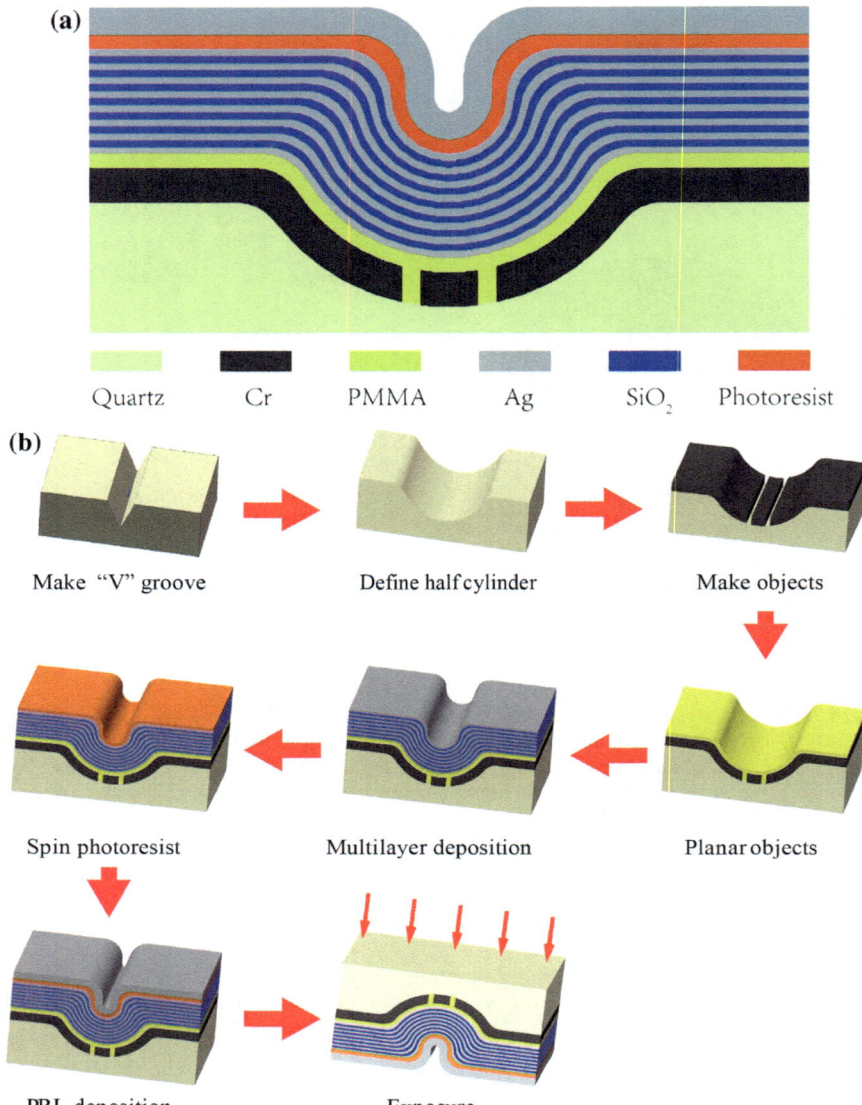

Fig. 4.23 **a** Schematic of demagnifying hyperlens with a PRL. **b** Process flow of hyperlens fabrication and lithography. Reproduced from [30] with permission. Copyright 2016, The Royal Society of Chemistry

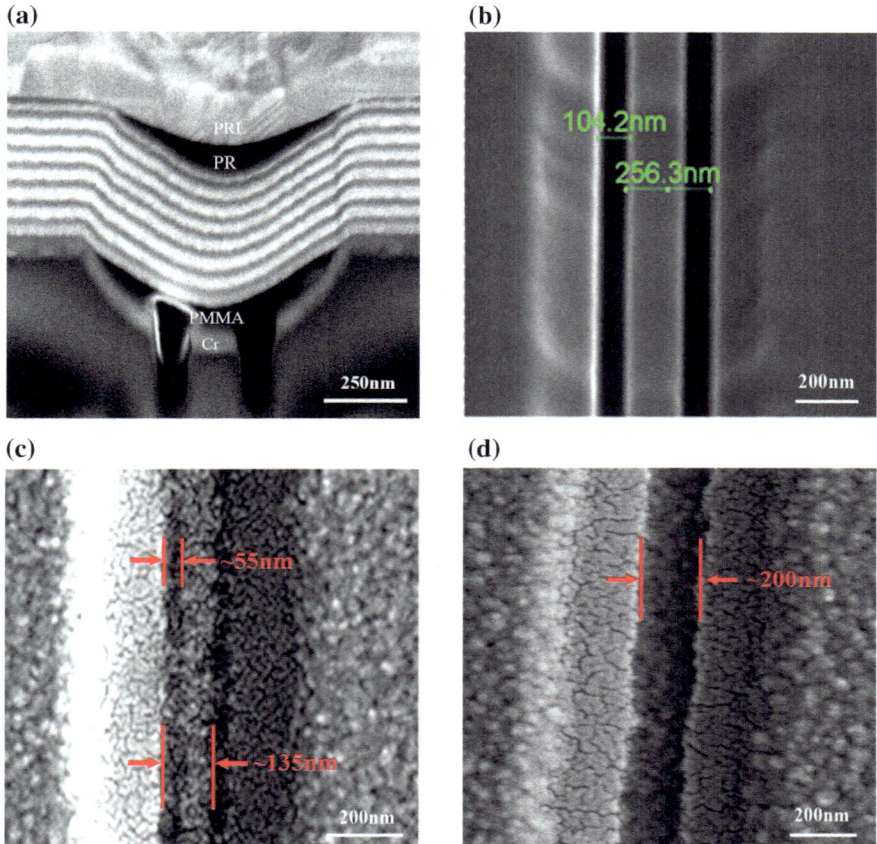

Fig. 4.24 **a** SEM image of the cross section of hyperlens. **b** The two slits mask pattern with 100-nm line width and 250-nm center-to-center distance. **c** Lithography result of resist pattern with hyperlens and PRL with 55-nm line width and 135-nm distance. **d** Resist pattern of hyperlens without a PRL. Reproduced from [30] with permission. Copyright 2016, The Royal Society of Chemistry

multiple films and resist layer, the thickness is less than radius with large groove opening aperture. 40-nm Cr film was deposited and milled by two slits with 100-nm line width and 250-nm center-to-center distance. Subsequently, a PMMA layer was spun on it and etched to about 45-nm thickness. Then 15 layers of Ag and SiO_2 films, each with a 20-nm thickness, were alternately deposited. 1–2-nm thickness Ge was pre-deposited beneath every Ag film as wetting material to improve surface smoothness (RMS: 1.2 nm). Then diluted SX AR-P3170 positive resist layer was spun over it with a 50-nm thickness. The final step is to deposit 120-nm-thick Ag reflector on the resist.

Figure 4.24a, b show the SEM images of hyperlens in cross-sectional view and the photomask before covered by hyperlens, showing a 250 nm center-to-center distance

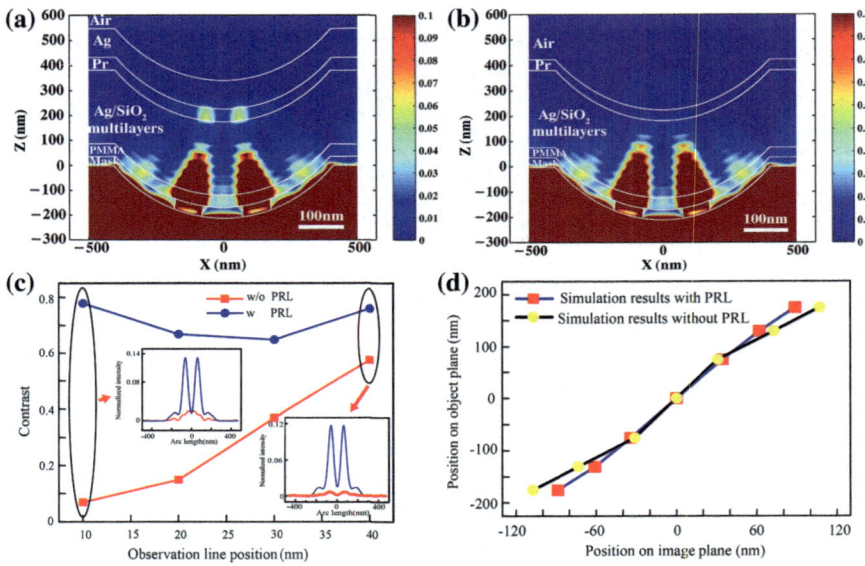

Fig. 4.25 Normalized electric field intensity distributions for hyperlens with (**a**) and without a PRL (**b**). **c** Variations of image contrast at positions 10–40 nm away from the hyperlens. The insets are the image profiles of hyperlens with/without a PRL, respectively. **d** Lateral positions relation between images and two silts on the mask with variant center-to-center distance. Reproduced from [30] with permission. Copyright 2016, The Royal Society of Chemistry

and a 100-nm line width. The sample with a PRL was exposed from the bottom side of the quartz substrate by a filtered and nearly collimated light (the central light wavelength is 365 nm and the line width is 10 nm). After peeling off the PRL, the exposed hyperlens sample was developed at room temperature.

The demagnifing lithography result with reflective hyperlens is presented in Fig. 4.24c, where two parallel lines could be clearly distinguished with a line width of ~55 nm and a separation distance of ~135 nm (about 1/2.7 times that of light wavelength), corresponding to a demagnification ratio of ~1.85. For the hyperlens without a PRL (Fig. 4.24d), the two slits could not be resolved and the image merges into one wide line with a width of ~200 nm.

To explain the experimental results in detail, a commercial software COMSOL Multiphysics is used to simulate the imaging performance of the designed hyperlens. The normalized electric field intensity distributions of the hyperlens with and without PRL are shown in Fig. 4.25a, b, respectively. As the light transmits through mask patterns and propagates inwards along the radial direction of hyperlens, angular wave vectors gradually increase and the sub-diffraction demagnified optical pattern is generated in the resist layer. Figure 4.25c shows the image contrast of two lines for the observation positions ranging from 10 to 40 nm below the output surface of the hyperlens. Obviously, the contrast of the hyperlens with a PRL is considerably high, usually larger than 0.6 throughout the 50 nm photoresist thickness, and much larger than those of the hyperlens without a PRL. Particularly, at position 30 nm and below away from the output surface of the hyperlens, the latter shows a rapid decrease in imaging property with contrast much smaller than 0.4. This helps to explain why no resolved resist pattern is observed in the lithography experiment shown in Fig. 4.24d.

The PRL scheme also helps to significantly enhance the electric intensity of images, about 4 times larger than that without it, as shown in inset of Fig. 4.25c. This is mainly due to the amplification of evanescent waves and resonance effect in the plasmonic cavity lens. On the other hand, the position relationship of object plane and image plane for the configuration of hyperlens with and without a PRL is shown in Fig. 4.25d, which demonstrates that more uniform demagnification ratio ~2.05 could be obtained for hyperlens with a PRL. Owing to the imperfection in the fabrication and lithography process, plus the fact that the multilayers have a catenary rather than circular shape, the achieved demagnification ratio in experiment is decreased to 1.85.

4.4 Interference Lithography of Periodic Patterns

In Abbe's imaging theory, all imaging processes are actually composed of interference of various spatial components. Consequently, interference is of fundamental importance for optical imaging. In optical lithography, it is also true that interference plays a more elementary role. For instance, the plasmonic imaging lithography is enabled by the observation of extraordinary Young's double slits interference [6, 12]. In this section, we focus on a more traditional definition of interference lithography, i.e., the fabrication of periodic patterns by the interference of finite coherent beams.

4.4.1 Normal Incidence

Interference lithography provides a simple way to fabricate periodic gratings with low-cost and high throughput. In the first experiment, interference of SPPs has led to a much smaller period beyond the diffraction limit [12]. To obtain higher resolution, higher order evanescent waves must be employed. In this section, large area and uniform deep subwavelength interference patterns are produced by squeezing bulk plasmon polaritons (BPPs) through a hyperbolic metamaterial (HMM) composed of metal and dielectric multilayers (Fig. 4.26). Once again, a reflective layer is incorporated to increase the quality of resist patterns. The grating masks used for BPPs excitation have period much larger than that of interference patterns, thus they can be fabricated easily [32, 33].

Numerical simulation with finite element method (FEM) for 1D mask was performed to demonstrate light intensity distribution inside the structure.

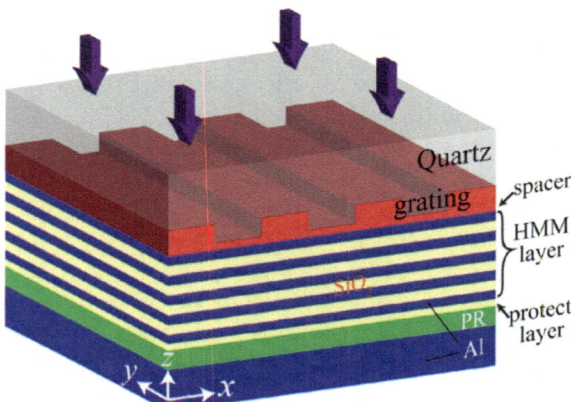

Fig. 4.26 Schematic of the BPP interference lithography with hyperbolic metamaterial. A 363.8 nm TM polarized light from the Ar-ion laser was vertically illuminated on the grating. Reproduced from [33] with permission. Copyright 2018, Optical Society of America

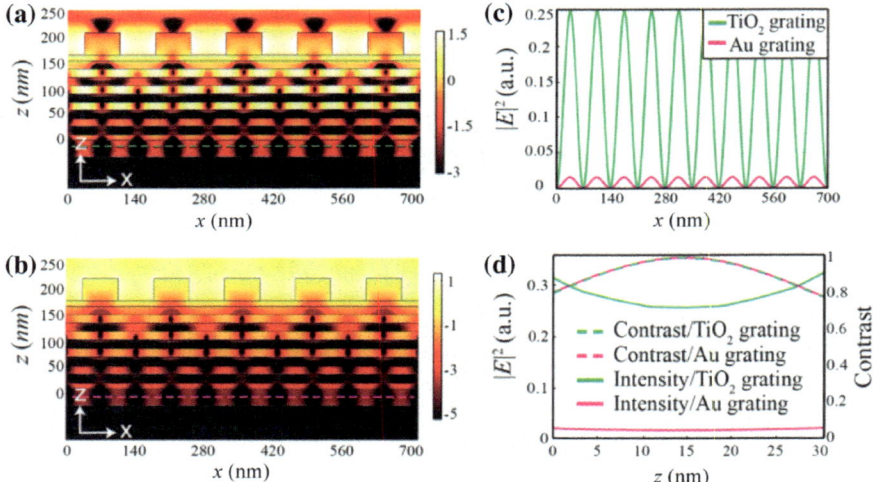

Fig. 4.27 **a, b** Simulated cross section of normalized intensity distribution in logarithm scale with **a** TiO₂ and **b** Au grating layer, respectively. The electric field intensity of the incident light is set 1 V/m. **c** Normalized $|E|^2$ electric field intensity along the horizontal lines at the middle of Pr layer for the two structures. **d** The image contrast and normalized intensity in the different depth of Pr layer. Note that the intensity curve follows a catenary function. Reproduced from [33] with permission. Copyright 2018, Optical Society of America

The normalized light intensity is defined as $|E|^2/|E_0|^2$, where $|E_0|^2$ is the incident light intensity. Meanwhile, the image contrast ratio is defined as $\left(|E_{max}|^2 - |E_{min}|^2\right)/\left(|E_{max}|^2 + |E_{min}|^2\right)$. Figure 4.27 shows the simulated normalized intensity distribution for two different grating structures (either TiO$_2$ or Au grating). Obviously, the light intensity of TiO$_2$/HMM/PR/Al is almost 16 times higher than that of the Au/HMM/PR/Al, which proves the higher energy efficiency of TiO$_2$ grating layer. Moreover, the image contrast for interference fringes in the whole Pr layer are both larger than 0.98 for both the two structures (Fig. 4.27d), which is attributed to the good filtering performance of HMM and the image enhancement of Al reflector. Owing to the catenary-like intensity distribution, the interfering fields are well maintained in the whole Pr layer.

In order to compare and analyze the efficiency of energy utilization, the above two structures were fabricated. Figure 4.28a, b shows the SEM images of TiO$_2$ grating and Au grating after etching process. The TiO$_2$ grating was planarized with a 7.5 nm peak-to-valley (PV) value and 0.6 nm RMS. Meanwhile, the top surface of Au grating layer, after the lift-off process, shows 4 nm PV profile and 0.37 nm RMS roughness. Figure 4.28c, d shows the SEM images of the resulting fringes obtained in the Pr layer after dissolving the back reflection Al film and developing process. Clearly, nearly uniform dense periodic line with 35 nm half-period was obtained, which is less than 1/10 of the incident light wavelength. The pattern area is about 20 mm × 20 mm. Figure 4.28e shows the cross-sectional SEM image of the BPP interference lithography structure. The ear-like structures on the top edge of the grating were caused by the assembling of Au atoms reflected at the etching process.

Both fringes show almost the same image contrast, as predicted by the simulation results in Fig. 4.27d. The exposure time of the structures with Au grating layer is almost 12 times longer than that with TiO$_2$ grating layer, which is consistent with the calculated intensity distribution. The slightly distortion of interference patterns in the SEM images depicted in Fig. 4.28c, d is mainly attributed to the film roughness, which leads to the random fluctuation of interference intensity and reduces fringe's uniformity.

4.4.2 Oblique Incidence

According to the diffraction equation, the transmitted wave vector through the metal–dielectric multilayers can be continuously changed by adjusting the incident light angle [34]. Figure 4.29 presents a schematic of the tunable interference lithography system. Two symmetrical TM polarized plane waves at 436 nm wavelength impinge on the Al grating from the fused silica substrate side. A thin TiO$_2$ dielectric spacer underneath the grating is used to enhance the coupling between the diffraction grating and the HMM, and thus improve the intensity of BPPs modes. Below the spacer is the HMM composed of alternately stacked 5 pairs 30-nm Ag and 35-nm SiO$_2$ thin

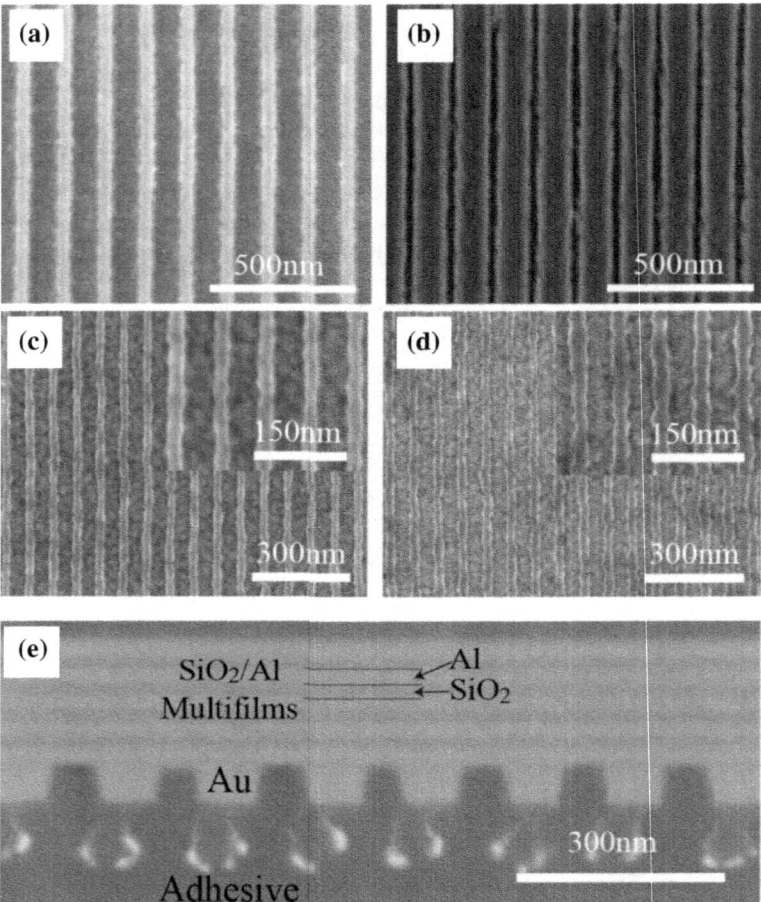

Fig. 4.28 **a**, **b** SEM images of TiO₂ and Au gratings before overturn. **c**, **d** SEM images of BPP interference fringes on Pr layer based on TiO₂ and Au gratings, respectively. **e** SEM cross section of the multilayers and grating. Reproduced from [33] with permission. Copyright 2018, Optical Society of America

films, followed by 30-nm Pr layer and 70-nm Al film, which acts as reflector for improving the imaging contrast.

Similar to traditional interference lithography [35], one can easily generate 2D periodic tunable patterns by combining multiple symmetrical illumination and 2D grating. The sketch of 2D structure with HMM and grating is shown in Fig. 4.30a. By considering the case where only the first diffraction order could pass the HMM, the upper and lower pitch of the interference pattern can be written as:

$$P_u = \frac{\lambda}{2} \frac{1}{m\lambda/p - n \sin \theta} \tag{4.4.1}$$

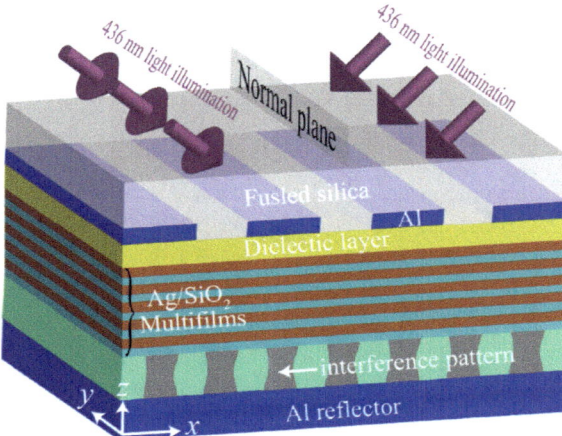

Fig. 4.29 Schematic of interference lithography with tunable period. The interference pattern in the resist layer shows an obvious catenary-like shape along the z-direction. Reproduced from [34] with permission. Copyright 2017, Optical Society of America

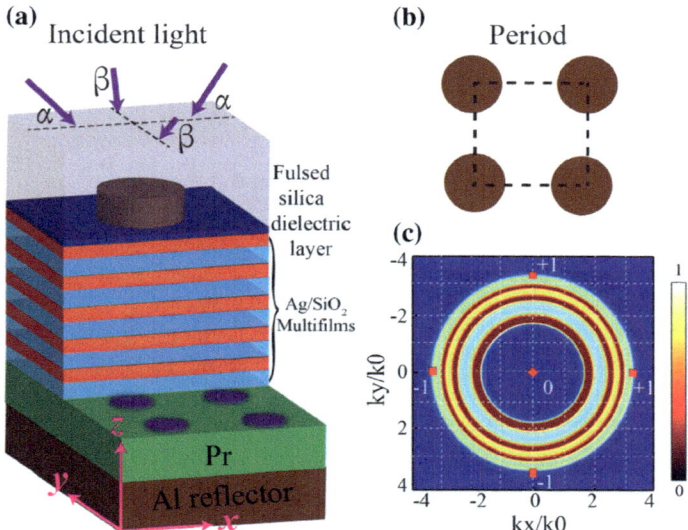

Fig. 4.30 Schematic for tunable BPPs interference lithography with 2D grating. **b** Square grating for BPPs excitation. **c** The position of diffraction light orders and optical transmission amplitude band for 5 pairs Ag (30 nm)/SiO$_2$ (35 nm) films. Here, the $\pm$ 1st diffraction orders are approximately locate at the positions of $(0, 3.4k_0)$, $(0, -3.4k_0)$, $(3.4k_0, 0)$ and $(-3.4k_0, 0)$. Reproduced from [34] with permission. Copyright 2017, Optical Society of America

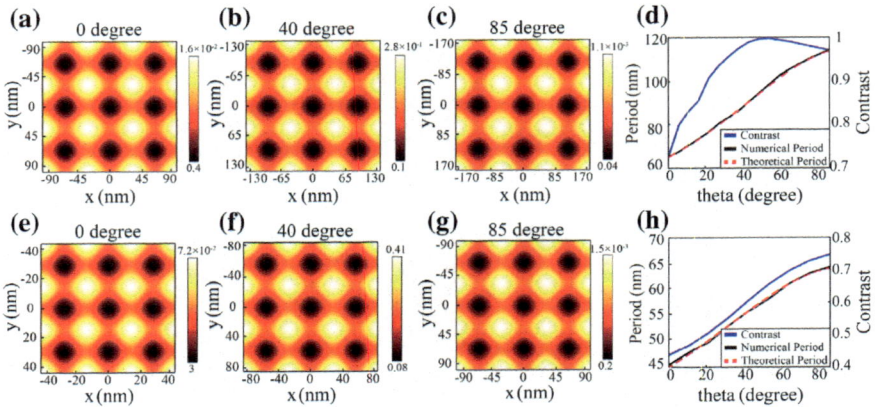

Fig. 4.31 **a–d** Structure with 5 pairs Ag (30 nm)/SiO$_2$ (35 nm) films. The period of the grating is 130 nm. **a** Normalized electric field intensity distributions (3 × 3 periods) in the *xy* plane at 0°, **b** 40° and **c** 85°. **d** The image contrast, numerical, and theoretical pitch resolution of the interference array dots for variant incident angles. **e–h** Structure with 5 pairs Ag (30 nm)/SiO$_2$ (15 nm). The period of the grating is 90 nm. Reproduced from [34] with permission. Copyright 2017, Optical Society of America

and

$$P_1 = \frac{\lambda}{2} \frac{1}{m\lambda/p + n \sin \theta}, \qquad (4.4.2)$$

where *m* is the diffraction order, *n* the refractive index of background material, and *p* the period of the excitation grating. The two equations are applied for the instances that the required diffraction orders located at the upper and lower boundary of the OTF. For the upper (lower) case, an additional incidence angle would reduce (increase) the wavevector. By adjusting the incident angle, these four diffraction orders excited by two couples of TM light can be transmitted. The angles between incident light with the *x*- and *y*-axis are defined as α and β.

At 436 nm wavelength, the pitch of the patterns can be continuously tuned from 45 nm (~λ/10) to 115 nm (~λ/4). As shown in Fig. 4.31a–c, e–g, normalized electric field intensity distributions (3 × 3 periods at 0°, 40° and 85°) in the *xy*-plane at the middle of the Pr layer for given incident angles are calculated for configurations of 5 pairs Ag (30 nm)/SiO$_2$ (35 nm) films and Ag (30 nm)/SiO$_2$ (15 nm) films, respectively. It can be seen that the 2D interference array dots with pitch resolution ranging from 65 to 115 nm and from 45 to 65 nm are produced. The contrast in Fig. 4.31d increases until the peak value around 50° and slightly decreases with the increase of incident angle, while the contrast in Fig. 4.31h increases among the entire incident angle range from 0° to 85°. The simulated period of the interference dots array increases and agrees well with the theoretical values.

4.5   Interference Lithography of Aperiodic Patterns

4.5.1   Polarization-Dependent Catenary Optical Fields

Besides the deep-subwavelength interference period, the interference patterns could be controlled by introducing more complex metal–dielectric structures such as multilayers and meta-mirrors [36, 37]. As shown in Fig. 4.32a, an additional aluminum film is added below circular patches array to form a meta-mirror [6]. The geometric parameters of $p = 200$ nm, $t = 20$ nm, $d = 30$ nm and $r = 45$ nm are used throughout the following discussions. Owing to the complex evanescent coupling schematically illustrated in the bottom right panel, both the E_z component below the patches and the E_x component between two patches follow the catenary function.

The interplay of vertical and horizontal catenary fields determines to some extent the optical properties (e.g., the reflection and absorption) of the MIM structures [38]. Figure 4.32b illustrates that there are two reflection troughs locating at $\lambda = 333$ and 500 nm. By comparing the spectra of lossless and lossy dielectrics, it can be seen that the energy loss mainly comes from metal, accompanying with enhanced fields localized at the patches [39]. If we hope energy is not absorbed by the electrons in metal, the horizontal coupling between adjacent patches should be suppressed thus electric fields are confined in the dielectric layer, which is the case at the wavelength of 375 nm. One additional benefit of operating at 375 nm is the interference patterns may possess better uniformity along the z-direction and higher contrast in the x-direction, as shown in the insets.

4.5.2   Interference of Circular Polarizations

The above catenary optical fields can be exploited for many important applications such as super-resolution lithography and spin-controlled photonics. First of all, since the MIM configuration is similar to the plasmonic cavity lens for nanolithography [14, 21], the intensity pattern can be directly recorded by the photoresist. In this case, the circular patches act as photomask while the dielectric acts as photoresist. Since the peak-to-peak distance between one pair of interference patterns is smaller than the diameter of the circular patch (90 nm in the simulations), the resolution of nanolithography is much smaller than the diffraction limit. Moreover, such structures can be easily fabricated using self-assembly of nanospheres [40].

Figure 4.33 shows the intensity and phase distribution for a metallic patches array. Under circularly polarized light (CPL) illumination at normal incidence, the E_z component forms ring patterns as a result of the symmetry of circular polarization. Note that the horizontal components E_x and E_y act as background for the ring-shaped patterns, which should be minimized by reducing the horizontal coupling between adjacent patches. The spiral phase shown in Fig. 4.33c indicates that there is a vertically polarized optical vortex induced by photonic spin–orbit interaction.

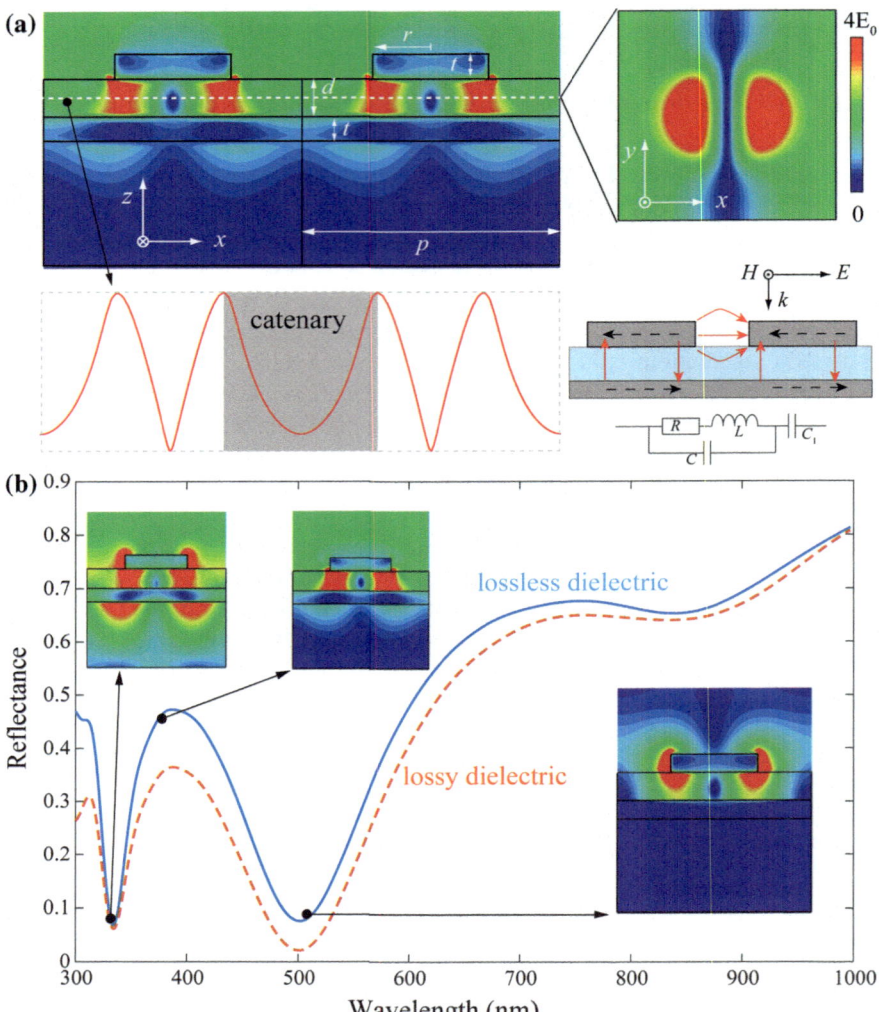

Fig. 4.32 Averaged electric field distribution induced by metallic patches backed by a reflective metallic layer. Incident light is polarized along the *x*-direction. **a** Fields in the *xz*- and *xy*-plane as well as that indicated by the dashed line ($\lambda = 375$ nm). The bottom right panel shows the equivalent circuit model composed of parallel and serial circuits. **b** Reflectance spectrum. The dashed curve corresponds to the case when a loss tangent of 0.05 is added into the dielectric. By comparing the two curves, it can be seen that the energy loss mainly occurs in the metallic parts. The inset shows the fields at corresponding wavelengths. Reproduced from [6] with permission. Copyright 2018, American Chemical Society

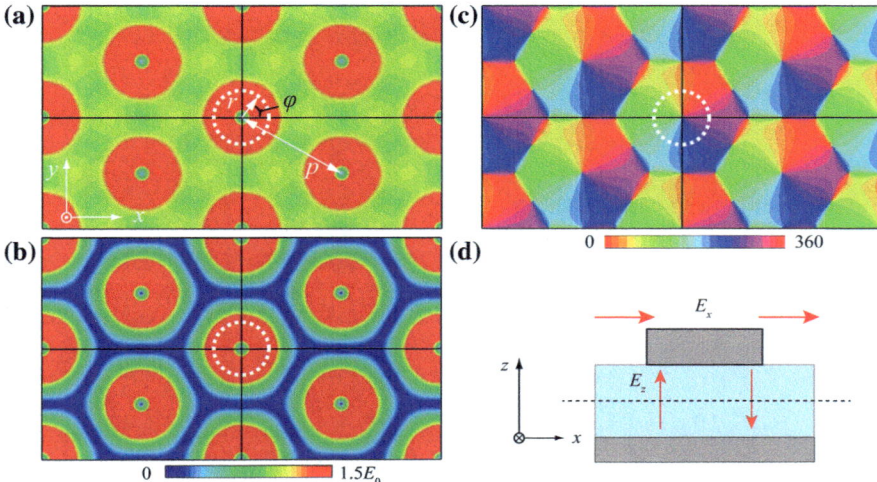

Fig. 4.33 Electric fields distribution in the middle of the photoresist layer under CPL illumination. **a** Averaged amplitude for the electric fields. The dashed circle indicates the circular patch. **b**, **c** Averaged amplitude and phase for E_z. **d** Schematic of the spin–orbit interaction. Note that there is always a phase shift of π between E_z at the two sides. The dashed line indicates the middle plane shown in (**a**)–(**c**). Reproduced from [6] with permission. Copyright 2018, American Chemical Society

However, different from the geometric phase resulting from polarization conversion from one circular state to its opposite one [41], the phase shift should reduce as $\Phi = \pm\varphi$ [42], where $\pm$ denotes LCP or RCP incidence, φ is the azimuthal angle. A simple explanation of the physical process is schematically illustrated in Fig. 4.33d: considering CPL as a rotating linearly polarized wave, at each instantaneous time, the phase of E_z at the two sides of a patch would have a shift of π, which means that the absolute value of phase shift is coincident with the azimuthal angle.

To further extend the concept, the interference of CPL is exploited to obtain more complex patterns. As shown in Fig. 4.34a, the coherent addition of CPL would induce space-variant linear polarization, leading to inhomogeneously oriented anisotropic patterns in the photoresist layer. As a result, this structure can convert polarization insensitive photoresist to be dependent on polarization, which is of particular importance in the recording of vectorial light fields. Strikingly, these anisotropic patterns can also generate geometric phase and redirect the incident CPL to one particular direction. Figure 4.34b, c illustrates the coherently added fields illuminating at $\pm10°$ within the yz-plane. Owing to the limited incidence angle, the asymmetric intensity distribution is not obvious, rendering a rotating dipole array with characteristic dimension smaller than 50 nm. Note that the horizontal electric fields shown in Fig. 4.34c are negligible because the horizontal coupling is small.

It is interesting to mention that the fields in Fig. 4.33 are similar to an optical skyrmion, which is a type of topological defect with promising applications in

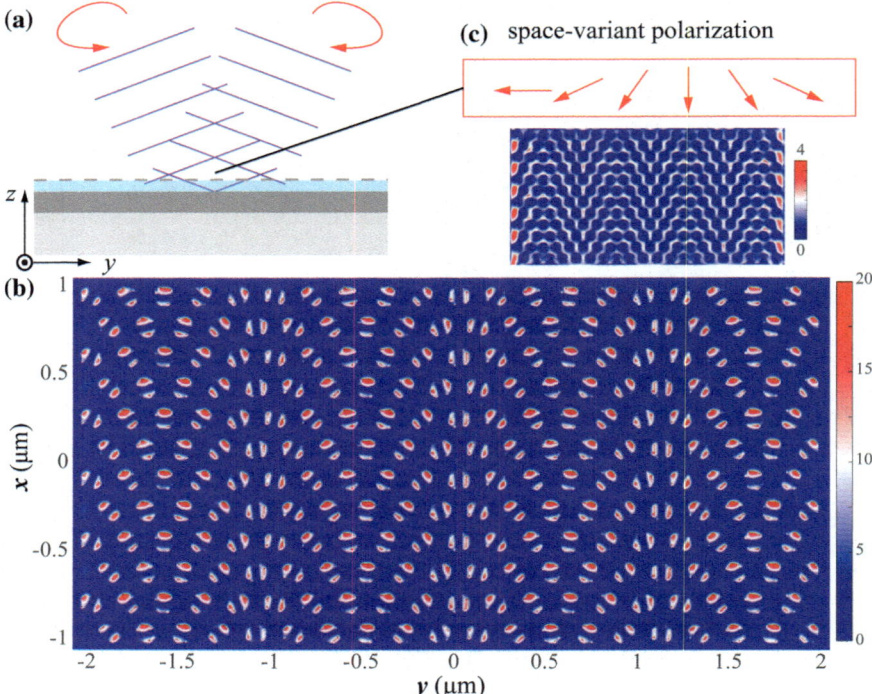

Fig. 4.34 Interference of CPL in the photoresist. **a** Schematic of the interference configuration. The space-variant polarization is shown in the right panel. **b** Averaged intensity in the middle of the photoresist layer calculated by $\left(|E_x|^2 + |E_y|^2 + |E_z|^2\right)/2$. **c** Averaged horizontal intensity calculated by $\left(|E_x|^2 + |E_y|^2\right)/2$. Since the fields were calculated by adding two separate results obtained in CST MWS, the color map is shown in a way different from previous figures. Reproduced from [6] with permission. Copyright 2018, American Chemical Society

magnetic storage and spintronics. It is shown that optical skyrmion lattices can be generated using evanescent electromagnetic fields such as SPPs [43]. Figure 4.35a shows the structure used to generate the SPPs, which consists of six gratings creating a hexagon on a 200 nm Au layer. The periodicity of the grating corresponds to the plasmonic wavelength (636 nm) and the bottom grating is displaced by half the plasmonic wavelength, to enable the a phase shift of π. The optical skyrmion lattice is created at the center of the slit, through the SPPs excited under circular polarization incidence. Figure 4.35b–e shows the axial and transverse electric fields distribution. The normalized three-dimensional electric field confirms the formation of a skyrmion lattice, i.e., each lattice site exhibits the distinct features of a Néel-type skyrmion, with their calculated skyrmion number being $S = 1$. Interestingly, if the single metal–dielectric surface is replaced by a MIM waveguide, the properties of skyrmion can be further modulated by the catenary plasmons.

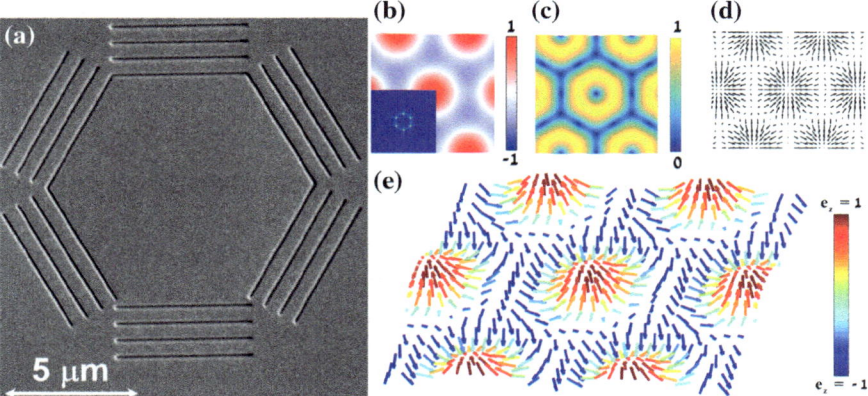

Fig. 4.35 Optical skyrmion lattice in evanescent electromagnetic fields. **a** SEM image of the slits array used to generate SPPs. **b** Axial (out-of-plane) electric field with the Fourier decomposition in the inset. **c** Amplitude of the transverse (in-plane) electric field. **d** Vector representation of the transverse electric field, showing polarization singularities at the center of each lattice site. **e** Vector representation of the local unit vector of the electric field (color coded for the value of its axial component), showing that each lattice site is a Néel-type skyrmion. Reproduced from [44] with permission. Copyright 2018, The Authors

4.6 Plasmonic Direct Writing Based on Catenary Plasmons

Direct writing is another corner stone of optical lithography since it is indispensable for the fabrication of photomasks. It has been shown that the catenary plasmons in MIM structures could greatly improve the performance of plasmonic direct writing. In the following, three kinds of structures are briefly discussed.

4.6.1 Metallic Tip

Inspired by the plasmonic reflection lithography [24], an improvement of apertureless tip is proposed by adding metal layer below the tip and photoresist [45]. The tip, photoresist, and metallic layer form a tip–insulator–metal (TIM) structure, in which a highly confined mode of plasmons both in transversal and longitude (z) directions occurs. Evanescent waves with large transversal spatial frequency components could be amplified and reduce the enlargement of light spot below the tip. In addition to the resolution improvement, the spot intensity inside photoresist along the z-direction becomes nearly uniformly distributed and results in an elongated depth of focus. Numerical simulations show that the thickness of photoresist plays a dominating role in determining the focusing spot size. Full width at half maximum (FWHM) of the spot size could reach sub-10 nm by optimizing geometrical parameters of

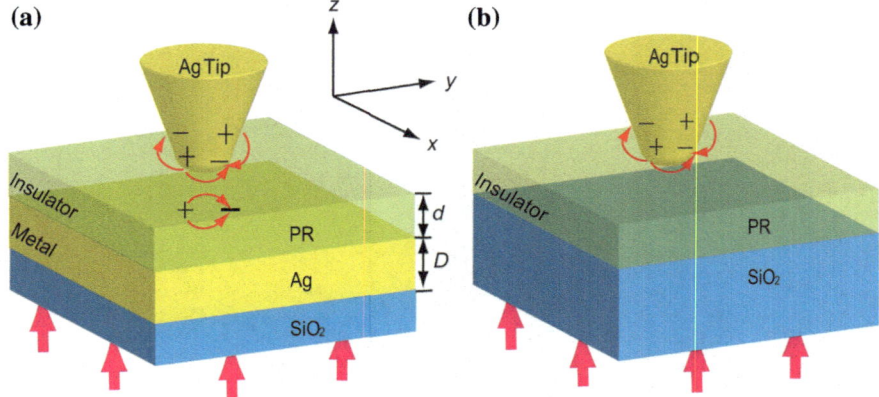

Fig. 4.36 **a** Schematic of the plasmon direct writing lithography with TIM structure. **b** A control case based on TI structure. Reproduced from [45] with permission. Copyright 2013, Springer Science + Business Media New York

the TIM structure. Moreover, circularly polarized light illumination in the normal direction beneath the metal layer delivers regular spot in circular shape.

The schematic of the TIM structure is shown in Fig. 4.36a, which consists of a silver sharp tip positioned above photoresist insulator film. Beneath the photoresist is silver metal layer deposited on the SiO_2 substrate. The distance between the tip and the surface of photoresist is assumed to be zero here, which could be realized in practice by using the tapping mode of an atomic force microscope. As a control case, the tip–insulator (TI) structure is presented in Fig. 4.36b by removing the metal film.

Monochromatic circularly polarized plane wave is incident beneath the substrate. The thickness of Ag film is assumed to be small enough for light transmission, thus in both structures, hot spots could be observed around the apex of metal tip. In some cases, this hot spot is attributed to the waveguiding effect [46], which concentrates a great number of charges around the tip's apex region and results in local field enhancement. In the static condition, the charges around apex could be viewed as a dipole.

Two major light concentration mechanisms would contribute to the formation of hot spot in Fig. 4.36a. The first mechanism is related to the localized surface plasmon resonance (LSPR) around the tip apex. To simplify the analysis, let us treat the tip apex as a nanoparticle with effective dipole moment $P = 4\pi\varepsilon_0\varepsilon_r\alpha E_0$, here α is the polarizability, and ε_r is the relative permittivity of the medium around. The polarizability of a small sphere of subwavelength diameter in the electrostatic approximation can be represented by equation $\alpha = 4\pi a^3(\varepsilon_m - \varepsilon_r)/(\varepsilon_m + 2\varepsilon_r)$, a and ε_m stand for the radius and relative permittivity of the particle. The polarizability experiences a great enhancement at resonance provided that the Fröhlich condition

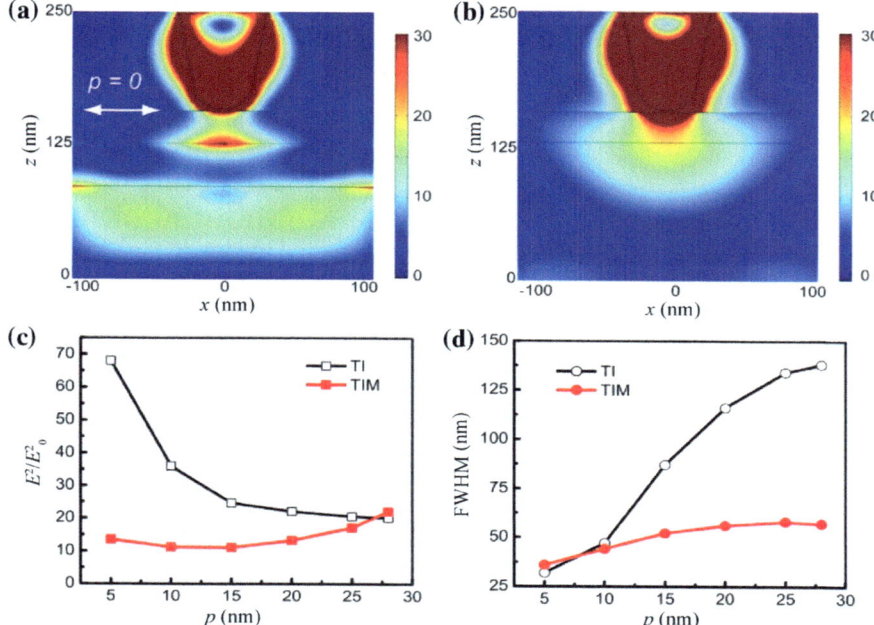

Fig. 4.37 Simulated electric field intensity distribution of **a** TIM and **b** TI structures. **c, d** Intensity and FWHM for different positions along the vertical direction in photoresist. The top surface of Pr is set as $p = 0$. The electric fields intensity in the TIM are featured by a catenary. Reproduced from [45] with permission. Copyright 2013, Springer Science + Business Media New York

$|\varepsilon_m + \varepsilon_r| \approx 0$ is valid. For the structure shown in Fig. 4.36, this condition could be approximately realized.

On the other hand, the additional metal layer would introduce another mechanism of surface plasmon resonance on the planar Ag film to reshape the hot spot. In this way, the SP resonance excitation for wide range of transversal wavevector occurs when $|\varepsilon_m + \varepsilon_r| \approx 0$. So the charges and light field at apex would be coupled and imaged near the metal film and help to form a narrower and elongated nano focus between the tip and metal film.

To illustrate the improvement of the hot spot associated to the TIM structure for nanolithography, 3D numerically simulation are performed using COMSOL Multiphysics 4.2a. The Ag tip apex radius is assumed to be 25 nm and shaped with a cone angle of 10°. The thickness of photoresist and the Ag film are $d = 30$ nm and $D = 40$ nm, respectively. At a wavelength of 365 nm, the relative permittivities of Ag, photoresist, and SiO$_2$ substrate are $\varepsilon_{Ag} = -2.4012 + 0.2448i$, $\varepsilon_{Pr} = 2.56$, $\varepsilon_{SiO_2} = 2.13$, respectively. As shown in Fig. 4.37, compared with the TI mode, the TIM structure has led to a catenary-shaped intensity distribution along the vertical direction, so the FWHM is suppressed in the entire range.

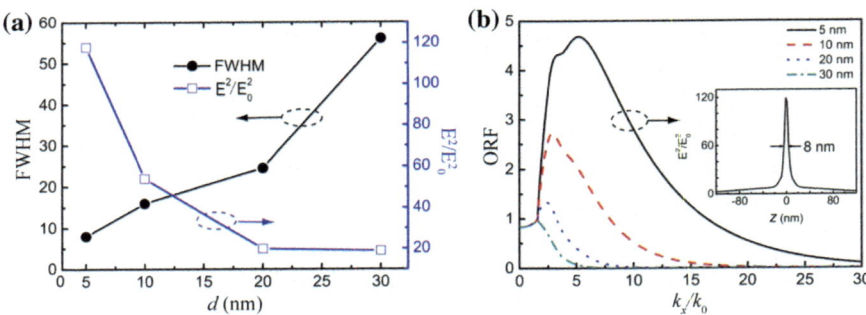

Fig. 4.38 **a** Simulated spot size (black line) and electric field enhancement (blue line) as a function of photoresist thickness d when the metal film is set as 15 nm. **b** ORF at different d. The inset shows the cross-section profile of the electric field distribution at 5 nm below the tip/Pr interface for the photoresist layer and the Ag film are both 5 nm. Reproduced from [45] with permission. Copyright 2013, Springer Science + Business Media New York

Thickness of the photoresist layer plays another key role in minimizing the spot size. As demonstrated in Fig. 4.38a, the spot size (FWHM) can be reduced by decreasing the thickness of Pr layer, and the corresponding electric field intensity increases sharply, which is caused by the enhanced resonant coupling between the tip and Ag film. This feature can also be seen from the optical reflection function (ORF) shown in Fig. 4.38b, which is obtained by calculating the reflection coefficient of different wave vector when the Pr thickness is set at 5, 10, 20, and 30 nm. It is shown that as the thickness of the Pr layer decreases, the ORF increases over a wide range of k_x. A wider spatial spectrum means a stronger resonant coupling between the metallic tip and Ag film, which results in a stronger electric field enhancement at the hot spot (blue line in Fig. 4.38a). Meanwhile, as the ORF becomes wider, the waves of larger k_x can be coupled into the TIM cavity so that a higher resolution and a smaller hot spot is achieved (black line in Fig. 4.38a). Especially, the optimal coupling occurs when the thickness of Pr layer is decreased to about 5 nm and thickness of Ag film is 5 nm as well. In such case, the waves of k_x smaller than $25k_0$ can be coupled into the TIM cavity efficiently (the black line in Fig. 4.38b). This implies that a hot spot size of 7.3 nm ($\lambda/50$) can be achieved, which is consistent with the simulation result of 8 nm as shown in the inset of Fig. 4.38b.

4.6.2   Bowtie-Shaped Nanoapertures

Besides metallic tip, bowtie-shaped aperture is also widely utilized to realize direct writing. Figure 4.39 shows the bowtie–metal–insulator–metal (BMIM) and bowtie–insulator (BI) structures [47]. A plane wave linearly polarized perpendicular to the aperture gap at $\lambda = 365$ nm irradiates the mask normally, which would induce the redistribution of the free electrons in the Al mask and excite local surface

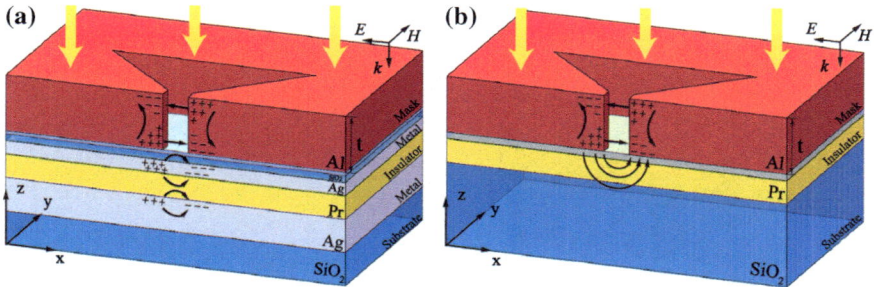

Fig. 4.39 Schematic configurations of the **a** BMIM and **b** BI nanolithography. Reproduced from [47] with permission. Copyright 2015, Springer Science + Business Media New York

plasmons. For the BI configuration, the mask and the Pr are separated by a 5-nm air gap. For the BMIM configuration, the recording structure is composed of 20-nm Ag superlens, 30-nm Pr layer, and 50-nm Ag cladding layer in the bottom. The Ag superlens amplifies evanescent waves scattered by the bowtie aperture, while the Ag cladding acts as a plasmonic mirror which could modulate the distribution of the electric field intensity in the Pr layer. Therefore, a strong resonance is formed in the silver-Pr-silver (Ag–Pr–Ag) structure, which gives rise to the enhancement of electromagnetic components. Note that there is a 10-nm SiO_2 film on the Ag superlens, which is used to protect the superlens from oxidation or scuffing.

As shown in Fig. 4.40, the electric field intensity and FWHM of the spots in two schemes show different behaviors inside the Pr layer. For the BI structure, the spot size extends from ~45 nm on the top surface to ~125 nm on the bottom surface along z-direction. Meanwhile, the electric field intensity decreases exponentially, which inevitably causes the spot with a shallow profile. On the contrary, in the BMIM structure, FWHM on the cross-section profile of xz-plane keeps ~30 nm on the top of the Pr layer and reaches a minimum value of 28 nm at the central plane, which is reduced by 67% compared with the value of 85 nm at the same position in BI structure. Moreover, the electric field intensity in BMIM structure along the z-direction is enhanced at least 10 times than that in the BI structure. This is crucial for the elongation of the aspect profile for the plasmonic lithography with bowtie aperture. However, FWHM on the cross-section profile of the yz-plane (inset of Fig. 4.40f) is hardly changed on the top surface of the Pr due to the lack of field confinement created by the aperture along the y-direction.

Since the size of the hot spots in the image plane is not identical on the cross-section profile of xz-plane and yz-plane, the scanning direction along the x-axis (as shown in Fig. 4.41a) during the exposure process would achieve higher resolution for the BMIM structure. By rotating both the bowtie structure and incidence polarization, higher resolution could be obtained (Fig. 4.41a, b). In order to evaluate the depth profile of the near-field recording in a BMIM nanolithography model, we refer to the theoretical near-field exposure model proposed in Ref. [48]. In the referenced model, the intensity distribution away from the exit plane of the bowtie aperture is described

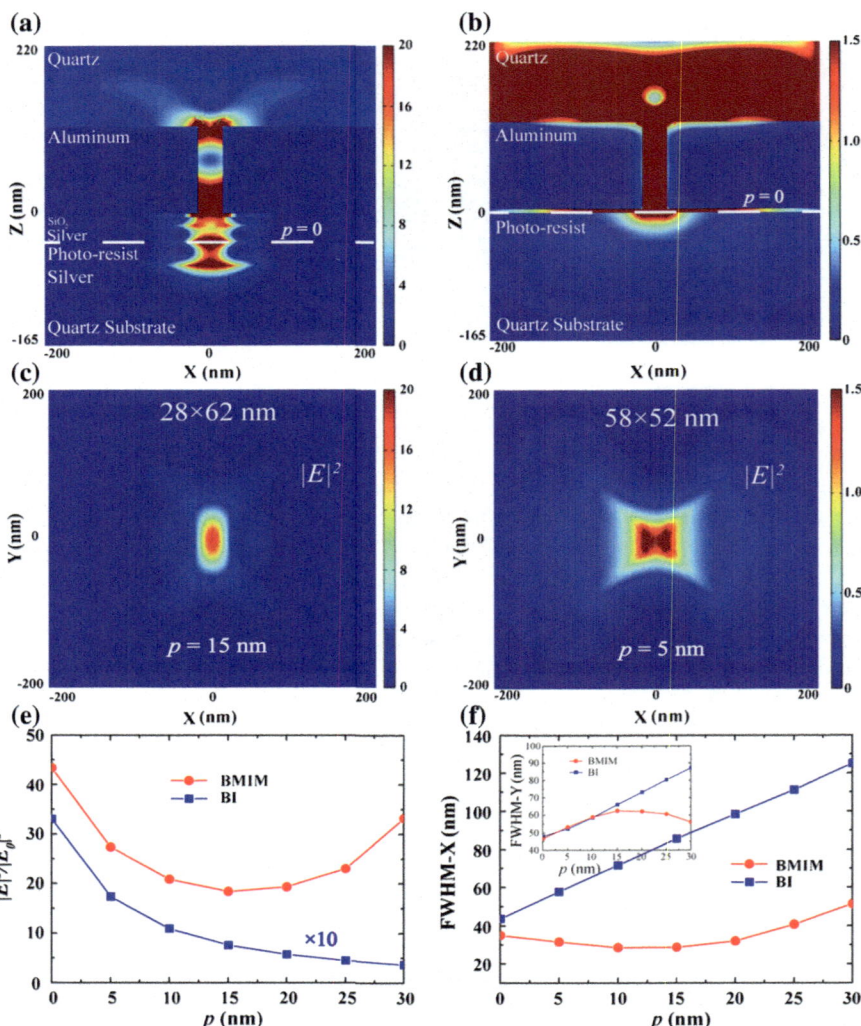

Fig. 4.40 Cross-section profiles of simulated electric field intensity distribution of **a** BMIM structure and **b** BI structure. Simulated electric field intensity distribution of **c** BMIM structure at 15 nm below the photoresist surface and **d** BI structure at 5 nm below the photoresist surface. **e** Intensity enhancement ratio and **f** FWHM of the hot spot for the two structure at different positions along the z-direction in the photoresist (the top surface of the photoresist is set as $p = 0$). The axis of FWHM-X and FWHM-Y (inset) corresponds to the FWHM on the cross-section profile of xz-plane and yz-plane, respectively. Reproduced from [47] with permission. Copyright 2015, Springer Science + Business Media New York

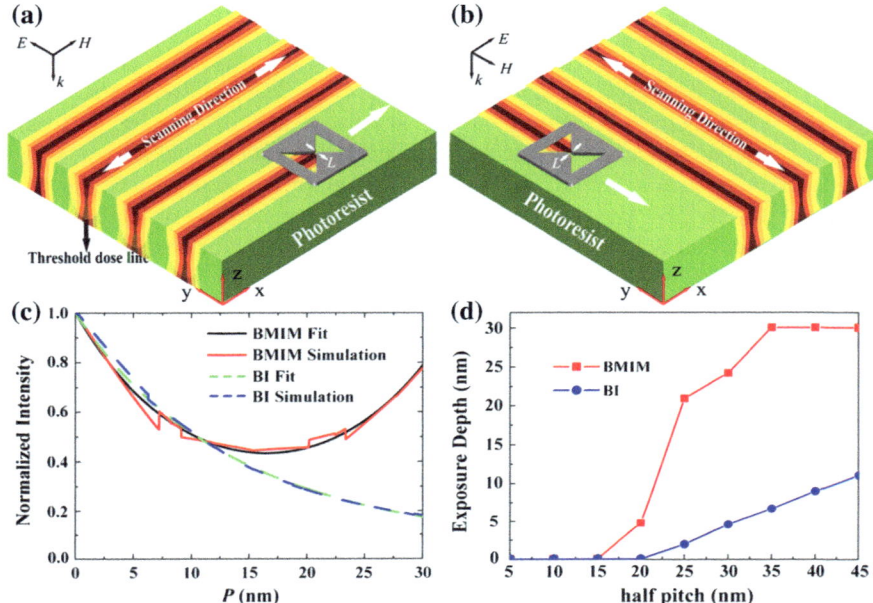

Fig. 4.41 Line array patterns recorded on the photoresist in **a** x-direction line scanning and **b** y-direction line scanning. **c** Normalized electric field intensity along z-direction in the photoresist in simulation and fitting results for BMIM and BI structure, respectively. The fitting equation for BI structure is $I(z) = I_i(1 + bz/a)^{-1/b}$ with $a = 14.054$ and $b = 0.217$. **d** Calculated exposure depth as functions of the half-pitch resolution of lithography patterns. Reproduced from [47] with permission. Copyright 2015, Springer Science + Business Media New York

using a hyperbola curve. This case is consistent with the BI structure, and the dashed lines in Fig. 4.41c correspond to the simulation (green) and fitting (blue) results. However, for BMIM nanolithography, the intensity distribution from the bowtie aperture exit would be modified due to the introduced Ag–Pr–Ag structure. The simulation results denoted by the red solid line show that the intensity distribution in the Pr layer is an even function which is attributed to multiple reflection of Ag–Pr–Ag structure. The sum of two simple exponential decay functions would be employed to approximately fit the intensity distribution. The black solid line in Fig. 4.41c represents the fitting data by an asymmetric equation $I(z) = I_i[a \exp(-bz) + (1 - a)\exp(bz)]$, with $a = 0.951$ and $b = 0.089$. According to the model with the catenary intensity distribution, the pattern profile depth for BMIM nanolithography model could be obtained as shown in Fig. 4.41d. In the case of BMIM nanolithography with the 30-nm ridge gap of bowtie aperture, the exposure depth is remarkably improved and the resolution limit reaches 15 nm. In addition, the exposure depth would reach 30 nm with a 35-nm half-pitch lithography pattern.

4.6.3 Virtual Scanning Tip

In this section, a novel evanescent Bessel beam with sub-diffraction characteristic is generated by utilizing HMM composed of alternative metal and dielectric layers, along with the concentric metasurface and plasmonic cavity lens in the form of Ag-Photoresist-Ag structure [36]. This method is based on the launch of high spatial frequency BPP modes, the spatial frequency filtering characteristic of HMM as well as the vectorial fields modulation ability provided by the plasmonic cavity lens. Numerical simulations indicate that an evanescent Bessel beam with a central spot size of 62 nm ($0.17\lambda_0$) could be maintained for a distance as large as 100 nm, and the experiment results also verify that the focusing spot could be compressed to about 65 nm beyond the diffraction limit. Compared with previous approaches, the method is featured by greatly increased working distance, thus may find potential applications in plasmonic direct writing and high-density optical storage, etc.

Figure 4.42 is a schematic configuration to obtain deep-subwavelength evanescent Bessel beam by two structures. One is the Bessel beam generating structure composed of the HMM and the concentric annular grating, which behaves as a metasurface to couple the propagating light to evanescent waves. The other is the photoresist recording structure in the form of a plasmonic cavity lens. The abovementioned structures are separated by a spacer layer, whose thickness could be set as the diffraction-free distance of the evanescent Bessel beam in experiments. Circularly polarized plane wave with a wavelength of 365 nm impinges on the 40-nm-thick Cr grating from the quartz substrate. The concentric annular grating serves as a generator of high spatial frequency evanescent waves, which excites multiple orders of evanescent waves with a radial wavevector of $k_r = nk_0 \sin(\theta_i) + 2m\pi/p$ in all azimuthal directions, where k_0 and θ_i are the incident wavevector in vacuum and incident angle, respectively; n is the refractive index of substrate; p is the period of grating and m is an integral diffraction order. In the case of normal illumination, k_r is reduced as $2\pi m/p$, and it means that the diffraction wavevector only depends on the diffraction order and the period of grating. Thus, six-ring annular patterns with the innermost ring diameter of $w = 100$ nm, the slit width of $d = 61$ nm, and the period of $p = 122$ nm would generate diffraction waves with the radial wavevector of $k_r = 3mk_0$. The Cr grating is planarized by the adhesive with a thickness of 30 nm. Adjacent to the planarization layer is the HMM, which is composed of two layers of Al_2O_3 films centrally embedded in three Al films (the thickness is 15 nm for each layer). The Pr recording structure is composed of 20-nm-thick Ag, 30-nm-thick Pr, and 70-nm-thick Ag. The values of the diameter of the innermost ring and the thickness of metal and dielectric film are determined by optimization.

It is well known that Bessel beam is described by the superposition of a set of plane waves with the wavevectors lying on the surface of cone in the Fourier space [49, 50]. The diffraction-free characteristic of Bessel beam emanates from the intensity distribution in the plane normal to z-axis proportional to $J_0^2(k_r r)$, where J_0 is the zero-order Bessel function of the first kind. In principle, the FWHM of Bessel beam is only decided by the spatial wavevector k_r. Compared with common SP modes, the

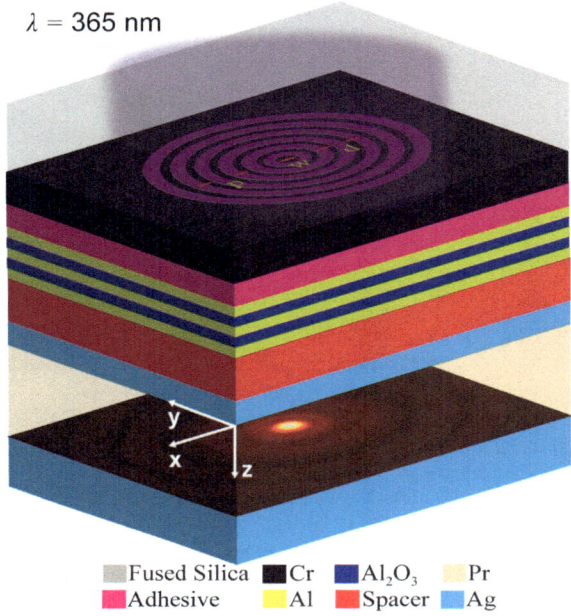

$\lambda = 365$ nm

Fused Silica Cr Al$_2$O$_3$ Pr
Adhesive Al Spacer Ag

Fig. 4.42 Schematic of the Al/Al$_2$O$_3$ hyperbolic metamaterial and Ag/Pr/Ag plasmonic cavity lens for the evanescent Bessel beam generation under circular polarization illumination. Reproduced from [36] with permission. Copyright 2017, The Royal Society of Chemistry

BPP modes generated by HMM could provide higher spatial frequency. The HMM based on the Al/Al$_2$O$_3$ multilayer structure acts like a homogeneous electromagnetic medium with a highly anisotropic, hyperbolic spatial frequency dispersion. Figure 4.43a is the dispersion relation of the multilayer structure, which represents the relationship between tangential wavevector k_x and longitudinal wavevector k_z. The real part (black curve) of k_z exhibits a hyperbolic profile with no cutoff spatial frequency, thus evanescent waves with infinitely large wavevector could be launched when ignoring the light absorption. However, due to the quick growth of the imaginary part of k_z, the ultrahigh spatial frequencies could not be supported, which makes the fact that the light modes with a specified k_x range of spatial frequencies would go through the metamaterial. Also, it clearly shows that a filtering window with low absorption for the specific BPP waves around the wavevector $3k_0$ is created through the multilayer metamaterial, resulting in that the BPP wave with the wavevector $3k_0$ could be coupled out the metamaterial and other diffraction waves outside the window range are damped. As expected, the calculated two-dimensional OTF for the proposed multilayer structure exhibits a filtering window around the wavevector $3k_0$ as shown in Fig. 4.43b.

Higher radial wavevector promises a narrower FWHM of Bessel beam. Thus, further analysis is performed to tune the filtering window of multilayer structure

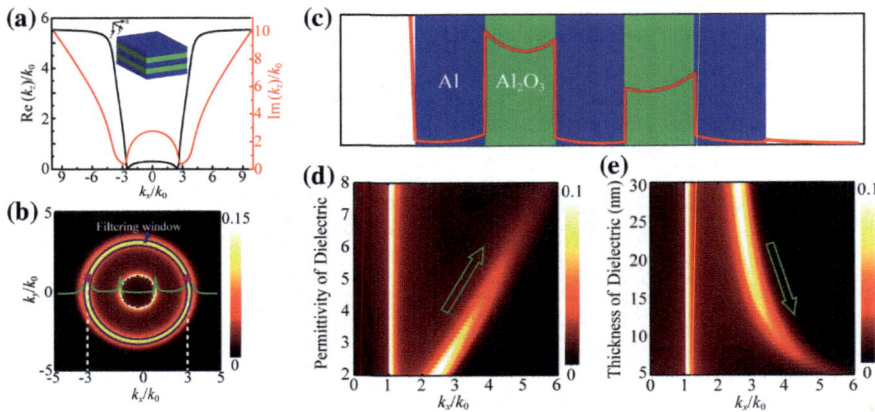

Fig. 4.43 **a** Calculated real (black curve) and imaginary (red curve) part of longitudinal wavevector k_z for variant tangential wavevector k_x in the proposed Al/Al$_2$O$_3$ metamaterial (Inset in **a**) at the incidence light wavelength of 365 nm. **b** Two-dimensional optical transmission function (OTF) versus the tangential wavevector k_x and k_y of HMM. The olive curve in (**b**) is the OTF distribution along the position of $k_y = 0$. The stars in (**b**) correspond to the center position of specific BPPs coupling window in multilayer metamaterial. **c** Catenary-shaped electric field intensity distribution (red curve) inside multilayer structure at $3k_0$. **d**, **e** Calculated OTF distributions versus **d** permittivity and **e** thickness of dielectric layer in the multilayer structure. When tuning the permittivity and thickness, other parameters are kept unchanged. Reproduced from [36] with permission. Copyright 2017, The Royal Society of Chemistry

towards higher spatial frequency. For the multilayer structure, the magnitude of electric field intensity in the dielectric layer is much higher than that in the metal layer. When the resonance effect of BPP wave occurs, the electric fields mainly concentrate in the dielectric layer, which is shown in Fig. 4.43c. This phenomenon has been testified by the mode analyses in the MIM structure, which presents the high confining capability of waveguide mode for electromagnetic energy.

The simulation is rendered by calculating the OTF distribution of multilayer versus the permittivity and thickness of dielectric layer, which shows that increasing the permittivity or decreasing the thickness could push the center wavevector of filtering window towards higher spatial frequency. The polarization of incidence light plays another crucial role in the generation of deep-subwavelength evanescent Bessel beam, which could be clarified by analytically calculating the field distribution after passing the HMM. If the illumination is circularly polarized, the components in Cartesian coordinate system are expressed as

$$
\begin{aligned}
E_x &= -\frac{1}{2}i|t(k_r)|k_z\big[\exp(i2\varphi)J_2(k_rr) - J_0(k_rr)\big]\exp(ik_zz) \\
E_y &= -\frac{1}{2}|t(k_r)|k_z\big[\exp(i2\varphi)J_2(k_rr) + J_0(k_rr)\big]\exp(ik_zz) \\
E_z &= -|t(k_r)|k_r\exp(i\varphi)J_1(k_rr)\exp(ik_zz)
\end{aligned}
\tag{4.6.1}
$$

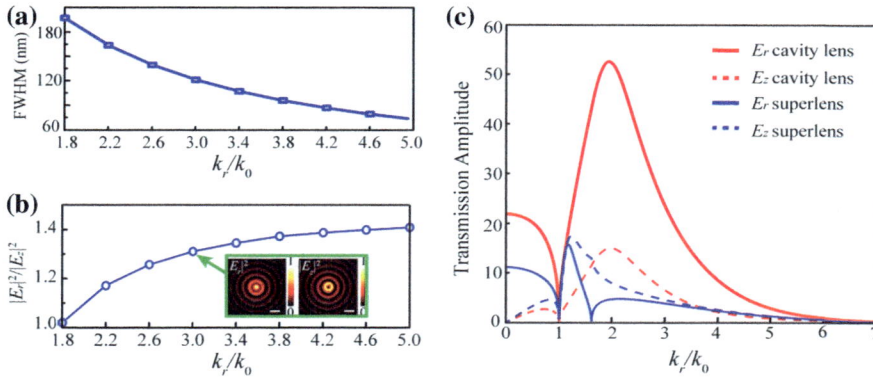

Fig. 4.44 **a** FWHM of the intensity profile of evanescent Bessel beam. **b** Ratio of electric intensity components as a function of the radial wavevector k_r under circularly polarized illumination. The insets in (**b**) are the distribution of $|E_r|^2$ and $|E_z|^2$ with $k_r = 3k_0$. The scale bar in insets is 100 nm. **c** Transmission amplitude of electric field components for the transverse wave vector through the structure of Ag–Pr–Ag (cavity lens) and Ag–Pr (superlens) in the Pr region. Reproduced from [36] with permission. Copyright 2017, The Royal Society of Chemistry

where r, φ, and z are cylindrical coordinates, k_r and k_z is the radial and longitudinal wavevector with $k_r^2 + k_z^2 = k_0^2$, J_m is the mth-order Bessel function of the first kind, and $t(k_r)$ is the transmission coefficient of diffraction wave.

According to Eq. (4.6.1), Fig. 4.44a illustrates the relationship between the resolution of evanescent Bessel beam and the radial wavevector k_r under circular polarization illumination. As expected, the FWHM of beam could be significantly reduced by increasing the radial wavevector. At the same time, the vectorial electric field components have different contributions to the evanescent Bessel beam. The radial component intensity $|E_r|^2$, defined as $|E_r|^2 = |E_x|^2 + |E_y|^2$, presents a spot with the zero-order Bessel function, while the longitudinal component intensity $|E_z|^2$ gives negative contribution with a doughnut distribution. For the radial component E_r, these two BPP waves would interfere constructively at the center because all BPP waves emerging from the grating arrive at the center with the same amplitudes and phases. However, the focusing of longitudinal components E_z would lead to a destructive interference and a null at the optical axis (Fig. 4.44b). Since the magnitude $|E_z|^2$ is comparable to $|E_r|^2$, the overall spot is greatly extended.

Fortunately, the transmission amplitudes of E_r and E_z could be effectively modulated in a broad spatial spectra by the Ag cladding, as illustrated in Fig. 4.44c. When an Ag cladding is introduced, the reflection effect of Ag cladding would modulate the electric field components and make the electric field component E_r to be dominant in the Pr region. As a result, the transmission amplitude of electric field component E_r is about 3.6 times than that of electric field component E_z at $k_r = 3k_0$. Figure 4.44c also shows that with traditional superlens structure, i.e., single Ag layer deposited on a Pr layer, the electric amplitude of E_r and E_z is approximately equal.

Figure 4.45a is a detailed comparison of the beam propagating properties for three different configurations. The top panel is the near-field diffraction limit case for the metallic slab (Cr film) with 60 nm diameter hole, which shows that the FWHM of a normal light beam quickly increases after leaving the metallic slab. The stray light field around the main spot is caused by periodic boundary conditions in both the x and y directions used in simulation. The simulated results of Bessel-BPPs focusing structure without cavity lens in the middle panel show that there is a hollow pattern existing in the main lobe of the transverse profiles, confirming that the longitudinal component $|E_z|^2$ gives more negative contribution as the distance in the z-direction becomes larger. At the same time, the intensity of side lobe increases with distance due to the interference of other diffraction waves outside the filtering window ($k_r = 3k_0$), especially the interference of the propagation wave components. When the plasmonic cavity lens is introduced, the hollow around the focus disappears, the side lobe is further suppressed, and the transverse profile of beam is remarkably compressed to about 62 nm owing to the modulation effect of electric field components and the ability to transfer the evanescent waves in a wider wavevector range compared with the Bessel-BPPs focusing structure with Pr medium.

Figure 4.45b shows the FWHM of the transverse profiles for different propagating distances along the z-axis for the above three excitation structures. In the Bessel-BPPs focusing structure without cavity lens, the transverse intensity profile of beam after the HMM remains unchanged over a small distance in near field region (distance from 0 to 20 nm), indicating the diffraction-free property of evanescent Bessel beam. However, the FWHM of diffraction-free beam is about 120 nm, which is consistent with the theoretical results. For Bessel-BPPs focusing structure incorporating the plasmonic cavity lens, the FWHM of the central spot of evanescent Bessel beam holds 62 nm in the distance of 0–100 nm. In the control case of a transparent hole with diameter 60 nm in the Cr film, the FWHM of the transverse profiles after the film rapidly increases and reaches to 350 nm at the distance of 100 nm, which is expanded about 5.7-folds than the case with plasmonic cavity lens. This result testifies the remarkable advantage of evanescent Bessel beam as a virtual tip in near-field probing over the near-field scanning microscopes. Additionally, Fig. 4.45c shows the exponential trend of the electric field intensity over the distance. Although the HMM would lead to relatively high absorption, the plasmonic cavity lens presents higher intensity due to the cavity resonance effect.

The recording patterns in Pr layer with the optimum exposure dose at different working distances in lithography processing are shown in Fig. 4.46. The experimental results with the working distance of 0 nm exhibit a spot size of 67 nm. When the distance is elongated to 40 nm, the size of spot recorded reaches 65 nm. Figure 4.46d indicates that even the distance arises to 80 nm, the spot could maintain the size of about 70 nm. Within the processing tolerant error, it can be concluded that the experimental results have successfully demonstrated the diffraction-free characteristic of the evanescent Bessel beam in the near field.

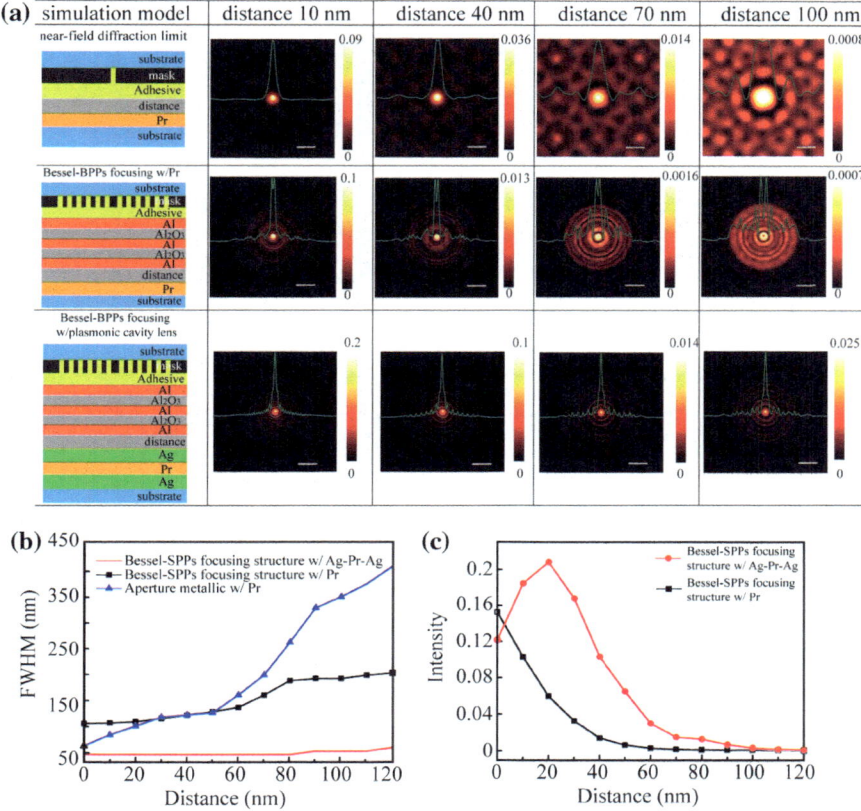

Fig. 4.45 a Cross-section of electric field intensity distributions on the plane, 15 nm away from the Pr surface for different distance. The configurations correspond to the near-field diffraction limit for a metallic slab with a 60 nm diameter hole; Bessel-BPPs focusing structure with Pr recording layer and Bessel-BPPs focusing structure with plasmonic cavity recording structure, respectively. The scale bar is 300 nm. **b** FWHM of the transmitted beams for different configurations versus different distances. **c** Total field intensity $|E|^2$ as a function of the distance along the z-axis. Reproduced from [36] with permission. Copyright 2017, The Royal Society of Chemistry

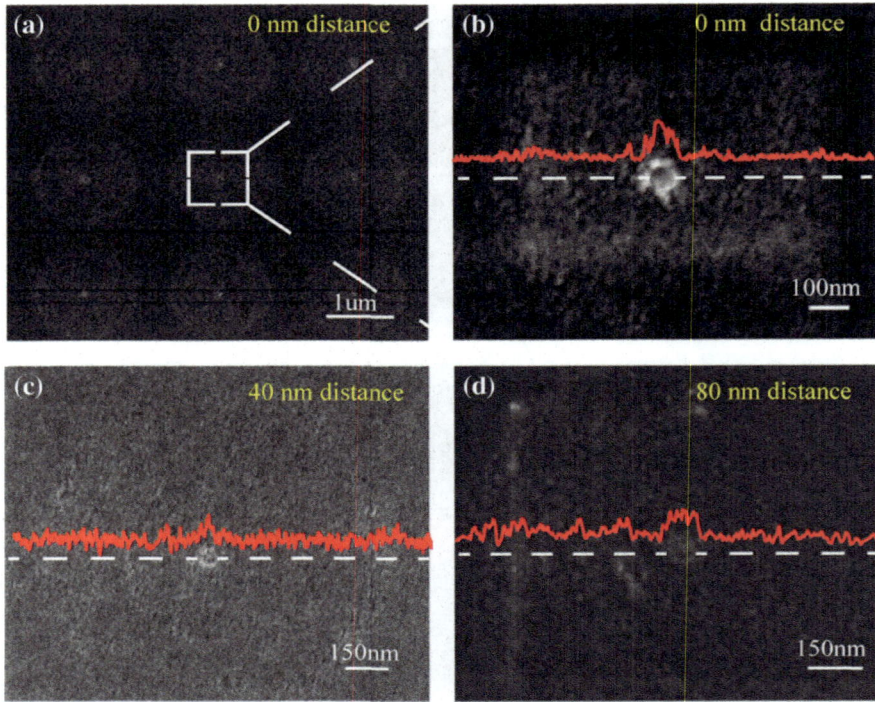

Fig. 4.46 **a** The lithography result by Bessel-BPPs focusing structure combined with plasmonic cavity lens at the distance of 0 nm. **b** The magnified SEM picture in (**a**). **c** and **d** Lithography results by Bessel-BPPs focusing structure combined with plasmonic cavity lens at the distance of 40 nm and 80 nm, respectively. The red curves in (**b**), (**c**), and (**d**) represent the corresponding SEM image gray values along the white dashed lines. Reproduced from [36] with permission. Copyright 2017, The Royal Society of Chemistry

References

1. P. Yeh, *Optical Waves in Layered Media*, 2nd edn. (Wiley, Hoboken, 2005)
2. X. Luo, Principles of electromagnetic waves in metasurfaces. Sci. China Phys. Mech. Astron. **58**, 594201 (2015)
3. X. Luo, M. Pu, X. Ma, X. Li, Taming the electromagnetic boundaries via metasurfaces: from theory and fabrication to functional devices. Int. J. Antennas Propag. **2015**, 204127 (2015)
4. G. Ren, C. Wang, G. Yi, X. Tao, X. Luo, Subwavelength demagnification imaging and lithography using hyperlens with a plasmonic reflector layer. Plasmonics **8**, 1065–1072 (2013)
5. X. Wu, C. Hu, M. Wang, M. Pu, X. Luo, Realization of low-scattering metamaterial shell based on cylindrical wave expanding theory. Opt. Express **23**, 10396–10403 (2015)
6. M. Pu, Y. Guo, X. Li, X. Ma, X. Luo, Revisitation of extraordinary Young's interference: from catenary optical fields to spin-orbit interaction in metasurfaces. ACS Photonics **5**, 3198–3204 (2018)
7. S.A. Maier, *Plasmonics: Fundamentals and Applications* (Springer Science & Business Media, Berlin, 2007)
8. J.B. Pendry, Negative refraction makes a perfect lens. Phys. Rev. Lett. **85**, 3966–3969 (2000)
9. B. Wood, J.B. Pendry, D.P. Tsai, Directed subwavelength imaging using a layered metal-dielectric system. Phys. Rev. B **74**, 115116 (2006)
10. S.A. Ramakrishna, T.M. Grzegorczyk, *Physics and Applications of Negative Refractive Index Materials* (CRC Press, Boca Raton, 2009)
11. X. Luo, T. Ishihara, Subwavelength photolithography based on surface-plasmon polariton resonance. Opt. Express **12**, 3055–3065 (2004)
12. X. Luo, T. Ishihara, Surface plasmon resonant interference nanolithography technique. Appl. Phys. Lett. **84**, 4780–4782 (2004)
13. R.P. Crease, The most beautiful experiment. Phys. World **15**, 19 (2002)
14. P. Gao, N. Yao, C. Wang, Z. Zhao, Y. Luo, Y. Wang, G. Gao, K. Liu, C. Zhao, X. Luo, Enhancing aspect profile of half-pitch 32 nm and 22 nm lithography with plasmonic cavity lens. Appl. Phys. Lett. **106**, 093110 (2015)
15. L. Liu, Y. Luo, Z. Zhao, W. Zhang, G. Gao, B. Zeng, C. Wang, X. Luo, Large area and deep sub-wavelength interference lithography employing odd surface plasmon modes. Sci. Rep. **6**, 30450 (2016)
16. N. Fang, H. Lee, C. Sun, X. Zhang, Sub-diffraction-limited optical imaging with a silver superlens. Science **308**, 534–537 (2005)
17. H. Lee, Y. Xiong, N. Fang, W. Srituravanich, S. Durant, M. Ambati, C. Sun, X. Zhang, Realization of optical superlens imaging below the diffraction limit. New J. Phys. **7**, 255 (2005)
18. D. Melville, R. Blaikie, Super-resolution imaging through a planar silver layer. Opt. Express **13**, 2127–2134 (2005)
19. P. Chaturvedi, W. Wu, V.J. Logeeswaran, Z. Yu, M.S. Islam, S.Y. Wang, R.S. Williams, N.X. Fang, A smooth optical superlens. Appl. Phys. Lett. **96**, 043102 (2010)
20. T. Taubner, D. Korobkin, Y. Urzhumov, G. Shvets, R. Hillenbrand, Near-field microscopy through a SiC superlens. Science **313**, 1595 (2006)
21. X. Luo, Plasmonic metalens for nanofabrication. Natl. Sci. Rev. **5**, 137–138 (2018)
22. D.B. Shao, S.C. Chen, Surface-plasmon-assisted nanoscale photolithography by polarized light. Appl. Phys. Lett. **86**, 253107 (2005)
23. T. Xu, L. Fang, J. Ma, B. Zeng, Y. Liu, J. Cui, C. Wang, Q. Feng, X. Luo, Localizing surface plasmons with a metal-cladding superlens for projecting deep-subwavelength patterns. Appl. Phys. B **97**, 175–179 (2009)
24. C. Wang, P. Gao, Z. Zhao, N. Yao, Y. Wang, L. Liu, K. Liu, X. Luo, Deep sub-wavelength imaging lithography by a reflective plasmonic slab. Opt. Express **21**, 20683–20691 (2013)
25. Z. Zhao, Y. Luo, N. Yao, W. Zhang, C. Wang, P. Gao, C. Zhao, M. Pu, X. Luo, Modeling and experimental study of plasmonic lens imaging with resolution enhanced methods. Opt. Express **24**, 27115–27126 (2016)

26. C. Wang, Y. Zhao, D. Gan, C. Du, X. Luo, Subwavelength imaging with anisotropic structure comprising alternately layered metal and dielectric films. Opt. Express **16**, 4217–4227 (2008)
27. L. Liu, X. Zhang, Z. Zhao, M. Pu, P. Gao, Y. Luo, J. Jin, C. Wang, X. Luo, Batch fabrication of metasurface holograms enabled by plasmonic cavity lithography. Adv. Opt. Mater. **5**, 1700429 (2017)
28. C. Wang, P. Gao, X. Tao, Z. Zhao, M. Pu, P. Chen, X. Luo, Far field observation and theoretical analyses of light directional imaging in metamaterial with stacked metal-dielectric films. Appl. Phys. Lett. **103**, 031911 (2013)
29. Z. Liu, H. Lee, Y. Xiong, C. Sun, X. Zhang, Far-field optical hyperlens magnifying sub-diffraction-limited objects. Science **315**, 1686–1686 (2007)
30. L. Liu, K. Liu, Z. Zhao, C. Wang, P. Gao, X. Luo, Sub-diffraction demagnification imaging lithography by hyperlens with plasmonic reflector layer. RSC Adv. **6**, 95973–95978 (2016)
31. M. Hirano, M. Aniya, A rational explanation of cross-profile morphology for glacial valleys and of glacial valley development. Earth Surf. Process. Landf. **13**, 707–716 (1988)
32. G. Liang, C. Wang, Z. Zhao, Y. Wang, N. Yao, P. Gao, Y. Luo, G. Gao, Q. Zhao, X. Luo, Squeezing bulk plasmon polaritons through hyperbolic metamaterial for large area deep sub-wavelength interference lithography. Adv. Opt. Mater. **3**, 1248–1256 (2015)
33. H. Liu, Y. Luo, W. Kong, K. Liu, W. Du, C. Zhao, P. Gao, Z. Zhao, C. Wang, M. Pu, X. Luo, Large area deep subwavelength interference lithography with a 35 nm half-period based on bulk plasmon polaritons. Opt. Mater. Express **8**, 199–209 (2018)
34. H. Liu, W. Kong, K. Liu, C. Zhao, W. Du, C. Wang, L. Liu, P. Gao, M. Pu, X. Luo, Deep subwavelength interference lithography with tunable pattern period based on bulk plasmon polaritons. Opt. Express **25**, 20511–20521 (2017)
35. Z. Zhang, J. Luo, M. Song, H. Yu, Large-area, broadband and high-efficiency near-infrared linear polarization manipulating metasurface fabricated by orthogonal interference lithography. Appl. Phys. Lett. **107**, 241904 (2015)
36. L. Liu, P. Gao, K. Liu, W. Kong, Z. Zhao, M. Pu, C. Wang, X. Luo, Nanofocusing of circularly polarized Bessel-type plasmon polaritons with hyperbolic metamaterials. Mater. Horiz. **4**, 290–296 (2017)
37. M. Pu, P. Chen, Y. Wang, Z. Zhao, C. Huang, C. Wang, X. Ma, X. Luo, Anisotropic meta-mirror for achromatic electromagnetic polarization manipulation. Appl. Phys. Lett. **102**, 131906 (2013)
38. M. Pu, C. Hu, M. Wang, C. Huang, Z. Zhao, C. Wang, Q. Feng, X. Luo, Design principles for infrared wide-angle perfect absorber based on plasmonic structure. Opt. Express **19**, 17413–17420 (2011)
39. Q. Feng, M. Pu, C. Hu, X. Luo, Engineering the dispersion of metamaterial surface for broadband infrared absorption. Opt. Lett. **37**, 2133–2135 (2012)
40. E.S. Kim, Y.M. Kim, K.C. Choi, Surface plasmon-assisted nano-lithography with a perfect contact aluminum mask of a hexagonal dot array. Plasmonics **11**, 1337–1342 (2016)
41. M. Pu, X. Li, X. Ma, Y. Wang, Z. Zhao, C. Wang, C. Hu, P. Gao, C. Huang, H. Ren, X. Li, F. Qin, J. Yang, M. Gu, M. Hong, X. Luo, Catenary optics for achromatic generation of perfect optical angular momentum. Sci. Adv. **1**, e1500396 (2015)
42. Y. Guo, M. Pu, Z. Zhao, Y. Wang, J. Jin, P. Gao, X. Li, X. Ma, X. Luo, Merging geometric phase and plasmon retardation phase in continuously shaped metasurfaces for arbitrary orbital angular momentum generation. ACS Photonics **3**, 2022–2029 (2016)
43. S. Tsesses, E. Ostrovsky, K. Cohen, B. Gjonaj, N. Lindner, G. Bartal, Optical skyrmion lattice in evanescent electromagnetic fields. Science **361**, 993–996 (2018)
44. S. Tsesses, E. Ostrovsky, K. Cohen, B. Gjonaj, N. Lindner, G. Bartal, Optical skyrmion lattice in evanescent electromagnetic fields. Arxiv:1805.11839 (2018)
45. J. Zhou, C. Wang, Z. Zhao, Y. Wang, J. He, X. Tao, X. Luo, Design and theoretical analyses of tip–insulator–metal structure with bottom–up light illumination: formations of elongated symmetrical plasmonic hot spot at sub-10 nm resolution. Plasmonics **8**, 1073–1078 (2013)
46. M.I. Stockman, Nanofocusing of optical energy in tapered plasmonic waveguides. Phys. Rev. Lett. **93**, 137404 (2004)

47. Y. Wang, N. Yao, W. Zhang, J. He, C. Wang, Y. Wang, Z. Zhao, X. Luo, Forming sub-32-nm high-aspect plasmonic spot via bowtie aperture combined with metal-insulator-metal scheme. Plasmonics **10**, 1607–1613 (2015)
48. S. Kim, H. Jung, Y. Kim, J. Jiang, J.W. Hahn, Resolution limit in plasmonic lithography for practical applications beyond 2x-nm half pitch. Adv. Mater. **24**, OP337–OP344 (2012)
49. X. Li, M. Pu, Z. Zhao, X. Ma, J. Jin, Y. Wang, P. Gao, X. Luo, Catenary nanostructures as highly efficient and compact Bessel beam generators. Sci. Rep. **6**, 20524 (2016)
50. D. McGloin, K. Dholakia, Bessel beams: diffraction in a new light. Contemp. Phys. **46**, 15–28 (2005)

Chapter 5
Catenary Plasmons for Flat Lensing, Beam Deflecting, and Shaping

Abstract As discussed in the Chap. 4, surface plasmons are collective excitations of free electrons and photons. The electric force line of surface plasmons at a metal–dielectric surface follows the function defined by a catenary of equal strength. When surface plasmons in adjacent interfaces are coupled together, the evanescent tails would lead to catenary optical fields described by hyperbolic cosine and sine functions. These catenary optical fields help to increase the focal depth of surface plasmon imaging and nanolithography. Here, we show that another unique property of the plasmonic catenary fields can be used to locally modulate the phase retardation. Based on the Young's double slits interference with unequal widths, the plasmonic propagating phase shift is revealed, and various functional flat plasmonic devices are designed and experimentally demonstrated. Since the gradient phase shift could introduce an additional horizontal wavevector, the classic Snell's law has also been generalized. Besides propagating phase shift, this chapter also describes the geometric phase induced by the rotated plasmonic nanoslits. Owing to the anisotropic field distribution and dispersion described by two catenary functions, the transmission of both metallic grating and rectangular nanoapertures depend on the polarization of incident light. Consequently, under circularly polarized illumination (with a spin angular momentum of $\pm\hbar$ for each photon), a space-variant surface structure would generate a polarization-dependent phase retardation. This geometric phase has been investigated to realize both flat lens and spin-controlled beam shaping.

Keywords Catenary plasmon · Flat lens · Extraordinary Young's interference · Structural color · Holography

5.1 Young's Double Slits Interference with Unequal Widths

Young's double slits interference is a fundamental optical phenomenon in wave optics, which is often treated as a turning point from geometric optics to wave optics. It is selected as one of the most beautiful experiments in physics [1], owing to its universal existence in electromagnetic, mechanic, and matter waves. Although double slits interference has been intensively investigated for centuries, recent studies

© Springer Nature Singapore Pte Ltd. 2019

X. Luo, *Catenary Optics*, https://doi.org/10.1007/978-981-13-4818-1_5

in subwavelength optics and electromagnetics have revealed some new properties not discovered before.

In the subwavelength scale, i.e., when the width and distance between these slits are smaller than the wavelength, metallic double slits have been proved to be an excellent platform to investigate the subwavelength light–matter interaction [2], to reveal the physical mechanism of extraordinary optical transmission (EOT) through sub-wavelength holes arrays [3], and to analyze the near-field characteristics of surface plasmon polaritons (SPPs) [4]. In Chap. 4, it has already been shown that the interference period on a metallic screen may be much smaller than the value predicted by traditional theory [2, 5]. This extraordinary Young's double slits interference (EYI) serves as the basis for plasmonic sub-diffraction-limited nanolithography.

After the original discovery of EYI phenomenon on a periodic slits array, it was soon found that for a metallic double nanoslit with different widths, the Young's interference order could be shifted via changing the width of the two slits [6, 7]. One could even get a dark fringe at the perpendicular bisector of the two slits, which is contrary to the classic Young's interference pattern. This abnormal effect reveals that the width of metallic slits could result in a tunable phase change of electromagnetic waves, which forms the basis of a series of functional flat optical devices. In the following, we shall discuss the far-field and near-field EYI phenomena produced by double slits with unequal widths.

5.1.1 Far-Field EYI

Figure 5.1 illustrates the plasmonic Young's double slits interference with equal and unequal slit widths. When the two slits are identical, symmetry analysis shows that the interference pattern would be symmetric too. Under transverse magnetically (TM, or p-) polarized illumination, the summed magnetic fields in the normal direction is coherently enhanced because there is no difference in the optical paths between the two slits. However, when we change the slit widths as 100 and 25 nm, respectively, the interference peaks horizontally shift by about 6°, resulting in a dark interference fringe at the normal direction. This anomalous interference effect indicates that light passing through the two slits have a relative phase shift of π. Nevertheless, this seems to be contrary to classic optical theory, which assumes that the width of slits only changes the transmitted intensity, but not the phase shift. Apparently, the phase difference can only be induced by the different propagation characteristics of guided waves in the two slits. According to the relation $\Delta\varphi = \Delta n k_0 h$, the mode index should have a strong dependence on the slit width.

Recalling the guiding modes in the form of catenary plasmons in the metal–insulator–metal (MIM) waveguides, the effect of mode index β/k_0 of antisymmetric mode (defined by the symmetry of the parallel electric component) can be calculated from the following equation:

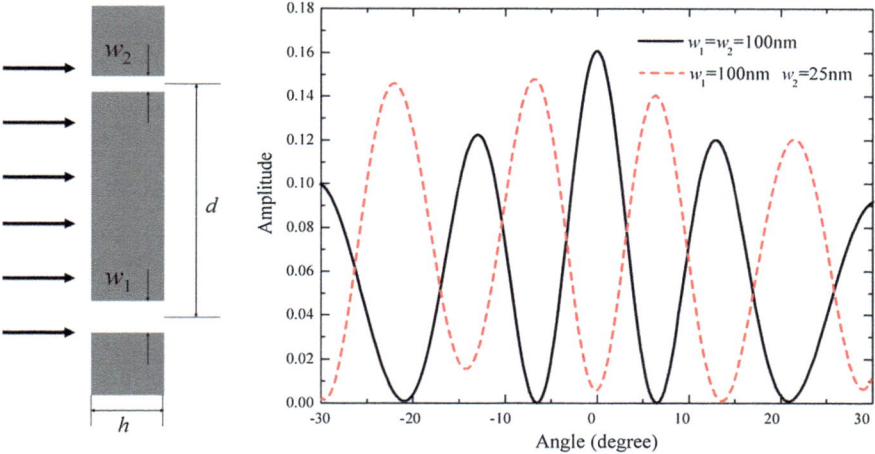

Fig. 5.1 Schematic of Young's interference effect with unequal slits widths. The right panel shows a comparison of the far-field diffraction pattern for the symmetric and asymmetric configurations. The thickness of the metal film is $h = 700$ nm and the distance between the two slits is $d = 2.6$ μm. Reprinted from [6] with permission. Copyright 2007, Optical Society of America

$$\tanh\left(\sqrt{\beta^2 - \varepsilon_d k_0^2}\, w/2\right) = -\frac{\varepsilon_d \sqrt{\beta^2 - \varepsilon_m k_0^2}}{\varepsilon_m \sqrt{\beta^2 - \varepsilon_d k_0^2}}, \tag{5.1.1}$$

where w is the width of the dielectric core, ε_d and ε_m are the permittivity of dielectric and metal, respectively. It can be easily seen that in this equation the propagation constant is directly related to the slit width. For extremely small width, the above equation can be approximated as

$$\left(\beta^2 - \varepsilon_d k_0^2\right) w/2 = -\frac{\varepsilon_d \sqrt{\beta^2 - \varepsilon_m k_0^2}}{\varepsilon_m}. \tag{5.1.2}$$

As we know, the propagation constant is much larger than the vacuum wavenumber for very small slit width. In this case, the following relation could be obtained:

$$\beta \approx \left|\frac{2\varepsilon_d}{\varepsilon_m w}\right|. \tag{5.1.3}$$

As discussed in Chap. 4, this width-dependent propagation constant can be understood using the catenary optical fields. When the width is smaller, the attenuation factor α of evanescent wave in the dielectric core could be larger. According to the dispersion relation, the propagation constant would increase correspondingly. The above EYI effect with two unequal slits is a direct result of the width-dependent propagation constants of catenary plasmons. Since the propagation constant varies

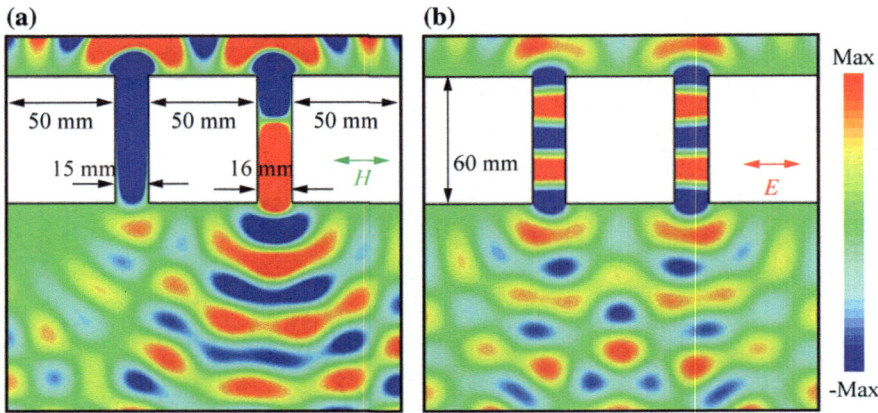

Fig. 5.2 Young's double slits interferences of microwaves with perfect electric conductors. **a** TE polarization. **b** TM polarization. Incident waves at a frequency of 10 GHz are launched from the top port. The widths of the two slits are 15 and 16 mm, with a center-to-center distance of 65.5 mm

quickly as the width changes, a small width difference may introduce a considerable phase difference.

In the microwave regime, high conductivity metals do not support SPP mode, thus TM-polarized wave in a parallel plate waveguide becomes transverse electromagnetic (TEM) polarization and the propagation constant is independent of the slit width. However, if the polarization is changed as transverse electric (TE, or *s*-) polarization, i.e., when the electric field is parallel to the boundary, the propagation constant has a strong width-dependence. As shown in Fig. 5.2, two slits with width (*w*) of 15 and 16 mm are perforated in a 60-mm-thick metal film. The center-to-center distance is 65.5 mm. In the numerical simulations, periodic boundary conditions are used with a periodicity of 181 mm. Using waveguide theory, it is easy to express the propagation constant of TE mode as

$$\beta = k_0\sqrt{1 - \left(\frac{\lambda}{2w}\right)^2}. \tag{5.1.4}$$

At the cutoff frequency, the propagation constant is zero thus no phase retardation occurs. When the frequency goes down further, the wave cannot propagate anymore. For instance, the propagation constants at 10 GHz ($\lambda = 30$ mm) are zero and $0.348k_0$ for $w = 15$ and 16 mm, corresponding to a relative phase difference of 4.37. Based on this principle, metallic lenses were proposed to construct large-aperture antennas in the microwave band [8].

The above microwave metallic lens works with the TE mode closing to the cutoff frequency, which means that the propagation constant is smaller than that in vacuum. To obtain a particular phase shift, rather long propagation distance larger than the wavelength is required, making such approach not practical in the high-frequency

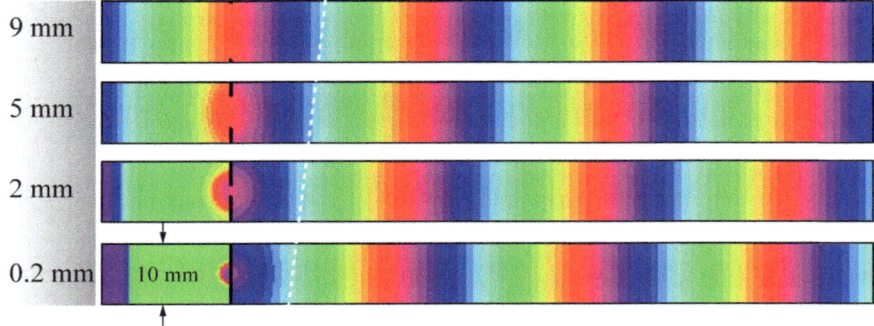

Fig. 5.3 Phase profile for a single slit with variable width. The period is 10 mm, the thickness of the sheet is 0.1 mm. The metal is assumed to be a perfect electric conductor. It can be seen that a smaller gap would produce a larger phase retardation, the effect of which is equal to let the wave propagate in a small distance. The white dashed line indicates the effective wavefront

band such as millimeter wave, infrared and visible range. Consequently, the plasmonic metal slits shown in Fig. 5.1 are completely different from the microwave metal lens and show great advantages such as lower profile and wider bandwidth.

Figure 5.3 shows another approach to modulate the phase shift with a single metallic slit perforated in an ultrathin metallic sheet. When the slit width is reduced from 9 to 0.2 mm, there is a transmission phase shift of 90° for TM-polarized wave. By stacking multilayers together, a full phase modulation in the range of $[0, 2\pi]$ could be realized [9]. Indeed, the physical origin of this effect is related to the catenary optical fields excited inside the slits. It should be noted that complex metallic structures may be treated as effective materials with tunable material parameters [10–12]. This concept is actually one important origin of the so-called metamaterials [13]. We would like to discuss these phenomena and their applications in the next chapter.

5.1.2 Near-Field EYI

Following the above discussion, if the metallic film supports surface wave, the Young's double slits interference can be used as a unidirectional near-field source, i.e., surface wave is excited in one direction and suppressed in the other one. As shown in Fig. 5.4, when the phase matching condition is met, the asymmetric double slits could excite surface plasmons to the left side, making the intensity in the right side to be virtually zero [7]. In general, the total SPPs at the left and right side can be decomposed into two parts, one directly generated by the slit and the other coming from the contribution of the other slit. Supposing the phase of SPPs directly generated by the left and right side are $\pi/2$ and 0, and the distance between the two slits is $\lambda_{SPP}/4$, the two SPP components at the left side would have the same phase

Fig. 5.4 **a** Near-field time-average intensity distribution. The thickness of the silver layer is 500 nm with two perforated slits, both filled with air. Widths of left and right slits are 50 and 200 nm, respectively. The TM-polarized, 632.8 nm plane wave is incident on the top side of the silver layer. **b** The spatial frequency spectrum along the *x*-direction, 20 nm away from the bottom side of the silver layer, showing that most energy is directed to the left side. Reproduced from [7] with permission. Copyright 2007, Optical Society of America

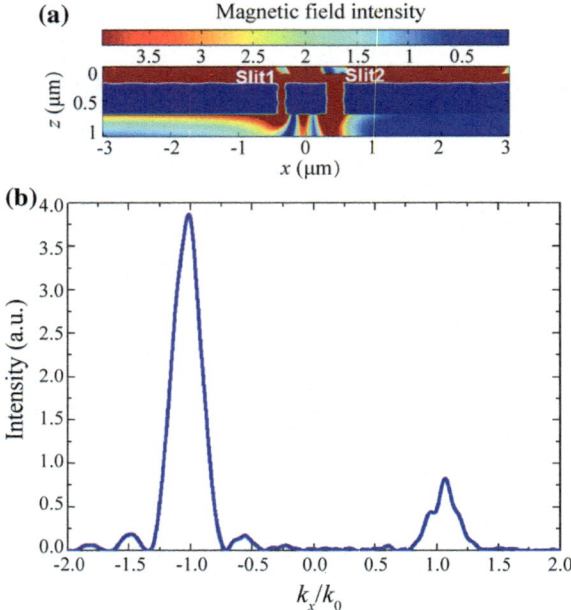

($\pi/2 = 0 + \pi/2$), leading to an enhanced intensity. In contrary, the two components at the right side are out of phase ($\pi/2 + \pi/2 = \pi$), leading to a vanishing intensity.

Compared with classic methods for plasmon generation, the intensity at the left side would be doubled. The proportion of the SPP field intensity along two opposite directions on an unilluminated surface excited by this setup is approximately 4:1. This effect has been utilized to increase the intensity of surface plasmon lithography [14]. Various variations have also been demonstrated to realize unidirectional surface plasmon sources [15, 16]. In particular, Zia and Brongersma performed a plasmonic double-slit experiment to reveal the strong analogy between SPP propagation along the surface of metallic structures and light propagation in conventional dielectric components. It justified the use of well-developed concepts from conventional optics and photonics in the design of new plasmonic integrated devices [17]. Once again, it should be noted that the EYI can also be modified to realize unidirectional excitation of microwave.

The near-field Young's interference provides a novel platform for the optical manipulation on the surface, which is critical for on-chip optical signal processing and other applications. In fact, some related experiments can be also treated in the framework of Young's interference [18, 19]. As shown in Chap. 4, the polarization of light has also been utilized to control the directional interference of SPPs via the concept of photonic spin–orbit interactions [20].

5.2 Wavefront Shaping via Plasmonic Slits

In this section, we show that plasmonic nanoslits could be used to realize almost arbitrary phase profile required for wavefront and beam shaping. Based on the gradient phase shift introduced across an ultrathin flat interface, classic Snell's law has been extended into a new form, i.e., the generalized law of reflection and refraction [21–23]. To emphasize the role of metasurface, this new optical law was also termed as metasurface-assisted laws of reflection and refraction (MLRR) [22]. With this formula, the propagation of light waves could be controlled at the surface directly, leading to a series of functional flat optical devices. One main advantage of the flat optical device is its ability to scale up without greatly increasing the size and weight, which breaks the fundamental barrier in traditional optical technologies. As a result, this technology is thought as disruptive and revolutionary [24–26].

5.2.1 Plasmonic Deflector and Generalized Snell's Law

Figure 5.5 shows the schematic of MLRR [22], where incident plane wave is reflected and refracted to directions away from that predicted by the classic geometrical optics. According to the continuity of tangential electric and magnetic fields at the boundaries, the following conditions must be satisfied [22, 26]:

$$n_1 k_0 \sin \theta_i + \nabla \Phi_r = n_1 k_0 \sin \theta_r,$$
$$n_1 k_0 \sin \theta_i + \nabla \Phi_t = n_2 k_0 \sin \theta_t, \tag{5.2.1}$$

where $\nabla \Phi$ is the phase gradient in the surface plane (may be different for reflection and transmission), which is determined by the arrangement of subwavelength structures, and may be changed with time by external stimuli such as electric and mechanic tuning. n_1 and n_2 are the refractive index of media at the incident and transmit sides. θ_i, θ_t and θ_r are the angles for incident, refracted, and reflected light. Combined with the generalized Fresnel's equations [22], the MLRR has been widely adopted in the generation and transformation of arbitrary wavefront across a thin sheet. It should be noted that similar equations have also been proposed with other structures such as V-shaped metallic antennas [27, 28] and rotated rectangular dielectric rods [29]. Indeed, the propagation phase in the nanoslits is one of the three kinds of phase modulating schemes [23, 26, 30].

According to Eq. (5.2.1), the key to realize MLRR is to construct subwavelength structures with gradient phase shift for either reflection or transmission. In the phase modulation process, the transmission and reflection amplitudes must be guaranteed to obtain sufficient high energy efficiency and low stray light (noise). In the following, we shall explain how to design a beam deflector based on the linear phase shift using gradient metallic nanoslits.

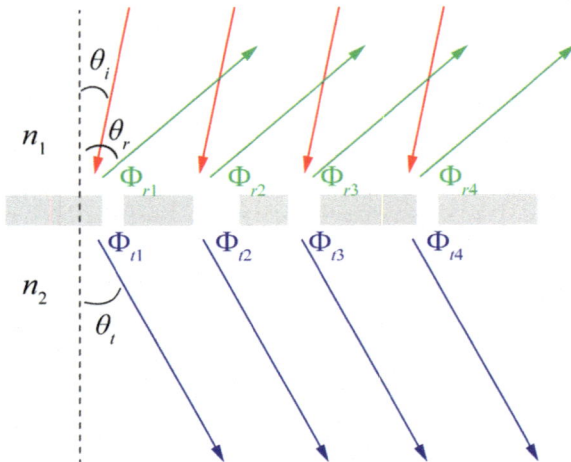

Fig. 5.5 Beam deflection via constructive interference in a given direction. The transmitted and reflected phase shift Φ_t and Φ_r are locally tuned by gradient subwavelength structures

As shown in Fig. 5.6, the plasmonic beam deflector is composed of a planar thin metal film perforated with gradient nanoslit arrays. The width of the slits is tuned to obtain the desired phase. With TM-polarized incidence (H field is parallel to the z-axis), SPPs can be excited at the slit entrance and funneled into the other side. Each slit may be treated as a MIM waveguide. Since the slit width is far less than the wavelength, only the fundamental mode dominates the propagation behavior. The complex propagation constant β for SPPs inside the slit region can be calculated by Eq. (5.1.1). The real and imaginary parts of β determine the phase velocity and the propagation loss of SPPs inside the slit, respectively. Given the thickness of metal film (i.e., slit depth) is h, the phase retardation of SPPs transmitted through the slit can be expressed as $\Delta\varphi = \Delta\varphi_1 + \Delta\varphi_2 + \mathrm{Re}(\beta h) + \Delta\varphi_3$, where $\Delta\varphi_1 = \arg[(n_1 - \beta/k_0)/(n_1 + \beta/k_0)]$ and $\Delta\varphi_2 = \arg[(\beta/k_0 - n_2)/(\beta/k_0 + n_2)]$ are the phase shifts occurring at the slit entrance and exit. When the medium on the illuminated and unilluminated side of metal film is the same, the two factors have the equal value but opposite sign. The last term $\Delta\varphi_3$ is originating from the multiple reflections between the entrance and exit interfaces, which can be calculated with the following equation:

$$\Delta\varphi_3 = \arg\left[1 - \left(\frac{1 - \beta/k_0}{1 + \beta/k_0} \right)^2 \exp(i2\beta h) \right]. \qquad (5.2.2)$$

Further calculation indicates that when the silt width is larger than 10 nm, the retardation contribution from $\Delta\varphi_3$ is less than 1 percentage of the factor (βd). That is to say, the phase retardation is mainly dominated by the real part of the propagation constant, which is approximated as $\Delta\varphi = \mathrm{Re}(\beta h)$. Therefore, the phase retardation $\Delta\varphi$ can be directly tuned by varying the slit width w if the other parameters are fixed.

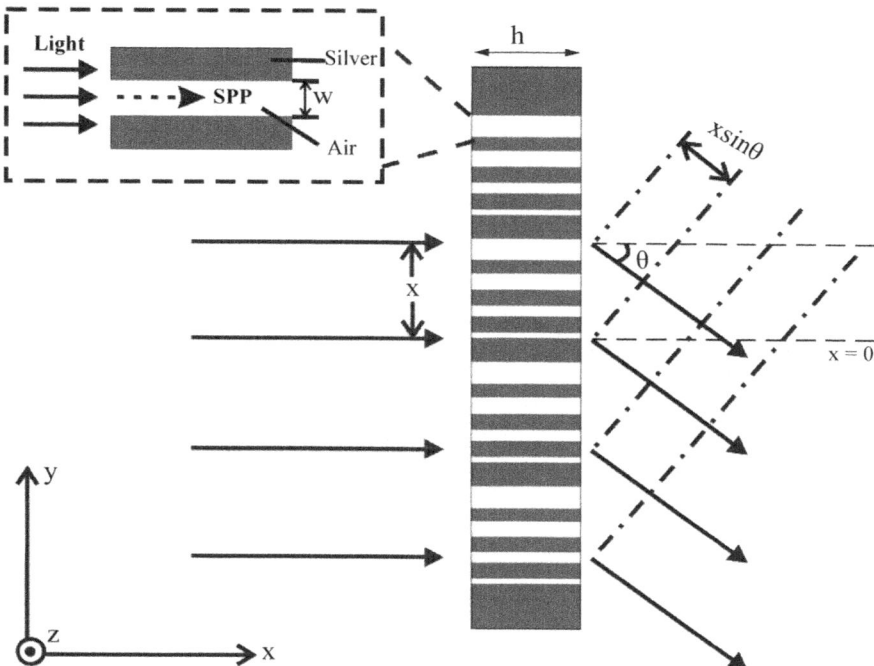

Fig. 5.6 Schematic of the metallic beam deflector. h and θ represent the thickness and deflection angle. The inset depicts the excitation and propagation of SPP in the nanoslit when the incident light impinges on the metal surface. Reproduced from [21] with permission. Copyright 2008, Optical Society of America

To realize beam deflection at a specific angle θ, as shown in Fig. 5.6, the phase retardation of light transmitted through the slits along the x-direction should take a linear form as [21]

$$\Delta\varphi = 2m\pi - \frac{2\pi}{\lambda}x \sin\theta, \tag{5.2.3}$$

where m is an integer number. Obviously, the key point of designing a plasmonic beam deflector is to determine the width and position of each nanoslit for desired phase retardation. Four deflectors with deflection angles of $30°, 45°, 60°$, and $80°$ are designed to illustrate this method and the wide range of defection angle. The thickness and aperture size are fixed as 500 nm and 6 μm, respectively. The wavelength of incident light is 650 nm, and the corresponding permittivity of used metal silver is $\varepsilon_m = -17.36 + 0.715i$. The medium surrounding the deflector and inside silts are assumed to be air. The designed slits' width ranges from 10 to about 50 nm, corresponding to a phase retardation modulation range of 2π. The thickness of the metal walls between any two adjacent slits is larger than the skin depth (24 nm for silver at

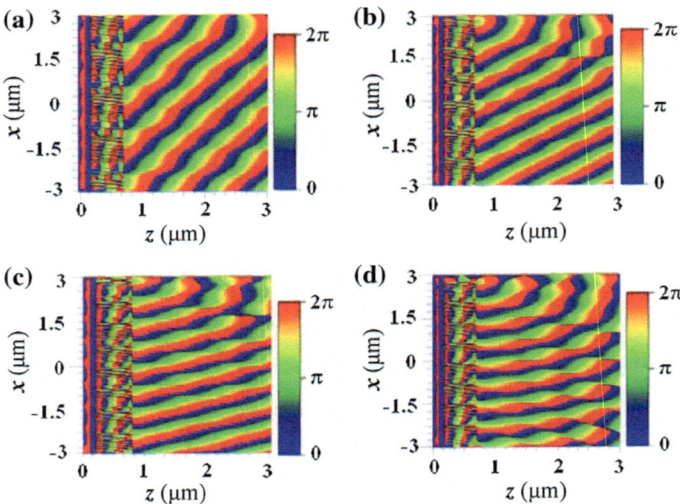

Fig. 5.7 Phase distribution of electric field for the four deflectors. **a–d** Correspond to the designed deflection angles of −30°, −45°, −60°, and −80°, respectively. Reproduced from [21] with permission. Copyright 2008, Optical Society of America

wavelength 650 nm) to reduce the plasmonic coupling and crosstalking effect. This localized phase modulation is different from traditional dielectric structures [31] and therefore enables wide-angle operation (large angle deflection under both normal and oblique incidences).

Finite difference time domain (FDTD) method was employed to simulate the structure with a size of 6 μm × 3 μm, surrounded with a boundary of the perfect matched layer (PML). TM-polarized plane wave is incident from the left side. The simulated phase distributions of the light field for the four deflectors are presented in Fig. 5.7, where the deflection angle can be approximately calculated by $\theta = \arcsin(N\lambda/D)$. Here, N is the number of phase fringes at the deflector's exit side, D is the aperture size of the deflector.

Figure 5.8 presents a detailed evaluation of the deflection effect with the angular spectrum in the far field. As expected, the maximum angular spectrum peak is localized at the angle of −30.12°, −45.26°, −60.40°, and −81.55° for the four deflectors. The slight angular deviation probably arises from the coupling effect of SPPs at both slits' entrance and exit sides. Due to the diffraction effect, larger deflection angle delivers greater divergence.

The energy efficiency, defined as the percentage of incoming power distributed in the central lobe of angular spectrum confined by the angles of two closely adjacent minimum, is one important parameter for the beam deflector. The calculated efficiencies for the four deflectors' are 47.04, 52.72, 51.74, and 53.38%. The energy loss mainly arises from the reflection at the illuminating side and the absorption loss during the propagation. In principle, higher efficiency can be obtained by choosing

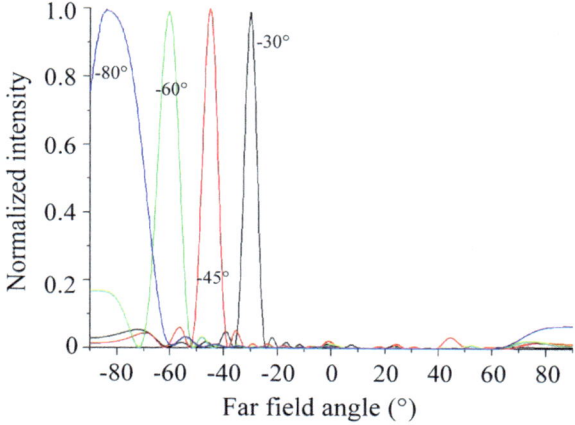

Fig. 5.8   Far-field angular spectrum of the deflectors for $-30°$, $-45°$, $-60°$, and $-80°$, respectively. Reproduced from [21] with permission. Copyright 2008, Optical Society of America

better plasmonic materials with smaller losses. Note that although narrow slits could induce larger propagation constant and thinner device profile, the Fresnel reflection at the entrance side would increase correspondingly. Consequently, a compromise between the thickness and efficiency must be made. Furthermore, by using reflective geometry, a higher efficiency may be obtained with a smaller thickness.

If the light beam is incident obliquely, deflection phenomenon resembling "negative refraction" can be achieved. Figure 5.9 gives an example of such case, in which the incident angle is set to be $30°$ and the refractive angle is $-30°$. The deflector is designed with an aperture of 4 µm and a thickness of 800 nm. From the phase distribution and far-field angular spectrum, it can be seen that the deflection angle agrees well with the designed one.

Owing to the broadband characteristic of SPPs, the deflector can also operate in a broad wavelength range. Figure 5.10 presents the calculated deflection angles for light ranging from 550 to 750 nm with a step of 10 nm, while the designed working wavelength and deflection angle are 650 nm and $45°$. There is a nearly linear angular shift with incident wavelength, approximately $0.9°/10\,\text{nm}$, as predicted by Eq. (5.2.3). This implies that in addition to manipulating beam, the plasmonic deflector can also be used for spatial and spectral multiplexing. Note that this deflector is similar to the blazed grating [31], but completely different from the periodic grating.

5.2.2   Flat Lens Based on Plasmonic Nanoslits

Similar to the plasmonic beam deflector, a planar metallic lens can be easily designed with nanoslits array for subwavelength focusing and imaging in the far field. In

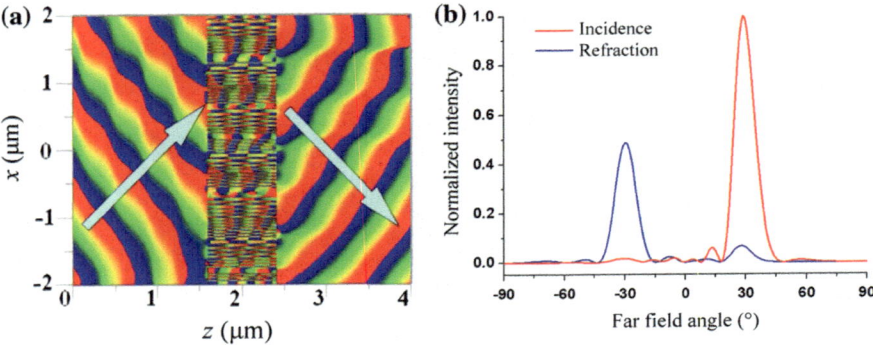

Fig. 5.9 **a** Calculated phase distribution of electric field for the designed deflector. Incident and deflection angles are designed as 30° and −30°, respectively. **b** Far field angular spectrum for the incident and emitted beams, respectively. Reproduced from [21] with permission. Copyright 2008, Optical Society of America

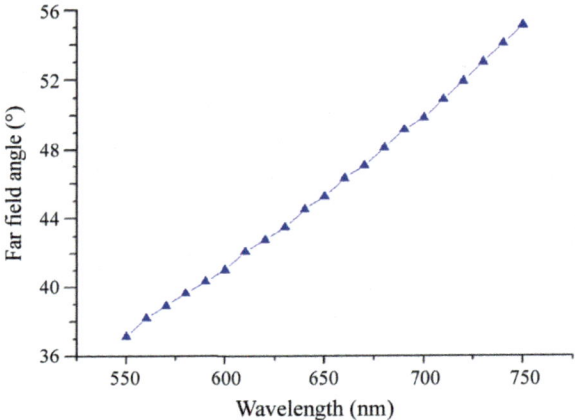

Fig. 5.10 The dependence of deflection angle on incident wavelength for the deflector designed for 45° at a wavelength of 650 nm. Reproduced from [21] with permission

contrary to the super-resolving superlens based on a thin metallic film [32], the metallic planar lens can realize imaging at any distance from the lens, provided that appropriate phase shift are designed with gradient nanoslits.

Figure 5.11 is the schematic of the optical imaging with a metallic slab lens composed of an Ag slab of thickness h perforated with a series of slits arranged symmetrically with respect to the central plane $x = 0$ [33]. The object, denoted as a point source, is positioned at the left side of the lens with a distance a and an optical image appears on the right side with a distance b. For simplicity of simulation, the 2D case is considered, i.e., both the lens and object extend uniformly and infinitely in the y-direction. When a TM-polarized light impinges on the surface of the lens,

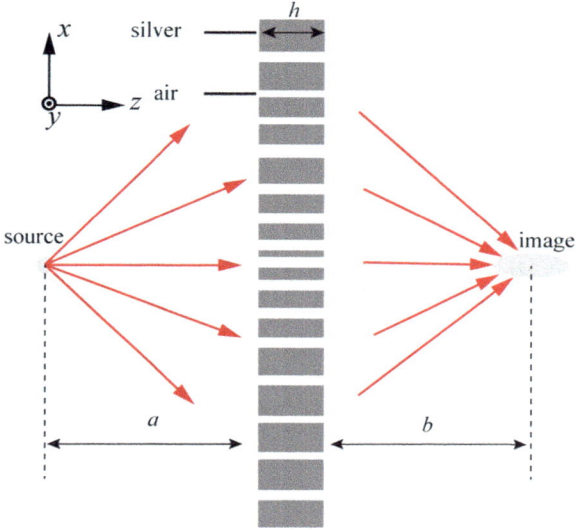

Fig. 5.11 A schematic of far-field optical imaging by a metallic slab lens composed of nanoslits. Reproduced from [33] with permission. Copyright 2007, American Institute of Physics

SPPs can be excited at the slit entrance and transmitted with a phase shift determined by the geometric parameters. For the imaging of object localized on the axis $x = 0$, the phase retardation of light transmitted through the lens should be

$$\Delta\varphi = 2m\pi + \frac{2\pi}{\lambda}\left(a + b - \sqrt{a^2 + x^2} - \sqrt{b^2 + x^2}\right), \qquad (5.2.4)$$

where m is an integer number. Clearly, the focusing behavior can be regarded as a special case of image formation. For plane wave illumination, a would be infinite and b should be the focal length f, thus Eq. (5.2.4) becomes

$$\Delta\varphi = 2m\pi + \frac{2\pi}{\lambda}\left(f - \sqrt{x^2 + f^2}\right). \qquad (5.2.5)$$

In the paraxial regime, i.e., $\max(x) \ll f$, the equation can be further approximated as

$$\Delta\varphi = 2m\pi - \frac{\pi x^2}{\lambda f}. \qquad (5.2.6)$$

In 2009, Verslegers et al., experimentally demonstrated the far-field 1D lens using a combination of thin film deposition and focused ion beam milling (FIB) [34]. The confocal experimental results show excellent agreement with the full electromagnetic field simulations. It was thought as a crucial step in the realization of plasmonic flat lenses, which offer a range of processing and integration advantages over conventional curved dielectric lenses. By further optimizing the geometry and parameters, almost arbitrary phase front may be constructed.

Fig. 5.12 Experimental demonstration of the flat lens based on nanoslits. The two samples show the effect of lens size on focusing for **a** a lens with 13 slits (80–150 nm wide) and **b** a lens with 11 slits (80–120 nm wide). The white lines show the lens position. Both scanning electron microscope (SEM) images are on the same scale. Reproduced from [34] with permission. Copyright 2009, American Chemical Society

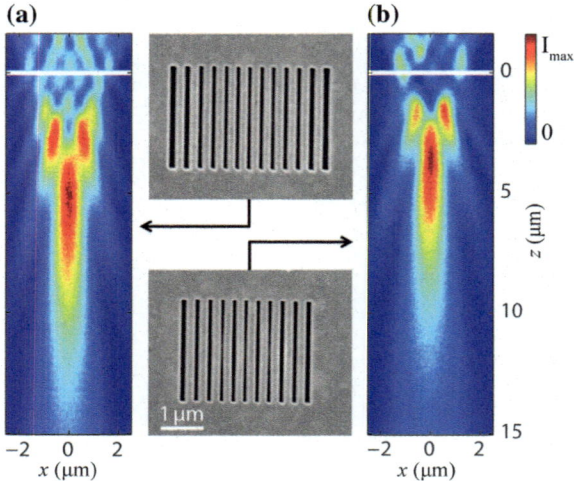

In the experiment, it was found that the effect of lens size can be exploited to control the focusing behavior as shown in Fig. 5.12. Both lenses introduce the same curvature to the incident plane wave. By omitting one outer slit on each side for the lens in Fig. 5.12a, one gets the lens in Fig. 5.12b. Even though these structures have the same phase front curvature, the actual focus moves closer to the gold film as the lenses get smaller, in good agreement to what the simulations predict. This confirms that, as lenses are scaled down to the size of only a couple of wavelengths and especially for long focal lengths, the simple Eqs. (5.2.5) and (5.2.6) are no longer rigorously valid. Instead, the diffraction behavior of the finite sized aperture should be taken into consideration. This effect will also be discussed in Sect. 5.3.1 for a modified aperture shape.

5.2.3   Tunable Plasmonic Nanoslits Lens

One advantage of the nanoslits lenses is that they can be dynamically tuned when active materials such as Kerr nonlinear materials [35] and phase-change materials [36] are filled in theses nanoslits. When the refractive index changes, the mode index of the SPP modes and the phase shift would change correspondingly.

In 2007, Min et al., investigated a metallic lens consisting of slits filled with Kerr nonlinear media [35]. Each slit is designed to transmit light with specific phase retardation controlled by the intensity of incident light, owing to the nonlinear response. Figure 5.13 shows FDTD simulation of time-average electric-field intensity distribution of beam focusing with a five-slit structure. The positions of focal spots are indicated by vertical white lines. The thickness of the silver film is 570 nm and the slit widths in the five-slit array are 100, 70, 60, 70, and 100 nm from up to down.

In Fig. 5.13a, a clear focus appears at about 2.7 μm away from the exit surface; however, it drops to about 0.6 μm with the increased intensity of incident light in Fig. 5.13b. When the incident amplitude grows from 1×10^8 to 2×10^8 V/m, only the central slit with $w = 60$ nm achieves F-P resonance, hence its increased effective refractive index is larger than the slits far from the center, which makes the focus closer to the exit surface. According to the analysis above, if the incident amplitude continues increasing, the central slit may be out of F-P resonance and make the focus far away from the exit surface again.

The experimental realization of tunable flat lens based on phase-change material is shown in Fig. 5.14 [36]. The structures were fabricated by milling 21 identical nanoslits on a 100 nm-thick gold film using FIB milling. 100 nm GST was then sputtered on the samples covering the entire surfaces. To achieve different focus patterns, GST in each slit was supposed to be selectively crystallized to tune the phase. Since GST crystallization inside the slits could not be precisely monitored and controlled in the current stage, the GST is binarized to either amorphous or crystalline phase. A 4 mW CW laser operating at 532 nm was focused on the samples by a 100× objective and scanned along the slits at a speed of 0.2 mm/s to ensure that GST is thoroughly crystalized inside the slits.

As a proof-of-concept demonstration, the far-field patterns are measured for three samples, namely a control sample before any GST crystallization, followed by two samples with focusing or deflecting behaviors achieved by selectively crystallizing the GST. Confocal scanning optical microscopy was used to characterize the far-field patterns of the structures. A laser beam at 1.55 μm with a spot size much wider than the structures illuminated the samples through the substrate. The transmitted field was probed on the other side of the samples as shown in Fig. 5.14b. For each case, the calculated phase front and the anticipated discrete phase distributions from the sample are shown in Fig. 5.14c. The far-field intensity distribution of the lens before GST crystallization is shown in the left panel of Fig. 5.14b. Since no GST crystallization variation is introduced among the slits, a nearly flat phase front is expected. The focusing effect is attributed to the diffraction of a finite size beam. The middle panel of Fig. 5.14b shows the lens with GST symmetrically crystallized about the lens axis, which provides a simplified phase front for the lens. The right panel of Fig. 5.14b illustrates the beam deflection or off-axis focus leaning toward the right-hand side. Simulation results based on the experimental conditions, namely using only 0% and 90% crystallized GST are shown in the insets of Fig. 5.14b. A relative good agreement was observed between the experimental results and the adapted simulations.

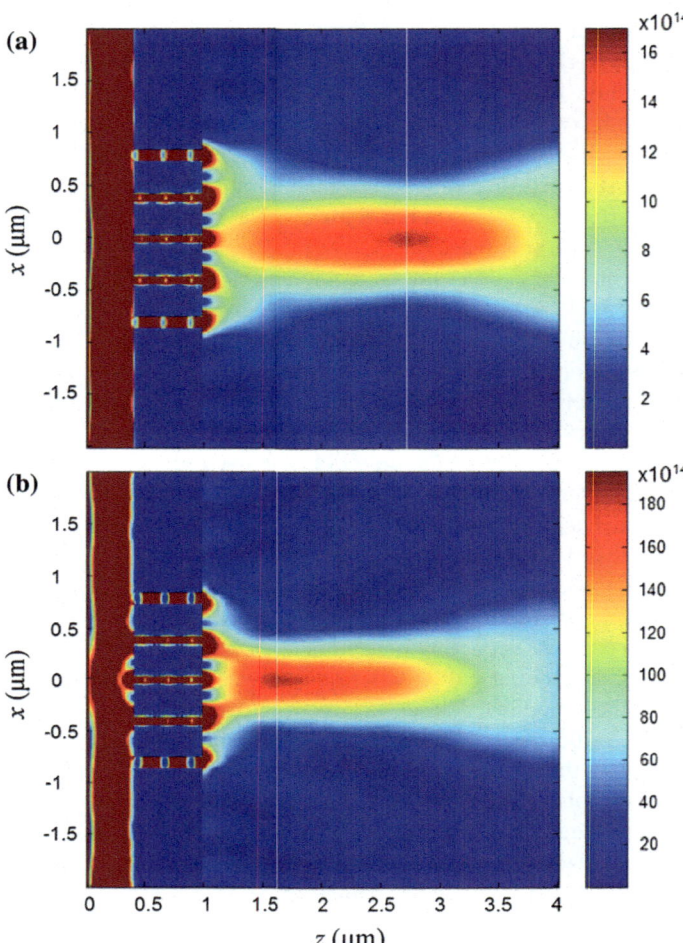

Fig. 5.13 The FDTD simulation of electric-field intensity $|E|^2$ with a five-slit metallic lens. The thickness of the Ag film is 570 nm, the distance between two silts is 400 nm (center to center). A TM-polarized plane wave ($\lambda = 850$ nm) is incident from the left side of the slit array. The electric-field amplitude of the incident light is 1×10^8 V/m in (**a**) and 2×10^8 V/m in (**b**). The vertical white lines indicate the positions of focal spots in the x-axis. Reproduced from [35] with permission. Copyright 2007, Optical Society of America

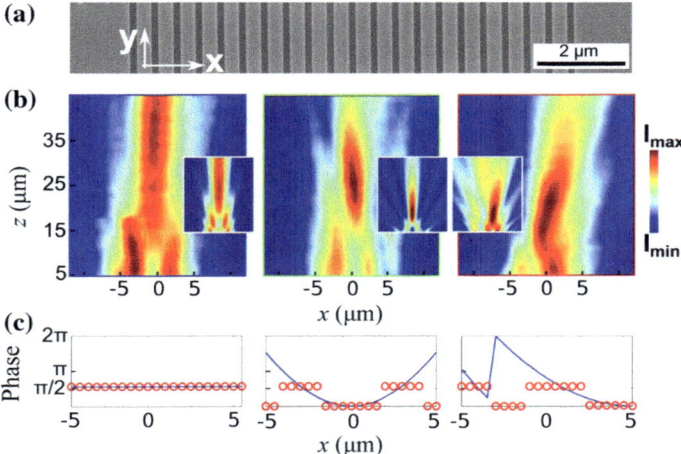

Fig. 5.14 a SEM image of the fabricated planar lens before sputtering GST. **b** Focusing patterns measured in the *xz* plane by confocal scanning optical microscopy for amorphous GST in all slits without crystallization, GST being crystallized in the selected slits to construct focusing and deflecting phase fronts, respectively. Insets show the corresponding simulation results of the planar lens in each case using the binarized GST crystallization levels. **c** The calculated phase fronts (blue curves) and the binarized discrete phase distributions (red circles) for the samples shown in (**b**). Reproduced from [36] with permission. Copyright 2015, The Authors

5.3 Plasmonic Hole Lens

5.3.1 Rectangular Holes

As polarization-dependent structures (only TM-polarized light could excite SPPs), 1D nanoslits can just manipulate incident beam with electric fields perpendicular to the long axis. To realize polarization-independent optical response, a square hole with a width of $a_x = a_y = a$ perforated in the thick metal film may be taken as the building element (Fig. 5.15).

Similar to the nanoslits, the phase retardation can be modulated from 0 to 2π (or $-\pi$ to π) by varying the hole size. As illustrated in Fig. 5.16, the solid curve shows that the phase retardation of a single square hole changes from $-\pi$ to π, while *a* changes from ~0.52π to 0.89π at the thickness of $h = 1.8\ \lambda$. Meanwhile, the normalized-to-area transmittance *T* of a single square hole in the inset figure can exceed 1 for a wide range of *a* due to the coupling of the surface waves and waveguide modes.

As shown in Fig. 5.17, a structured lens is designed with an aperture of $D = 280\ \mu m$, a focal length of $f = 240\ \mu m$, a film thickness of $h = 19.08\ \mu m$, and an incident wavelength of $\lambda = 10.6\ \mu m$. Similar to the 1D case, the required phase profile of the lens can be calculated according to the equal optical path principle as

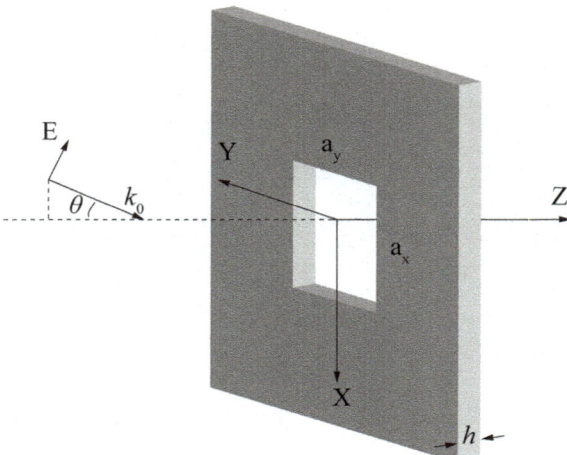

Fig. 5.15 Schematic of a single rectangular hole with side a_x and a_y perforated on a thin metal film of thickness h. The film is illuminated by a TM-polarized plane wave with its incidence angle of θ (E is in the XZ plane). Adapted from [37] with permission. Copyright 2008, Optical Society of America

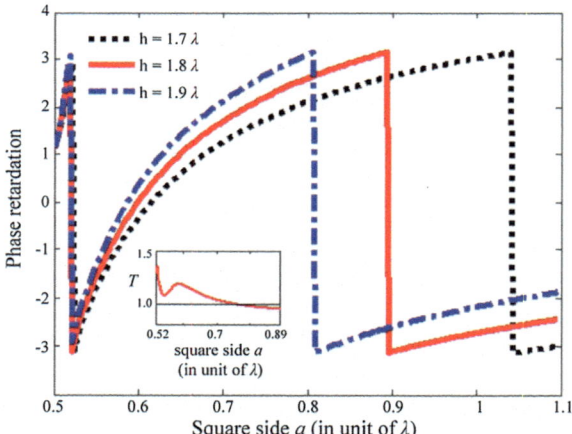

Fig. 5.16 Phase retardation of a single square hole in a metallic film as a function of the hole size a for several values of the thickness h (both a and h are in units of wavelength λ). Normal incidence and TM polarization are assumed for the incident plane wave. The inset shows the normalized-to-area transmittance (T) versus hole size a when the thickness is $h = 1.8\lambda$. Reproduced from [37] with permission. Copyright 2008, Optical Society of America

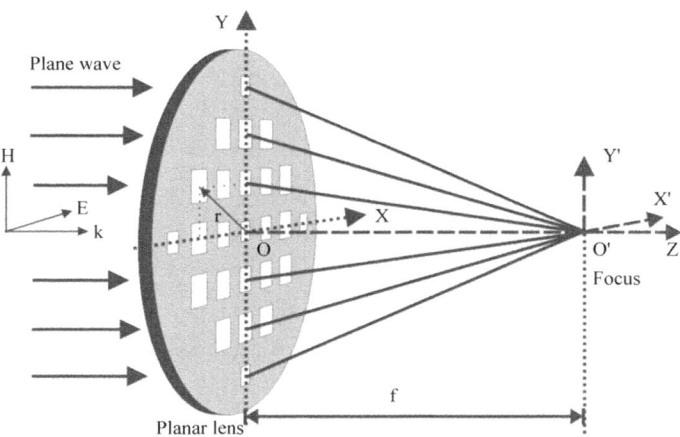

Fig. 5.17 Schematic of the 2D structured lens with a focal length of f. The film is perforated with square holes of different sizes. The gray part indicates metal, and the white part is air. It is illuminated by a plane wave with electric field polarized in the X-direction. Reproduced from [37] with permission. Copyright 2008, Optical Society of America

$$\Delta\varphi = 2m\pi + \frac{2\pi}{\lambda}\left(f - \sqrt{r^2 + f^2}\right), \qquad (5.3.1)$$

where r is the radius. Since the phases at the corresponding center of the holes can be directly determined, the distribution of the holes in the lens may follow either a uniform (located at the cross points of either the Cartesian or polar coordinates) or an arbitrary regulation. To verify the principle, the holes are arranged regularly in the Cartesian coordinates and the distance between adjacent holes is chosen as 1.75λ. Note that unlike the 1D slit which has no cutoff frequency under TM polarization, the modes in rectangular holes may be evanescent when the hole is too small compared with the wavelength. As a result, the distance between each hole may be larger than the wavelength.

When an incident plane wave with E polarized along X-direction illuminates on the metal lens, the phase retardation in different holes varies and a focal spot can be formed on the Z-axis. To reduce the calculation difficulty for large-scale lens, a hybrid method was used: First, numerical simulation using the FDTD approach was performed to obtain the E field in the vicinity of the exit surface. Second, a program using the Fresnel–Kirchhoff diffraction integral method was used to transform the near field to the far field.

Figure 5.18 shows the intensity distributions at the cross-sections in different planes, indicating an elliptic focal spot in the Z direction and the maximum intensity appears at $Z = 232$ μm, i.e., the focal length is 232 μm, which is close to the designed value (240 μm). Meanwhile, a clear circular focal spot with the full width at half-maximum (FWHM) of ~11.2 μm is observed in the XY plane ($Z = 232$ μm).

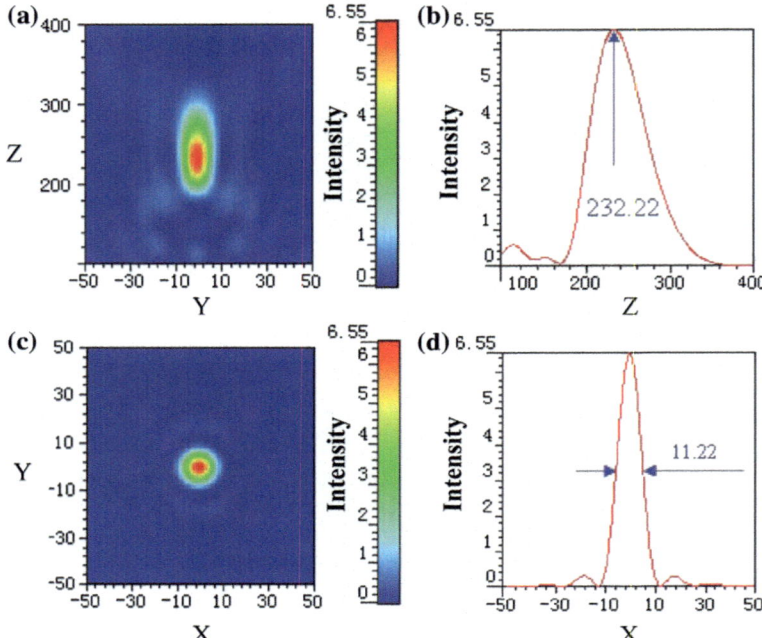

Fig. 5.18 a Numerical calculation of intensity distribution in the YZ plane at a distance ranging from 180 to 420 μm from the exit surface of the metal film. **b** Intensity distribution along the Z-axis. The focal position is $Z = 232$ μm. **c** Intensity distribution in the XY plane at $Z = 232$ μm. **d** Cross-section of (**c**) at $Y = 0$. The FWHM is marked. Reproduced from [37] with permission. Copyright 2008, Optical Society of America

Compared with the diffraction-limited FWHM of 11.08 μm, there is only an increase of 1.08%.

The transmittance of the rectangular holes array is evaluated according to the integral of the time-averaged Poynting vector S over the entrance surface and the exit surface for achieving the total energy E_{in} input on the entrance surface and the total energy E_{out} output from the exit surface, respectively. The transmittance $t = E_{out}/E_{in}$ is calculated to be about 20%. In principle, the transmittance can be further increased with the optimization for the distance of the square holes.

Compared with simple square or rectangular apertures, it has been proved that more complex apertures, such as coaxial and cross-shaped apertures, generally offer higher coupling efficiency of incident light in the scenario of EOT [38]. Owing to the localized surface plasmon resonance, it is reasonable to expect that the phase shift can be easily modulated by changing the geometric structures. In order to demonstrate this, two polarization-independent lenses with focal lengths of 15 and 25 μm have been designed using cross-shaped holes array as shown in Fig. 5.19. To facilitate comparison between devices, the two lenses have the same array size. Since the structures are more complex than the rectangular holes, the fabrication difficulty may

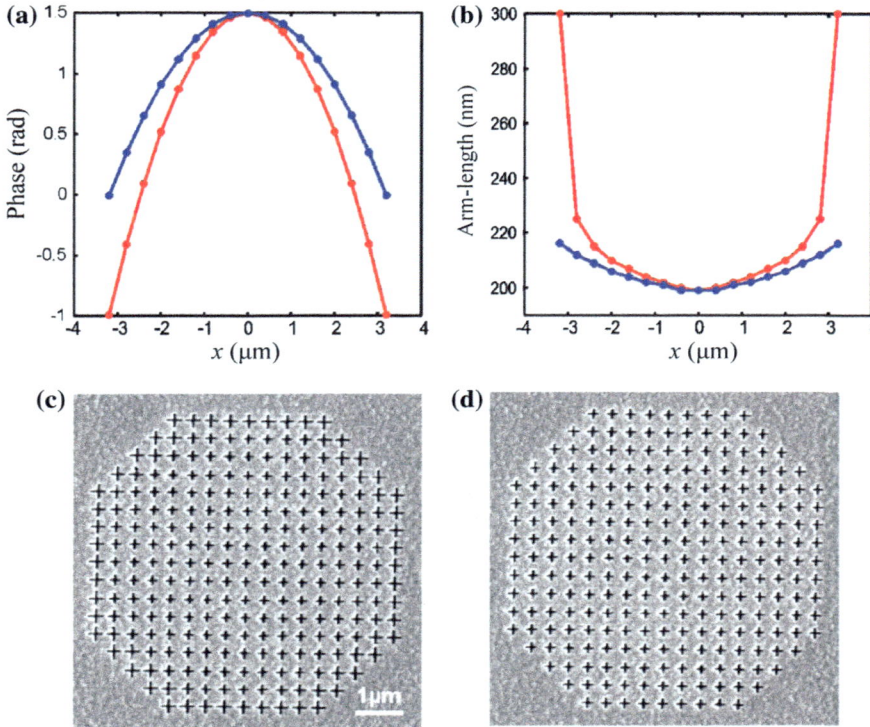

Fig. 5.19 **a** The calculated phase and **b** arm length along one of the symmetric axes (*x*-axis) of the 2-D aperture array lenses operating at 850 nm wavelength: red line, $f = 15$ μm; blue line, $f = 25$ μm. The thickness of the metallic film is 140 nm. The period and arm width are kept as 400 nm and 40 nm. **c, d** SEM images of the fabricated devices with $f = 15$ μm and $f = 25$ μm. Reproduced from [38] with permission. Copyright 2010, American Chemical Society

increase. Nevertheless, this is not a big problem since the micro-fabrication technologies have evolved greatly along with the rapid development of microelectronics [39]. One advantage of the cross-shaped structure is that the propagation constant of SPPs therein may be larger, thus a smaller thickness is sufficient to introduce the required phase shift. For instance, with a thickness of 140 nm, the structures could induce a phase shift of π at a wavelength of 850 nm by changing the length of the arms, as demonstrated in Fig. 5.20a, b. Furthermore, since the unit cell can be smaller, the lattice constant could be reduced to be smaller than the wavelength, thus high-order diffraction orders could be suppressed.

As displayed in Fig. 5.20b, c, the behaviors of the two designed lenses are rather different. For the lens with $f = 15$ μm, the transmitted intensity has its maximum at the exit interface of the structures and the intensity of the transmitted beam decreases steadily along the *z*-direction. In contrast, for the lens with $f = 25$ μm the axial

Fig. 5.20 Measured axial intensity profile (on the *yz* plane) of light passing through: **a** a reference structure consisting of apertures with fixed arm lengths (250 nm); **b** a flat lens with $f = 15$ µm; **c** a flat lens with $f = 25$ µm. Reproduced from [38] with permission. Copyright 2010, American Chemical Society

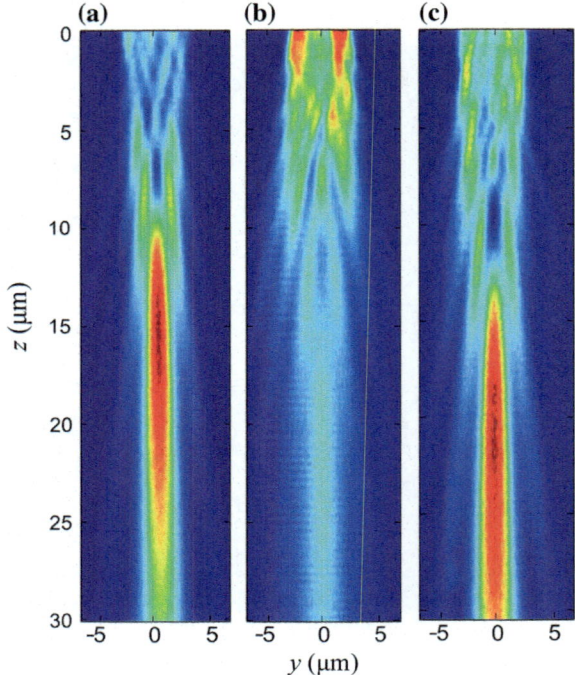

intensity maximum of the transmitted beam occurs at around $z = 22$ µm, which is close to its designed focal position.

The difference in the optical responses of the designed lenses can be understood by considering the diffraction pattern of an aperture of radius a. Along the axis of the beam path, the intensity distribution of the diffracted light at a distance z from the aperture can be estimated using the equation [38]

$$I(z) = I_0 \sin^2\left(\pi a^2/(2\lambda z)\right), \tag{5.3.2}$$

where I_0 is the intensity maximum. Equation (5.3.2) gives an intensity maximum at $z = 15$ µm for 850 nm light passing through an aperture of diameter 7.2 µm. As shown in Fig. 5.20a, the position of the axial intensity maximum of the reference structure agrees well with this theory. Considering light passing through a micrometer-scale lens, the existence of the diffraction at the boundary of the devices implies that Huygens' spherical wavelets centered on these regions carry a divergent wavefront. Consequently, the intensity profile of light transmitted through the fabricated lens is a result of interference between focusing and diffraction effects along the beam path. When the locations of intensity maxima arising from the focusing and the diffraction coincide, the two effects cancel to produce a low intensity in the designed focal position, as illustrated in Fig. 5.20b. By employing a slowly varying phase profile to shift the focal length away from the axial intensity maximum produced by diffraction alone, the behavior of the device could become more close to the theoretical prediction.

It should be remembered that the above effect only occurs for small flat lenses with a diameter only several times larger than the wavelength. For large-aperture plasmonic flat lenses, the influence of diffraction on the focal distance can be ignored.

5.3.2 Circular Holes

Although rectangular and cross-shaped holes are efficient in phase modulation, circular holes are easier to fabricate. In principle, the propagation characteristics of SPPs in a circular hole can be understood using waveguide theory. In cylindrical coordinates, the waveguide modes for both electric and magnetic fields can be written as $\psi(r, \phi, z) = \psi_n(r)\exp(im\phi)\exp(i\beta z)$, where ω and β are the angular frequency in free space and the propagation constant of SPP that is determined from the boundary conditions. From the requirement of continuity of the tangential components of the electric field and the magnetic field at the cylindrical surface $r = a$, the dispersion relation can be obtained by solving the equation:

$$\left[\frac{J'_m(au)}{uJ_m(au)} - \frac{H^{(1)'}_m(av)}{vH^{(1)}_m(av)}\right]\left[\frac{k^2_d J'_m(au)}{uJ_m(au)} - \frac{k^2_m H^{(1)'}_m(av)}{vH^{(1)}_m(av)}\right]$$
$$= \frac{m^2\beta^2}{a^2}\left(\frac{1}{v^2} - \frac{1}{u^2}\right)^2,$$

(5.3.3)

where $J_m(au)$ and $H^{(1)}_m(av)$ are the mth-order first kind Bessel function and Hankel function ($m = 0$ for the fundamental mode), $k_d = (\omega/c)(\varepsilon_d)^{1/2}$ and $k_{Ag} = (\omega/c)(\varepsilon_m)^{1/2}$ are the wavenumbers in the dielectric and metal. The propagation constants in the dielectric and metal are given by $u = (k^2_d - \beta^2)^{1/2}$, $v = (k^2_m - \beta^2)^{1/2}$, respectively. In principle, this equation can be solved to find the phase change βh, in a way similar to the case of the 1D MIM waveguide. Considering the coupling of adjacent holes, the theoretical phase shift is only an approximation, and rigorous simulations are required to obtain better designs.

As shown in Fig. 5.21, a polarization-independent flat lens with a focal length of 10 μm at 531 nm was designed with circular holes array (the geometric parameters are shown in Table 5.1) [40]. These holes were milled through the gold film by FIB and then covered with a 200 nm PMMA film by spin-coating. The PMMA filled the holes and remained as a 200 nm-thick uniform film on the gold surface. At the positions of the holes, only slight (~20 nm) pits in the PMMA layer were observed. One advantage of filling the holes with the polymer is it decreases the cutoff radius of the circular waveguide, thus enabling larger phase shift with the same radius and length.

Fig. 5.21 SEM image of the holey metal lens before the PMMA spin-coating process. Reproduced from [40] with permission. Copyright 2013, American Chemical Society

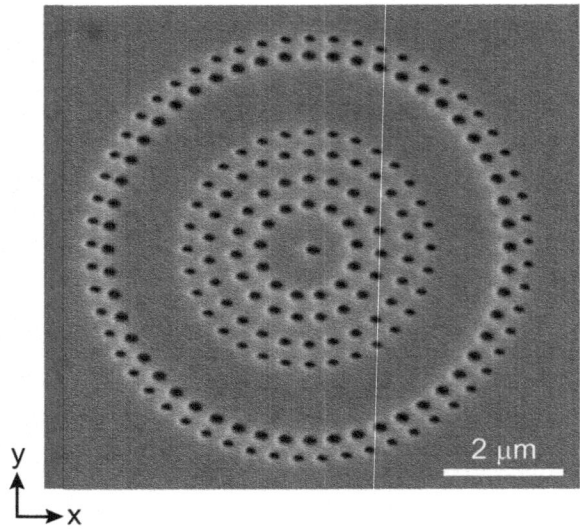

Table 5.1 Geometric parameters of the flat lens based on circular holes array [40]

Concentric ring number	1	2	3	4	5	6	7
Distance from the origin (μm)	0	0.8	1.2	1.6	2.0	3.3	3.63
Hole radius (nm)	83	76	69	62	56	83	56
Number of holes	1	13	19	26	32	52	57

Figure 5.22 illustrates the transmission intensities of the fabricated sample at 488, 531, and 647 nm in two orthogonal linear polarizations which were recorded using a microscope and a CCD camera. The focal lengths for both polarizations at 488, 531, and 647 nm are 12, 10, and 8 μm, respectively. Although this chromatic dispersion is disadvantageous for white-light imaging, it provides a tunability via wavelength. At all these wavelengths, the focus profiles show little dependence to the incident polarizations.

The holes array can also be arranged to realize much more complex phase distribution such as helical phase featured by beams carrying orbital angular momenta (OAM). As schematically shown in Fig. 5.23, an array of nanoholes with a circular graded size distribution along the azimuthal direction can be used to generate optical vortex [41]. Each nanohole introduces a specific phase change determined by its radius. Therefore, by carefully choosing the spatial distribution of the nanohole radii, a total phase change of multiple of 2π can be imposed on the wavefront of the beam passing through such an array. Such structures are compact, versatile, and can be readily integrated with optical fibers or on a chip.

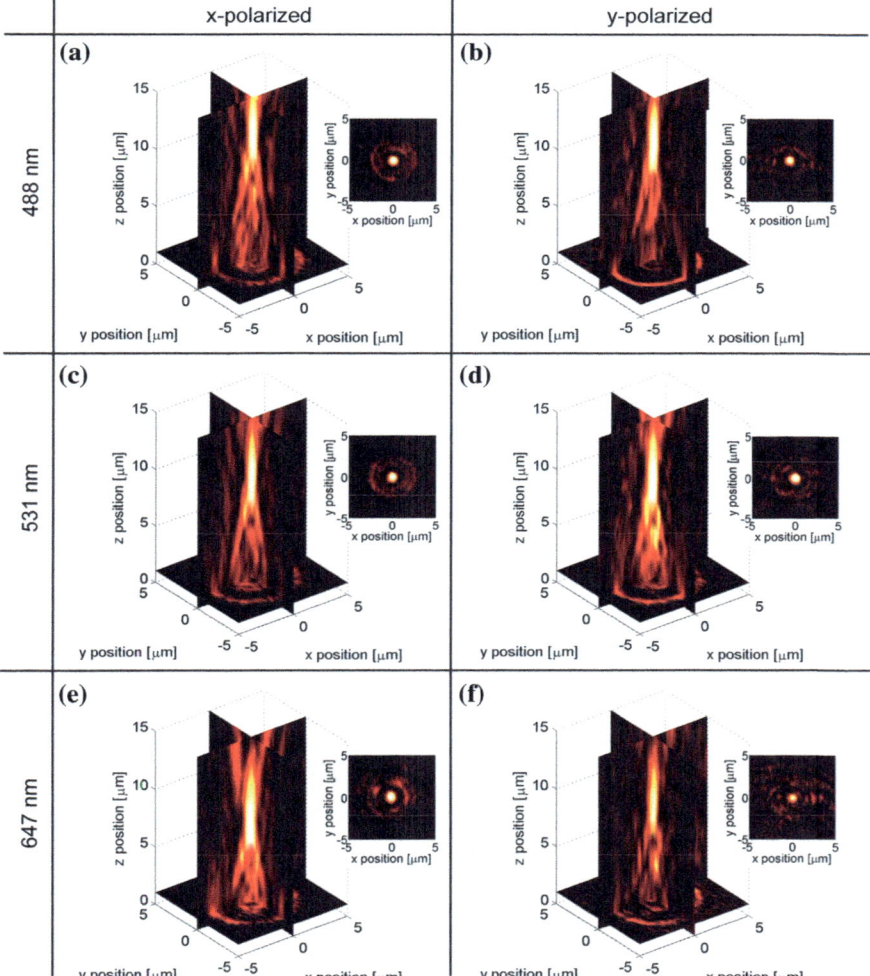

Fig. 5.22 Pseudocolor cross-section intensity maps measured at various heights above the sample surface for x-polarized and y-polarized incident light at 488 nm, 531 nm, and 647 nm. The xy-planes are at $z = 1$ μm, the yz-planes are at $x = 0$, and the xz-planes are at $y = 0$. The insets show the intensity at xy-planes at the focus for each wavelength. Reproduced from [40] with permission. Copyright 2013, American Chemical Society

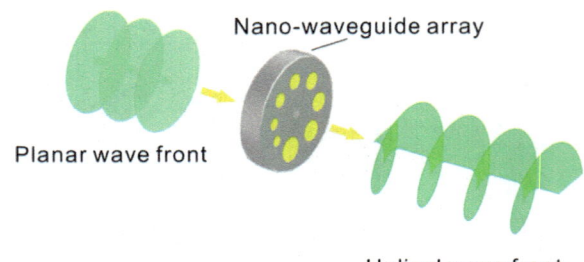

Fig. 5.23 Schematic of a nanoholes array that induces wavefront shaping. Reproduced from [41] with permission. Copyright 2014, American Chemical Society

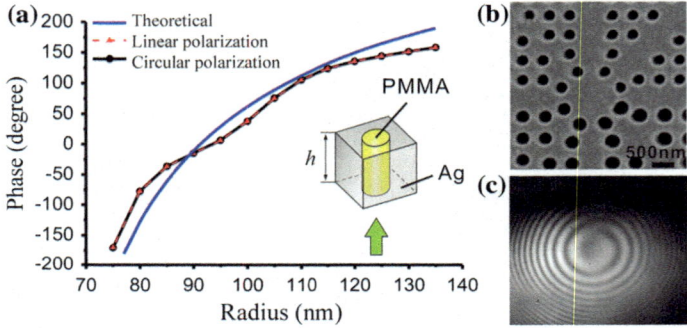

Fig. 5.24 **a** Simulated phase distribution. The solid blue line corresponds to theoretical predictions using dispersion equation. The dashed red line corresponds to results from numerical simulations for the case of a linearly polarized input beam; the black solid-dotted line corresponds to the case of circular polarization. **b** SEM image of the fabricated sample. **c** Measured interference pattern with a Gaussian beam. Reproduced from [41] with permission. Copyright 2014, American Chemical Society

Table 5.2 Calculated phase change for various hole radius [41]

Phase change	−160°	−105°	−50°	−10°	45°	90°	128°	165°
Hole radius (nm)	76	80	85	90	100	106	115	132

Figure 5.24 shows the phase change as a function of the nano-hole radius for an 800 nm-thick silver film, indicating that a change from 76 to 132 nm in a circular array is enough to obtain the required phase distribution. The calculated phase for eight hole radii is listed in Table 5.2. The fabricated sample (Fig. 5.24b) is designed by the fitting these results with the required phase shifts.

The spiral pattern in Fig. 5.24c is resulting from the interference between the probe beam and a coaxial Gaussian beam (the reference beam), revealing the presence of the helical wavefront of the optical vortex with topological charge $l = 1$. Compared with the catenary OAM generators [42], this design is independent of polarization and sensitive to the operational wavelength. Consequently, the choice of design would rely on the specific requirements on the optical performances. For broadband optical communication involving OAM and polarization multiplexing, the catenary structures based on geometric phase seem to be preferable.

5.4   Achromatic Optical Lens Based on Nanoslits Array

One common feature of flat optical devices based on local phase manipulation is the chromatic dispersion. As shown in the plasmonic beam deflector, the deflection angle for a normal incidence plane wave is

$$\theta = \sin^{-1}\left(\frac{\lambda}{2\pi}\frac{\partial\varphi}{\partial x}\right). \tag{5.4.1}$$

If the phase shift is a constant independent of frequency (or wavelength), the deflection angle would increase for larger wavelength, leading to strong wavelength-dependent behavior. When these structures are used as flat lenses, the focal length for a paraxial lens could be approximately written as

$$f = \frac{\lambda_0 f_0}{\lambda}, \tag{5.4.2}$$

where f_0 is the focal length at a wavelength of λ_0. This wavelength-dependent focal length would lead to strong chromatic aberration in images. Over the past several years, many different approaches have been investigated to overcome this issue [43–47]. As one representative example, here we show that properly tuned nanoslits could eliminate this aberration and realize achromatic beam deflecting, lensing, and imaging [48].

In general, to realize any desired functionality, the phase retardation of the optical component is required to compensate the phase retardation of propagation in free space. For example, in the traditional refractive lens, the light path in the lens is used to compensate the light path in free space propagation from the lens to the focal spot. The dispersion of required phase φ at the point r can be generally written as

$$\varphi(r, \lambda) = -\frac{2\pi}{\lambda}l(r) + C(\lambda), \tag{5.4.3}$$

where $l(r)$ is the physical distance between the interface at position r and the desired wavefront. The free parameter $C(\lambda)$ can be set as an arbitrary wavelength-dependent value to optimize the elements. Therefore, relative phase distribution $\Delta\varphi(r, \lambda) = \varphi(r, \lambda) - C(\lambda)$ may be used to substitute the absolute phase distribution $\varphi(r, \lambda)$. For an achromatic component, $l(r)$ should be a constant at the same point r so that the desired phase at an arbitrary point r is only a function of wavelength, so that Eq. (5.4.3) can be written as

$$\Delta\varphi(r,\lambda)\cdot\lambda = -2\pi l(r) = const. \tag{5.4.4}$$

Now, there is an assist parameter $\Delta\varphi\lambda$ to design the achromatic components. For an achromatic flat component, $\Delta\varphi\lambda$ should be dispersionless at any point.

When a MIM slit with subwavelength width is illuminated by TM-polarized radiation, the complex propagation constant β of SPP mode in MIM waveguide is given by the Eq. (5.1.1):

$$\tanh\left(\frac{\sqrt{\beta^2 - k_0^2\varepsilon_d}\,w}{2}\right) = -\frac{\varepsilon_d\sqrt{\beta^2 - k_0^2\varepsilon_m}}{\varepsilon_m\sqrt{\beta^2 - k_0^2\varepsilon_d}}, \tag{5.4.5}$$

where w is the width of slit, k_0 is the wave vector of light in free space, ε_d and ε_m are the permittivities of the dielectric medium filled in the slit and the metal. The dispersion of metal permittivity $\varepsilon_m(\omega)$ can be written in the Drude form:

$$\varepsilon_m(\omega) = \varepsilon_\infty - \frac{\omega_p^2}{\omega^2 + i\omega\gamma}, \tag{5.4.6}$$

where ω is the angular frequency of the incident electromagnetic radiation, ε_∞ is the permittivity at infinite angular frequency, ω_p is the bulk plasma frequency which represents the natural frequency of the oscillations of free conduction electrons, and γ is the collision frequency. At the frequency $\omega < \omega_p$ and $\omega \gg \gamma$, $\varepsilon_m(\omega)$ can be approximated as

$$\varepsilon_m(\omega) \approx -\frac{\omega_p^2}{\omega^2} \ll -1. \tag{5.4.7}$$

The transversal wavevector in the metal layer can be then simplified as

$$k_{\|m} = \sqrt{\beta^2 - k_0^2\varepsilon_m} \approx k_0\frac{\omega_p}{\omega}. \tag{5.4.8}$$

The right side of Eq. (5.4.5) is converted to

$$-\frac{\varepsilon_d\sqrt{\beta^2 - k_0^2\varepsilon_m}}{\varepsilon_m\sqrt{\beta^2 - k_0^2\varepsilon_d}} \approx \frac{\omega}{\omega_p}\frac{\varepsilon_d k_0}{\sqrt{\beta^2 - k_0^2\varepsilon_d}} \ll 1. \tag{5.4.9}$$

Since it is much smaller than 1, the left side of Eq. (5.4.5) can be expanded using Taylor series:

$$\tanh^{-1}(z) = z + \frac{z^3}{3} + \frac{z^5}{5} + \cdots + \frac{z^{2k-1}}{2k-1}. \tag{5.4.10}$$

By choosing only the first term, there is

$$\frac{\sqrt{\beta^2 - k_0^2 \varepsilon_d} w}{2} = \tanh^{-1}\left(\frac{\omega}{\omega_p} \frac{\varepsilon_d k_0}{\sqrt{\beta^2 - k_0^2 \varepsilon_d}}\right) \approx \frac{\omega}{\omega_p} \frac{\varepsilon_d k_0}{\sqrt{\beta^2 - k_0^2 \varepsilon_d}}. \qquad (5.4.11)$$

Now, the $\beta\lambda$ can be written as a function of the permittivity of dielectric ε_d and slit width w:

$$\beta\lambda = 2\pi \sqrt{\varepsilon_d \left(\frac{2c}{\omega_p w} + 1\right)} = const. \qquad (5.4.12)$$

Since the waveguide length h is a constant, there is

$$\beta h\lambda \approx \Delta\varphi \cdot \lambda = const., \qquad (5.4.13)$$

which implies an approximately achromatic performance. Different from Eq. (5.1.3) which predicts a small wavelength variation, here the propagation constant is inversely proportional to the wavelength. The large dispersion of metal permittivity is actually the reason why the nanoslits flat lens can be transformed from chromatic to achromatic. From another point of view, the propagation constant can be written as

$$\beta = k_0 \sqrt{\varepsilon_d \left(\frac{2c}{\omega_p w} + 1\right)}, \qquad (5.4.14)$$

while the mode index is a constant value:

$$n_{\text{eff}} = \sqrt{\varepsilon_d \left(\frac{2c}{\omega_p w} + 1\right)}. \qquad (5.4.15)$$

Once again, it can be seen that a larger propagation constant could be obtained with small w. However, when β becomes too large, the approximation made in Eq. (5.4.8) is not valid anymore. Moreover, when the imaginary part is taken into account, the approximation would be worse. Consequently, it seems that the above theory cannot be simply extended to the microwave regime.

The dispersive behavior of the MIM waveguide is demonstrated both theoretically and numerically. The basic unit of the lens based on MIM waveguide is shown by the cross section in the inset of Fig. 5.25a. The slit width w is varied from 20 to 100 nm, and the length of waveguide h is fixed at 3 μm. The relative permittivity of the material filled in the slit is assumed to be $\varepsilon_d = 1$ for air. Silver is chosen as the metal in this model due to its lower optical loss.

The dispersion of the MIM silts is shown as Fig. 5.25b, where $\Delta\varphi$ is defined as phase shift compared with the phase for a slit width of $w = 20$ nm. The propagation constant β gets the maximum when slit width w gets the minimum, so that

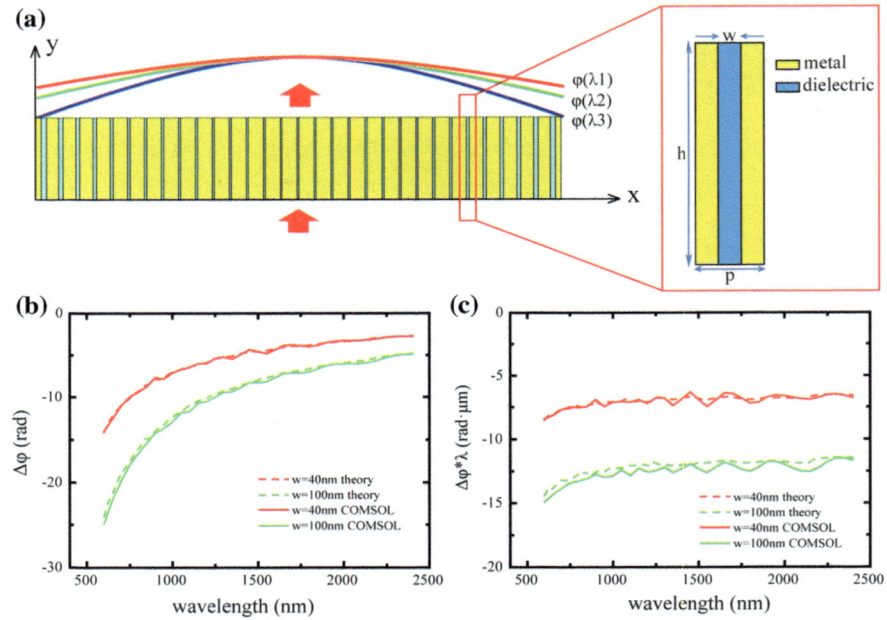

Fig. 5.25 Structural geometry and phase shift dispersion. **a** Side view of the achromatic lens, including 51 unit cells as shown in the inset. **b** Relative phase shift $\Delta\varphi$ through MIM structure at different wavelengths at varied slit widths in theoretical and numerical simulation. **c** Theoretical and numerical calculated $\Delta\varphi\lambda$ at different wavelengths for different slit widths. Reproduced from [48] with permission. Copyright 2016, The Authors

$\Delta\varphi$ is negative at $w > 20$ nm. Figure 5.25c shows the theoretically calculated and numerically simulated $\Delta\varphi\lambda$ with different silt widths at different wavelengths. Obviously, $\Delta\varphi\lambda$ varies rapidly for the wavelength smaller than 800 nm and becomes stable at a longer wavelength. This phenomenon matches the theoretical analysis. The slight oscillation can be explained by Fabry–Perot effect.

For a 1D achromatic lens, the required phase retardation as a function of spatial distance x can be calculated using Eq. (5.2.5). In a proof-of-concept experiment, the phase distribution is designed for $f = 5$ μm. The number of slits is 51 and the period of the structure is chosen to be 200 nm. As shown in Fig. 5.26a, the focus length is very close to 5 μm at different wavelengths. Insets of Fig. 5.26a show the electric-field intensity distribution of plasmonic lens illuminated by light at wavelengths of 1000, 1500, and 2000 nm, respectively. Although the sizes of the focal spot are different, which is determined by diffraction limit, the focus length is the same so this flat lens is achromatic.

To analyze the achromatic performance, the simulated phase distributions at 200 nm above the output plane of the flat lens are shown as solid lines in Fig. 5.26b. The dashed lines represent the theoretical results for target focal length at different wavelengths. The simulated phase distributions of output light with different wave-

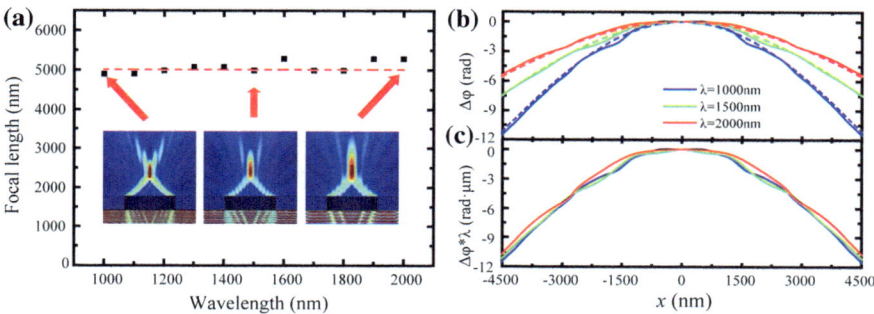

Fig. 5.26 Achromatic plasmonic lens based on nanoslits array. **a** Focal lengths of the designed achromatic lens at different wavelengths. Electric-field intensity distributions at the wavelength $\lambda = 1000, 1500,$ and 2000 nm are shown in the inset. **b** Numerical (solid line) simulated and ideal (dashed line) phase distribution at 200 nm above the output surface at the wavelength $\lambda = 1000$ nm (blue), 1500 nm (green) and 2000 nm (red). **c** Simulated spatial distribution of $\Delta\varphi\lambda$ at the wavelength $\lambda = 1000$ nm (blue), 1500 nm (green) and 2000 nm (red). Reproduced from [48] with permission. Copyright 2016, The Authors

lengths show good agreement with the theoretical prediction. Figure 5.26c shows the normalized $\Delta\varphi\lambda$ as a function of the x-position. Simulated $\Delta\varphi\lambda$ is almost the same, which leads to the same focal length. The slight focal shift can be explained by the little deviation of $\Delta\varphi\lambda$.

Compared with metasurface based on rectangular dielectric resonator [43] which gets achromatism at only a few discrete wavelengths and encounter much greater deviation at the other wavelengths, this nanoslits design can hold achromatic performance in a continuously broadband range $\lambda = 1000–2000$ nm. In addition, this kind of achromatic flat components can be simply designed without complex parameters scanning.

5.5 Super-Oscillatory Metalens

It is well known that the resolution of the optical system is dependent on the wavelength of the illuminating light and the effective aperture and numerical aperture (NA), as defined by the Rayleigh criterion [49]. Consequently, extensive efforts have been made to break this barrier for centuries by exploiting the evanescent wave containing fine details of objects [50–52]. These methods allow sub-diffraction imaging at the cost of complex near-field operation. Besides, some super-resolution fluorescent microscopes, including photo-activated localization microscope and stimulated emission depletion (STED) [53, 54], deliver impressive achievement in obtaining nanofeatures of bio-samples. Due to the pre-labeling of samples by specific dyes, it has limited applications. Therefore, it would be imperative to have an access to far-field noninvasive super-resolution imaging.

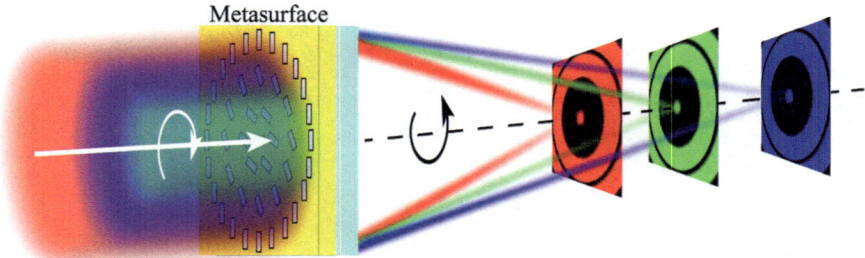

Fig. 5.27 Schematic of the ultra-broadband super-oscillatory lens based on rotated nanoapertures perforated in a thin metallic film. The red, green and blue components are focused at different longitudinal positions with similar super-oscillatory behaviors. Reproduced from [61] with permission. Copyright 2015, WILEY-VCH Verlag GmbH & Co. KGaA, Weinheim

In 1952, Giuliano Toraldo di Francia [55] extended the concept of super-gain antenna in the microwave to demonstrate that there is fundamentally no theoretical limit on the resolving power. Instead of being a theoretical limit, the Rayleigh criterion may be only a practical limit for uniform pupil functions. By changing the pupil functions, the width of the main focal spot could be suppressed to an arbitrarily small value at the sacrifice of brighter side lodes. This approach is well known as pupil filtering technology [56] and now termed as optical super-oscillation [57]. It was found that the super-oscillatory behavior occurs in a region, where the band-limited functions are able to oscillate faster than their highest Fourier components, which is somewhat similar to the weak measurements in quantum system. Based on the super-oscillatory interference, super-oscillatory lens (SOL) provides a promising approach for far-field noninvasive super-resolution imaging. SOL has been successfully applied to the sub-diffraction focusing and imaging as well as the heat-assisted magnetic recording in confocal scanning system [58, 59]. The key point of super-oscillatory imaging is the appropriate pupil filter, which is usually realized by binary amplitude elements, spatial light modulators or specially designed nanostructures with strong phase dispersion [60]. In 2015, an ultra-broadband super-oscillatory lens (UBSOL) was proposed and experimentally verified to achieve subwavelength focusing behavior for wavelengths over the whole visible and near-infrared light [61]. Figure 5.27 shows the operational principle, where space-variant anisotropic nanoapertures are used to combine the phases required by focusing and super-oscillatory wavefront manipulation in a single metasurface. The mechanism of the geometric phase is the same as that shown in Chap. 2, which means that under circular polarization incidence, a rotated anisotropic aperture would generate a cross-polarized component with an additional phase shift twice as the rotation angle. In the equivalent circuit theory, the impedance of the apertures array (or so-called metallic mesh) is described by the catenary of equal strength [9, 62] combined with optical nanocircuits [63].

Although geometric phase guarantees the dispersionless super-oscillatory phase, the focal length of a common metalens is dependent on the wavelength. In the following, a white-light super-oscillatory imaging system is realized by using the

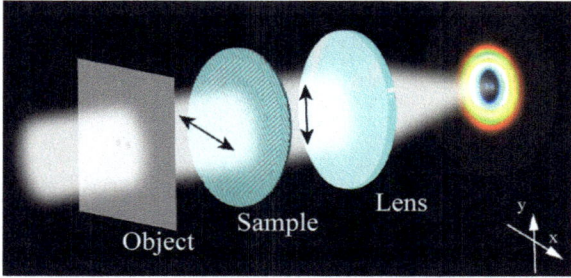

Fig. 5.28 Schematic of achromatic broadband super-resolution imaging by super-oscillatory meta-surface. By combining the metasurface sample and achromatic lens, the details otherwise missing can be observed in the imaging plane. Reproduced from [64] with permission. Copyright 2018, WILEY-VCH Verlag GmbH & Co. KGaA, Weinheim

dispersionless spin–orbit interactions in geometric metasurfaces, which provides perfect 0 and π phase-only manipulation for the white light. Super-oscillatory imaging patterns for extended targets were successfully obtained for visible light ranging from 400 to 700 nm and the sub-diffraction resolution is about 0.64 times of the Rayleigh limitation [64].

As illustrated in Fig. 5.28, the metasurface filter is positioned at the exit-pupil plane of the imaging optical system, where objects illuminated with white incoherent light could be imaged at the image plane away from the metasurface filter. For the simplicity of analysis, the image performance could be regarded as an achromatic lens combined with the metasurface filter with a phase function $\varphi_{MF}(r)$. The point spread function (PSF) of the system is

$$I(\rho) \propto \left(\frac{1}{\lambda f}\right)^2 \left| \int_0^R \exp[i\varphi_{MF}(r)] J_0\left(\frac{2\pi r\rho}{\lambda f}\right) r \, dr \right|^2, \tag{5.5.1}$$

where λ is the light wavelength, f the focal length of the achromatic lens, and R the metasurface radius. The goal of designing a metasurface filter (MF) is to yield a PSF with a minimum width of the main lobe and small side lobe levels within the local field of view (FOV). This could be performed by optimization methods like linear programming methods and particle swarm optimization. Instead of using phase retardations in conventional binary microstructures etched on transparent substrates or spatial light modulators, the geometric phase is utilized here. The polarization conversion efficiency is dependent on the difference of the transmission efficiency for two orthogonal polarizations. In principle, they are both related to the width and period of the grating, which can be described by a catenary of equal strength [9, 62].

As a demonstrative example, a metasurface is designed to contain arrays of customized rectangular metallic gratings with $\xi = 45°$ and $\xi = 135°$ etched on a glass substrate, where ξ is the angle between axis x and the orientation of the gratings. The

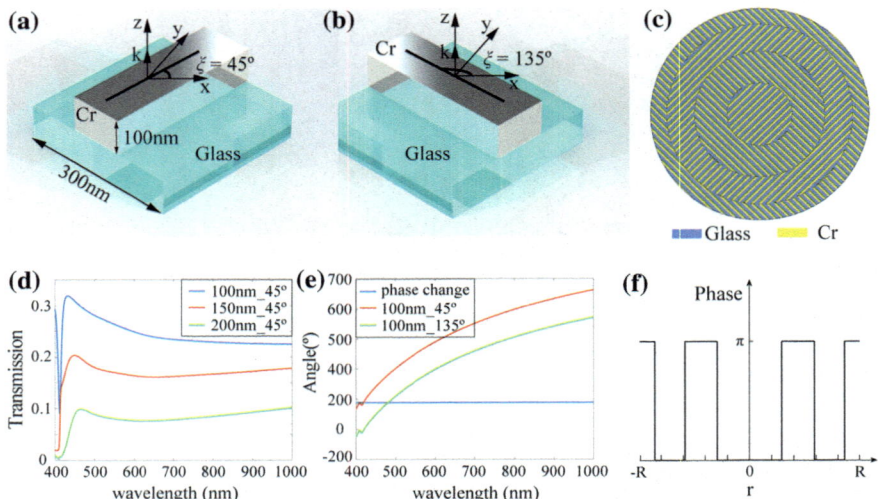

Fig. 5.29 Schematic of the unit cell with **a** $\xi = 45°$ and **b** $\xi = 135°$. **c** Structure of the metasurface whose diameter is 9 mm. **d** Transmission efficiency of the unit cell with different widths of Cr. **e** Phase shift of unit cell with different orientations. **f** Radial phase distribution of metasurface. Reproduced from [64] with permission. Copyright 2018, WILEY-VCH Verlag GmbH & Co. KGaA, Weinheim

rectangular nanometallic grating with two orthogonal orientations, 300×300 nm period, 100 nm thickness are shown in Fig. 5.29a, b. Since the efficiency has a strong dependence on the width of the grating theoretically, the light-field manipulation ability of the unit cells with the width of 100, 150 and 200 nm are simulated by the commercial software CST Microwave Studio. For simplicity of fabrication, the chromium (Cr) is used as the grating material. As illustrated in Fig. 5.29d, the Cr grating with the width of 100 nm is finally chosen to achieve higher energy efficiency in a broad spectrum from 400 to 1000 nm. Furthermore, the modulated phase change of the transmitted light field is approximately fixed to be 180°.

In Fig. 5.29c, f, the metasurface is designed to contain four regions with the normalized radius of each phase-jump position being $r_1 = 0.297$, $r_2 = 0.594$ and $r_3 = 0.85$. These parameters have been demonstrated to obtain 0.6 times the full width of the Airy pattern ($\lambda = 532$ nm, $R = 4.5$ mm and $f = 500$ mm). The unit cells in Fig. 5.29a, b are distributed alternately in the neighboring annulus. As the orientation of the units in neighboring circles are orthogonal to each other, the metasurface could provide nearly perfect 0 and π dispersionless phase manipulation for incident white light. In the local coordinates, the transmission and reflection property can be described by the generalized Fresnel's equation combined with the equivalent impedances. The unique dispersionless phase-only manipulation property of the metasurface for white light plays a significant role in the broadband super-oscillatory imaging.

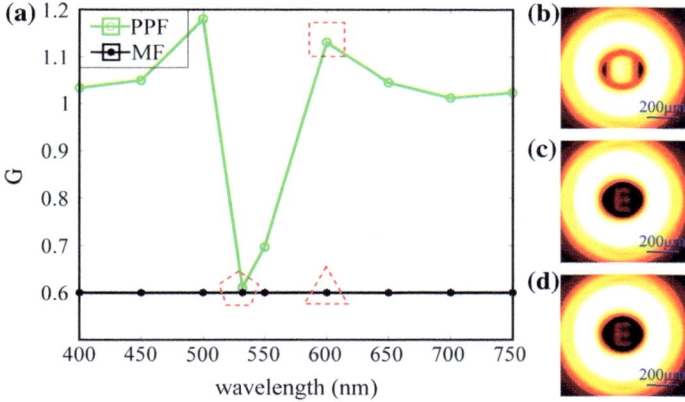

Fig. 5.30 **a** Comparison of G between PPF and MF at wavelengths from 400 nm to 750 nm. **b–d** Simulation results for imaging of "E", corresponding to the cases indicated by the square, triangle, and pentagon, respectively. The "E" has a size of 100 μm × 124 μm. Reproduced from [64] with permission. Copyright 2018, WILEY-VCH Verlag GmbH & Co. KGaA, Weinheim

To show the advantage of MF over classic 0–π phase pupil filter (PPF), these two super-oscillatory imaging performances are compared. Figure 5.30a illustrates the resolution factor G of these two methods for light wavelength ranging from 400 to 750 nm. Obviously, the PSF width of PPF method shows a strong dependence on the wavelength, where the super-oscillatory effect occurs closely at the working wavelength 532 nm and G increases greatly as wavelength goes slightly away from it. This phenomenon indicates that super-oscillation is sensitive to the change of phase modulation, even for 10% variance generated by a binary phase structure. According to the phase retardation mechanism, the phase shift of a binary structure can be written as

$$\Delta\Phi = \frac{2\pi}{\lambda}\Delta d(n - 1), \tag{5.5.2}$$

where Δd and n are the etched depth and refraction index of the glass, which are about 578 nm and 1.46, thus the phase shift is π at $\lambda = 532$ nm. When the wavelength is shifted to 582 nm, the phase shift becomes 0.914 π. Therefore, the PPF can only be applied to the imaging within a narrow bandwidth. This can be further demonstrated by the simulation images of "E" object in Fig. 5.30b.

Owing to the dispersionless phase shift of the designed MF, it could overcome the wavelength dependence of PPF. As illustrated in Fig. 5.30c, d, the images of "E" could be resolved clearly and have approximately the same imaging quality and contrast for both wavelengths at 600 and 532 nm. It is worth noting that the resolution enhancement factor G is scaled with respect to the Airy spot size at the correspondent wavelength. The resolving ability would scale linearly with the wavelength and the resolution of a MF imaging system is a sum for all the wavelength

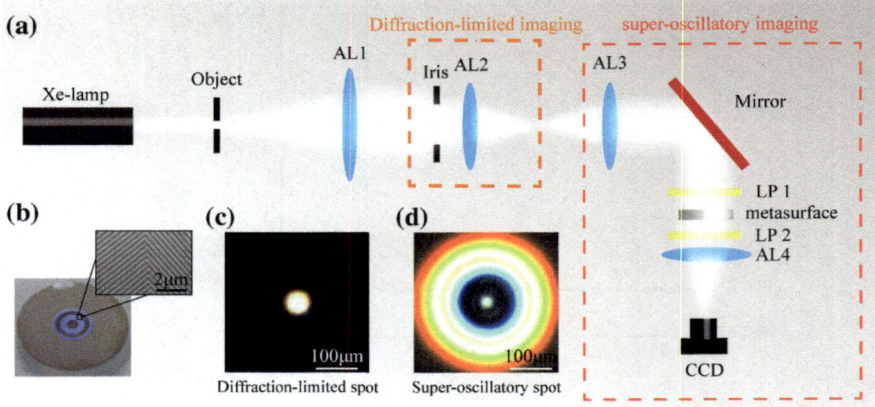

Fig. 5.31 **a** Experimental setup comprising a Xe-lamp, an optical collimator (AL1, f1 = 1000 mm), an iris, three same achromatic lenses (AL2, AL3, AL4, f = 500 mm), a mirror, two linear polarizers (LP1, LP2), a designed metasurface and a CCD. **b** The proposed metasurface. The inset indicates the distribution of the Cr gratings between two annuluses. **c** Diffraction-limited image of a 20 μm hole. **d** Super-oscillatory image of a 20 μm hole. Here the positions and distances of elements are not in accurate scale with the experiment optic setup for a good visualization. Reproduced from [64] with permission. Copyright 2018, WILEY-VCH Verlag GmbH & Co. KGaA, Weinheim

range, which is approximately equal to that of the central wavelength. On the other hand, it is believed that an arbitrarily high resolution would be theoretically possible for MF if the negative issues like fabrication challenges, ultra-low energy efficiency could be overcome. From these points of view, MF could significantly promote the development of broadband super-resolution imaging system.

The experimental setup consists of a diffraction-limited imaging section and a super-resolution imaging section. As shown in Fig. 5.31, the object is placed at the front focal plane of the optical collimator (compound achromatic lens, AL1), acting as the infinite distant target. Firstly, it is imaged at the focal plane of lens (AL2) with finite aperture and the image here is diffraction-limited. Following it is the relayed super-oscillatory imaging section, composed by a 4*f* system with two lenses AL3 and AL4. At the exit pupil of the system (before the lens AL4), there is a pupil filter group with a specially designed metasurface filter sandwiched by two orthogonal linear polarizers. The super-oscillatory image is recorded in the CCD at the second imaging plane.

First of all, the PSF is measured by using a 20 μm size transparent circular hole on an opaque screen. For the case without the filter, the measured PSF at CCD plane is an Airy spot with the full width of 82.8 μm, as shown in Fig. 5.31c. When the filter group is inserted in the optical path, the PSF in Fig. 5.31d exhibits an obvious super-oscillatory pattern with a much smaller bright central spot surrounded by a bright ring about few spot sizes away. The super-oscillatory central spot size is about

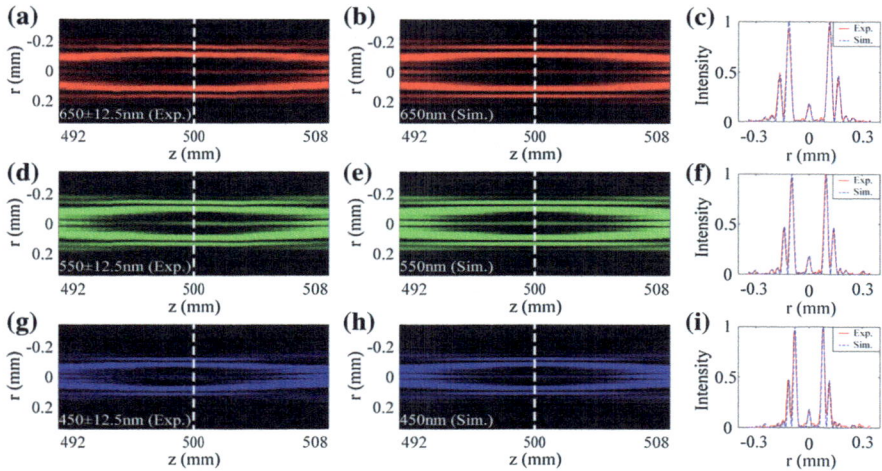

Fig. 5.32 **a, d, g** Measured axial PSF with narrow band of 650 ± 20, 550 ± 20 and 450 ± 20 nm. **b, e, h** Simulated axial PSF with fixed wavelength of 650, 550, and 450 nm. **c, f, i** Normalized PSF profiles along the radial direction in the focal plane of the theoretical and experimental results. Reproduced from [64] with permission. Copyright 2018, WILEY-VCH Verlag GmbH & Co. KGaA, Weinheim

51.75 μm, being 0.625 times of the Airy spot. This is a bit larger than the design result $G = 0.6$ owing to some inevitable errors induced by fabrication and measurement.

To precisely evaluate the super-oscillatory PSF in the experiment, three monochromatic light performances for the wavelengths of 650 ± 20, 550 ± 20, and 450 ± 20 nm are measured in variant image planes and compared with simulation results. As shown in Fig. 5.32, the measured and simulated axial PSFs show nearly identical profiles, which demonstrate the good achromatic ability of this method. The PSFs in the focal plane are also plotted along the radial direction in Figs. 5.32c, f, i.

Figure 5.33 shows the experimental imaging results for two transparent holes with 20 μm diameter and 60 μm center-to-center distance. The two-point objects are totally unresolved for the case without MF, and are clearly distinguished with MF. The cross-section profiles of the two images are plotted in Fig. 5.33c, demonstrating that the super-oscillatory image clearly resolved the two holes. Figure 5.33d plots the contrast of measured images when the distance between the two holes is ranging from 20 to 120 μm. By assuming the resolving contrast criterion of about 0.153, the minimal resolvable distance of two points is about 60 μm and the minimal resolving ability is demonstrated to be approximately 0.64 times of the Airy pattern.

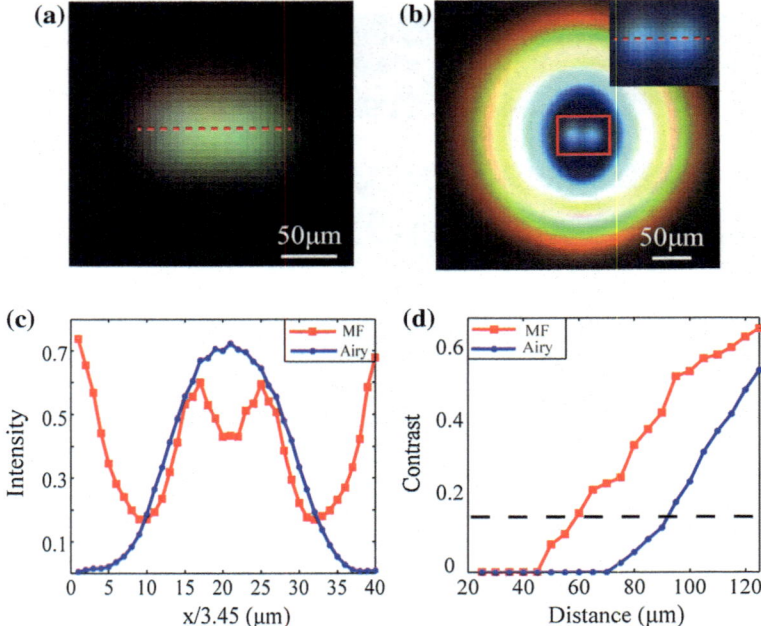

Fig. 5.33 **a** Diffraction-limited image. **b** Super-oscillatory image. **c** Intensity distribution of the central place marked by dashed lines in (**a**) and the insert of (**b**). **d** Experiment imaging contrast for variant center-to-center distances of two-hole targets. The dashed line is the contrast of Rayleigh criterion. Here the contrast is defined as $(Ip - Iv)/(Ip + Iv)$, where Ip is the peak intensity and Iv the central valley intensity. When the two-point objects are unresolved, the Ip is equal to Iv. Reproduced from [64] with permission. Copyright 2018, WILEY-VCH Verlag GmbH & Co. KGaA, Weinheim

5.6 Structural Colors and Color Holography

5.6.1 Structural Colors Based on Linear Dispersion of Catenary Plasmons

The manipulation of phase delay in subwavelength structures has important application in the generation of structural color based on destructive and constructive interferences. As early as the discovery of EOT in 1998 [3], subwavelength holes and gratings have been widely utilized as high-purity color filters [65].

To further increase the transmission efficiency, Xu et al. proposed a nanoresonator structure based on aluminum and zinc selenide to filter white light into individual colors [66]. Based on the linear dispersion of antisymmetric catenary plasmons, the photon–plasmon–photon conversion can be realized efficiently at any specific resonant wavelength. Compared with traditional color-filtering methods, the new design significantly improved the absolute transmission, bandwidth, and compactness. In

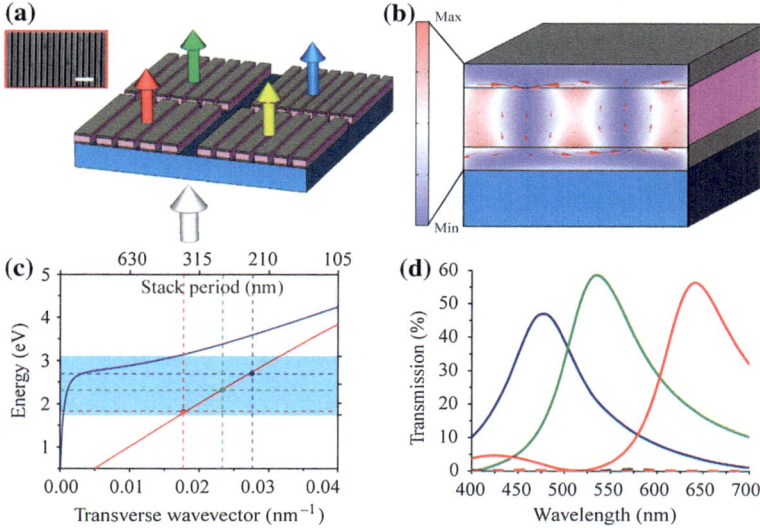

Fig. 5.34 **a** Schematic diagram of the proposed plasmonic nanoresonators on a magnesium fluoride substrate. Inset is the scanning electron microscopy (SEM) image of the fabricated device. Scale bar, 1 μm. **b** Cross-section of the catenary shaped magnetic field intensity and electric displacement distribution (red arrow) inside the MIM stack at a wavelength of 650 nm with 360 nm stack period. **c** Plasmon dispersion in MIM stack array. The shaded region indicates the visible range. **d** Simulated transmission spectra for the RGB filters. The solid and dash curves correspond to TM and TE illuminations respectively. Reproduced from [66] with permission. Copyright 2010, Macmillan Publishers Limited

addition, the filtered light is naturally linearly polarized, making it attractive for direct integration in liquid crystal displays without a separate polarizer layer.

Figure 5.34 shows the schematic of the operating principle. The color filter is designed based on periodic grating perforated in MIM stacks. When the incident light is coupled into the gap region as catenary plasmon, an interference field (Fig. 5.34b) will be formed and contribute to increase or decrease the transmission efficiency determined by the phase shift Φ. According to the dispersion curve shown in Fig. 5.34c, the linear dispersion of the antisymmetric mode indicates the effective mode index n is independent of the frequency and wavelength. Consequently, the resonant wavelength is directly determined by the grating period, which can be written as $p = \lambda \Phi / 2n\pi$. By linearly tuning the period, the transmission wavelength could be readily controlled. Figure 5.34d shows the transmission spectra for periods of 360, 270 and 230 nm.

As an experimental validation of the design, a gradient grating was fabricated as shown in Fig. 5.35. The optical microscopy image demonstrates that the transmission peak has been tuned from red to green and deep blue. One additional advantage of this method is the transmission peak may be dynamically tuned by adding tunable

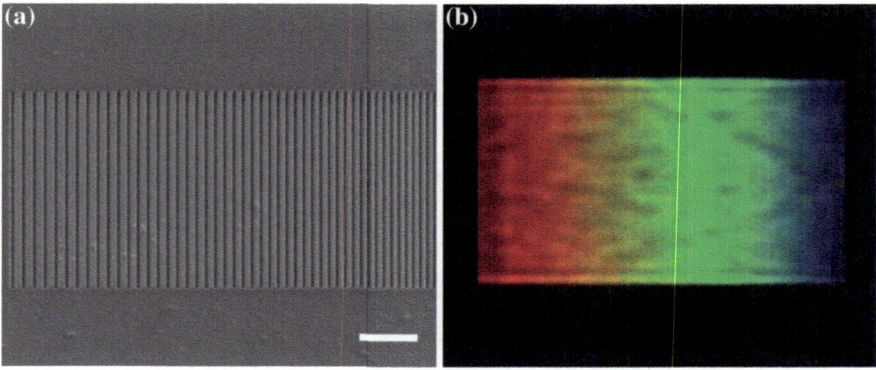

Fig. 5.35 a SEM image of the fabricated 1D plasmonic spectroscope with gradually changing periods from 400 to 200 nm (from left to right). Scale bar, 2 μm. **b** Optical microscopy image of the plasmonic spectroscope illuminated with white light. Reproduced from [66] with permission. Copyright 2010, Macmillan Publishers Limited

materials such as polyaniline (PANI) and poly(2,2-dimethyl-3,4 propylene dioxythiophene) (PolyProDOT-Me$_2$) in these slits [67].

5.6.2 Structural Colors Based on Polarization Conversion

In the previous discussion, the metallic gratings are either used to introduce width-dependent phase retardation or polarization-dependent transmission and geometric phase. In this section, the catenary optical fields in reflective metallic grating are used to realize polarization-dependent reflection and geometric phase. This mechanism is used to achieve high-purity red, green and blue (RGB) structural colors. The plasmonic shallow grating (PSG) produces colors by photon spin restoration, which reflects a circularly polarized light to its co-polarized state at specific wavelengths. FWHM of ~16 nm with high efficiency (~75%) was theoretically obtained and experimentally demonstrated.

The concept is schematically illustrated in Fig. 5.36a. To realize circular polarization inversion, a 180° phase difference between the two linearly polarized components along the x and y directions (Φ_x and Φ_y) is necessary. In the optical experiment setup, relatively narrow bandwidth of commercial quarter-wave plate makes it difficult to generate broadband circularly polarized light and measure the co-polarized counterpart. In order to have a clear vision of the spin restoration, a linearly polarized incident light polarized at 45° with respect to the grating is used instead, since the PSG will convert it to its cross-polarization when $|\Phi_x - \Phi_y| = 180°$ is achieved.

Silver was chosen here due to its low loss in the visible bands. With the simultaneous contributions of the propagating surface plasmons (PSP) resonance and localized surface plasmons (LSP) resonance, a sharp peak occurs at the reflective

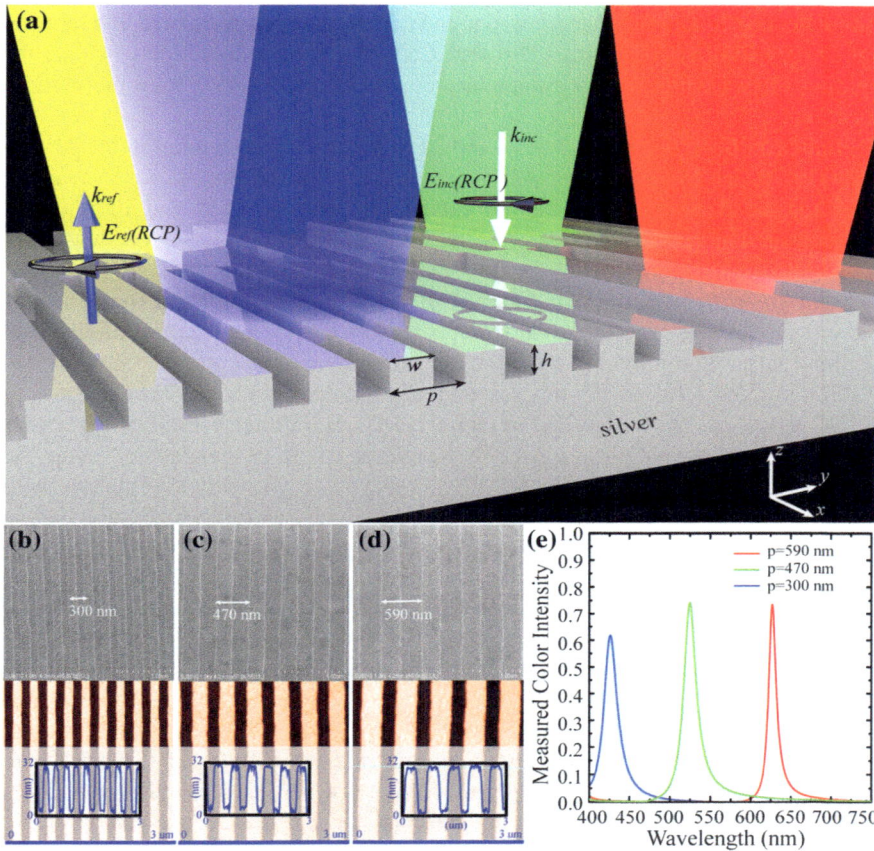

Fig. 5.36 Structural color filters formed by PSG. **a** Schematic diagram of the principle. The white and blue arrow indicates the incident and reflective right-handed circular polarization. The black arrows depict the spin state of the incident and reflective wave. The red, yellow, green, and blue beams represent the reflected waves with restored spin states. The duty cycle (w/p) is fixed to be 0.55. **b–d** The SEM and atomic force microscope (AFM) images of the samples with periods of 300, 470, and 590 nm. **e** Experimentally measured reflection spectra of the cross-polarized reflection for 45° polarized light at normal incidence. Reproduced from [69] with permission. Copyright 2018, Xiangang Luo et al.

cross-polarization spectra, so that a specific color can be efficiently filtered out. In order to realize a broad palette of colors besides the three primary colors of GRB, it is necessary to change the periods of the PSGs to make the spectral peak locations undergo a continuous blue-shift from 668 to 430 nm. The proposed PSG consists of a one-dimensional periodic arrangement of silver grooves etched on a silver film as shown in Fig. 5.36a. The dimensions are chosen as $w = 0.55p$, $h = 30$ nm, where p-w and h represent the width and depth of the groove, respectively. Here p represents the period, which is kept subwavelength to avoid unexpected high-order diffractions.

The dielectric function of silver is obtained from the data measured by Johnson and Christy [68]. A commercial software CST MWS was adopted to calculate the building blocks while unit cell boundary conditions were used along the x and y directions.

Figures 5.36b–d displays the SEM and AFM images of three samples (the periods are 300, 470, and 590 nm) fabricated by interference lithography. As observed in the SEM images, all three prepared filters exhibit uniform line width and high fidelity. The measured morphology images of a small part of the PSG indicate that the texture is patterned with 30 nm peak-to-valley profile and a surface roughness of about ~0.5 nm (root-mean-square) is achieved. Figure 5.36e depicts the experimentally measured reflective cross-polarization spectra. Three sharp peaks are observed at 627 nm (red), 525 nm (green), and 430 nm (blue), while the corresponding periods of PSGs are 590, 470 and 300 nm, respectively. Reflective efficiency peaks reach up to 75% with the smallest FWHM to be only ~16 nm. Furthermore, the magnitudes at off-resonant wavelengths are strongly suppressed, both of which help to improve the purity of the generated colors. It is remarkable that such fascinating performance is obtained by the PSGs consisting of only 30 nm-thick functional layer mounted on the silver mirror (~100 nm), which enables the construction of ultrathin chromatic display devices.

The grating is fabricated on silicon via laser interference lithography and transferred to silver film by template stripping [70]. It is of great importance to smooth the PSG since the surface roughness will deteriorate the performance by inducing scattering loss, which broadens the resonant bandwidth, especially when the groove depth is only 30 nm. To experimentally measure the performance of the PSG color filters with a size of 30 mm × 30 mm, a commercial spectrophotometer is utilized. For the proposed PSG color filters, a broad palette of RGB colors can be continuously produced by varying the periods of the gratings.

Figure 5.37a shows the calculated and measured reflective cross-polarization spectra with the period varying from 300 to 630 nm. All the values have been normalized to their maxima. The line shapes of the measured data closely resemble those of simulated results, except for some slight differences when the period is 530, 550, 560, and 630 nm. These discrepancies mainly result from structural defects caused by fabrication errors. Another reason leading to the difference is that the incident light is not perfectly normal to the sample. Figure 5.37b depicts the locations of reflective peaks obtained from the spectral results. Interestingly, the experimental data (red circles) exhibits considerable agreement to the simulated ones (black squares). Figure 5.37c reveals the experimentally measured optical images of the samples yielded by the PSGs with periods of 300, 390, 410, 470, 510, and 590 nm.

To have a better understanding of the high-purity colors and tunable region, the corresponding chromatic coordinates are calculated and plotted in CIE 1931 chromaticity diagram as shown in Fig. 5.37d [71]. All the corresponding chromatic coordinates are depicted as black dots located near the CIE boundary, which indicate the region of monochromatic colors. The experimentally obtained chromatic coordinates reasonably agree with their simulation counterparts. Slight variations are attributed to the fact that asymmetric reflection line shapes decrease the purity of colors [72].

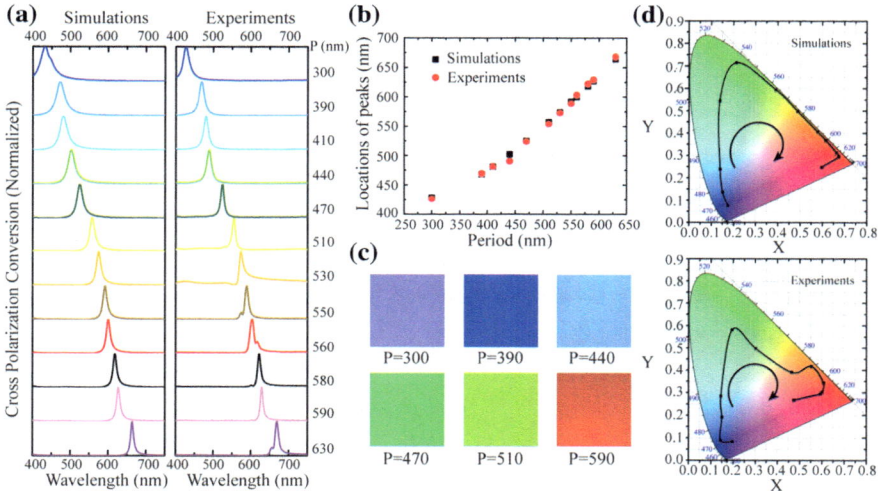

Fig. 5.37 **a** Normalized cross-polarized spectra of the PSG when the period varies from 300 to 630 nm. **b** Reflective peak locations of the simulated (black squares) and measured (red circles) results shown in (**a**). **c** Experimental images of the colors. A broad palette of color with high contrasts is realized. **d** Chromatic coordinates in CIE 1931 diagram obtained from the simulated and measured spectra. The black arrows indicate the variable tendency according to the increasing of the grating period. Reproduced from [69] with permission. Copyright 2018, Xiangang Luo et al.

Anyway, the color gamut in CIE 1931 diagram achieved by this design is larger than that in many previous works [66, 73, 74]

Besides the realization of high-purity color filters, the feasibility of high-resolution display of arbitrary chromatic patterns is investigated. As an example, three red characters "IOE" in a blue background are experimentally demonstrated. The logo fabricated by nested interference lithography spans 30 mm × 30 mm with a group of silver grooves exhibiting two periods: 600 nm for the red characters "IOE" and 400 nm for the blue background. Figures 5.38a, b show the SEM images of the different periodic building blocks constituting characters and background, where the insets display zoom-in views of the arrays. Uniform dense line-patterns imply that this PSG can be readily applied in high-resolution chromatic display. The reflective cross-polarization photographs taken for incident light polarized at 45° are shown in Fig. 5.38c. It can be seen that a bright red "IOE" sharply contrasts the blue background. Furthermore, the two distinct colors are still observed even at the corners and edges of the logo.

Subsequently, encrypted color display was experimentally demonstrated. When rotating the linear polarizer in front of the CCD camera to polarize the reflective light at −20° with respect to the grooves, a dim red acronym "IOE" and green background emerge as depicted in Fig. 5.38d. Figure 5.38e indicates that all the colors vanish when the reflective light and incident light possess the same polarization state. The sample looks white and maps to the central point in the CIE 1931 diagram. These

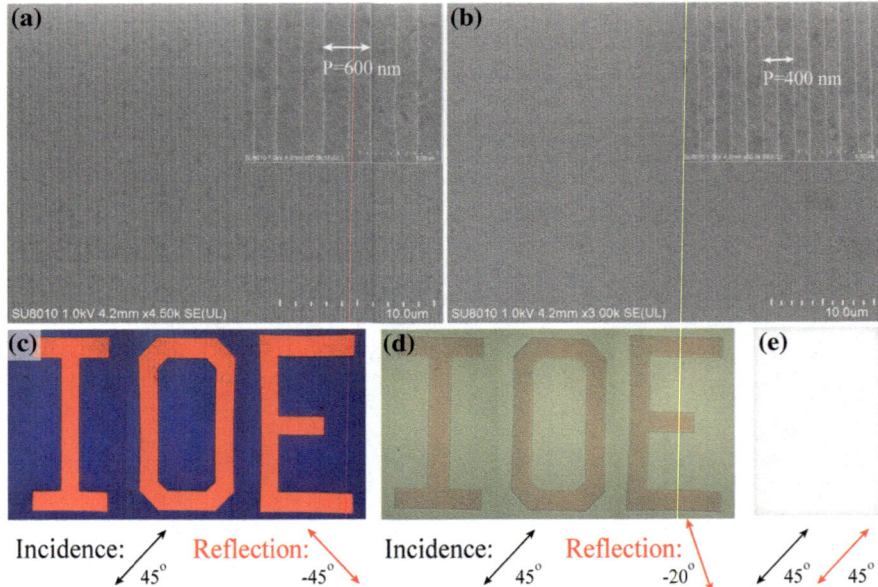

Fig. 5.38 SEM images of the building blocks for the logo with a period of **a** 600 nm and **b** 400 nm. **c–e** Microscopic images of the sample for 45° polarized incident light. The polarization direction of reflection light is (**c**) −45°, **d** −20°, and **e** 45°. Reproduced from [69] with permission. Copyright 2018, Xiangang Luo et al.

fascinating phenomena offer platforms for encryption applications. The encrypted information recorded in the PSG can be only unscrambled by a couple of specific incident polarizations and/or reflective helicities. In addition, PSG composed of silver makes the security tags or encrypted information readily disappear by natural oxidation, presenting a difficult-to-reuse guarantee and enhancing the safety.

In order to explore the physical mechanism inside the spin restoration for circularly polarized light or cross-polarization conversion for 45° polarized wave, the reflective amplitude and the phase difference between two linearly polarized components along the x and y directions are calculated as depicted in Fig. 5.39a. For a one-dimensional PSG with a period of 400 nm illuminated by a normally incident linearly polarized light at a polarization angle of 45°, the reflective amplitude of the x-component electric field remains flat while the y-component spectra line undergoes a dip at a wavelength of 474 nm. Interestingly, the blue curve in the inset spectrum reaches a sharp peak at the same wavelength, which indicates that a phase difference of 180° is achieved between the orthogonal linear polarizations ($|\Phi_y - \Phi_x| = 180°$). This is striking since the thickness of the grating is only 30 nm, which is ∼1/16 of the operating wavelength.

The electromagnetic response of the PSG is analyzed with CST MWS, and compared with the analytic PSP dispersion equation [2]:

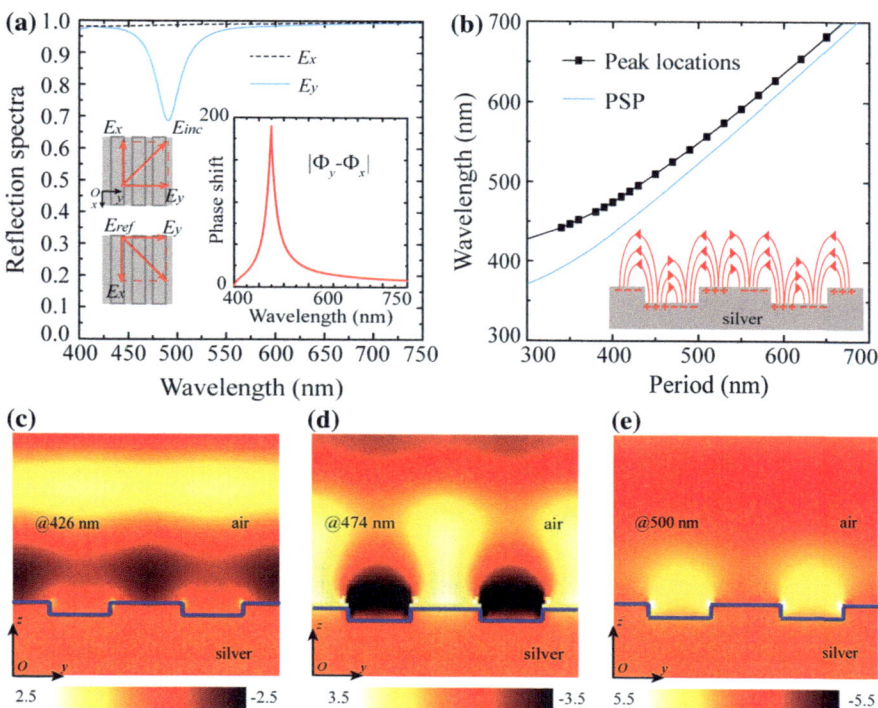

Fig. 5.39 **a** Reflective spectral amplitudes for x- and y-polarized waves. The inset shows the phase difference between the two orthogonal polarizations. **b** Comparison between the reflective peaks and PSP resonant wavelengths corresponding to variable grating periods. The inset depicts the collective oscillation of free electrons at silver–air interface. **c–e** 2D color maps of the electric-field (E_y) distribution with period of 400 nm at **c** $\lambda = 426$ nm, **d** $\lambda = 474$ nm, and **e** $\lambda = 500$ nm, respectively. Reproduced from [69] with permission. Copyright 2018, Xiangang Luo et al.

$$\lambda_{PSP} = p \sqrt{\frac{\varepsilon_m \varepsilon_d}{\varepsilon_m + \varepsilon_d}}, \qquad (5.6.1)$$

where λ_{PSP} is the vacuum wavelength, p the period of the PSG and equals to the effective wavelength for PSP, ε_d and ε_m the dielectric functions of the air and silver, respectively. We stress that this equation describes the resonant vacuum wavelength, rather than the wavelength of SPP on a single metal–dielectric interface, which is usually written as

$$\frac{2\pi}{\lambda_{SPP}} = \frac{2\pi}{\lambda} \sqrt{\frac{\varepsilon_m \varepsilon_d}{\varepsilon_m + \varepsilon_d}}. \qquad (5.6.2)$$

As plotted in Fig. 5.39b, the blue curve corresponding to the PSP dispersion nearly overlaps the calculated peak locations, which signifies that PSP resonance dominates the spin restoration. The less-than-ideal superposition may result from the LSP tightly confined at the edges of the grooves, which alters the dispersion diagram of the surface plasmons. In fact, Eq. (5.6.1) is only correct when the shallow grating does not change the propagation constant of PSP. Note that both LSP and PSP show obvious catenary properties.

In an effort to get a clear visualization of the contributions made by the plasmonic resonances, the electric-field distributions at the vertical cross section with a period of 400 nm are plotted in Fig. 5.39c–e, where the field amplitudes (E_y) are normalized to the incidence at the wavelengths of 426, 474 and 500 nm, respectively. As depicted in Figs. 5.39c, d, the electric field is enhanced at the silver–air interface and coupled with that in the adjacent elements, verifying the excitation of PSP resonance. However, Fig. 5.39e shows that the reinforced electromagnetic fields mainly locate at the corners and sidewalls of silver groove, which is a feature of LSP mode [75].

5.6.3   Color Holography

Based on catenary optical fields and anisotropic impedance dispersion, metallic gratings and anisotropic nanoapertures are promising candidates for both the manipulation polarization and geometric phase. Besides useful in flat lensing and orbital angular momentum generation, these structures have enabled flexible design of computer-generated holograms. Although these meta-holograms provide sufficient viewing range (covering the entire space), the new challenge is how to multiplex multi-wavelengths into one metasurface simultaneously and eliminate the crosstalk among different wavelengths. Recently, researchers demonstrated multicolor spectral modulation by integrating 3 plasmonic pixels with different sizes into one metasurface [76]. These 3 plasmonic pixels independently respond to RGB lights. Since the spectral selectivity is not perfect, the pixel designed for one wavelength may respond to other unintended wavelengths.

To solve the cross-talks between different channels, a space division method was developed to create multicolor and even 3D meta-hologram using one single type of plasmonic pixel [77]. This approach represents a great advance for meta-holography because of the following reasons. First, it provides a methodology to overcome the fundamental crosstalk limitation for meta-hologram. A schematic diagram of this methodology and a measured color image are shown in Fig. 5.40a, b. In the design, the imaging patterns of different colors are separated apart spatially. Laser beams are irradiated at an angle from the vertical axis, which shifts the imaging patterns and merges into the final hologram image. By this method, the crosstalk among different wavelengths can be eliminated through recording the imaging patterns separately on a broadband hologram. Second, different from conventional multicolor meta-holograms, which rely on several types of plasmonic pixels, this approach uses one plasmonic pixel to modulate broad bandwidth wavelengths, which results in an increased viewing range of the transmitted light due to the reduced pixel period on the metasurface. Third, a full-color meta-holographic image in the 3D space was

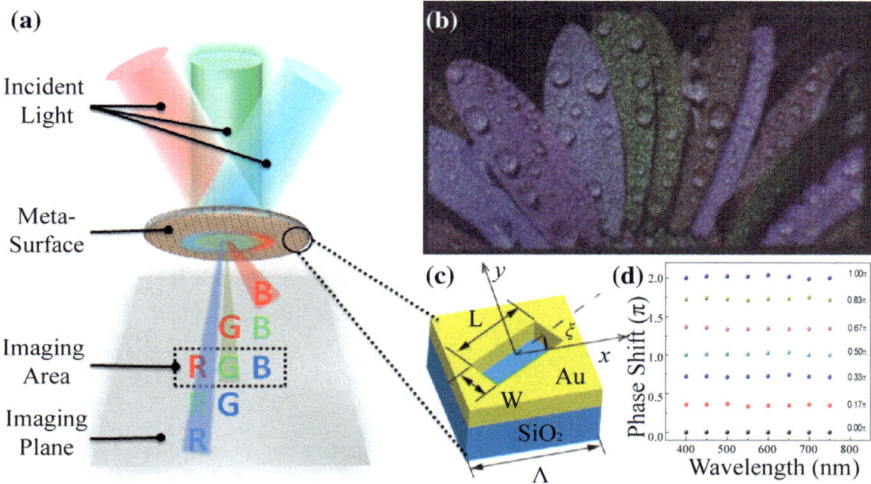

Fig. 5.40 Schematic of the methodology. **a** Space division method for meta-hologram. **b** Experimental holographic image of a "flower" RGB image. **c** Structure of a plasmonic nano-slit aperture. **d** Simulation results for the phase shift with different ξ (in radians) from 400 to 750 nm wavelength. (Simulation settings: material of the metal is Au; the thickness of the Au film is 120 nm, period $\Lambda = 200$ nm, length $L = 140$ nm, width $W = 60$ nm). Reproduced from [77] with permission. Copyright 2016, The Authors

experimentally demonstrated. The virtual object was displayed in the space which has the volume of 324,000 μm^3. It was the first realization of multicolor 3D meta-holography based on plasmonic metasurface, which may be one big step toward the realization of matrix-free 3D display with a broad viewing angle observed with naked eyes. Furthermore, this approach is also promising for a wide range of other applications such as data storage, security, and authentication.

As shown in Fig. 5.40c, the building block to construct the meta-hologram is a set of space-variant nanoslit apertures perforated in a metallic film on the quartz substrate, which is based on the geometric phase to modulate the transmitted light. According to the geometric phase or the so-called photonic spin–orbit interaction, the phase of the cross-polarized light (Φ) depends on the orientation angle (ξ) of the nanoslit by the relation $\Phi = \pm 2\xi$ for circular polarization incidence. It is worth to mention that such nanoslit is efficient for a broadband wavelength covering the visible range (380–780 nm). Figure 5.40d shows the simulation results of the phase shift, which differs from the theory slightly as a result of the contribution from the propagating phase (the thickness of the aperture is not zero). This type of nanoaperture was utilized to design one single holographic image that contains all the imaging patterns corresponding to different colors. These images are separated and positioned at different spatial locations. Consequently, these individual imaging patterns are independent of each other so that the crosstalk among colors is eliminated.

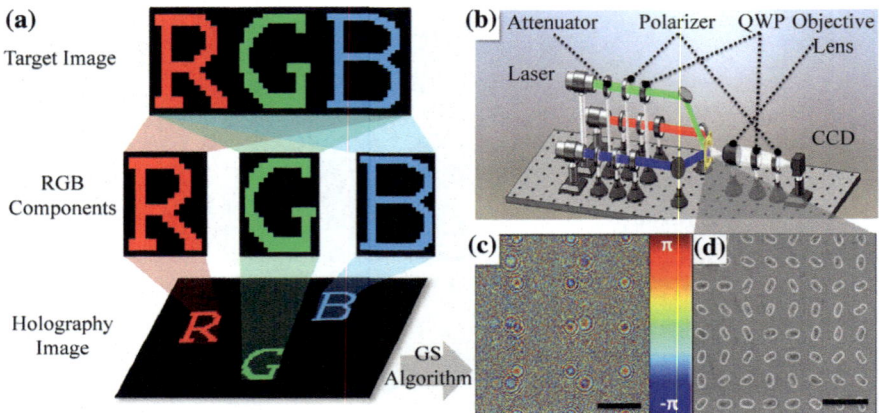

Fig. 5.41 **a** Principle for the holographic imaging. **b** Experimental setup for the reconstruction process (QWP stands for quarter-wave plate). **c** Phase profile of the meta-hologram calculated by the GS algorithm corresponding to the "flower" holographic image, scale bar: 500 μm; color bar: phase shift. **d** SEM image of the nanoslits, scale bar: 1 μm. Reproduced from [77] with permission. Copyright 2016, The Authors

The meta-hologram is capable to reconstruct both multicolor 2D and 3D images. For a multicolor 2D meta-hologram image based on three primary (RGB) colors, the recording and reconstruction principle are shown in Fig. 5.41a, b. To design the arrangement of the nanoslits, a colorful image is first divided into its RGB components. The corresponding RGB components are shifted to different locations of the imaging plane to separate themselves from each other. These components then make one new hologram image. The phase distribution of the hologram is calculated by the Gerchberg–Saxton (GS) algorithm [78], with the termination condition defined by

$$SSE = \frac{\iint (|g(u, v)| - |G(u, v)|)^2 du dv}{\iint |G(u, v)|^2 du dv} < \varepsilon, \tag{5.6.3}$$

where *SSE* is the mean square error between the target image and the holographic image, $g(u, v)$ the Fourier transform of the incident light wave function, $G(u, v)$ the Fourier transform of the target image, and ε the preset tolerance. In the calculation, the local minimum effect of the GS algorithm was not particularly taken care and the initialized phase of every pixel of the image is set to be zero.

The phase distribution of the hologram calculated by the GS algorithm is shown in Fig. 5.41c. In the reconstruction process, the individual color component, for example, the red light, obtains not only the red patterns, but also the "unwanted" patterns designed for green and blue light. The selection of the correct imaging patterns can be then achieved by the Shifted–Fraunhofer diffraction for 2D holography. As depicted in Fig. 5.42a, it is possible to tune the k_x and k_y components of the transmitted light

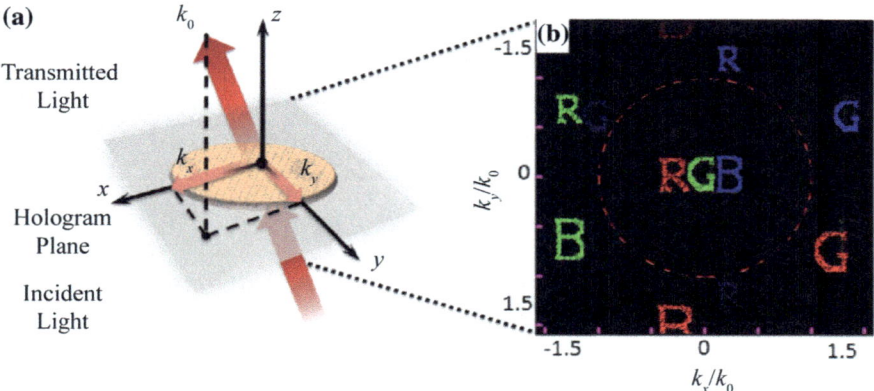

Fig. 5.42 **a** Schematic of the k space modulation of incident light. **b** Simulation results for the hologram design that shifts all the "unwanted" image into the k space corresponding to evanescent waves. The red dotted line marks the k space boundary between the evanescent wave and the propagating wave. Reproduced from [77] with permission. Copyright 2016, The Authors

wave vector without affecting the holographic reconstruction. For example, the wave vector in the x-direction can be calculated as: $k_x = k_0\cos\theta_x$, where k_0 is the transmitted light wave vector and θ_x the incident angle respect to the x-axis. During the tuning of k_x and k_y, the holographic images are shifted on the imaging plane. The correct holographic images can be shifted and superimposed into the final image while the "unwanted" patterns are moved out of the imaging region.

In principle, the maximal viewing angle can be even larger than 180° for the metasurface plasmonic pixels which has the period smaller than the wavelength. In this case, the radial component is larger than $|k_0|$ and k_z becomes an imaginary number, which means that the transmitted light is converted into evanescent waves. With such a large k space, it is possible to design the "unwanted" patterns in the k space corresponding to the evanescent waves. The "unwanted" patterns then "disappear" in far field and are hidden from the observer. Simulation results are shown in Fig. 5.42b to demonstrate such a design. It is clear that the unwanted patterns corresponding to each color locate outside the k space of the propagating wave. The observer does not capture the near-field waves and these patterns are hidden. This design strategy makes the imaging area only consist of the desired imaging patterns.

In this design, 1.6×10^7 numbers of the nanoslits were fabricated on a total area of 4 mm². Figure 5.41d shows the SEM image of the holographic metasurface. The simulated and experimental holographic images for a multicolor "flower" pattern are shown in Fig. 5.43. It is clear that the experimental results restore the features of the simulated images. The images corresponding to RGB colors are independent of each other and the crosstalk among different wavelengths is eliminated. Therefore, the image details are clear and the final image is vivid. It should be noted that in this experiment, a 405 nm laser is used to represent the blue color. However, the real blue

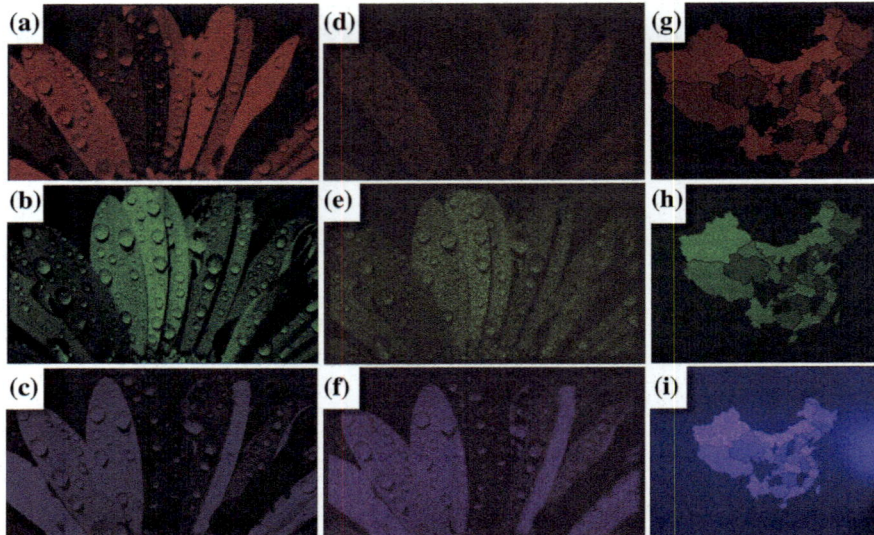

Fig. 5.43 Multi-image meta-holography. **a–c** Simulation of the "flower" holography for the red, green, and blue imaging patterns. **d–f** Experimental results corresponding to the red, green, blue imaging patterns, respectively. **g–i** RGB holographic images of "map of China", which are reconstructed by the same hologram metasurface of the "flower" with different incident angles. Reproduced from [77] with permission. Copyright 2016, The Authors

color represents a wavelength range from 450 to 495 nm. This wavelength mismatch is accountable for the difference between the simulation and experimental results.

The signal-to-noise ratio (SNR) is 126.3, more than 5 times higher than the previous meta-hologram design [79]. The SNR is defined as the ratio between the peak intensity in the image to the standard deviation of the background noise. Such high-quality images are mainly attributed to the neglectable cross-talk between different colors. The off-axis illumination also reduces the noise from the co-polarized light and contributes to the SNR. The above off-axis illumination method has also been used to design various types of photonic devices with different functionalities, including the following examples:

First, multi-image holography can be achieved. As the position of the patterns on the image plane depends on the $|k_x|$ and $|k_y|$ of the illumination light, multi-image holograms can be designed on one single sample. Switching among different images is achieved by tuning the incident angle of the illumination light. With this method, we designed and fabricated a dual-image hologram consisting of the "flower" (Fig. 5.43d–f) and "map of China" images (Fig. 5.43g–i). When the incident angle of the light is set as $\theta_x = 71.55°$ and $\theta_y = 71.55°$, the red "flower" appears in the imaging area. The image changes into the red "map of China" as the incident angle is tuned to be $\theta_x = 71.58°$ and $\theta_y = 90°$. It is worth to mention that by this method more images can be compiled into the meta-hologram. With proper design, superposition

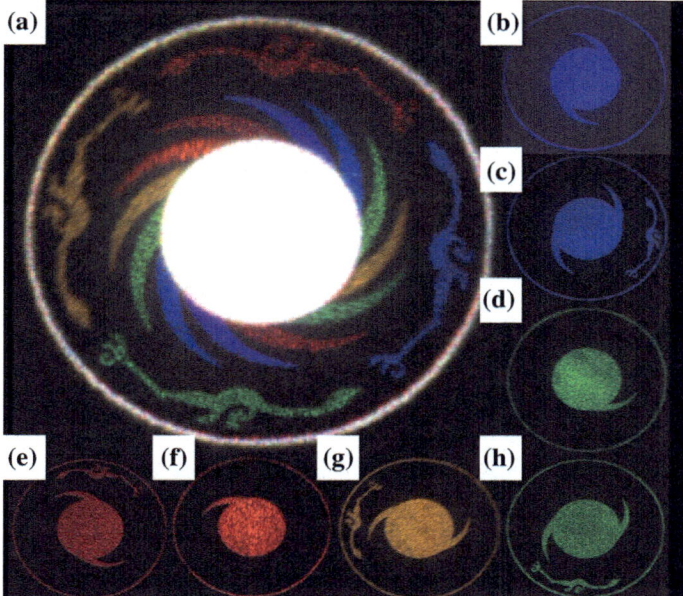

Fig. 5.44 Seven color meta-holography. **a** Seven color holographic image for the "Sun Phoenix", a pattern discovered on an ancient artifact gold coil, made in the Chinese Shang Dynasty 3000 years ago. **b–h** Patterns corresponding to purple, blue, green, red, orange, yellow and cyan colors, respectively. Reproduced from [77] with permission. Copyright 2016, The Authors

of even more images is achievable. A dynamic meta-hologram is even possible if the frames of dynamic images are compiled into the metasurface and each frame is switched by tuning the incident angle.

Second, a 7 color meta-hologram image is designed and fabricated for the first time, as shown in Fig. 5.44. It follows the similar design principle of the RGB hologram except that lasers at 7 different wavelengths were used to reconstruct the image. As discussed previously, introducing more wavelengths increases the data capacity in the imaging area. Compared with the RGB design, the key feature for such full-color holography is that the color gamut is increased by 1.39 times. The mixing of 7 colors extends the range of the colors available for the holography and provides a much improved capability to display more colorful and superior image. Furthermore, the light sensors and retina of animals are sensitive to a wide range of wavelengths, which requires a much larger color gamut than the one mixed by RGB. For the topics of the authentication, it is possible to fabricate a super-thin meta-hologram including both the visible and "invisible" imaging patterns at a specific wavelength, so the holographic image contains "secret" information invisible for the human eyes. Furthermore, for the data security applications, introducing more wavelengths is similar to increasing the length of the password, which increases the order of the complexity and greatly enhances the security level.

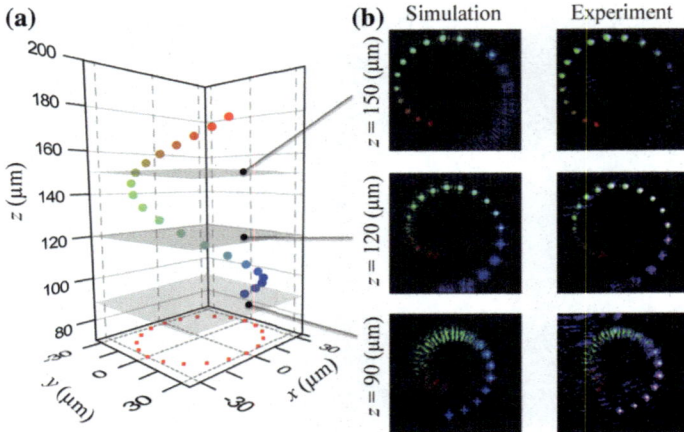

Fig. 5.45 Multicolor 3D meta-holography based on plasmonic nanoapertures. **a** Schematic diagram of the 3D object in the spatial coordinate. **b** Experimental results for the cross-sections at different z-positions. Reproduced from [77] with permission. Copyright 2016, The Authors

Finally, a multicolor 3D object was also designed and fabricated. A modified point source algorithm is adapted to design the metasurface. The 3D object is represented by a collection of RGB point sources. The complex amplitude at the hologram plane is then calculated as the superposition of the light fields from the entire 3D object consisting of all point sources. Different from the 2D holographic characterization setup, the 3D object needs to scan through the vertical direction of the imaging space. The experimental setup is modified to scan the xy imaging plane across the z-axis. As presented in Fig. 5.45, a spiral helix pattern composed of 20 light spots is designed. At different positions along the z-axis, the holographic image is captured by the CCD camera. These isolated points are arranged in a space of 60 μm × 60 μm × 90 μm, with a volume of 324,000 μm^3.

As a final remark, we note that similar metasurfaces with anisotropic apertures have been widely utilized in the design of color and/or 3D holography. This section only presents a basic design principle. The readers may find more detailed results in recent literature [80–82].

References

1. R.P Crease, The most beautiful experiment. Phys. World **15**, 19 (2002)
2. X. Luo, T. Ishihara, Surface plasmon resonant interference nanolithography technique. Appl. Phys. Lett. **84**, 4780–4782 (2004)
3. T.W. Ebbesen, H.J. Lezec, H.F. Ghaemi, T. Thio, P.A. Wolff, Extraordinary optical transmission through sub-wavelength hole arrays. Nature **391**, 667–669 (1998)
4. W.L. Barnes, A. Dereux, T.W. Ebbesen, Surface plasmon subwavelength optics. Nature **424**, 824–830 (2003)

5. M. Pu, Y. Guo, X. Li, X. Ma, X. Luo, Revisitation of extraordinary Young's interference: From catenary optical fields to spin-orbit interaction in metasurfaces. ACS Photonics **5**, 3198–3204 (2018)
6. H. Shi, X. Luo, C. Du, Young's interference of double metallic nanoslit with different widths. Opt. Express **15**, 11321–11327 (2007)
7. T. Xu, Y. Zhao, D. Gan, C. Wang, C. Du, X. Luo, Directional excitation of surface plasmons with subwavelength slits. Appl. Phys. Lett. **92**, 101501 (2008)
8. W.E. Kock, Metal-lens antennas. Proc. IRE **34**, 828–836 (1946)
9. M. Pu, X. Ma, Y. Guo, X. Li, X. Luo, Theory of microscopic meta-surface waves based on catenary optical fields and dispersion. Opt. Express **26**, 19555–19562 (2018)
10. W.E. Kock, Metallic delay lenses. Bell Syst. Tech. J. **27**, 58–82 (1948)
11. J. Pendry, A. Holden, W. Stewart, I. Youngs, Extremely low frequency plasmons in metallic mesostructures. Phys. Rev. Lett. **76**, 4773–4776 (1996)
12. R. Shelby, D. Smith, S. Schultz, Experimental verification of a negative index of refraction. Science **292**, 77–79 (2001)
13. V.G. Veselago, E.E. Narimanov, The left hand of brightness: Past, present and future of negative index materials. Nat. Mater. **5**, 759–762 (2006)
14. T. Xu, L. Fang, B. Zeng, Y. Liu, C. Wang, Q. Feng, X. Luo, Subwavelength nanolithography based on unidirectional excitation of surface plasmons. J. Opt. -Pure Appl. Opt. **11**, 085003 (2009)
15. G. Lerosey, D.F.P. Pile, P. Matheu, G. Bartal, X. Zhang, Controlling the phase and amplitude of plasmon sources at a subwavelength scale. Nano Lett. **9**, 327–331 (2009)
16. C. Lu, X. Hu, H. Yang, Q. Gong, Ultrawide-band unidirectional surface plasmon polariton launchers. Adv. Opt. Mater. **1**, 792–797 (2013)
17. R. Zia, M.L. Brongersma, Surface plasmon polariton analogue to Young's double-slit experiment. Nat. Nanotechnol. **2**, 426 (2007)
18. X. Luo, M. Pu, X. Li, X. Ma, Broadband spin Hall effect of light in single nanoapertures. Light Sci. Appl. **6**, e16276 (2017)
19. B. Gjonaj, A. David, Y. Blau, G. Spektor, M. Orenstein, S. Dolev, G. Bartal, Sub-100 nm focusing of short wavelength plasmons in homogeneous 2D space. Nano Lett. **14**, 5598–5602 (2014)
20. F.J. Rodríguez-Fortuño, G. Marino, P. Ginzburg, D. O'Connor, A. Martínez, G.A. Wurtz, A.V. Zayats, Near-field interference for the unidirectional excitation of electromagnetic guided modes. Science **340**, 328–330 (2013)
21. T. Xu, C. Wang, C. Du, X. Luo, Plasmonic beam deflector. Opt. Express **16**, 4753–4759 (2008)
22. X. Luo, Principles of electromagnetic waves in metasurfaces. Sci. China-Phys. Mech. Astron. **58**, 594201 (2015)
23. Y. Xu, Y. Fu, H. Chen, Planar gradient metamaterials. Nat. Rev. Mater. **1**, 16067 (2016)
24. P. Lalanne, P. Chavel, Metalenses at visible wavelengths: Past, present, perspectives. Laser Photonics Rev. **11**, 1600295 (2017)
25. F. Capasso, The future and promise of flat optics: A personal perspective. Nanophotonics **7**, 953 (2018)
26. X. Luo, Subwavelength optical engineering with metasurface waves. Adv. Opt. Mater. **6**, 1701201 (2018)
27. N. Yu, P. Genevet, M.A. Kats, F. Aieta, J.-P. Tetienne, F. Capasso, Z. Gaburro, Light propagation with phase discontinuities: Generalized laws of reflection and refraction. Science **334**, 333–337 (2011)
28. X. Ni, N.K. Emani, A.V. Kildishev, A. Boltasseva, V.M. Shalaev, Broadband light bending with plasmonic nanoantennas. Science **335**, 427–427 (2012)
29. M. Khorasaninejad, W.T. Chen, R.C. Devlin, J. Oh, A.Y. Zhu, F. Capasso, Metalenses at visible wavelengths: Diffraction-limited focusing and subwavelength resolution imaging. Science **352**, 1190–1194 (2016)
30. S. Chen, Z. Li, Y. Zhang, H. Cheng, J. Tian, Phase manipulation of electromagnetic waves with metasurfaces and its applications in nanophotonics. Adv. Opt. Mater. **6**, 1800104 (2018)

31. P. Lalanne, S. Astilean, P. Chavel, E. Cambril, H. Launois, Blazed binary subwavelength gratings with efficiencies larger than those of conventional échelette gratings. Opt. Lett. **23**, 1081–1083 (1998)
32. J.B. Pendry, Negative refraction makes a perfect lens. Phys. Rev. Lett. **85**, 3966–3969 (2000)
33. T. Xu, C. Du, C. Wang, X. Luo, Subwavelength imaging by metallic slab lens with nanoslits. Appl. Phys. Lett. **91**, 201501 (2007)
34. L. Verslegers, P.B. Catrysse, Z. Yu, J.S. White, E.S. Barnard, M.L. Brongersma, S. Fan, Planar lenses based on nanoscale slit arrays in a metallic film. Nano Lett. **9**, 235–238 (2009)
35. C. Min, P. Wang, X. Jiao, Y. Deng, H. Ming, Beam manipulating by metallic nano-optic lens containing nonlinear media. Opt. Express **15**, 9541–9546 (2007)
36. Y. Chen, X. Li, Y. Sonnefraud, A.I. Fernández-Domínguez, X. Luo, M. Hong, S.A. Maier, Engineering the phase front of light with phase-change material based planar lenses. Sci. Rep. **5**, 8660 (2015)
37. Y. Chen, C. Zhou, X. Luo, C. Du, Structured lens formed by a 2D square hole array in a metallic film. Opt. Lett. **33**, 753–755 (2008)
38. L. Lin, X.M. Goh, L.P. McGuinness, A. Roberts, Plasmonic lenses formed by two-dimensional nanometric cross-shaped aperture arrays for Fresnel-region focusing. Nano Lett. **10**, 1936 (2010)
39. M. Totzeck, W. Ulrich, A. Gohnermeier, W. Kaiser, Semiconductor fabrication: Pushing deep ultraviolet lithography to its limits. Nat. Photonics **1**, 629–631 (2007)
40. S. Ishii, V.M. Shalaev, A.V. Kildishev, Holey-metal lenses: Sieving single modes with proper phases. Nano Lett. **13**, 159–163 (2013)
41. J. Sun, X. Wang, T. Xu, Z.A. Kudyshev, A.N. Cartwright, N.M. Litchinitser, Spinning light on the nanoscale. Nano Lett. **14**, 2726–2729 (2014)
42. M. Pu, X. Li, X. Ma, Y. Wang, Z. Zhao, C. Wang, C. Hu, P. Gao, C. Huang, H. Ren, X. Li, F. Qin, J. Yang, M. Gu, M. Hong, X. Luo, Catenary optics for achromatic generation of perfect optical angular momentum. Sci. Adv. **1**, e1500396 (2015)
43. F. Aieta, M.A. Kats, P. Genevet, F. Capasso, Multiwavelength achromatic metasurfaces by dispersive phase compensation. Science **347**, 1342–1345 (2015)
44. Z. Zhao, M. Pu, H. Gao, J. Jin, X. Li, X. Ma, Y. Wang, P. Gao, X. Luo, Multispectral optical metasurfaces enabled by achromatic phase transition. Sci. Rep. **5**, 15781 (2015)
45. W.T. Chen, A.Y. Zhu, V. Sanjeev, M. Khorasaninejad, Z. Shi, E. Lee, F. Capasso, A broadband achromatic metalens for focusing and imaging in the visible. Nat. Nanotechnol. **13**, 220–226 (2018)
46. S. Wang, P.C. Wu, V.-C. Su, Y.-C. Lai, M.-K. Chen, H.Y. Kuo, B.H. Chen, Y.H. Chen, T.-T. Huang, J.-H. Wang, R.-M. Lin, C.-H. Kuan, T. Li, Z. Wang, S. Zhu, D.P. Tsai, A broadband achromatic metalens in the visible. Nat. Nanotechnol. **13**, 227–232 (2018)
47. O. Avayu, E. Almeida, Y. Prior, T. Ellenbogen, Composite functional metasurfaces for multispectral achromatic optics. Nat. Commun. **8**, 14992 (2017)
48. Y. Li, X. Li, M. Pu, Z. Zhao, X. Ma, Y. Wang, X. Luo, Achromatic flat optical components via compensation between structure and material dispersions. Sci. Rep. **6**, 19885 (2016)
49. L. Rayleigh, XXXI. Investigations in optics, with special reference to the spectroscope. Philos. Mag. Ser. **5**(8), 261–274 (1879)
50. N.I. Zheludev, What diffraction limit? Nat. Mater. **7**, 420–422 (2008)
51. M. Pu, C. Wang, Y. Wang, X. Luo, Subwavelength electromagnetics below the diffraction limit. Acta Phys. Sin. **66**, 144101 (2017)
52. F. Qin, M. Hong, Breaking the diffraction limit in far field by planar metalens. Sci. China Phys. Mech. Astron. **60**, 044231 (2017)
53. S.W. Hell, J. Wichmann, Breaking the diffraction resolution limit by stimulated emission: Stimulated-emission-depletion fluorescence microscopy. Opt. Lett. **19**, 780–782 (1994)
54. B. Huang, M. Bates, X. Zhuang, Super resolution fluorescence microscopy. Annu. Rev. Biochem. **78**, 993 (2009)
55. G.T. di Francia, Super-gain antennas and optical resolving power. G Suppl Nuovo Cim **9**, 426–438 (1952)

56. M. Born, E. Wolf, *Principles of Optics: Electromagnetic Theory of Propagation, Interference and Diffraction of Light*, 7th ed. (Cambridge University Press, 1999)
57. E.T.F. Rogers, N.I. Zheludev, Optical super-oscillations: Sub-wavelength light focusing and super-resolution imaging. J. Opt. **15**, 094008 (2013)
58. E.T.F. Rogers, S. Savo, J. Lindberg, T. Roy, M.R. Dennis, N. Zheludev, Super-oscillatory optical needle. Appl. Phys. Lett. **102**, 031108 (2013)
59. F. Qin, K. Huang, J. Wu, J. Teng, C. Qiu, M. Hong, A supercritical lens optical label-free microscopy: Sub-diffraction resolution and ultra-long working distance. Adv. Mater. **29**, 1602721 (2017)
60. C. Wang, D. Tang, Y. Wang, Z. Zhao, J. Wang, M. Pu, Y. Zhang, W. Yan, P. Gao, X. Luo, Super-resolution optical telescopes with local light diffraction shrinkage. Sci. Rep. **5**, 18485 (2015)
61. D. Tang, C. Wang, Z. Zhao, Y. Wang, M. Pu, X. Li, P. Gao, X. Luo, Ultrabroadband super-oscillatory lens composed by plasmonic metasurfaces for subdiffraction light focusing. Laser Photonics Rev. **9**, 713–719 (2015)
62. L.B. Whitbourn, R.C. Compton, Equivalent-circuit formulas for metal grid reflectors at a dielectric boundary. Appl. Opt. **24**, 217–220 (1985)
63. N. Engheta, Circuits with light at nanoscales: Optical nanocircuits inspired by metamaterials. Science **317**, 1698–1702 (2007)
64. Z. Li, T. Zhang, Y. Wang, W. Kong, J. Zhang, Y. Huang, C. Wang, X. Li, M. Pu, X. Luo, Achromatic broadband super-resolution imaging by super-oscillatory metasurface. Laser Photonics Rev. **12**, 1800064 (2018)
65. C. Genet, T.W. Ebbesen, Light in tiny holes. Nature **445**, 39–46 (2007)
66. T. Xu, Y.-K. Wu, X. Luo, L.J. Guo, Plasmonic nanoresonators for high-resolution colour filtering and spectral imaging. Nat. Commun. **1**, 59 (2010)
67. T. Xu, E.C. Walter, A. Agrawal, C. Bohn, J. Velmurugan, W. Zhu, H.J. Lezec, A.A. Talin, High-contrast and fast electrochromic switching enabled by plasmonics. Nat. Commun. **7**, 10479 (2016)
68. P.B. Johnson, R.W. Christy, Optical constants of the noble metals. Phys. Rev. B **6**, 4370–4379 (1972)
69. M. Song, X. Li, M. Pu, Y. Guo, K. Liu, H. Yu, X. Ma, X. Luo, Color display and encryption with a plasmonic polarizing metamirror. Nanophotonics **7**, 323–331 (2018)
70. X. Zhu, Y. Zhang, J. Zhang, J. Xu, Y. Ma, Z. Li, D. Yu, Ultrafine and smooth full metal nanostructures for plasmonics. Adv. Mater. **22**, 4345–4349 (2010)
71. T. Smith, J. Guild, The C.I.E. colorimetric standards and their use. Trans. Opt. Soc. **33**, 73 (1931)
72. V.R. Shrestha, S.-S. Lee, E.-S. Kim, D.-Y. Choi, Aluminum plasmonics based highly transmissive polarization-independent subtractive color Filters exploiting a nanopatch array. Nano Lett. **14**, 6672–6678 (2014)
73. Y. Shen, V. Rinnerbauer, I. Wang, V. Stelmakh, J.D. Joannopoulos, M. Soljačić, Structural colors from Fano resonances. ACS Photonics **2**, 27–32 (2015)
74. A.F. Kaplan, T. Xu, L.J. Guo, High efficiency resonance-based spectrum filters with tunable transmission bandwidth fabricated using nanoimprint lithography. Appl. Phys. Lett. **99**, 143111 (2011)
75. L.J. Sherry, S. Chang, G.C. Schatz, R.P. Van Duyne, B.J. Wiley, Y. Xia, Localized surface plasmon resonance spectroscopy of single silver nanocubes. Nano Lett. **5**, 2034–2038 (2005)
76. Y.-W. Huang, W.T. Chen, W.-Y. Tsai, P.C. Wu, C.-M. Wang, G. Sun, D.P. Tsai, Aluminum plasmonic multicolor meta-hologram. Nano Lett. **15**, 3122–3127 (2015)
77. X. Li, L. Chen, Y. Li, X. Zhang, M. Pu, Z. Zhao, X. Ma, Y. Wang, M. Hong, X. Luo, Multicolor 3D meta-holography by broadband plasmonic modulation. Sci. Adv. **2**, e1601102 (2016)
78. R.W. Gerchberg, W.O. Saxton, A practical algorithm for the determination of phase from image and diffraction plane pictures. Optik **35**, 237–250 (1972)
79. X. Ni, A.V. Kildishev, V.M. Shalaev, Metasurface holograms for visible light. Nat. Commun. **4**, 2807 (2013)

80. Z.-L. Deng, G. Li, Metasurface optical holography. Mater. Today Phys. **3**, 16–32 (2017)
81. S. Wang, X. Ouyang, Z. Feng, Y. Cao, M. Gu, X. Li, Diffractive photonic applications mediated by laser reduced graphene oxides. Opto-Electron. Adv. **1**, 170002 (2018)
82. X. Zhang, M. Pu, J. Jin, X. Li, P. Gao, X. Ma, C. Wang, X. Luo, Helicity multiplexed spin-orbit interaction in metasurface for colorized and encrypted. Ann. Phys. **529**, 1700248 (2017)

Chapter 6
Beam Shaping via Microscopic Meta-surface-wave

Abstract In previous chapter, we discussed the theory, design principle, and application of phase modulation based on plasmonic nanoslits, nanoholes, and other nanoapertures. The coupling of SPPs at the interfaces forms catenary plasmons featured by catenary-liked intensity profile. This can be understood from two aspects: first, the analytic mathematical description of plasmonic modes in metal–insulator–metal layered waveguide takes the form of hyperbolic cosine and sine functions; second, the summation of evanescent tails of waveguide modes would form a catenary. In this chapter, we show a generalized concept of catenary optical fields. The interference fields of two subwavelength scatters would follow a catenary shape. For instance, the two sides of a subwavelength slit perforated in a thin metallic screen could generate strong localized fields featured by a catenary function. This effect can be also observed in periodic slits, i.e., 1D grating. Interestingly, the equivalent impedance of such grating is described by the catenary of equal strength, which is termed catenary dispersion. Based on these properties, we proposed the concept of microscopic meta-surface-wave, which forms one important basis to discuss the light–matter interaction in subwavelength structures.

Keywords Meta-surface-wave · Wavefront shaping · Beam steering · Cloak · Virtual shaping

6.1 Microscopic Meta-surface-wave

As one special waveform, surface wave is widespread in optics, electromagnetics, and acoustics, among others. According to Rayleigh's definition, surface waves must propagate along one particular interface, which are completely different from the "space waves" that spread in three dimensions [1]. In principle, the behaviors of surface waves are closely associated with the materials and geometries. By changing these parameters, one could generate a large variety of surface waves such as surface plasmon polaritons (SPPs), Zenneck surface waves, Dyakonov waves as well as Tamm states [2, 3]. The control of surface waves has become one critical goal of integrated optical devices and systems [4, 5].

© Springer Nature Singapore Pte Ltd. 2019

X. Luo, *Catenary Optics*, https://doi.org/10.1007/978-981-13-4818-1_6

The emerging metasurfaces are also related to surface waves. As one kind of two-dimensional metamaterials, metasurfaces have attracted increasing attentions [6–9]. While they have a reduced dimension, the almost infinite subwavelength surface structures provide us great flexibility in the control of electromagnetic waves [9, 10]. In 2015, we term the waves in metasurfaces as meta-surface-waves (M-waves) since they are propagating along surfaces but have unusual properties beyond classic surface waves [3, 9]. Historically, the concept of M-waves can date back to the discovery of extraordinary Young's double slits interference (EYI) [11, 12].

For quite a long time, the physics of M-waves has not been fully addressed. It is often wrongly thought that surface waves must propagate along the "macroscopic" surface. In 2018, we revisited the concept of M-waves from a microscopic view. Three notable conclusions can be drawn from these results: First, M-waves are not limited to propagate along the apparent macroscopic surface. In the microscopic scale, the interface inside the structured materials also supports the propagation of interfacial waves. Second, the amplitudes of these surface waves decay exponentially away from the microscopic surface, and the fields of the coupled modes behave like catenaries [12], which lead to a much shorter and tunable effective wavelength along the propagating direction. Third, the impedance dispersion of thin nanoslits array is described using a mathematic model similar to the "catenary of equal strength," which presents an additional link between the catenary function and subwavelength engineering optics or the so-called engineering optics 2.0 (EO 2.0) [9, 13].

6.1.1 Catenary Theory of the Microscopic M-Waves

From a microscopic point of view, there are nearly infinite numbers of interfaces between the constitutive materials in metamaterials and metasurfaces. Although these interfaces may not support surface waves when the dimension is much larger than the wavelength, strong modifications must be considered in the deep-subwavelength scale. This can be understood by using the generalized Helmholtz equation [9, 12]:

$$\nabla^2 \mathbf{E} + \nabla \left[\mathbf{E} \cdot \frac{\nabla \varepsilon}{\varepsilon} \right] + k_0^2 \varepsilon \mu \mathbf{E} = 0, \qquad (6.1.1)$$

where $\mathbf{E}$ is the electric field, ε and μ are the permittivity and permeability, and k_0 is the vacuum wavenumber. At the boundaries, $\nabla \varepsilon$ approaches infinity and is responsible for the coupling between free-space waves and localized modes. To understand Eq. (6.1.1), waves propagating through thin slits perforated in perfect electric conductors (PEC) are investigated. As shown in Fig. 6.1, the 2D metallic slab waveguide (the dimension along the y-axis is infinite) supports transverse electromagnetic (TEM) waves without cutoff frequency, and the propagation constant is equal to that in vacuum. When the thickness of waveguide is reduced to much smaller than the operational wavelength, however, strong scattering would occur at the edges, which makes the fields distribution change dramatically. Similar to the SPP fields at the slit

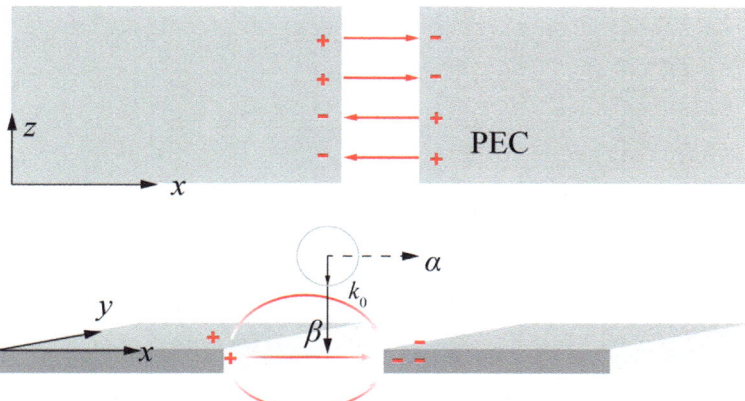

Fig. 6.1 Waves in the microscopic regime. The top panel shows the electric fields in a thick slit cut in PEC. The bottom panel shows the M-wave in a thin slit. Reproduced from [1] with permission. Copyright 2018, Optical Society of America

edges [14], this new kind of field takes a form of hyperbolic cosine catenary function as a result of the evanescent coupling, which means that the localized wave has a large vertical propagation constant β along z-direction and an imaginary horizontal component α along x-direction. As will be expatiated in the following, this propagation constant is dependent on the slit width, which resembles the SPP effect once again [15].

These catenary-shaped interfacial waves may be treated as one special vertically propagating M-wave which deviates from traditional diffraction theory. In the rigorous coupled wave analysis (RCWA), the in-plane wavevector of the (m, n)-order mode inside the grating layer is often expressed as [16]

$$\vec{k}_{m,n} = \vec{k}_{||} + m\frac{2\pi}{\Lambda_x}\vec{i} + n\frac{2\pi}{\Lambda_y}\vec{j}, \tag{6.1.2}$$

where Λ_x and Λ_y are the periods along x- and y-directions. Obviously, the in-plane wavenumber can only be a real number, in contrary to the imaginary components shown in Fig. 6.1. Since the RCWA has been demonstrated to be correct for periodic gratings with finite thickness, it is logical to attribute the discrepancy to the small thickness of the metasurface. If one wants to make the RCWA to be applicable, the thin metasurface may be considered as an effective medium with larger refractive index and proper thickness. Alternatively, the generalized boundary condition of metasurfaces can be used to calculate the electromagnetic response [17].

As shown in Fig. 6.2, the electromagnetic properties of simple metallic gratings can be obtained using impedance theory. To investigate the influence of gap width on the optical properties, the electric fields along the central line are illustrated in Fig. 6.2b. Since the scattering fields are mainly evanescent waves, the amplitude

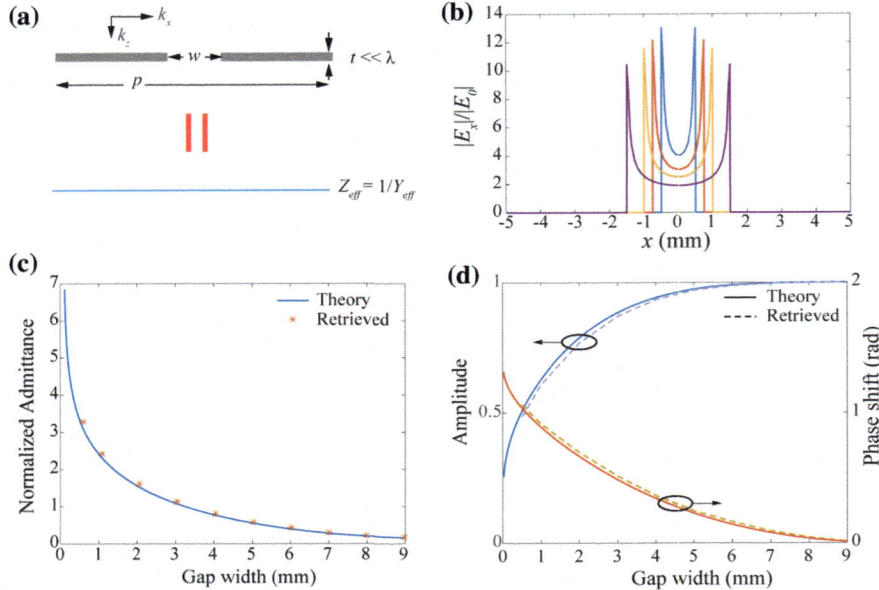

Fig. 6.2 Vertical M-waves associated with catenary optical fields. **a** Equivalent of a thin sheet perforated with subwavelength slits. The period p and thickness t are 10 and 0.1 mm. **b** E_x distribution at 10 GHz for slit widths of 1, 1.5, 2, and 3 mm. **c** Normalized admittance for different gap widths. The curve is half of a catenary curve of equal strength. **d** Transmission amplitude and phase calculated using FEM and impedance theory. Reproduced from [1] with permission. Copyright 2018, Optical Society of America

distributions follow a hyperbolic cosine catenary shape. Using equivalent circuit model (a stationary approach was given in Appendix D of this book) [18], the normalized admittance Y_{eff} can be written as

$$Y_{\text{eff}} = \frac{1}{Z_{\text{eff}}} = -i \frac{2p\left(n_1^2 + n_2^2\right)}{\lambda} \ln \csc \frac{\pi w}{2p}, \qquad (6.1.3)$$

where Z_{eff} is the normalized surface impedance, p is the period of grating, w is the width of the slit, λ is the wavelength, and n_1 and n_2 are the refractive indexes for the background materials. Using the generalized Fresnel's equations involving equivalent impedances [3, 17], the transmission and reflection coefficients can be easily obtained. Alternatively, the impedance sheet may be treated as a homogeneous thin film with effective permittivity of [19, 20]

$$\varepsilon_{\text{eff}} = 1 + i \frac{Y_0 Y_{\text{eff}}}{\varepsilon_0 \omega d_{\text{eff}}}, \qquad (6.1.4)$$

where d_{eff} is the effective thickness, ε_0 is the permittivity of vacuum, Y_0 is the admittance of vacuum, and ω is the angular frequency. Interestingly, the above two equations possess a form of "catenary of equal strength," which has been utilized to generate photonic spin–orbit interaction and achromatic geometric phase [21, 22]. To highlight this interesting property, it is termed as catenary optical dispersion. Figure 6.2c shows the effective admittance calculated by Eq. (6.1.3) and retrieved from finite element method (FEM) calculation. To further demonstrate the agreement, the calculated transmission amplitudes and phases are illustrated in Fig. 6.2d.

6.1.2 Application of M-Wave in Amplitude and Phase Modulation

The above catenary theory is useful in the design of various functional metasurfaces. First of all, we investigate the performance of multilayered metallic gratings as spectral filters. When the distance between the layers is large enough, the evanescent coupling between them could be ignored and the electromagnetic properties may be directly calculated using transfer matrix or interference theory [23, 24]. As can be seen in Fig. 6.2c, the retrieved admittance is a bit larger than the theory. In order to make the theory more accurate, in the following discussion the normalized admittance is revised to be

$$Y_{\text{eff}} = -i\frac{2p\left(n_1^2 + n_2^2\right)}{\alpha\lambda} \ln \csc \frac{\pi w}{2p}, \tag{6.1.5}$$

where α is an additional experiential term.

Without loss of any generality, we shall compare the modified theory with numerical simulations for a multilayered structure with five dielectric spacers and six metasurfaces. The permittivity of the dielectric spacer is set to be either 1 or 3.5 while the gap width is 0.5 mm or 2.5 mm. Figure 6.3 shows a comparison of the theoretical and numerical results for four combinations of geometric and electric parameters, implying that the theories are accurate enough. The blue regions indicate the bandgap formed by the multilayer.

Note that when the left and right boundaries of the unit cell are set as PEC by adding two metallic sheets, each unit cell would act as a metallic waveguide. In this case, the propagation constant is almost independent of the incidence angle, thus the above band-stop filters could operate in all angle of incidences. To demonstrate this intriguing phenomenon, the reflectance for the case shown in Fig. 6.3a was calculated for various incidence angles as illustrated in Fig. 6.4. It is shown that the reflectance in 10–14 GHz can be maintained to be higher than 99% even for incident angle up to 88°. This exotic behavior can be interpreted using the transfer matrix formalism: Unlike the case of classic multilayered structures where the phase shift is

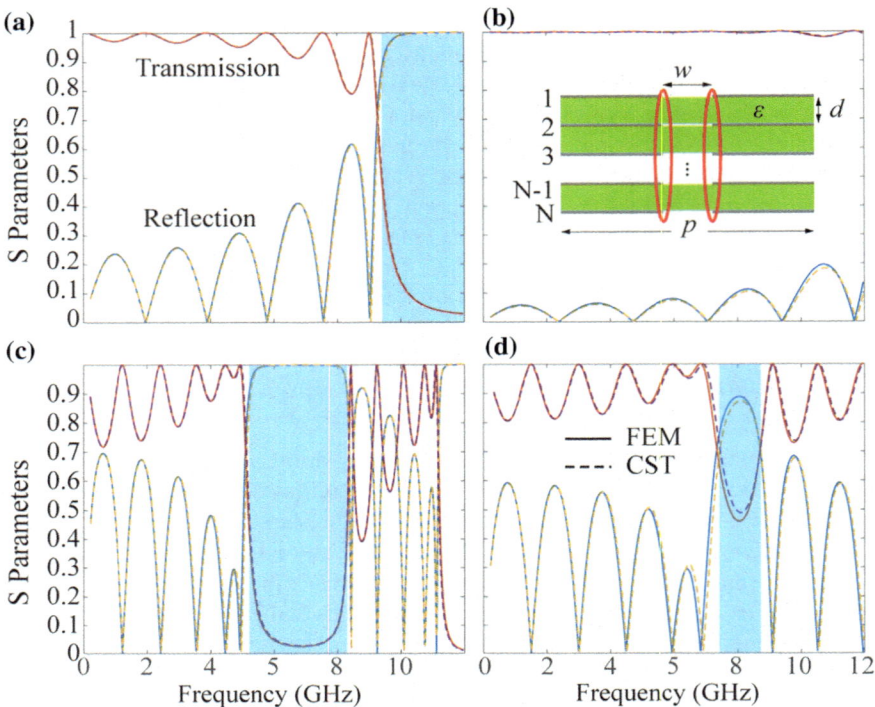

Fig. 6.3 Simulated transmission and reflection coefficients under various conditions. The period, spacer thickness, and layer number are fixed to be $p = 5$ mm, $d = 10$ mm, and $N = 6$. The other parameters are **a** $w = 0.5$ mm; $\varepsilon = 1$; **b** $w = 2.5$ mm; $\varepsilon = 1$; **c** $w = 0.5$ mm; $\varepsilon = 3.5$; and **d** $w = 2.5$ mm; $\varepsilon = 3.5$. The blue regions indicate the transmission bandgap, while the red ellipses in (**b**) show the effective interfaces supporting the vertical M-waves. Reproduced from [1] with permission. Copyright 2018, Optical Society of America

angle dependent, the waves in this structure always propagate along the z-direction, thus the bandgap and transmission peak are almost independent of incident angle.

For strongly coupled multilayers, the catenary model shall only provide initial values for the geometric parameters. In what follows, rigorous simulations are used to design thin and high-efficient beam deflectors comprised of gradient slits array. According to the generalized Snell's law, proper phase modulations with constant amplitude are required to deflect beam efficiently [3, 9]. Like the case for SPPs in metal–insulator–metal (MIM) waveguides [25], the current structures rely on the change of width to modulate the phase shift. Figure 6.5 shows three designs (Designs A, B, and C) with deflection angles equal to 32.4°, 45.6°, and 59° at 14 GHz, respectively. The scattering peaks in the calculated radar cross sections (RCS) are in good agreement with the theory, which demonstrates that the zeroth order and opposite order reflections are well suppressed. Although the amplitude of opposite deflection increases for larger angles, it may be still much better than previous designs based on

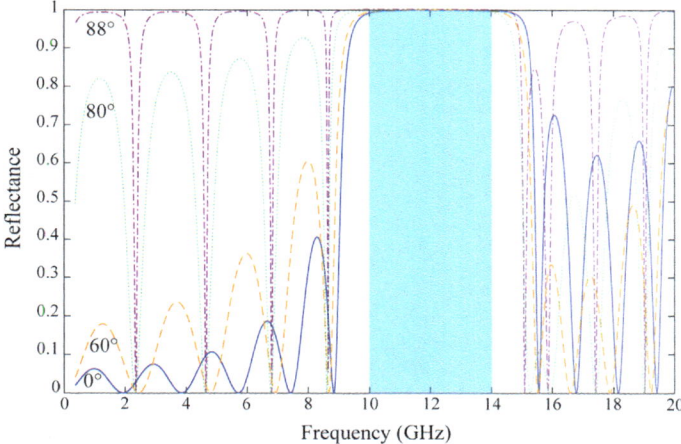

Fig. 6.4 Reflectance at incidence angles ranging from 0° to 88°. The blue region indicates the omnidirectional bandgap. Reproduced from [1] with permission. Copyright 2018, Optical Society of America

dielectric resonances [26, 27]. In fact, we expect the deflection angle can be as high as 88° for larger samples. When used as focusing lenses, this indicates a numerical aperture (NA) being close to 0.999.

Since a reflective layer is added in the bottom of the device, the reflectance of the meta-mirror can be maintained as high as 99%. Figure 6.5b shows the z-component of electric fields for Design A. In order to suppress the horizontal coupling between neighboring unit cells, vertical metallic sheets have been added at the two sides of each unit. Consequently, each unit acts as a localized waveguide, ensuring large-angle filtering and beam steering [28, 29].

6.2   All-Metallic Surface Structure for Virtual Shaping

The catenary optical fields in the metallic structures provide important insight to the physical mechanism of localized phase modulation. In this section, an all-metallic phase-gradient metasurface is used to simultaneously reduce the specular reflection and infrared emission in broad wavebands and wide incident angles [30]. The metasurface is composed of a monolayer of uniaxial birefringent metal gratings with space-variant orientation. The core idea behind this design is to combine the low-emission nature of metal and geometric phase originated from photonic spin–orbit interaction in inhomogeneous anisotropic material. Additional advantages of these all-metallic structures are their good mechanical properties, such as high strength, high stiffness, and good ductility, which are essential to some engineering applications.

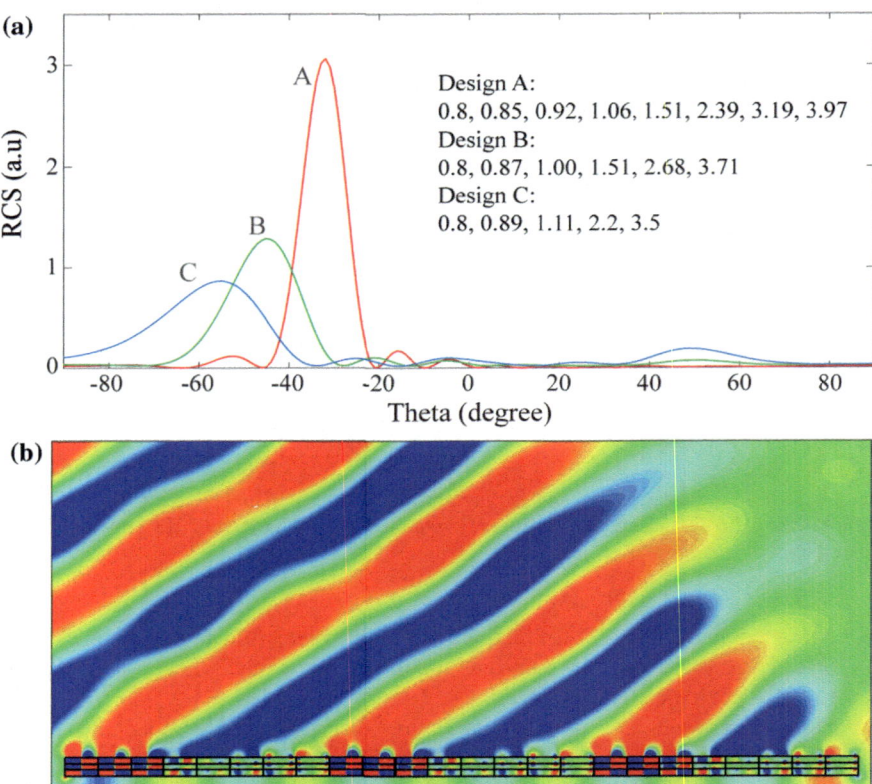

Fig. 6.5 Large-angle beam deflectors based on slits array operating at $f = 14$ GHz. The width parameters (in mm) for different slits are shown in the inset of (**a**). Other parameters are $p = 5$ mm, $d = 1$ mm, N = 3, and $\varepsilon = 3.5$. **b** Distribution of E_z for Design A. Reproduced from [1] with permission. Copyright 2018, Optical Society of America

According to spin–orbit interaction in inhomogeneous structures, under circularly polarized light illumination, the anisotropic reflection would result in a geometric phase for the cross-polarized light, which is twice the orientation angle of the metallic strips (ξ) and can be written as $\Phi = 2\sigma\xi$ [21], where $\sigma = \pm 1$ denotes the left-handed circular polarization (LCP) and right-handed circular polarization (RCP). By designing the phase distribution, the reflected beam will be forced to propagate in well-defined ways with respect to the specular reflection direction [31]. In the simplest case, a chessboard-like configuration of two orthogonal metallic gratings (Fig. 6.6) will induce an abrupt phase change of $\Delta\Phi = \pi$. In this circumstance, the metasurface will be virtually shaped just like a chessboard pattern with different heights [32]. Under normal incidence, the backscattered energy is redirected to four diagonal directions where $\varphi = 45°$, $135°$, $225°$, and $315°$, thus the reflected wave

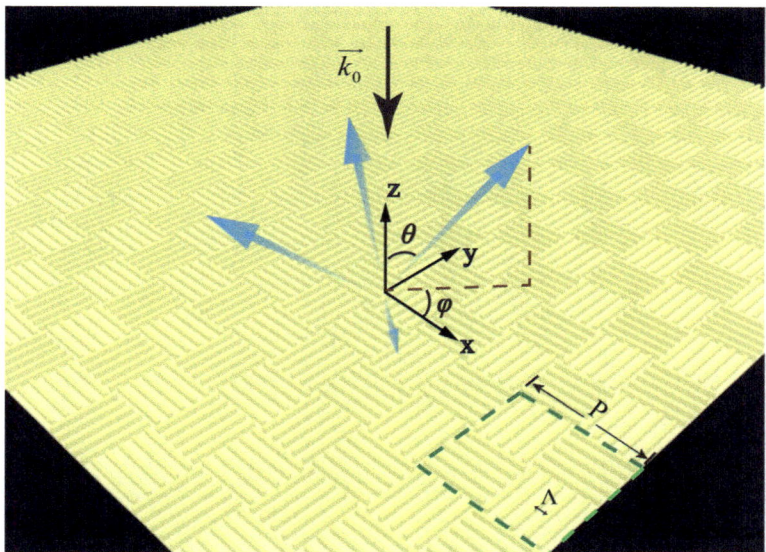

Fig. 6.6 Schematic of the all-metallic metasurface, showing the electromagnetic scattering in the upper half-space. The dashed box indicates the super unit of the metasurface. Reproduced from [30] with permission. Copyright 2018, WILEY-VCH Verlag GmbH & Co. KGaA, Weinheim

along the specular direction is canceled out. The angle between each beam with the z-axis is calculated by

$$\theta = \arcsin\left(\frac{\lambda}{P}\right), \tag{6.2.1}$$

where λ is the wavelength of incident wave, P is the period of the metasurface satisfying $P = 2n\Lambda$, Λ is the grating period, and n is the number of metallic strips. Due to its fourfold geometric symmetry, the metasurface not only performs well for circularly polarized incidence, but also for linearly polarized incidence, because a linear polarization can be treated as a superimposition of two circularly polarized light with opposite handedness but same amplitude [31].

To understand the electromagnetic properties of the basic elements and gain more physical insights into the underlying mechanism of the metasurface, numerical simulations are performed with commercial software CST Microwave Studio (MWS). Unit cell boundary is used in the simulation to take the mutual coupling of adjacent elements into account.

As shown in Fig. 6.7, when the geometric parameters are $\Lambda = 6\,\mu m$, $d = 1.3\,\mu m$, $h = 3.3\,\mu m$, incident circular polarization could be converted to its cross polarization (the definition is the same as that in CST MWS) in a broadband range. When the materials are changed from PEC to realistic gold, the energy of cross polarization is

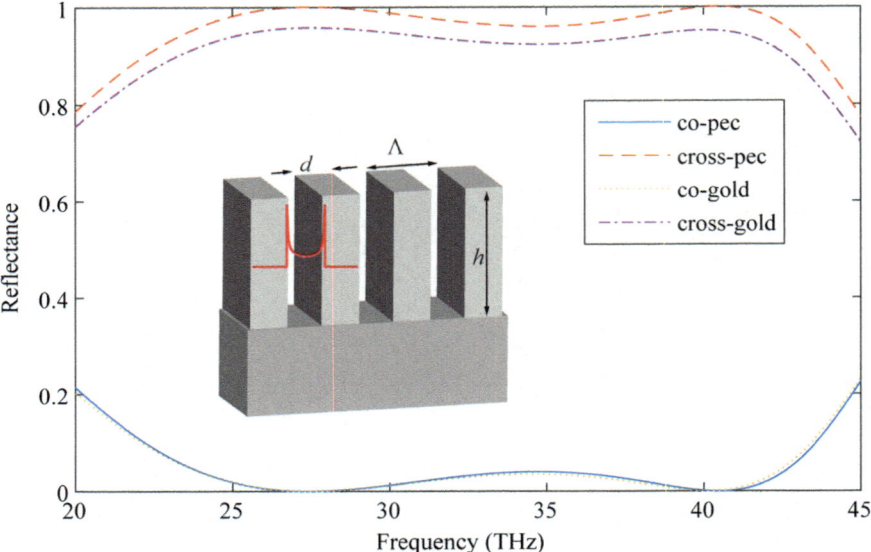

Fig. 6.7 Calculated reflectance of metallic grating composed of PEC and gold under circular polarization incidence. In the legends, co- and cross-mean the co- and cross-polarization components

reduced as a result of the ohmic loss. Nevertheless, the intensity of co-polarization is almost unchanged.

Interestingly, the localized waves in the metallic grating show some similarities with the SPPs [1]. According to the equivalent circuit theory, the grating behaves like inductors and capacitors for transverse electric (TE) and transverse magnetic (TM) polarizations, respectively. The effective admittances can be written in a form of catenary of equal strength for both polarizations as [18]

$$Y_x = -iY_0 \frac{2p(n_1^2 + n_2^2)}{\lambda} \ln \csc \frac{\pi(\Lambda - d)}{2\Lambda} \tag{6.2.2}$$

and

$$Y_y = iY_0 \frac{\lambda}{\Lambda} \left(\ln \csc \frac{\pi d}{2\Lambda} \right)^{-1}, \tag{6.2.3}$$

where $Y_0 = 1/377$ S is the admittance of vacuum.

Besides ultrathin gratings, Eqs. (6.2.2) and (6.2.3) can also be used to analyze the thick grating shown in Fig. 6.7 [30]. Using transfer matrix method (TMM) and considering the waveguide effect in the grating, the reflection phase for both polarizations can be easily obtained. As depicted in Fig. 6.8a, the phase shifts of two linearly polarized waves in a metallic grating in the microwave regime are simulated. The two curves are not linear and a near constant phase difference could be obtained in a

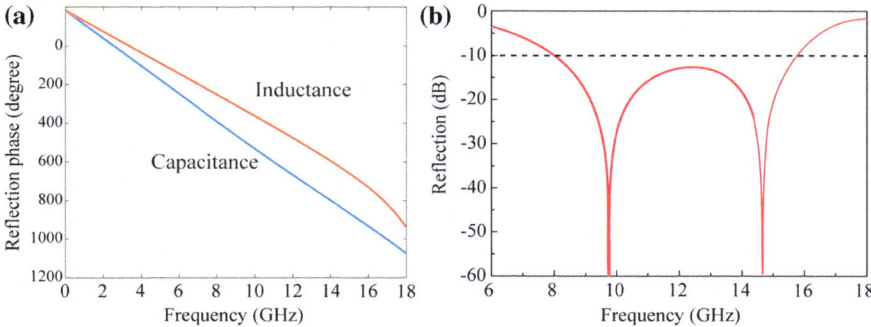

Fig. 6.8 **a** Theoretically calculated reflection phase for the two polarizations. The geometric parameters are set as $\Lambda = 17$ mm, $d = 4$ mm, and $h = 9$ mm. The distance between the port the top surface of the grating is 20 mm. **b** Calculated specular reflection for a chessboard-like metasurface. Adapted from [30] with permission. Copyright 2018, WILEY-VCH Verlag GmbH & Co. KGaA, Weinheim

wide frequency band. When such structures are illuminated with circularly polarized light, a broadband polarization conversion could be realized, which finally lead to broadband geometric phase.

In the following discussion, we focus on the virtual shaping in infrared band to obtain compatibility against infrared detection and CO_2 laser detection. To guarantee the performance at 10.6 μm, the optimized geometric parameters are chosen as $\Lambda = 5$ μm, $d = 1.6$ μm, $h = 2.7$ μm, where d is the width and h is the thickness of each gold strip (see the inset in Fig. 6.9a). For TE and TM polarized incidences, the reflectivity for different incident angles is calculated and shown in Fig. 6.9b. It is clearly observed that the reflectivity is more than 95% at wavelengths beyond 8 μm for all the incident cases, indicating that the metallic gratings possess ultralow absorption loss and infrared emissivity. Figure 6.9c shows the relative phase differences between the x- and y-directions ($\Delta \Phi = \Phi_x - \Phi_y$) for different incident angles. One can see that the phase shift at the wavelength of 10.6 μm is about π, thus the corresponding conversion efficiency is up to nearly 100%.

The cross-polarized and co-polarized reflectivities under circularly polarized illumination for different incident angles are calculated as shown in Fig. 6.9d. According to the generalized Snell's law, the co-polarized component corresponds to the specular reflectivity which means that the reflected wave will propagate in accordance with the traditional reflection law, while the cross-polarized component corresponds to the reflected wave that will be steered to pre-designed directions. It can be seen that the co-polarized reflectivity is less than 0.1 from 8.5 to 12.5 μm and the minimum reflection (nearly 0) occurs at about 10.6 μm. Although the wavelength for minimum reflectivity is slightly shifted toward a larger value at oblique incidence, the mean values of the co-polarized reflectivity are smaller in the whole infrared window of 8–14 μm. To investigate the physical mechanism, the simulated electric field distributions in the gratings under TE and TM polarizations at the wavelength of

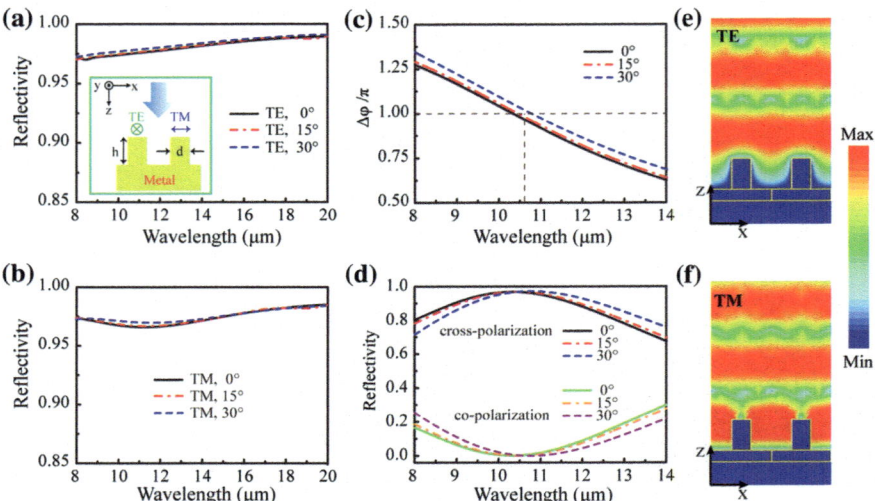

Fig. 6.9 **a, b** Simulated reflectivities of the unit cell under TE and TM polarized illumination for different incident angles. The inset in **a** illustrates the geometry of the element array. **c** Relative phase differences between x- and y-polarizations for different incident angles. **d** Simulated reflectivities for cross polarization and co-polarization under circularly polarized illumination for different incident angles. **e, f** Simulated electric field distributions of the gratings under TE and TM polarizations for normal incidence at 10.6 μm. Reproduced from [30] with permission. Copyright 2018, WILEY-VCH Verlag GmbH & Co. KGaA, Weinheim

10.6 μm are shown in Fig. 6.9e, f. Obviously, the two orthogonal waves are reflected at different interfaces to induce a relative phase shift.

Full-wave simulations were performed to validate the virtual shaping performance of the designed metasurface. A subgroup made of 2×2 gratings with orthogonal orientations (see the dashed box in Fig. 6.6) was used in the simulations with periodic boundaries. Every grating is comprised of five identical metallic strips ($n = 5$) to increase the geometrical similarity for each element, which complies with the unit cell boundary hypothesis in previous simulation. In fact, since the subgroup could be optimized rigorously, the final details of the structure model may be different from the original unit cells, especially for limited numbers of stripes. In order to verify the polarization-independent property, both TM- and TE-polarized plane waves were considered to illuminate on the structure at normal incidence. The calculated reflectivities under both polarizations are identical, as shown by the blue line in Fig. 6.10a. For comparison, an unpatterned gold plate with the same dimension was simulated (red line). It is obvious that the specular reflectivity for the metasurface is smaller than the unpatterned gold plate. The reflectivity from 8.5 to 12.5 μm is less than 0.1 and the minimum value occurs at about 10.6 μm. The scattering patterns of the metasurface and the metallic plate for normal incidence at 10.6 μm are also compared in Fig. 6.10b–e. According to the geometric theory, highly directed specular reflection is observed for the metallic plate. However, when it is covered by

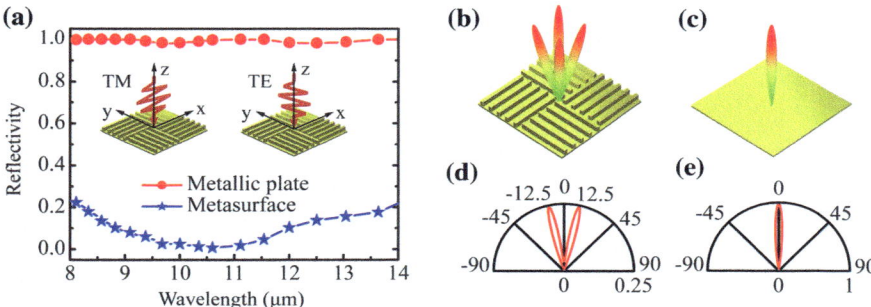

Fig. 6.10 **a** Full-wave simulated reflectivity of the metasurface and the metallic plate for TE and TM polarizations at normal incidence. **b, c** 3D scattering patterns of the metasurface and the metallic plate for normal incidence at 10.6 μm, respectively. **d, e** Scattering patterns of the metasurface and the metallic plate in the $\varphi = 45°$ plane. Reproduced from [30] with permission. Copyright 2018, WILEY-VCH Verlag GmbH & Co. KGaA, Weinheim

the all-metallic metasurface, the reflected energy is mainly split into four diagonal directions, resulting in significant reduction of reflection in the specular direction, as shown in Fig. 6.10b. The scattering pattern of the metasurface at $\varphi = 45°$ (or 225°) plane is presented in Fig. 6.10d, from which one can see that the scattering angle θ is about 12.5° and in good agreement with the theoretical calculation (12.2°). Note that according to Eq. 6.2.1, the scattering angle may increase by reducing the number of metallic strips (n).

Following the theoretical and numerical investigations, the simultaneous low-reflection and low-emission properties were investigated experimentally. A metasurface sample was fabricated with an area of 12×12 mm^2 using laser direct writing combined with conventional photolithography techniques. First, a layer of photoresist AZ1500 with a thickness of 2.55 μm was coated on a 1 mm-thick quartz substrate and then a 100 nm-thick SiO$_2$ film was deposited upon the photoresist layer using magnetron sputtering. Subsequently, a 500 nm-thick photoresist AZ1500 was coated on the SiO$_2$ layer. The target pattern was formed by laser direct writing on the photoresist AZ1500. Then, the pattern was transferred into the SiO$_2$ film and the photoresist AZ1500 successively. Finally, the exposed SiO$_2$ was etched off in the areas where the target pattern locates, and an Au layer with a thickness of 200 nm was sputtered on the patterned photoresist, resulting in the metasurface sample. Although the fabricated metasurface is not an ideal all-metallic structure as simulated, the same optical functionality can be still achieved.

The reflectivity of the fabricated sample was measured using the Fourier transform infrared spectrometer (FTIR spectrometer). Three specular angles (15°, 20°, and 30°) were considered for measurements. The measured reflectivity of the metasurface under TE and TM polarized incidences in the 8–14 μm was recorded in Fig. 6.11a, b in comparison with a same-sized Au plate without subwavelength gratings. The specular reflectivity is less than 0.1 from 10 to 14 μm for all the incidences, indicating

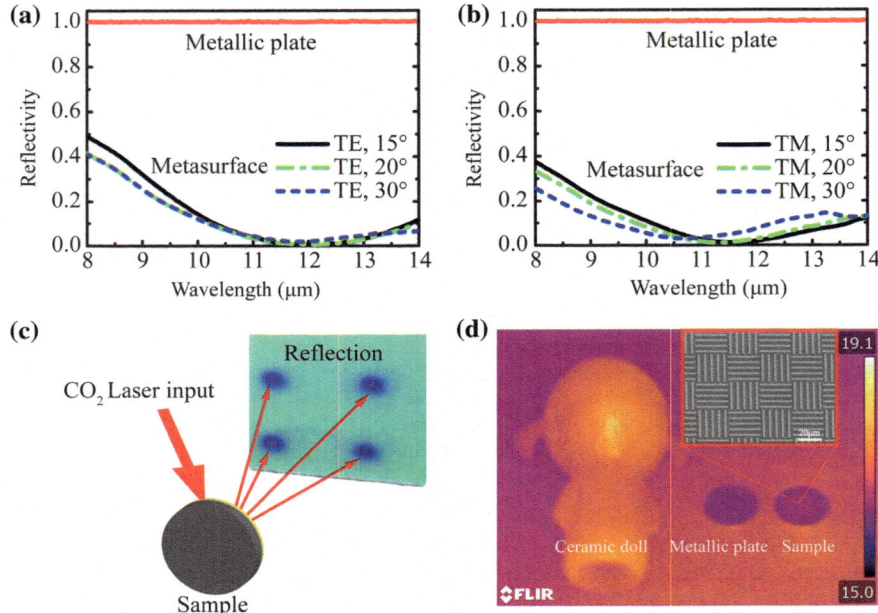

Fig. 6.11 **a, b** Measured reflectivity of the metasurface under oblique incidences for TE and TM polarizations. **c** Measured scattering pattern for the metasurface sample using a CO_2 Laser. **d** Thermal infrared images of a ceramic doll, a metallic plate and the fabricated sample. The inset is the scanning electron microscope (SEM) image of a part of the fabricated metasurface. The color bar indicates the temperature. Scale bar: 20 μm. Reproduced from [30] with permission. Copyright 2018, WILEY-VCH Verlag GmbH & Co. KGaA, Weinheim

the metasurface has excellent broadband and wide incidence angles properties in scattering the laser beam.

To check the scattering pattern, a CO_2 Laser (10.6 μm) was illuminated on the sample with a small oblique angle, and the reflection signal was collected with an infrared color plate. As shown in Fig. 6.11c, four spots are observed on the color plate, indicating the four reflected beams. The low infrared emission property is characterized by comparing the thermal infrared image of the fabricated sample with a ceramic doll and a metallic plate. The measurement was performed at room temperature using a commercial thermal infrared imager (FLIR) and the results are shown in Fig. 6.11d. The temperature of the ceramic doll is much higher than that of the metallic plate and the metasurface sample. Comparing the measured results of the sample and metallic plate, the approximate equivalent temperatures indicate the metasurface has ultralow thermal infrared emissivity, which is only slightly larger than the metallic plate.

Although the above metasurface consists of metallic gratings with merely two orthogonal orientation angles, the design can be easily extended to create more complex optical elements. For example, a high-performance meta-hologram was

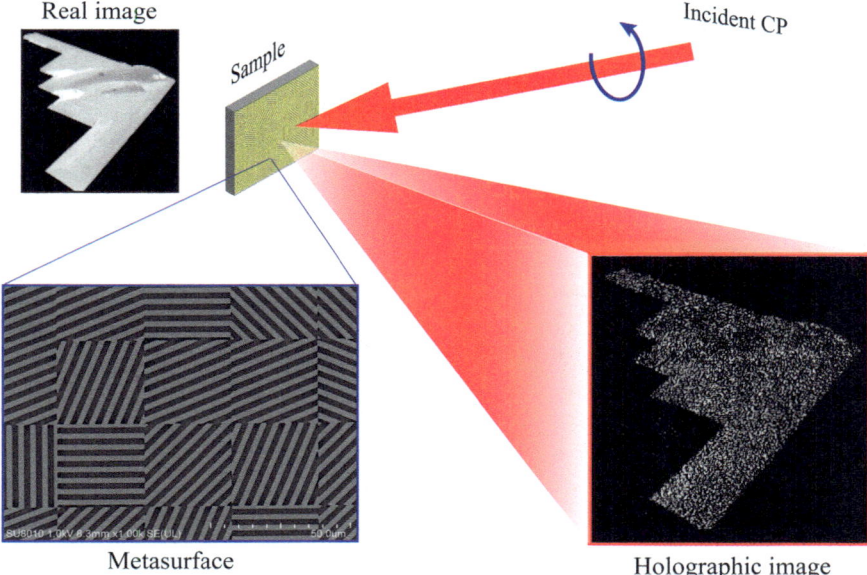

Fig. 6.12 SEM image of part of the fabricated metasurface and the measured holographic image. Reproduced from [30] with permission. Copyright 2018, WILEY-VCH Verlag GmbH & Co. KGaA, Weinheim

used to demonstrate the superb capability of phase manipulation of the metallic metasurface at the wavelength of 10.6 μm. The phase map of a holographic image, i.e., the binary B2 stealth aircraft model, was computed by means of the point source algorithm [33, 34]. The required phase was coded via rotation of the birefringent metallic gratings. An eight-level phase-gradient metasurface was fabricated and the measured results are shown in Fig. 6.12.

As a final remark, we note that the metallic metasurface can be shifted to short wavelength range by simply scaling the geometric size. In near-infrared region, the corresponding geometric parameters of the element are optimized as $\Lambda = 0.72$ μm, $d = 0.16$ μm, $h = 0.4$ μm. The simulated reflectivity for circularly polarized incidence is illustrated in Fig. 6.13a. The reflectivity is larger than 95% at the wavelengths beyond 0.9 μm, demonstrating an ultralow infrared emissivity. It can be seen that the co-polarized reflectivity (corresponding to specular reflectivity) is much less than 0.1 in 0.9–1.6 μm, covering a wide range of typical laser wavelengths. Figure 6.13b shows the full-wave simulated reflectivity of the metasurface in comparison to a metallic plate. One can see that the reflectivity from 0.9 to 1.6 μm is dramatically reduced. The scattering patterns for the metasurface and metallic plate at 0.93, 1.06 and 1.54 μm were also presented for comparisons, as shown in Fig. 6.13c–n. Excellent specular reflection suppression can be observed when the metallic plate is covered by metasurface.

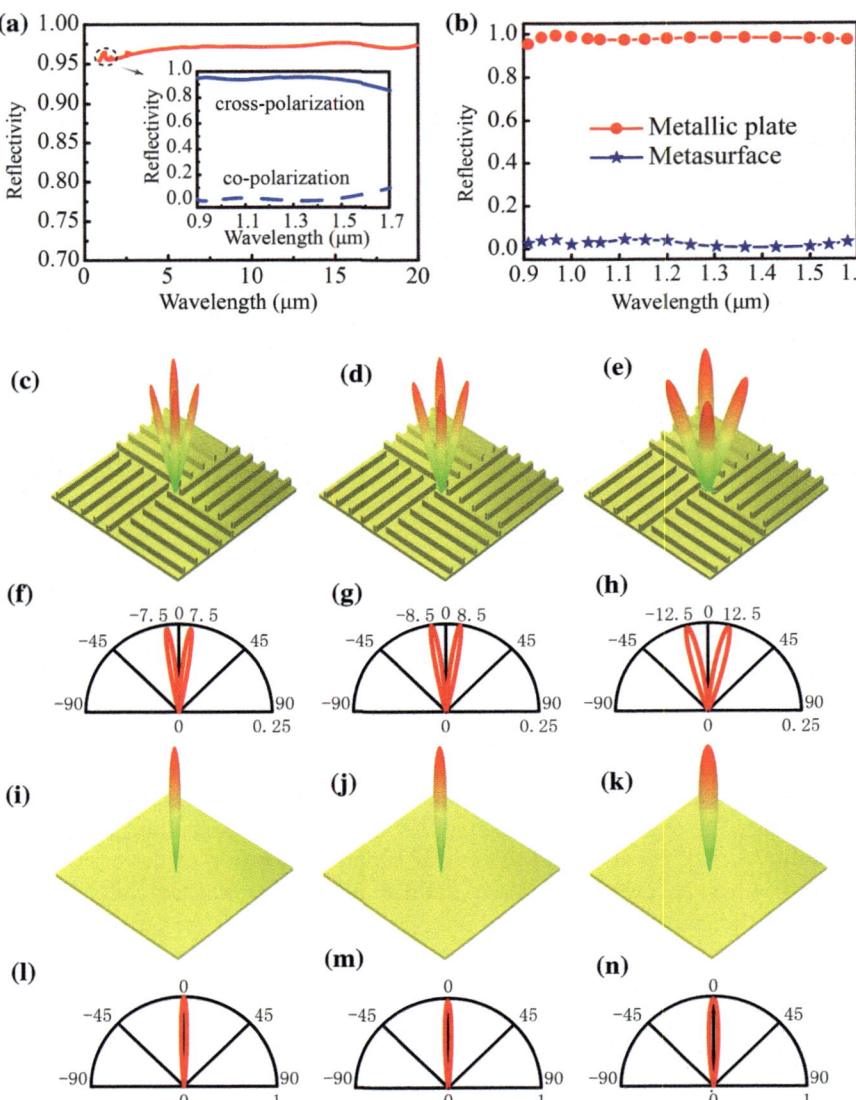

Fig. 6.13 a Simulated reflectivity under circularly polarized illumination. The inset illustrates the reflectivity for cross polarization and co-polarization. **b** Simulated reflectivity of the metasurface and the metallic plate for TE and TM polarizations at normal incidence. **c–e** 3D scattering patterns of the metasurface at 0.93, 1.06, and 1.54 μm, respectively. **f–h** Scattering patterns of the metasurface at 0.93, 1.06, and 1.54 μm in the $\varphi = 45°$ plane. **i–k** 3D scattering patterns of the metallic plate at 0.93, 1.06, and 1.54 μm, respectively. **l–n** Scattering patterns of the metallic plate at 0.93, 1.06, and 1.54 μm in the $\varphi = 45°$ plane. Reproduced from [30] with permission. Copyright 2018, WILEY-VCH Verlag GmbH & Co. KGaA, Weinheim

6.3 Broadband Virtual Shaping via Layered Metasurfaces

As shown in previous section, the geometric phase based on the photonic spin–orbit interaction is a good solution to control the reflection direction of electromagnetic beam. Since geometric phase is accompanied by spin conversion, the operation bandwidth of metasurface is ultimately limited by the polarization conversion efficiency, and thus an ultra-broadband polarization convertor is highly desired. As shown in Fig. 6.14a, a meta-mirror composed of bilayer metasurfaces separated from a metallic reflection plane was proposed to realize broadband polarization conversion [35]. Each metasurface is constructed by metallic cut-wire array in a hexagonal lattice with C6 symmetry to enhance the structural symmetry. The orientations of the cut wires in each unit are orthogonal to each other to ensure the dispersion engineering in both dimensions (u- and v-axes) [36]. Besides, to broaden the working bandwidth of the metasurface, the substrate thickness d is selected as 3 mm (one quarter of the wavelength at 17 GHz) to form a Fabry–Pérot-like cavity with low quality factor.

The polarization conversion performances of meta-mirror were simulated using CST Microwave Studio with unit cell boundary conditions. The dimensions of the unit cell were optimized as $a = 9$ mm, $l_1 = 0.9$ mm, $w_1 = 2.8$ mm, $l_2 = 8.5$ mm, and

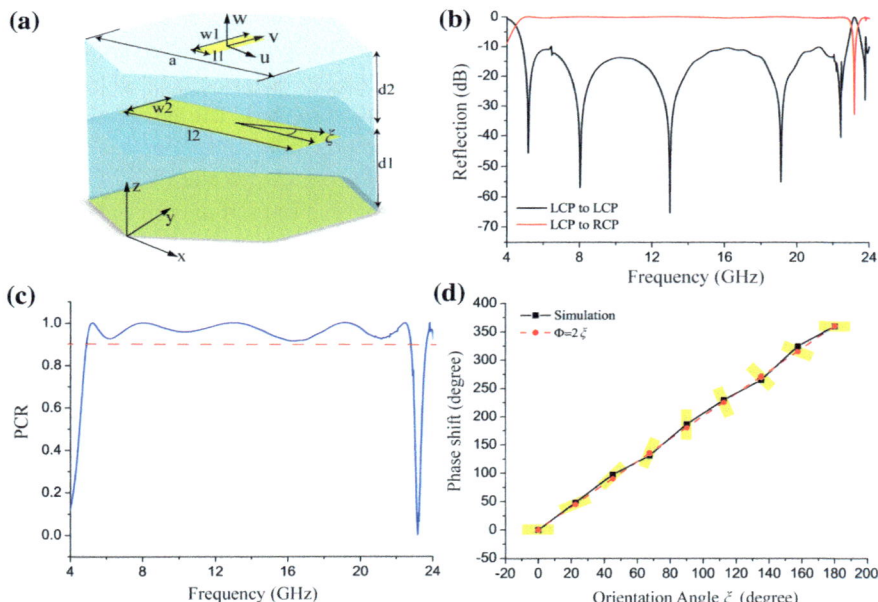

Fig. 6.14 Schematic of the bilayer metasurface. **a** A unit cell. **b** Simulated reflectance for cross polarization and co-polarization under normal circularly polarized illumination. **c** Polarization conversion ratio (PCR) of the unit cell. **d** Phase shift as a function of orientation angle ξ at 13 GHz. Reproduced from [35] with permission. Copyright 2018, The Royal Society of Chemistry

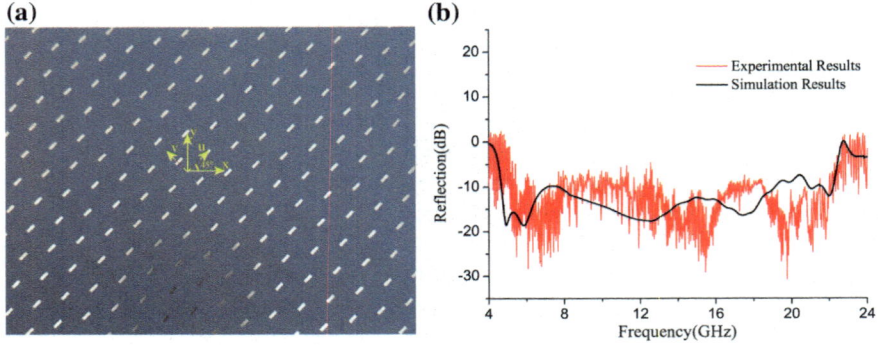

Fig. 6.15 **a** Top view of the sample. **b** Measured and simulated reflection coefficient of co-polarized component under *x*-polarized incidence. Reproduced from [35] with permission. Copyright 2018, The Royal Society of Chemistry

$w_2 = 2.4$ mm. The metal is assumed to be PEC and the permittivity of the dielectric spacer is 2.65. Obviously, the reflectance of the co-polarized wave is below -10 dB in an ultra-wide frequency band. The polarization conversion ratio (PCR) is over 90% from 4.9 GHz to 22.8 GHz, i.e., the relative bandwidth ratio is more than 4:1. Besides, when the orientation of subwavelength building varies from 0° to 180°, a linear phase shift from 0° to 360° can be obtained. In order to check the role of each metasurface played in the formation of the ultra-broad operation bandwidth, single MIM metasurface configurations have also been simulated. The upper and lower metasurfaces, respectively, exhibits high cross-polarization conversion efficiency in the higher frequency band (21–24 GHz) and lower frequency band (5–21 GHz) due to the geometry differences between them. The whole metasurface can be taken as a hybrid configuration of them.

To verify the ultra-wideband polarization conversion characteristics of the structure, a sample with dimension 300×300 mm^2 was fabricated with print circuit board (PCB) technique as shown in Fig. 6.15a. The circular polarization conversion (LCP-to-RCP) was replaced by the linear polarization conversion (*x*-to-*y*) in the experiment for simplification since they need to satisfy the same conversion condition. The *u*-axis of sample has an angle of 45° with respect to the *x*-axis so that the *x*-polarized wave is transformed to *y*-polarization. As depicted in Fig. 6.15b, the measured reflection coefficient (co-polarization) is below -10 dB in the frequency range between 5 and 22.4 GHz.

Here, equivalent circuit method is utilized to explain the principle of 2D dispersion engineering. According to the electric field distribution in Fig. 6.16, we can treat the metallic structure as a *LC* circuit element, where the inductance *L* origins from the metallic wires while the capacitance *C* stems from the metallic gap. Consequently, the anisotropic impedances of the metasurfaces can be expressed as

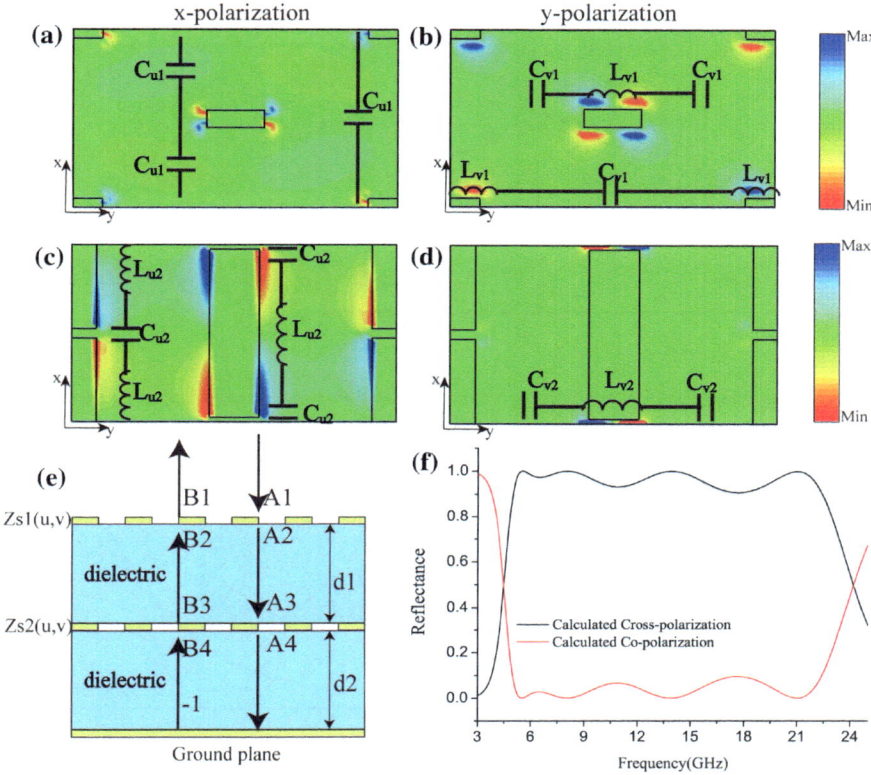

Fig. 6.16 **a**, **c** The electric field distribution of different layers under normal x-polarized light illumination at 14 GHz. **b**, **d** The electric field distribution of different layers under normal y-polarized light illumination. **e** Schematic of the transmission line model for the bilayer metasurfaces. **d** The calculated results based on transfer matrix method. Reproduced from [35] with permission. Copyright 2018, The Royal Society of Chemistry

$$
\begin{aligned}
Z_{s1,v} &= 1/Y_{s1,v} = 1/(i\omega C_{u1}) \\
Z_{s1,u} &= 1/Y_{s1,u} = i\omega L_{u1} + 1/(i\omega C_{u1}) \\
Z_{s2,v} &= 1/Y_{s2,v} = i\omega L_{v2} + 1/(i\omega C_{v2}) \\
Z_{s2,u} &= 1/Y_{s2,u} = i\omega L_{u2} + 1/(i\omega C_{u2})
\end{aligned}
\tag{6.3.2}
$$

where u and v represent the orthogonal main axes of the adopted bilayer metasurface, 1 and 2 denote the layer numbers. Generally speaking, the equivalent parameters may be obtained using quasi-static theory and connected with catenary function [37]. Alternatively, they can be retrieved by fitting the simulated reflected phases with the calculated results. In this case, there are $(L_{s1,u}, C_{s1,v}, C_{s1,u}) = (5 \times 10^{-11}$ H, 6×10^{-16} F, 1.5×10^{-14} F) and $(L_{s2,v}, L_{s2,u}, C_{s2,v}, C_{s2,u}) = (3.2 \times 10^{-9}$ H, 2×10^{-11} H, 4×10^{-16} F, 2×10^{-13} F). By utilizing the TMM shown in Fig. 6.16e, the

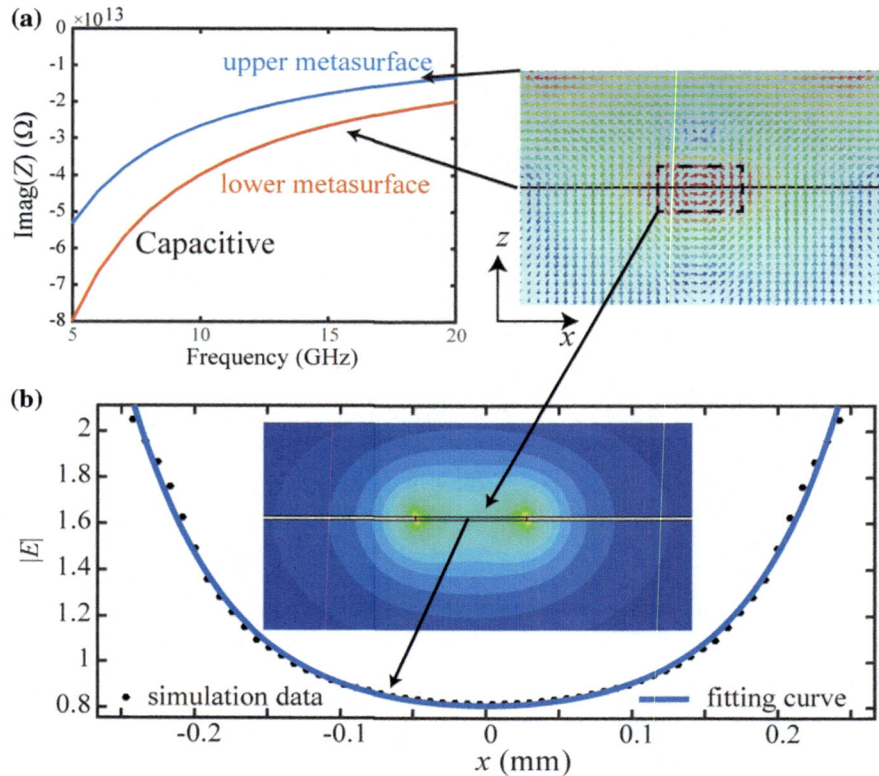

Fig. 6.17 **a** Calculated impedances of the metasurfaces as a function of frequency and corresponding electric field distribution at 12 GHz, which show strong local field enhancement in the metallic gap. **b** The magnitude of electric field distribution in the metallic gap and the evolution curve is fitted by a catenary function. Reproduced from [35] with permission. Copyright 2018, The Royal Society of Chemistry

reflectance of the proposed structure was calculated and displayed in Fig. 6.16f, which is consistent with the simulation results in Fig. 6.14b.

Figure 6.17a shows the impedances of the metasurface for $Z_{s1,v}$ and $Z_{s2,v}$. The local field enhancement in the metallic gap in the right panel of Fig. 6.17a gives a qualitative explanation. The magnitude of the electric field along the metallic gap is retrieved and shown in the inset of Fig. 6.17b in black dot. By fitting these data, it is found that the local electric field enhancement in the metallic gap follows a catenary function: $|E| = a \exp(bx) + c \exp(-dx) + e$, where x is the coordinate position and the fitting coefficients are $a = c = 359.8$, $b = d = 15.08$, and $e = 7367$ with a R-square of 0.9944.

The key of virtual electromagnetic wave shaping is to engineer the scattering through the proper phase distribution along the interface. Based on the principle of geometric phase, one can design arbitrary phase profiles along the interface by

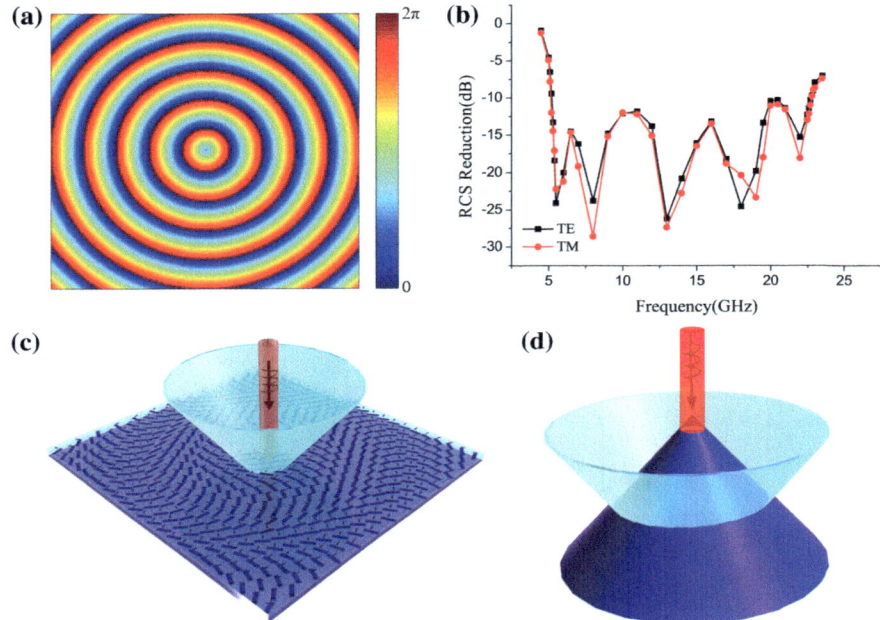

Fig. 6.18 Sketch map of electromagnetic illusion induced by Bessel-type phase distribution. **a** The phase profile of the Bessel-type phase distribution. **b** The simulated results of backward RCS reduction. **c** The scattered beams from the bilayer metasurface. **d** The scattered waves from a cone under normal incidence. Reproduced from [35] with permission. Copyright 2018, The Royal Society of Chemistry

simply rotating the local optical axis. As shown in Fig. 6.18a, a Bessel-type linear phase distribution is arranged along the radial direction $\Phi(x, y) = 2\pi r/P$ on an area of 280×280 mm^2, where r is the distance between the center of sample and each basic element, P is 70 mm. In this way, the reflected beam would spread like the ring with different radius, namely, the virtual shape is a cone. Under normal incidence of linearly polarized wave, the schematic 3D scattering patterns of the bilayer metasurface and a cone at 14 GHz are shown in Fig. 6.18c, d. The backward RCS in 5.2–22.8 GHz is reduced more than 10 dB compared with a bare metallic plate for both TE and TM polarizations (the maximum of the backward RCS reduction exceeds 25 dB around 12 GHz).

To verify the ultra-wideband RCS reduction characteristics of the structure, a sample with 300×300 mm^2 was fabricated and measured in microwave anechoic chamber. Two standard linearly polarized horn antennas were used for transmitting and receiving the reflected waves, and the orientation of linearly polarized horn antennas can be rotated to measure both the TE and TM polarizations. The measured RCS results of the fabricated sample and metallic flat plates with same dimensions

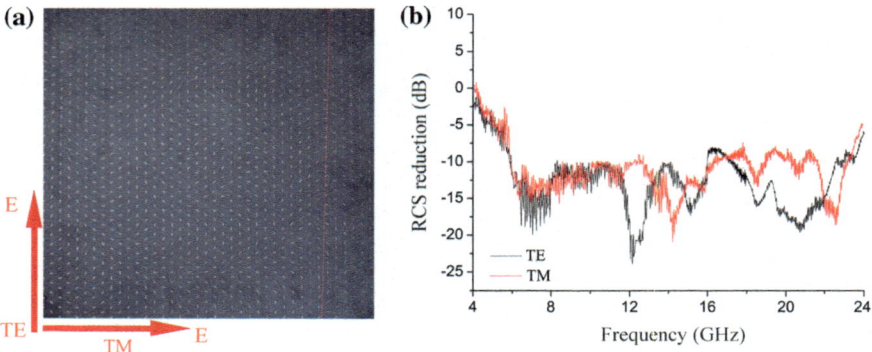

Fig. 6.19 Experimental results for the RCS reduction. **a** Top view of the sample. **b** Measured results for both TE and TM polarizations at an incidence angle of 5°. Reproduced from [35] with permission. Copyright 2018, The Royal Society of Chemistry

are presented in Fig. 6.19b, which display a reasonable agreement with the numerical results.

6.4   Achromatic Skin Cloak

As inspired by the concept of electromagnetic virtual shaping [31, 38], reflective phase-gradient metasurface can be utilized to replace traditional transformation-optics-based invisibility cloak [39]. Compared with bulky metamaterials, a thin layer of metasurface could realize on-demand control of the reflected amplitude and phase, thus the new type cloaking device may be termed as skin cloak [40, 41]. Although many impressive designs have been reported, broadband-metasurface-based skin cloaks are difficult to realize owing to the intrinsic chromatic dispersion. In this section, a nonlinear Gires–Tournois interference (GTI) model was employed to overcome this problem [42]. By delicately designing the dispersion of resonant structures composed of metallic rings, a broadband surface cloak was designed, as schematically depicted in Fig. 6.20. Although the geometries of the unit cells are very similar to traditional Gires–Tournois interferometer [43], their physical mechanisms are quite different. For traditional ones, the amplitude and phase shifts can be attributed to the multiple interferences between well-isolated interfaces. While in this case, as the distances between adjacent metasurfaces are in deep-subwavelength scale, strong magnetic coupling will complicate the interference effects.

In general, the role of phase-gradient metasurface in cloaking is to compensate the phase difference between the bump (the object that needs to be hidden) and the corresponding flat ground. To simplify the case, the bump is assumed to be two tilted flat planes with an inclination angle of θ_b (defined as the acute angle between the bump and the ground plane). As the surface cloak should possess a reversed phase

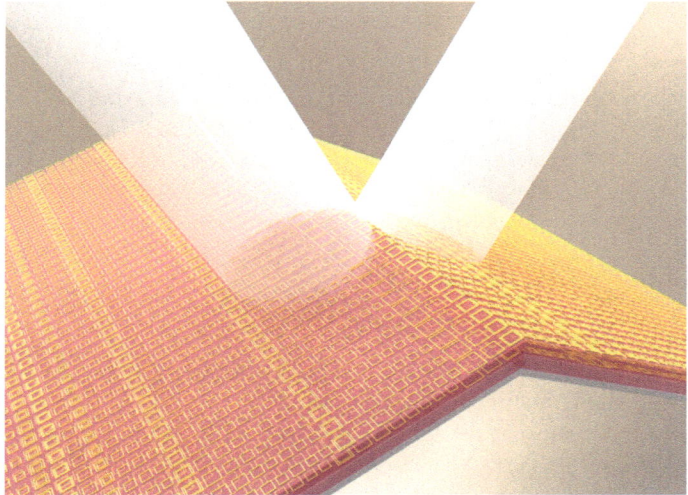

Fig. 6.20 Scheme of the broadband surface cloak. When a broadband incident wave impinges on the cloaked bump, it will be reflected as if the wave is incident on a flat ground plane. Reproduced from [42] with permission. Copyright 2019, WILEY-VCH Verlag GmbH & Co. KGaA, Weinheim

profile induced by the bump, the required phase distribution of the corresponding surface cloak can be denoted by

$$\varphi_1(x, f) = k_0 x \sin \theta + P_1(f), \tag{6.4.1}$$

where f is the operation frequency, k_0 is the wavenumber in vacuum, x is the spatial location in the metasurface, $\theta = \arcsin(2\sin\theta_b)$ corresponds to the deflection angle, and $P_1(f)$ is a frequency-dependent constant. For the two sides of the bump, the inclination angles have opposite signs, thus the deflection angle is also reversed.

In order to achieve achromatic cloaking in a wide band, there are two factors that need to be considered, i.e., the spatial location x and the operation frequency f, as both of them would vary the value of φ_1. In previous researches, the cloaks were designed for fixed frequencies, so the unit cell at certain location only needs to provide a fixed phase shift. For a broadband application, the required phase distribution is varied not only with spatial location but also with operation frequency, which makes the traditional design method invalid. It can be inferred that the simplest solution for Eq. (6.4.1) is that the frequency-dependent phase response φ_1 is linear with f and $P_1(f)$ is invariant ($\partial P_1(f)/\partial f = 0$) across the designed region. However, this situation can hardly be satisfied by metasurface as the typical phase response by such structure is dispersive owing to its resonant nature. With some mathematical manipulations, Eq. (6.4.1) can be rewritten as

$$S = \sin\theta = \frac{1}{k_0 x}[\varphi_1(x, f) - P_1(f)]. \tag{6.4.2}$$

Broadband cloaking can be achieved if the deflection function S is constant in a wide frequency range. Different from the aforementioned case that $P_1(f)$ was invariant [15, 44, 45], here a nonlinear $P_1(f)$ is employed to broaden the operating bandwidth. To further illustrate this issue, the following expression is derived by taking a derivative of Eq. (6.4.2):

$$\frac{\partial S}{\partial f} = \frac{1}{k_0 x}\left[\frac{\partial \varphi_1(x, f)}{\partial f} - \frac{\partial P_1(f)}{\partial f}\right] - \frac{1}{k_0 x f}[\varphi_1(x, f) - P_1(f)]. \tag{6.4.3}$$

Apparently, in order to let the left side of this equation near to zero, the metasurface need to provide highly nonlinear phase responses.

In order to precisely manage the phase responses of the unit cells, a database is set up to contain the performances and related geometries of all the unit cells. An optimization process based on particle swarm algorithm is proposed to obtain the required nonlinear $P_1(f)$ and select corresponding unit cells that can make φ_1 to desired pseudo-linear phases. Unlike the traditional design process that is achieved by a means of trial and error, this computer-assisted process can dramatically reduce the optimization complexity and enhance the robustness. A metasurface cloak (400 mm in length, 200 mm in width, 2 mm in thickness) with a tilted angle $\theta_b = 15°$ was designed to work across X band to Ku band. Similar to the case of multi-resonant broadband absorbers [17], although only discrete frequencies were optimized, the small dispersions between adjacent frequencies will lead to a broadband behavior across the whole region.

To further confirm that the cloaking performance can be realized in broadband, full-wave simulations were made by employing CST Microwave Studio. The simulated far-field radiation patterns for a PEC ground, a bare bump and a cloaked bump as a function of frequency and azimuthal angle ψ_1 under both polarizations across 10–17 GHz are shown in Fig. 6.21.

As expected, a bare bump has two reflection peaks at $\psi_1 = \pm 15°$ for all frequencies, while a flat PEC ground only has one peak at $\psi_1 = 0°$. Notably, after the bump is covered by the cloak, the original two peaks in the designed frequency region disappeared and rejoined to one peak at $\psi_1 = 0°$. To further illustrate this issue, the reflected electric field distributions and the far-field radiation patterns for these three cases at 15.5 GHz are plotted in the inset of Fig. 6.21. Apparently, a bare bump will induce a strong scattering for the incident wave and distort the original wavefronts for both TE and TM polarizations at all these frequencies. After covering the bump with the surface cloak, the distortion of the electric field is restored as if the incident wave is reflected by a flat PEC ground. These results indicate that the designed cloak could hide the bump perfectly in a wide bandwidth ranging from 10.7 to 16.6 GHz under normal incidence. The fractional bandwidth (defined as the ratio between the central frequency to the 3 dB bandwidth) is about 43.2% and much larger than its previous counterparts.

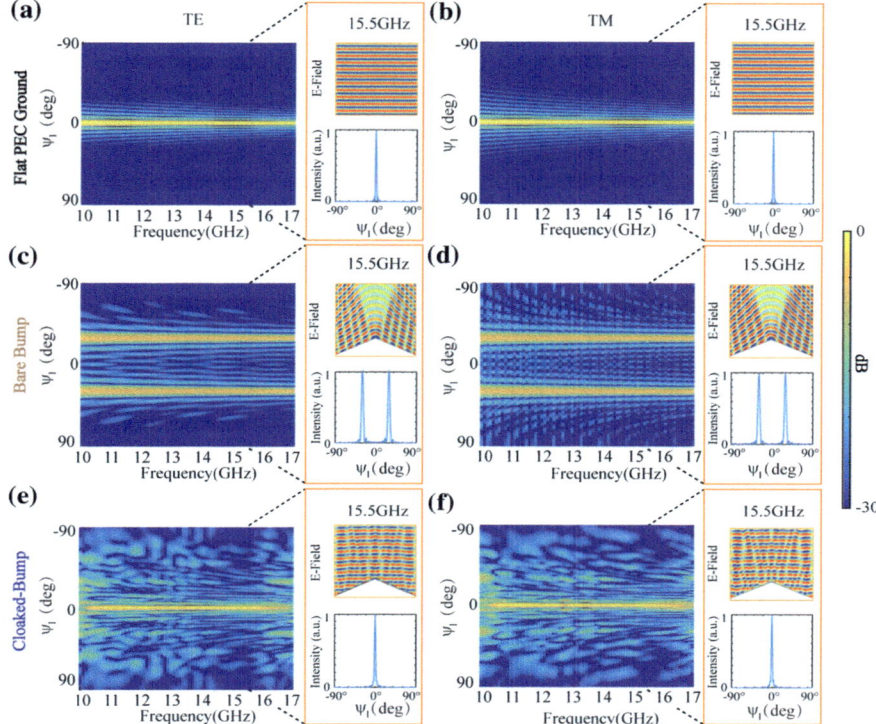

Fig. 6.21 The performance of the surface cloak under normal incidence. **a–f** The simulated far-field radiation patterns across 10–17 GHz at normal incidence for flat PEC ground (**a**, **b**), bare bump (**c**, **d**), and cloaked bump (**e**, **f**) under both polarizations. The insets are the reflected electric fields (upper) and far-field normalized radiation patterns (lower) at 15.5 GHz. Reproduced from [42] with permission. Copyright 2019, WILEY-VCH Verlag GmbH & Co. KGaA, Weinheim

Experiments were conducted to evaluate the performance of the cloak at oblique incidence with a vector network analyzer in an anechoic chamber, and the photography of the sample is shown in Fig. 6.22a. High-gain horn antennas were used as the transmitter and receiver, respectively. The distances from the horns to the sample are more than 2 m and sufficient to ensure the uniform illumination. The incident angle was fixed at $-10°$ and the reflection field was scanned from $0°$ to $90°$ with an interval of $2°$. As the operation bandwidth is across X band to Ku band, two pairs of horn antennas working at 8–12 and 12–18 GHz were employed. The measurements were performed with these antennas for three samples: the flat PEC ground, the bare bump, and the cloaked bump, under TE and TM polarizations. The measured radiation patterns as a function of frequency and azimuthal angle ψ_2 are depicted in Fig. 6.22b–g. The measured far-field patterns are broader than the simulated ones because that the waves emitted from the horns are not rigorous plane waves with wide beam. As observed, the reflection peak of the bare bump located at around

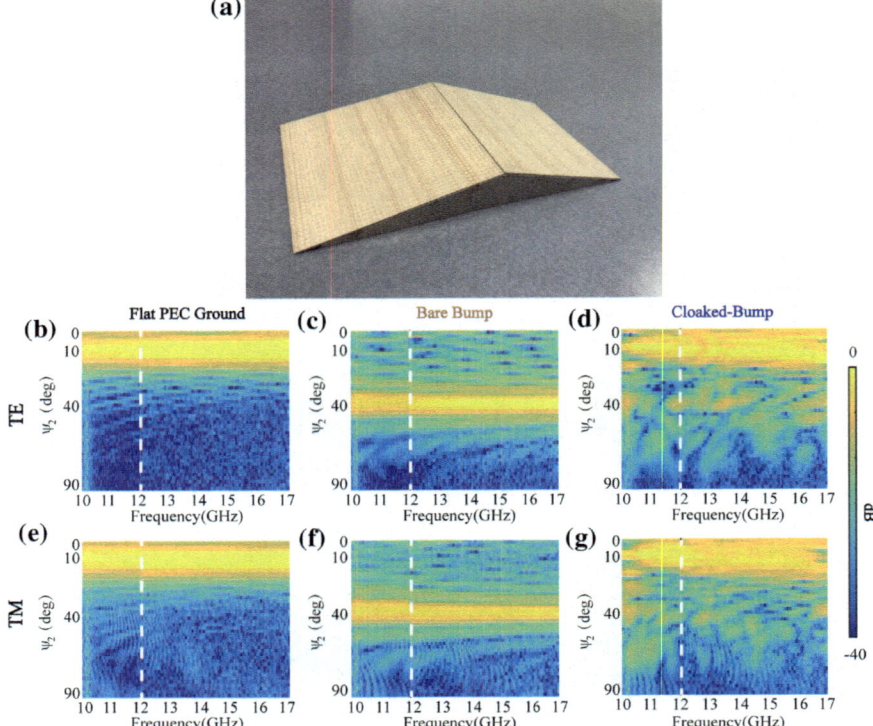

Fig. 6.22 Measured results under oblique incidence. **a** Photography of the cloaked bump. **b–g** Measured far-field radiation patterns across 10–17 GHz at −10° incidence for flat PEC ground (**b**, **e**), bare bump (**c**, **f**) and cloaked bump (**d**, **g**). The discontinuity (denoted in dotted white line) occurred at 12 GHz was due to the change of horn antennas. Reproduced from [42] with permission. Copyright 2019, WILEY-VCH Verlag GmbH & Co. KGaA, Weinheim

40° which accorded with the conventional law of reflection (another peak at −20° cannot be measured using the testing system). After it was wrapped by the cloak, the peak moved to around 10° for both polarizations in broadband that showed the same behavior as the flat PEC ground. A discontinuity occurs at 12 GHz due to the change of horn antennas.

It should be mentioned that although the profile of the bump is simple in the aforementioned case, the design method is suitable for bumps with arbitrary geometry as their needed phase responses can be treated as the superposition of several nonlinear phases.

6.5 Wide-Angle Beam Steering

In this section, we show that the catenary optical fields can be utilized to realize wide-angle beam steering [46]. It is well known that the field of view (FOV) of a lens determines the range of vision in various detecting/imaging devices ranging from microscopes to commercial cameras and microwave/optical radars. In classical optics, a stack of bulky and expensive lenses and off-axis techniques are generally used to obtain a large FOV. For example, the spherical Luneburg lens possesses theoretically full-space imaging capability [47–50]. Inspired by the wide FOV visual system found in arthropods and many vertebrates, scientists have also developed artificial analogues of compound eyes via advanced 3D fabrication technology [51]. Nevertheless, the requirement of nonhomogeneous refractive index distribution of Luneburg lens and the precise shape, fill factor, size of ommatidium pose great challenges to implementation. Although one can replace the spherical and nonhomogeneous index lenses by 2D geodesic surfaces with constant refractive index [52, 53], it is still not easy to be implemented in planar optical systems.

2D metasurface has emerged as a promising alternative to traditional bulky components. However, single-layered metasurface lenses usually suffer from the off-axis aberration. A direct method to realize wide FOV is adopting multilayered metasurface lenses [54, 55], in a way similar to traditional approaches [56] but with a more compact volume. For instance, a doublet-corrected 2D metasurface lens (metalens) operating in the near infrared is proposed with a FOV reaching $60° \times 60°$ [54]. Similar doublets have been demonstrated in the visible region with a FOV of $50°$ [55]. From a geometric perspective, the angle-induced image aberration could be considered as a consequence of the breaking of rotational symmetry of flat metalens. By transforming the rotational symmetry into translational symmetry, a wide-angle lens at visible band with FOV beyond $\pm80°$ has been experimentally demonstrated with catenary-inspired continuous nanoapertures [57], as shown in Chap. 2 of this book. Besides, the angle-invariant focusing behavior makes such a lens suitable for wide-angle Fourier analysis [58]. In the following, a high-efficiency wide-angle lens is demonstrated in the microwave regime [46]. The flat lens is sometimes termed as metalens to distinguish it from the plasmonic flat lens described in previous chapters.

The key to construct a wide FOV metalens is the realization of perfect conversion from the rotational symmetry to translational symmetry in light fields. For this purpose, the following relation should be met:

$$k_0 \sin\theta_x x + k_0 \sin\theta_y y + \Phi_m(x, y) = \Phi_m(x + \Delta_x, y + \Delta_y), \qquad (6.5.1)$$

where k_0 is the wavenumber in free space, $\Phi_m(x, y)$ is the phase shift profile carried by the flat lens, Δ_x and Δ_y correspond to the translational shift of $\Phi_m(x, y)$ at incidence angles of θ_x and θ_y.

With some mathematical manipulations, Eq. (6.5.1) can be rewritten as

$$\Phi_m(x + \Delta_x, y + \Delta_y) - \Phi_m(x, y) = k_x x + k_y y, \qquad (6.5.2)$$

and

$$\frac{\partial \Phi_m}{\partial x} = \frac{k_x}{\Delta_x} x,$$

$$\frac{\partial \Phi_m}{\partial y} = \frac{k_y}{\Delta_y} y, \tag{6.5.3}$$

where $k_x = k_0 \sin\theta_x$ and $k_y = k_0 \sin\theta_y$. Consequently, the phase profile of the flat lens should be

$$\Phi_m = \frac{k_x x^2}{2\Delta_x} + \frac{k_y y^2}{2\Delta_y}. \tag{6.5.4}$$

If the lens is circular symmetric, there are

$$\Delta_x = \Delta_y = \Delta, k_x = k_y = k_0 \sin\theta. \tag{6.5.5}$$

Then, we have

$$\Phi_m = \frac{k_0 \sin\theta}{2\Delta}(x^2 + y^2) = k_0 \frac{r^2}{2\Delta/\sin\theta}. \tag{6.5.6}$$

This is just the quadratic phase profile for a normal thin lens in the paraxial regime:

$$\Phi(r) = k_0 \frac{r^2}{2f} = \frac{\pi r^2}{\lambda f}, \tag{6.5.7}$$

where f is the focal length and the horizontal shift can be written as $\Delta = f\sin\theta$. For oblique incidence beam which lies in the xz-plane with an arbitrary angle of θ to the normal axis of the lens, the phase carried by the outgoing light should be

$$\Phi(r) = k_0 \frac{r^2}{2f} + k_0 x \sin\theta = \frac{k_0}{2f}\left((x + f\sin\theta)^2 + y^2\right) - \frac{f k_0 \sin^2\theta}{2}. \tag{6.5.8}$$

Since the last term in the right hand of Eq. (6.5.8) is independent of r and can be neglected, there is only a transversal shift of $f\sin\theta$ in the x-direction with respect to the Eq. (6.5.7). As indicated in Fig. 6.23a, the rotational effect of the oblique incidence light is perfectly converted to the translational symmetry of the output one.

For such a wide FOV metalens, the wavefront picked by a normal and off-axis collimated light is illustrated in Fig. 6.23b. Obviously, the phase changes rapidly with spatial location. For normal incidence, the effective transverse wave vector induced by the phase gradient is equal to the wave vector in free space k_0 at $r = f$. When the light fields pass through region where $r > f$, it will become evanescent, and

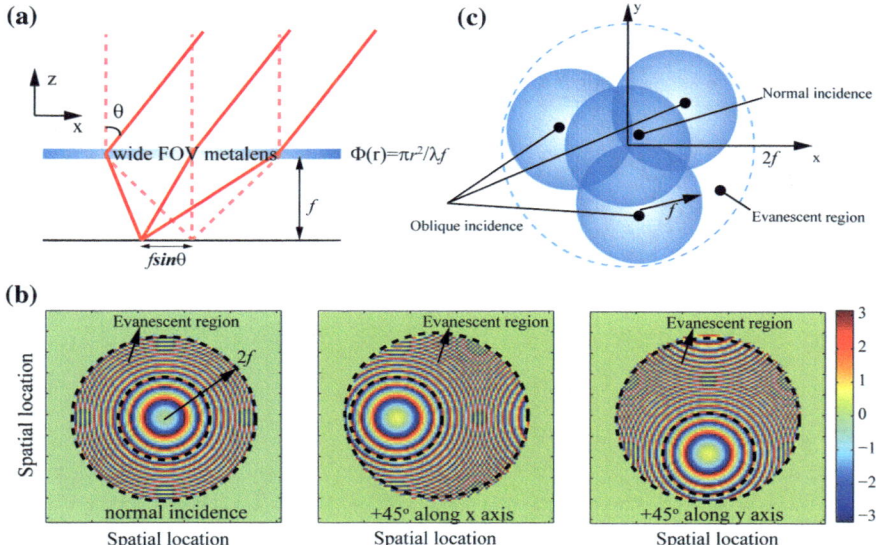

Fig. 6.23 Wide FOV metalens with a quadratic phase profile. **a** Illustration of the symmetry conversion from rotational symmetry of incidence to translational symmetry of outgoing wave based on a wide FOV metalens. **b** Wavefront of outgoing light when it is incident from different angles. **c** Illustration of angle-selective exaction region. Reproduced from [46] with permission. Copyright 2018, WILEY-VCH Verlag GmbH & Co. KGaA, Weinheim

therefore do not contribute to the focal spot. For oblique incidence of different angle, the evanescent zone would shift horizontally, as shown in Fig. 6.23c.

In order to construct such a phase profile, the geometric phase induced by photonic spin–orbit interaction (PSOI) is adopted [31]. It has been theoretically shown that the cross-polarization conversion efficiency of single layer ultra-thin transmissive metasurface is fundamentally limited to 25% with respect to the total incident energy [59]. Besides the reflective configurations presented in previous sections, here an ultra-thin bilayer metasurface is utilized for high-efficiency and wide-angle polarization conversion. Figure 6.24a presents the schematic of a single unit cell arranged in a hexagonal lattice with a period of P. This structure is composed of two identical metallic patterns printed on the two sides of a dielectric substrate. In order to obtain orientation-independent conversion efficiency and high structural symmetry, each rotating metallic cut wire is located at the center of a circular aperture to minimize the coupling between neighboring unit cells. The geometries of the unit cell are optimized using CST Microwave Studio, and the permittivity and thickness of the dielectric spacer are set as $\varepsilon = 4.6$ and $d = 2$ mm (~0.127λ). The metallic patterns are taken as perfect electric conductor (PEC) with a thickness of 0.017 mm. The period of unit cell is fixed as $P = 4$ mm and other geometries are optimized as $R = 1.9$ mm, $L = 3.3$ mm, and $W = 0.75$ mm.

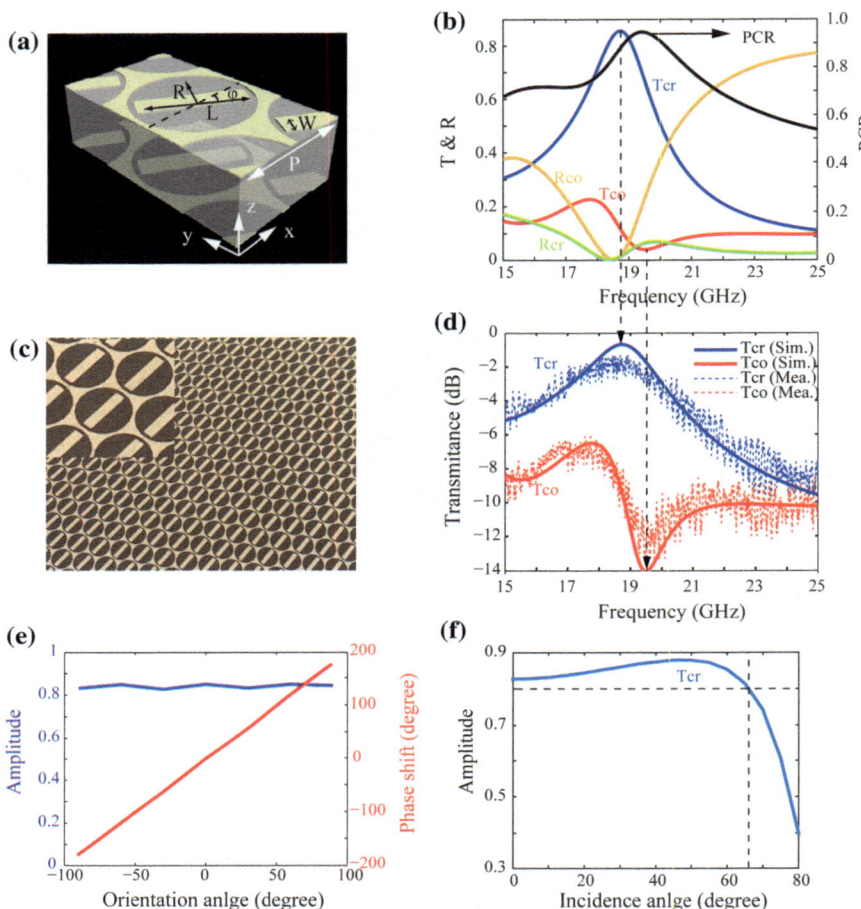

Fig. 6.24 **a** Schematic of the unit cell with defined geometries, **b** Simulated transmittance (T_{co} and T_{cr}) and reflectance (R_{co} and R_{cr}) of co-polarization and cross-polarization components under circularly polarized illumination. **c** Photos of the fabricated sample to characterize the polarization conversion performance. **d** Comparison between the simulated and measured transmission properties. **e** The transmission amplitude and phase shift of cross-polarization components as a function of orientation angle. **f** The transmission amplitude of cross-polarization components as a function of incidence angle. Reproduced from [46] with permission. Copyright 2018, WILEY-VCH Verlag GmbH & Co. KGaA, Weinheim

Figure 6.24b shows the simulated transmittance (T_{co} and T_{cr}) and reflectance (R_{co} and R_{cr}) of the co-polarization and cross-polarization components, where T and R denote the transmission and reflection. It can be seen that most energy of the incident circular polarization light is transmitted through the metasurface and transformed to the cross-polarization components (T_{cr}) with a peak efficiency up to 85% around 18.8 GHz, while other components (T_{co}, R_{co}, and R_{cr}) can be neglected at this point. The conversion efficiency of bilayer metasurface is significantly higher than that of single layer configuration. By cascading more metasurface layers, one may obtain a higher conversion efficiency [60], but at a cost of increased device thickness and fabrication difficulty. The polarization conversion ratio (PCR) of transmitted components is defined as $PCR = T_{cr}/(T_{cr} + T_{co})$. It can be seen that the PCR exhibits a maximum of 95% around 19.4 GHz. In order to verify this design, a sample with outer dimension of 350×350 mm^2 was fabricated on a FR4 substrate, as shown in Fig. 6.24c. The measured transmittances of the co- and cross-polarized components are denoted by dotted lines in Fig. 6.24d, which are consistent with the simulated ones denoted by solid lines. Subsequently, the transmittance and phase of T_{cr} as a function of the orientation angle φ are investigated and displayed in Fig. 6.24e. As expected, the geometric phase shift exhibits a twice relationship with the orientation angle φ while the amplitude is almost invariant. Besides, the proposed structure exhibits a high tolerance of incidence angle θ. As shown in Fig. 6.24f, the transmittance of T_{cr} is larger than 80% even when the incidence angle is titled by $60°$, which is a key factor to realize the wide FOV metalens.

The physical mechanism of PSOI is attributed to the anisotropy of the unit cell, which can be drawn from the electric field distribution within the unit cell excited by orthogonal linear polarizations. As shown in Fig. 6.25a, there is a strong localized field enhancement in the narrow metallic aperture for the x-polarized incidence, while this effect is quite weak for the y-polarized incidence. The field distribution within the narrow metallic aperture at 19 GHz is extracted and plotted in Fig. 6.25b, whose profile can be well described by a nonsymmetric catenary curve due to the multiple reflection between the nonsymmetric metallic aperture as

$$|E(x)| = a \, \exp(-bx) + c \, \exp(dx), \qquad (6.5.9)$$

where a, b, c, and d are the variables in the generalized catenary function. When these variables equal to $a = 26460$, $b = 12.48$, $c = 27820$, and $d = 18.07$, the catenary curve (solid curve) can well approach the simulated electric field profile (dotted curve) with a R-square of 0.993.

It is found that the catenary optical fields are important for the superb performance. On the one hand, the transversal catenary field results in highly anisotropic surface impedances, which promise high-efficiency polarization conversion and geometric phase modulation. On the other hand, the angle-independent PSOI is attributed to the strong localization of vertical catenary field since it gives rise to magnetic coupling between the top and bottom layers, as illustrated in Fig. 6.25d–f. When variables are $a = c = 1480$ and $b = d = 3.617$, the symmetric catenary curve (solid curve)

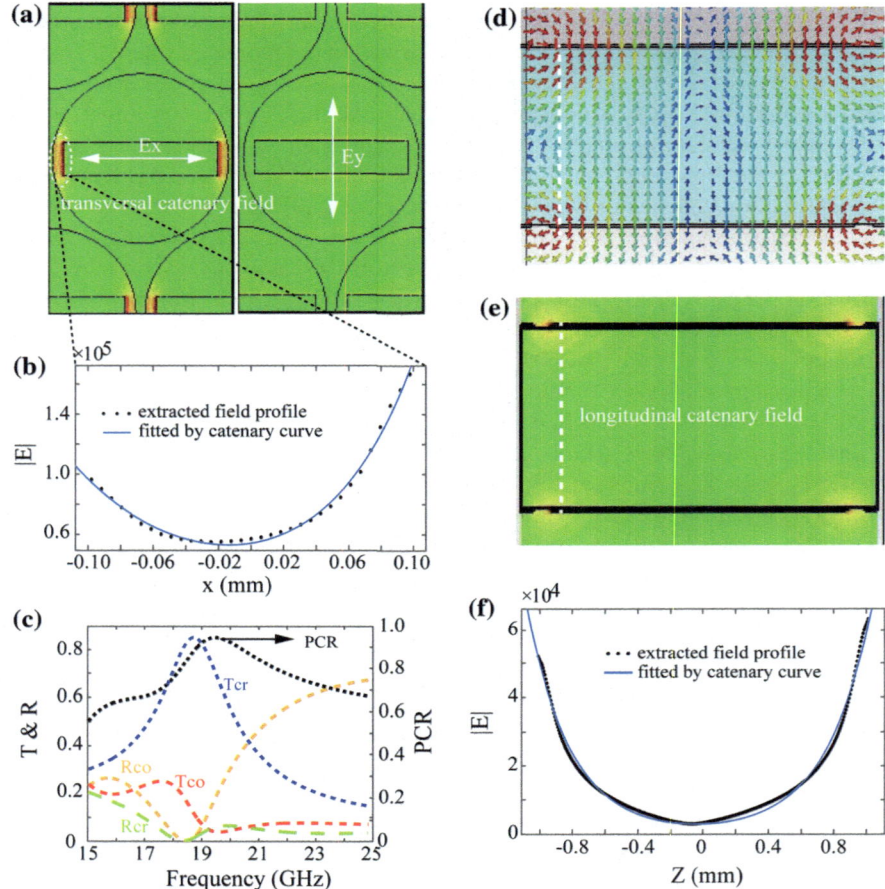

Fig. 6.25 Transversal and longitudinal catenary local fields. **a** Electric field distribution within the unit cell at 19 GHz under orthogonal linear polarizations. **b** The extracted field profile within the metallic aperture (indicated by the white elliptical circle in **a**), which is well fitted by a nonsymmetric catenary function. **c** Calculated transmission properties and PCR based on equivalent circuit theory and transfer matrix method. **d** The vectorial electric field distribution and **e** Electric field magnitude distribution at the *xoz* cross plane at 19 GHz. **f** The extracted field profile is well fitted by a symmetric catenary function. Reproduced from [46] with permission. Copyright 2018, WILEY-VCH Verlag GmbH & Co. KGaA, Weinheim

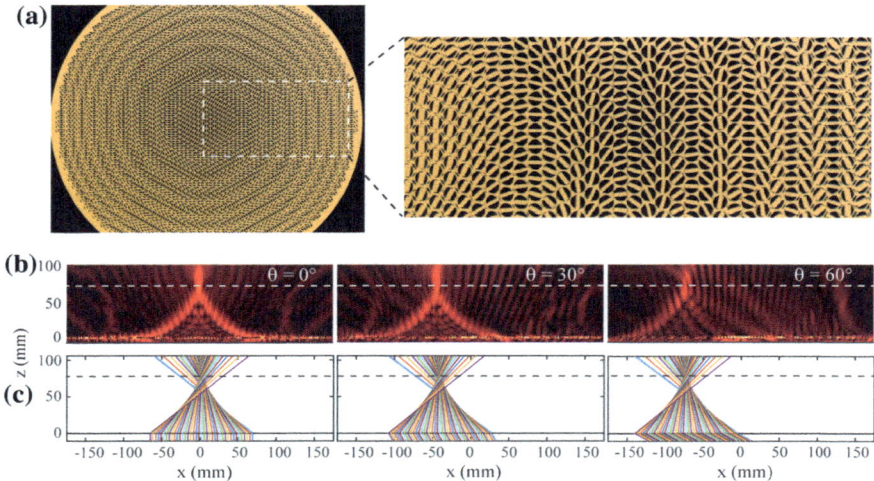

Fig. 6.26 Performance of the wide-angle flat lens. **a** Perspective and zoom view. **b** Simulated light intensity distributed on the *xoz*-plane at 19 GHz. **c** Ray trajectories of 19 GHz before and after propagating through the flat lens. Left, middle, and right panels of (**b**) (**c**), respectively, represent the case of for $\varphi = 0°$, 30° and 60°. Reproduced from [46] with permission. Copyright 2018, WILEY-VCH Verlag GmbH & Co. KGaA, Weinheim

approaches the simulated electric field profile (dotted curve) with a R-square of 0.990. Similar to the wide-angle absorbers [23], the magnetic resonance makes the effective refractive index very high. According to the Snell's law, the diffraction angle is much smaller than the incidence angle, thus the performance seems to be angle independent [8].

Based on the above designed unit cell, a wide FOV flat lens with a diameter of 350 mm and a focal length $f = 87.5$ mm (corresponding to a high numerical aperture (NA) of 0.89) was constructed by rotating the unit cells in a hexagonal lattice. As presented in Fig. 6.26a, the orientation of each unit cell is determined according to the principle of geometric phase:

$$\xi(r) = \frac{\Phi(r)}{2\sigma} = \frac{\pi r^2}{2\sigma \lambda f}. \tag{6.5.10}$$

Full-wave simulations with circularly polarized plane wave are carried out by employing the finite-difference time domain (FDTD) method. The simulated electric field distributions at 19 GHz within the *xoz*-plane are shown in Fig. 6.26b. It can be seen that the focal lengths and electric field patterns are nearly invariable for different incidence angles, $\theta = 0°$, 30°, and 60°, implying that the FOV of the high NA metalens is beyond ±60°. Compared with the doublet metalens reported in Refs. [54, 55], the proposed metalens has a relative thinner thickness and larger FOV. Note that there are some translational shifts corresponding to $\Delta = -f\sin\theta$. The values of Δ are 0,

Table 6.1 Beam steering performances of various antennas [46]

	Scanning range	Thickness (λ)	Side lobe (dB)	Fractional bandwidth (%)	Diffraction efficiency (%)
This design	±60°	0.127	−20	25	93
Ref. [65]	±30°	0.231	−15	23.7	87
Ref. [66]	±60°	0.13	−15	8.5	48
Ref. [67]	±50°	0.17	−11	10	not mentioned
Ref. [68]	±30°	0.46	−10	3.4	85

−43.75, and −75.8 mm, respectively, agreeing with the theoretical expectations. Based on the generalized Snell's law, the ray trajectories were simulated at 19 GHz before and after propagating through metalens, as displayed in Fig. 6.26c, which clearly demonstrate the nearly perfect conversion from rotational symmetry to translational symmetry. Different areas of metalens are excited for different incidence angles and the rays through other area become evanescent waves.

Based on the reciprocity principle, the proposed wide FOV metalens with a high NA can behave as a wide-angle beam steering antenna that can redirect the radiation of a circularly polarized point source toward different directions, when the source is transversely shifted along the focal plane. As a proof of concept, we measured the far-field power patterns of a circularly polarized horn through the fabricated wide FOV lens. Figure 6.27a shows the numerically calculated power pattern obtained from vectorial diffraction theory (VDT). The measured power patterns at 19 GHz with different transverse shifts Δ are shown in Fig. 6.27b, from which one can see that ±60° beam steering can be realized simply by transversely changing the location of the antenna within an area of $\Delta \in [−75.8, 75.8 \text{ mm}]$. The side lobe for the most tilted beam (60°) is smaller than −20 dB. The measured diffraction efficiency defined as the ratio between the power toward the desired direction and the total transmittance is high up to 93% when θ is 0°. The gain of the beam steering antenna is about 15 dB larger than the horn antenna.

The proposed wide FOV flat lens is competent for two-dimensional beam steering because of the C6 rotational symmetry of the metasurface. It is no doubt that the scanning range can be further improved by optimizing the metalens design. For example, by merging geometric phase and propagation phase within the same metalens [61–64], more flexible beam steering can be accessed due to the symmetry-breaking spin–orbit interaction. Although the proposed metalens is optimized at 19 GHz, it can also work well with a fractional bandwidth beyond 25%, as indicated in Fig. 6.27b–f.

Table 6.1 compares the beam steering performances (scanning range, device thickness, side lobe, operation bandwidth, and diffraction efficiency) of the proposed design and previous transmissive arrays, demonstrating obvious advantages.

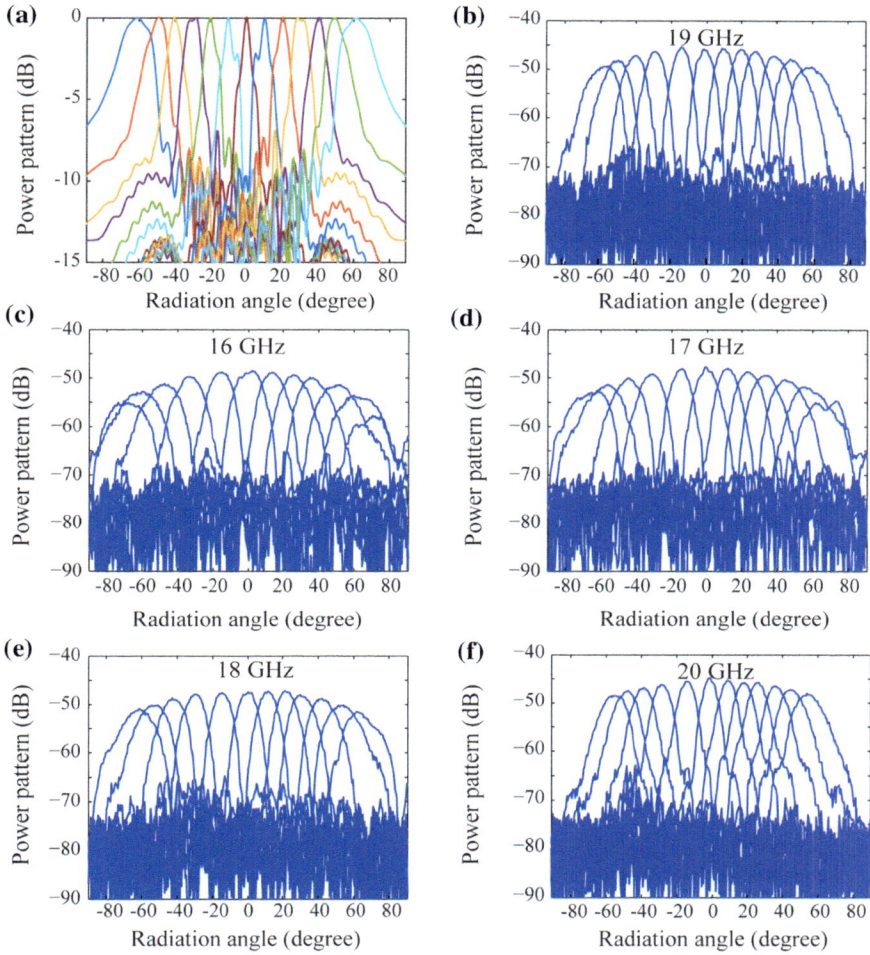

Fig. 6.27 Multi-wavelength behavior of the quadratic metalens antenna. **a** Calculated far-field power patterns. **b–f** Measured far-field power patterns at different frequencies when a circularly polarized antenna is transversely shift at a distance of $z = -87.5$ mm. **b** 19 GHz, **c** 16 GHz, **d** 17 GHz, **e** 18 GHz, and **f** 20 GHz. Reproduced from [46] with permission. Copyright 2018, WILEY-VCH Verlag GmbH & Co. KGaA, Weinheim

6.6   Switchable Beam Manipulation via Phase-Change Materials

Phase-change material (PCM) is a promising and earth-abundant alternative to the next-generation nonvolatile optical devices, offering a new avenue to realize switchable photonic devices [69]. As typical PCMs, GeSbTe (GST) alloys have been used for many years in optical disk storage and have been introduced to reconfigurable photonic devices recently [70, 71]. GST alloys can be switched repeatedly between amorphous and crystalline states or to an intermediate state by an appropriate thermal, optical, or electrical stimulus. Additionally, such materials exhibit high refractive index contrast between amorphous and crystalline states and low losses in the near- and middle-infrared spectral range. With these extraordinary properties, GST alloys are ideal materials for switchable or reconfigurable devices, such as thermal emitters, Fresnel zone plates, and absorbers [69–73]. In this section, switchable PSOI in the mid-infrared spectral range is experimentally demonstrated via the combination of plasmonic metasurface with GST [74]. To verify the proposed approach, three meta-devices based on a simple MIM configuration were fabricated by microfabrication techniques. These designed meta-devices turn on PSOI for spin Hall effect of light (SHEL) [22, 75], vortex beam generation, and optical holography when the GST is in amorphous state. When the GST layer turns into crystalline state, these effects are switched off. Experimental spectral results show that the cross-polarization reflectance of the unit cell approaches 60% in amorphous state and turns to approximately zero in crystalline state in a broadband wavelength range from 8.5 to 10.5 μm.

As depicted in Fig. 6.28a, an insulator layer consisting of $Ge_2Sb_2Te_5$ (GST-225) and MgF_2 film is sandwiched by the bottom gold ground plane and the top array of subwavelength plasmonic gold antennas. The GST-225 film acts essentially as a switchable dielectric medium that changes the optical response of the meta-devices. MgF_2 film was deposited on the top of the GST-225 layer to protect it against oxidation in the atmosphere. Furthermore, the MgF_2 layer serves as a refraction index-matching layer between high-index GST-225 and low-index air, resulting in a significant improvement of the polarization conversion efficiency.

Each anisotropic metallic antenna can be taken as a local wave plate. Under normal incidence, a circularly polarized beam is scattered into waves with the same polarization as that of the incident beam without phase shift, and waves of the opposite circular polarization with a spin-dependent geometric phase. By controlling the local orientation of the fast axes of the metallic elements between 0 and π, one can realize phase variation covering the full $0–2\pi$ range while maintaining equal reflection amplitude. Note that the amplitude of opposite circular polarization depends on the phase retardation and thus a wave plate with a phase retardation of π is desired to realize 100% conversion efficiency.

Based on the PSOI and geometric phase, three meta-devices were demonstrated with switchable functions at a wavelength of $\lambda = 9.6$ μm. As shown in Fig. 6.28b, when the GST layer is in amorphous state, PSOI occurs in the three meta-devices. Under the illumination of linear polarization, the devices 1 and 2 show an angular SHEL and generate two deflected vortex beams, respectively. Under the illumination of circularly polarized light, the third device generates one holographic image of abbreviation of Institute of Optics and Electronics (IOE). However, when the GST layer is switched into crystalline state, the PSOI-enabled phenomena disappear and there is only a bright spot. In this case, all of the three devices behave as conventional reflective mirrors.

(a)

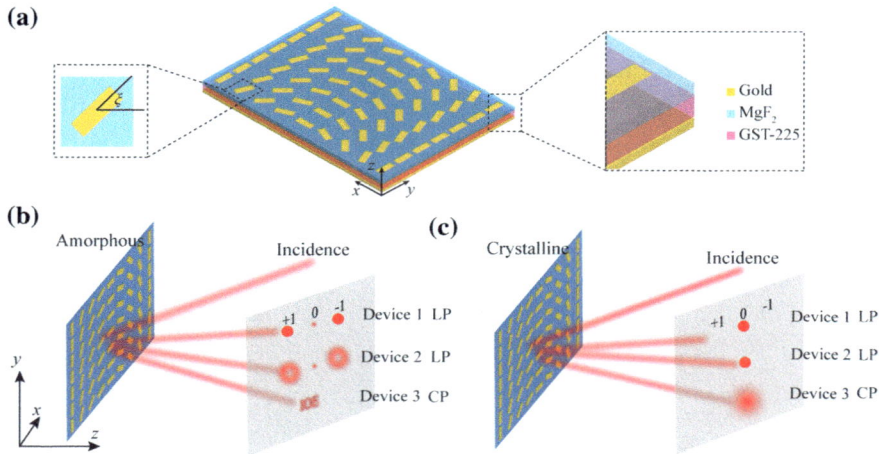

Fig. 6.28 Schematic of the switchable PSOI. **a** Structure and materials. **b, c** Optical performances of three designed meta-devices when the GST layer is in amorphous and crystalline states. LP and CP represent linear and circular polarizations. Reproduced from [74] with permission. Copyright 2018, The Authors

The unit cell of the devices is depicted in Fig. 6.29a. The thickness of GST h_{GST} is 600 nm to turn off the polarization conversion after crystallization. The thickness of MgF_2 ($h_{MgF2} = 150$ nm) is chosen to be thick enough to block the contact between air and GST. The antennas array is composed of rectangle patches with a length of l and a width of w, and the thickness of antennas ($h = 100$ nm) is chosen to be as thin as possible to reduce the fabrication difficulty. Figure 6.29b presents the SEM images of the fabricated antennas array.

Numerical simulations are performed using the frequency solver in CST Microwave Studio. The simulated results are shown in the upper part of Fig. 6.29c, d. As illustrated in the figures, the reflectance of cross polarization is over 70% in a broad spectral range from 9 to 12 μm in the amorphous state, while the co-polarized reflectance is approximately close to 0 (here, the circular polarization conversion of a common mirror is neglected for simplicity of discussion). When GST changes into crystalline state, the cross-polarized reflectance turns to be less than 5% in 7–11 μm. The performance of the sample was characterized with a Fourier transform infrared spectrometer (Bruker Vertex 80) in reflection mode with an incident angle of 15°. In order to characterize the broadband performance, two linear polarizers are utilized at 45° with respect to the patch antenna. The measured results are shown in the bottom part of Fig. 6.29c, d. In amorphous state, the cross-polarized reflectance approaches to 60% and the co-polarized reflectance is less than 10% in the spectral range from 8.5 to 11.5 μm. The sample was heated on a hotplate for 20 min to ensure the complete crystallization. In crystalline state, the cross-polarized reflectance decrease to be less than 2%. Besides, the PCR is calculated to characterize the working bandwidth. Figure 6.29e shows the simulated and measured PCR with respect to the wavelength

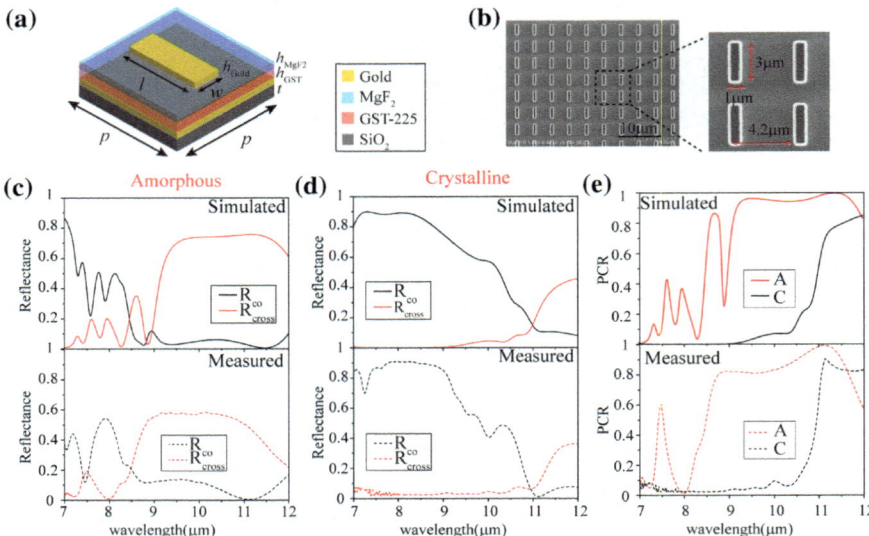

Fig. 6.29 **a** Schematic of the unit cell. **b** SEM images of fabricated patch antennas array with the width $l = 3$ μm, length $w = 1$ μm, and $p = 4.2$ μm. **c**, **d** The co-polarization and cross-polarization reflectance in amorphous and crystalline states. **e** The PCR with respect to the wavelength in amorphous (A) and crystalline (C) states. Reproduced from [74] with permission. Copyright 2018, The Authors

in two states, respectively. The simulated results depict that the PCR is beyond 90% in amorphous state and below 10% in crystalline state in a broadband spectral range from 9 to 11.5 μm, indicating a broadband feature. The measured results also show the broadband feature with PCR beyond 80% in amorphous state and below 10% in crystalline state in a broadband spectral range from 8.5 to 10.5 μm.

To reveal the physical mechanisms, the unit cells are analyzed at different orientation angles. Figure 6.30b shows the relationship of reflectance and phase shift with respect to the orientation angles. R_{cross_a} and PS represent the cross-polarized reflectance and phase shift at 9.6 μm in amorphous state. Note that the cross-polarized reflectance remains approximately the same as the angle varies. The calculated phase shift shows a good agreement with geometric phase. In addition, it can be seen that the unit cells have low cross-polarized reflectance in crystalline state.

Furthermore, to explore the physical mechanism of different responses of the unit cell in two states, the reflective phase difference between two linear components along the x- and y-directions are calculated as depicted in Fig. 6.30c. For the unit cell illuminated by a normally incident linearly polarized light with a polarization angle of 45°, a phase difference around π with a slight variation of $\pm 10\%$ ($0.9\pi - 1.1\pi$) is achieved from 8.8 to 11.5 μm in amorphous state, resulting in an "ON" state of PSOI. On the contrary, the phase difference in crystalline state is nearly equal to 0, implying an "OFF" state of PSOI.

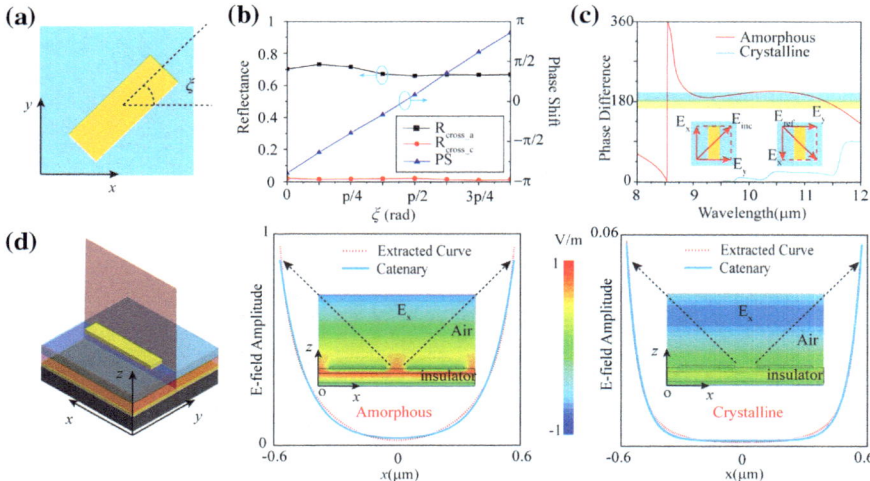

Fig. 6.30 **a** Top view of the unit cell with an orientation angle of ξ. **b** Simulated cross-polarization reflectance and phase shift as a function of ξ in two states. **c** The calculated phase difference between the two orthogonal polarizations in two states. **d** 3D view of the unit cell and the electric field distributions in the red cut plane. The right two panels show the extracted electric field amplitude (red dotted line) and fitted catenary curve (blue solid line) between two adjacent patches for amorphous state (left) and crystalline state (right) at 9.6 μm, respectively. The insets are the distributions of electric fields in the unit cell under circularly polarized illumination. Reproduced from [74] with permission. Copyright 2018, The Authors

In an effort to get a more clear visualization of this effect, the instantaneous electric field distribution under normal incidence is illustrated. Figure 6.30d shows the electric field E_x at the resonant wavelength of 9.6 μm in the yz-plane. The electric field in amorphous state (left) is obviously enhanced at the metal–insulator interface and coupled with that in the adjacent elements. The electric field profile between the adjacent elements can be described by the well-known catenary curve. Such field distribution stems from the near field interaction between adjacent unit cells. By tuning the coupling strength of catenary optical fields, ultra-broadband PSOI can be enabled or suppressed. In crystalline state, the refractive index of GST gets very large, and the transversal coupling is suppressed, resulting in a weak anisotropy and thus low polarization conversion efficiency.

In order to demonstrate the versatility of this approach, three flat meta-devices are fabricated and characterized. Experimental results in different states (amorphous and crystalline states) and SEM images of the three meta-devices are shown in Fig. 6.31.

The first device is fabricated to demonstrate SHEL. The linear phase gradient induced by a change of orientation with the coordinates produces a spin-dependent transverse wave vector. Thus, the anisotropic metasurface deflects right- and left-hand beams to opposite directions, which can be considered as an angular SHEL. The designed meta-device consists of periodic arrays of 16 unit cells with an incremental

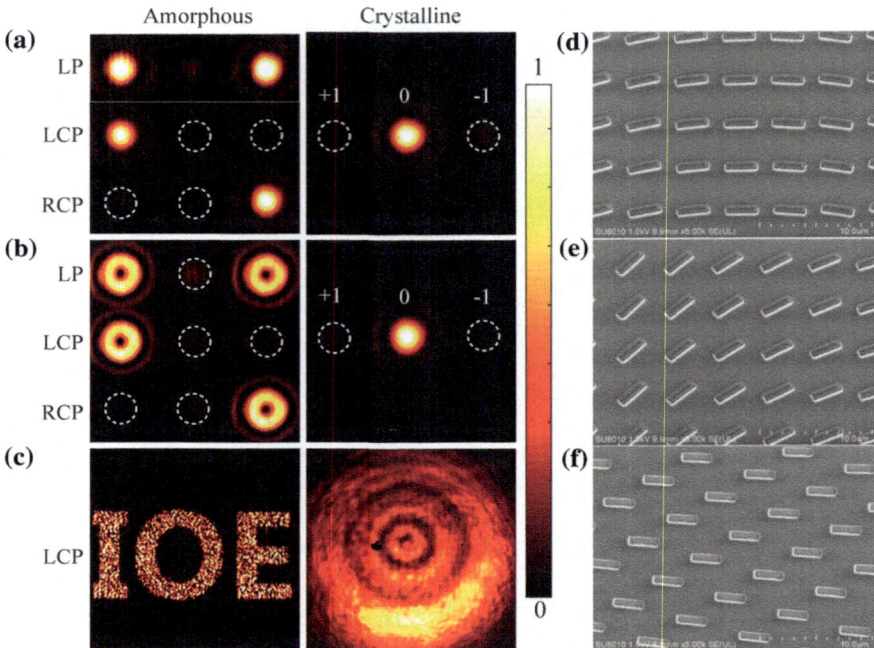

Fig. 6.31 The measured intensity and SEM images of three switchable meta-devices. **a–c** The intensity patterns for **a** SHEL, **b** vortex generation, and **c** holography in amorphous and crystalline states. The constant background produced by thermal radiation has been removed. **d–f** SEM images of the three meta-devices. Reproduced from [74] with permission. Copyright 2018, The Authors

rotation angle of $\pi/16$. The left panel of Fig. 6.31a shows the measured reflected diffraction patterns at the wavelength of 9.6 µm when the device is in amorphous state. The central dim spot (0th order) is attributed to the unmodulated spin component of reflected beam, while +1st- and −1st-order bright spots stem from the SHEL of two converted spin components. Under the illumination of linear polarized light, the measured efficiencies of the +1st-, 0th-, and −1st-order diffractions are 26.1, 2.4, and 26.1%, respectively. Besides, under the illumination of LCP or RCP, it is observed that a majority of the reflected light concentrated in the +1st- or −1st-order, thus the meta-device works as an anomalous deflector. When this device is heated at 200 °C for 20 min, the GST layer turns into crystalline state. In this case, the majority of reflected light is concentrated in the zeroth order whatever the polarization of incident light is, so the SHEL is "switched off" and the meta-device behaves as a specular mirror.

The second device is a vortex beam generator, which is crucial for many classic and quantum applications. To facilitate the measurement, an azimuthal phase dependence $\exp(il\varphi)$ is superimposed on the abovementioned deflector, where φ is the azimuthal angle and the l is the topological charge. As illustrated in Fig. 6.31b, the donut-shaped patterns present at +1st- and −1st-order (linear polarized incidence) and at +1st/−1st-order (LCP/RCP incidence). When the device turns into crystalline state, the donut-shaped patterns vanish and there is only a bright spot at the zeroth order.

The last device is a hologram that generates one holographic image of the abbreviation of the IOE. The holographic image is designed based on point source algorithm. The left panel of Fig. 6.31c displays the measured reflected intensity pattern under the illumination of a light beam with pure circular polarization (RCP or LCP) in amorphous state. As the meta-device turns into crystalline state, the designed holographic pattern disappears and there is only a bright spot, implying the information carried by the device is concealed. Such performance may provide a new route to dynamic holographic image generation and encrypted information storage.

Note that the functions of meta-devices can be switched between amorphous and crystalline states. The re-amorphization of GST can be achieved by several methods, such as thermal annealing, electrical stimulus, and laser pulse illumination [70, 73]. However, the fact that the re-amorphization temperature of the GST (~640 °C) is beyond the melting point of 100 nm-thick Au (~300 °C) hinders the reconfiguration of the above devices by thermal annealing process. Nevertheless, with proper design, electrical stimulus or laser pulses can be utilized to realize the reconfiguration of meta-devices. The typical switching time is several nanoseconds for electrical stimulus and even merely tens of femtoseconds for laser pulses.

References

1. M. Pu, X. Ma, Y. Guo, X. Li, X. Luo, Theory of microscopic meta-surface waves based on catenary optical fields and dispersion. Opt. Express **26**, 19555–19562 (2018)
2. J.A. Polo, A. Lakhtakia, Surface electromagnetic waves: a review. Laser. Photonics. Rev. **5**, 234–246 (2011)
3. X. Luo, Principles of electromagnetic waves in metasurfaces. Sci. China-Phys. Mech. Astron. **58**, 594201 (2015)
4. W.L. Barnes, A. Dereux, T.W. Ebbesen, Surface plasmon subwavelength optics. Nature **424**, 824–830 (2003)
5. P. Cheben, R. Halir, J.H. Schmid, H.A. Atwater, D.R. Smith, Subwavelength integrated photonics. Nature **560**, 565–572 (2018)
6. T. Xu, Y.-K. Wu, X. Luo, L.J. Guo, Plasmonic nanoresonators for high-resolution colour filtering and spectral imaging. Nat. Commun. **1**, 59 (2010)
7. M. Khorasaninejad, F. Capasso, Metalenses: versatile multifunctional photonic components. Science **358**, eaam8100 (2017)
8. M. Pu, X. Ma, X. Li, Y. Guo, X. Luo, Merging plasmonics and metamaterials by two-dimensional subwavelength structures. J. Mater. Chem. C **5**, 4361 (2017)
9. X. Luo, Subwavelength optical engineering with metasurface waves. Adv. Opt. Mater. **6**, 1701201 (2018)
10. S.B. Glybovski, S.A. Tretyakov, P.A. Belov, Y.S. Kivshar, C.R. Simovski, Metasurfaces: from microwaves to visible. Phys. Rep. **634**, 1–72 (2016)
11. X. Luo, T. Ishihara, Surface plasmon resonant interference nanolithography technique. Appl. Phys. Lett. **84**, 4780–4782 (2004)
12. M. Pu, Y. Guo, X. Li, X. Ma, X. Luo, Revisitation of extraordinary Young's interference: from catenary optical fields to spin-orbit interaction in metasurfaces. ACS Photonics **5**, 3198–3204 (2018)

13. X. Luo, Engineering optics 2.0: a revolution in optical materials, devices, and systems. ACS Photonics **5**, 4724-4738 (2018)
14. X. Luo, T. Ishihara, in *Sub 100 nm lithography based on plasmon polariton resonance*. 2003 International Microprocesses and Nanotechnology Conference (IEEE, 2003), pp. 138–139
15. T. Xu, C. Wang, C. Du, X. Luo, Plasmonic beam deflector. Opt. Express **16**, 4753–4759 (2008)
16. M.G. Moharam, T.K. Gaylord, Rigorous coupled-wave analysis of metallic surface-relief gratings. J. Opt. Soc. Am. A **3**, 1780–1787 (1986)
17. M. Pu, C. Hu, C. Huang, C. Wang, Z. Zhao, Y. Wang, X. Luo, Investigation of Fano resonance in planar metamaterial with perturbed periodicity. Opt. Express **21**, 992–1001 (2013)
18. L.B. Whitbourn, R.C. Compton, Equivalent-circuit formulas for metal grid reflectors at a dielectric boundary. Appl. Opt. **24**, 217–220 (1985)
19. Q. Feng, M. Pu, C. Hu, X. Luo, Engineering the dispersion of metamaterial surface for broadband infrared absorption. Opt. Lett. **37**, 2133–2135 (2012)
20. T. Senior, Approximate boundary conditions. IEEE Trans. Antennas Propag. **29**, 826–829 (1981)
21. M. Pu, X. Li, X. Ma, Y. Wang, Z. Zhao, C. Wang, C. Hu, P. Gao, C. Huang, H. Ren, X. Li, F. Qin, J. Yang, M. Gu, M. Hong, X. Luo, Catenary optics for achromatic generation of perfect optical angular momentum. Sci. Adv. **1**, e1500396 (2015)
22. X. Luo, M. Pu, X. Li, X. Ma, Broadband spin hall effect of light in single nanoapertures. Light. Sci. Appl. **6**, e16276 (2017)
23. M. Pu, C. Hu, M. Wang, Z. Zhao, C. Wang, Q. Feng, X. Luo, Design principles for infrared wide-angle perfect absorber based on plasmonic structure. Opt. Express **19**, 17413–17420 (2011)
24. H.-T. Chen, Interference theory of metamaterial perfect absorbers. Opt. Express **20**, 7165–7172 (2012)
25. Y. Li, X. Li, M. Pu, Z. Zhao, X. Ma, Y. Wang, X. Luo, Achromatic flat optical components via compensation between structure and material dispersions. Sci. Rep. **6**, 19885 (2016)
26. Z. Ma, S.M. Hanham, P. Albella, B. Ng, H.T. Lu, Y. Gong, S.A. Maier, M. Hong, Terahertz all-dielectric magnetic mirror metasurfaces. ACS Photonics **3**, 1010–1018 (2016)
27. R. Paniagua-Domínguez, Y.F. Yu, A.E. Miroshnichenko, L.A. Krivitsky, Y.H. Fu, V. Valuckas, L. Gonzaga, Y.T. Toh, A.Y.S. Kay, B.S. Luk'yanchuk, A.I. Kuznetsov, Generalized Brewster effect in dielectric metasurfaces. Nat. Commun. **7**, 10362 (2016)
28. D. Van Labeke, D. Gerard, B. Guizal, F.I. Baida, L. Li, An angle-independent frequency selective surface in the optical range. Opt. Express **14**, 11945–11951 (2006)
29. M. Pu, X. Li, Y. Guo, X. Ma, X. Luo, Nanoapertures with ordered rotations: symmetry transformation and wide-angle flat lensing. Opt. Express **25**, 31471–31477 (2017)
30. X. Xie, X. Li, M. Pu, X. Ma, K. Liu, Y. Guo, X. Luo, Plasmonic metasurfaces for simultaneous thermal infrared invisibility and holographic illusion. Adv. Funct. Mater. **28**, 1706673 (2018)
31. M. Pu, Z. Zhao, Y. Wang, X. Li, X. Ma, C. Hu, C. Wang, C. Huang, X. Luo, Spatially and spectrally engineered spin-orbit interaction for achromatic virtual shaping. Sci. Rep. **5**, 9822 (2015)
32. S. Simms, V. Fusco, Chessboard reflector for RCS reduction. Electron. Lett. **44**, 316–317 (2008)
33. L. Huang, X. Chen, H. Mühlenbernd, H. Zhang, S. Chen, B. Bai, Q. Tan, G. Jin, K.-W. Cheah, C.-W. Qiu, J. Li, T. Zentgraf, S. Zhang, Three-dimensional optical holography using a plasmonic metasurface. Nat. Commun. **4**, 2808 (2013)
34. X. Li, L. Chen, Y. Li, X. Zhang, M. Pu, Z. Zhao, X. Ma, Y. Wang, M. Hong, X. Luo, Multicolor 3D meta-holography by broadband plasmonic modulation. Sci. Adv. **2**, e1601102 (2016)
35. Y. Guo, J. Yan, M. Pu, X. Li, X. Ma, Z. Zhao, X. Luo, Ultra-wideband manipulation of electromagnetic waves by bilayer scattering engineered gradient metasurface. RSC Adv. **8**, 13061–13066 (2018)
36. Y. Guo, Y. Wang, M. Pu, Z. Zhao, X. Wu, X. Ma, C. Wang, L. Yan, X. Luo, Dispersion management of anisotropic metamirror for super-octave bandwidth polarization conversion. Sci. Rep. **5**, 8434 (2015)

37. G.G. Macfarlane, Quasi-stationary field theory and its application to diaphragms and junctions in transmission lines and wave guides. J. Inst. Electr. Eng. Part III A Radiolocation **93**, 703–719 (1946)
38. J.R. Swandic, *Bandwidth Limits and Other Considerations for Monostatic RCS Reduction by Virtual Shaping* (Naval Surface Warfare Center, Carderock Div., 2004)
39. J.B. Pendry, D. Schurig, D.R. Smith, Controlling electromagnetic fields. Science **312**, 1780–1782 (2006)
40. X. Ni, Z.J. Wong, M. Mrejen, Y. Wang, X. Zhang, An ultrathin invisibility skin cloak for visible light. Science **349**, 1310–1314 (2015)
41. C. Huang, J. Yang, X. Wu, J. Song, M. Pu, C. Wang, X. Luo, Reconfigurable metasurface cloak for dynamical electromagnetic illusions. ACS Photonics **5**, 1718–1725 (2018)
42. Y. Huang, M. Pu, F. Zhang, J. Luo, X. Li, X. Ma, X. Luo, Broadband functional metasurface: Achieving non-linear phase generation towards achromatic surface cloaking and lensing. Adv. Opt. Mater. 1801480 (2019)
43. F. Gires, P. Tournois, Interferometre utilisable pour la compression d' impulsions lumineuses modulees en frequence. C. R. Acad. Sci. Paris **258**, 6112–6115 (1964)
44. N. Yu, P. Genevet, M.A. Kats, F. Aieta, J.-P. Tetienne, F. Capasso, Z. Gaburro, Light propagation with phase discontinuities: generalized laws of reflection and refraction. Science **334**, 333–337 (2011)
45. X. Ni, N.K. Emani, A.V. Kildishev, A. Boltasseva, V.M. Shalaev, Broadband light bending with plasmonic nanoantennas. Science **335**, 427–427 (2012)
46. Y. Guo, X. Ma, M. Pu, X. Li, Z. Zhao, X. Luo, High-efficiency and wide-angle beam steering based on catenary optical fields in ultrathin metalens. Adv. Opt. Mater. **6**, 1800592 (2018)
47. R.K. Luneburg, *Mathematical Theory of Optics* (Brown University, 1944)
48. H. Ma, T. Cui, Three-dimensional broadband and broad-angle transformation-optics lens. Nat. Commun. **1**, 124 (2010)
49. N. Kundtz, D.R. Smith, Extreme-angle broadband metamaterial lens. Nat. Mater. **9**, 129–132 (2010)
50. Y.-Y. Zhao, Y.-L. Zhang, M.-L. Zheng, X.-Z. Dong, X.-M. Duan, Z.-S. Zhao, Three-dimensional Luneburg lens at optical frequencies. Laser Photonics Rev. **10**, 665–672 (2016)
51. D. Wu, J.-N. Wang, L.-G. Niu, X.-L. Zhang, S.Z. Wu, Q.-D. Chen, L.P. Lee, H.B. Sun, Bioin-spired fabrication of high-quality 3D artificial compound eyes by voxel-modulation femtosec-ond laser writing for distortion-free wide-field-of-view imaging. Adv. Opt. Mater. **2**, 751–758 (2014)
52. K. Liu, Y. Guo, M. Pu, X. Ma, X. Li, X. Luo, Wide field-of-view and broadband terahertz beam steering based on gap plasmon geodesic antennas. Sci. Rep. **7**, 41642 (2017)
53. J. L. McFarland, Catenary geodesic lens antenna. U.S. patent 3,383,691 (1968)
54. A. Arbabi, E. Arbabi, S.M. Kamali, Y. Horie, S. Han, A. Faraon, Miniature optical planar camera based on a wide-angle metasurface doublet corrected for monochromatic aberrations. Nat. Commun. **7**, 13682 (2016)
55. B. Groever, W.T. Chen, F. Capasso, Meta-lens doublet in the visible region. Nano Lett. **17**, 4902–4907 (2017)
56. T. Gissibl, S. Thiele, A. Herkommer, H. Giessen, Two-photon direct laser writing of ultracom-pact multi-lens objectives. Nat. Photon **10**, 554–560 (2016)
57. M. Pu, X. Li, Y. Guo, X. Ma, X. Luo, Nanoapertures with ordered rotations: symmetry trans-formation and wide-angle flat lensing. Opt. Express **25**, 31471–31477 (2017)
58. W. Liu, Z. Li, H. Cheng, C. Tang, J. Li, S. Zhang, S. Chen, J. Tian, Metasurface enabled wide-angle fourier lens. Adv. Mater. **30**, 1706368 (2018)
59. Y. Wang, M. Pu, Z. Zhang, X. Li, X. Ma, Z. Zhao, X. Luo, Quasi-continuous metasurface for ultra-broadband and polarization-controlled electromagnetic beam deflection. Sci. Rep. **5**, 17733 (2015)
60. W. Luo, S. Sun, H.-X. Xu, Q. He, L. Zhou, Transmissive ultrathin pancharatnam-berry meta-surfaces with nearly 100% efficiency. Phys. Rev. Appl. **7**, 044033 (2017)

61. Y. Guo, M. Pu, Z. Zhao, Y. Wang, J. Jin, P. Gao, X. Li, X. Ma, X. Luo, Merging geometric phase and plasmon retardation phase in continuously shaped metasurfaces for arbitrary orbital angular momentum generation. ACS Photonics **3**, 2022–2029 (2016)

62. F. Zhang, M. Pu, X. Li, P. Gao, X. Ma, J. Luo, H. Yu, X. Luo, All-dielectric metasurfaces for simultaneous giant circular asymmetric transmission and wavefront shaping based on asymmetric photonic spin–orbit interactions. Adv. Funct. Mater. **27**, 1704295 (2017)

63. F. Zhang, M. Pu, J. Luo, H. Yu, X. Luo, Symmetry breaking of photonic spin-orbit interactions in metasurfaces. Opto-Electron. Eng. **44**, 319–325 (2017)

64. J.P. Balthasar Mueller, N.A. Rubin, R.C. Devlin, B. Groever, F. Capasso, Metasurface polarization optics: independent phase control of arbitrary orthogonal states of polarization. Phys. Rev. Lett. **118**, 113901 (2017)

65. P. Zhang, S. Gong, R. Mittra, Beam-shaping technique based on generalized laws of refraction and reflection. IEEE Trans. Antennas Propag. **66**, 771–779 (2018)

66. C. Huang, W. Pan, X. Ma, B. Zhao, J. Cui, X. Luo, Using reconfigurable transmit array to achieve beam-steering and polarization manipulation applications. IEEE Trans. Antennas Propag. **63**, 4801–4810 (2015)

67. J.Y. Lau, S.V. Hum, Reconfigurable transmit array design approaches for beamforming applications. IEEE Trans. Antennas Propag. **60**, 5679–5689 (2012)

68. W. Pan, C. Huang, P. Chen, M. Pu, X. Ma, X. Luo, A beam steering horn antenna using active frequency selective surface. IEEE Trans. Antennas Propag. **61**, 6218–6223 (2013)

69. Q. Wang, E.T.F. Rogers, B. Gholipour, C.-M. Wang, G. Yuan, J. Teng, N.I. Zheludev, Optically reconfigurable metasurfaces and photonic devices based on phase change materials. Nat. Photonics **10**, 60–65 (2016)

70. Y. Qu, Q. Li, K. Du, L. Cai, J. Lu, M. Qiu, Dynamic thermal emission control based on ultrathin plasmonic metamaterials including phase-changing material GST. Laser Photonics Rev. **11**, 1700091 (2017)

71. Y. Chen, X. Li, Y. Sonnefraud, A.I. Fernández-Domínguez, X. Luo, M. Hong, S.A. Maier, Engineering the phase front of light with phase-change material based planar lenses. Sci. Rep. **5**, 8660 (2015)

72. P. Hosseini, C.D. Wright, H. Bhaskaran, An optoelectronic framework enabled by low-dimensional phase change films. Nature **511**, 206–211 (2014)

73. C.H. Chu, M.L. Tseng, J. Chen, P.C. Wu, Y.-H. Chen, H.-C. Wang, T.-Y. Chen, W.T. Hsieh, H.J. Wu, G. Sun, D.P. Tsai, Active dielectric metasurface based on phase-change medium. Laser Photonics Rev. **10**, 986–994 (2016)

74. M. Zhang, M. Pu, F. Zhang, Y. Guo, Q. He, X. Ma, Y. Huang, X. Li, H. Yu, X. Luo, Plasmonic metasurfaces for switchable photonic spin-orbit interactions based on phase change materials. Adv. Sci. **5**, 1800835 (2018)

75. A. Shaltout, J. Liu, A. Kildishev, V. Shalaev, Photonic spin Hall effect in gap–plasmon metasurfaces for on-chip chiroptical spectroscopy. Optica **2**, 860–863 (2015)

Chapter 7
Catenary Optical Fields and Dispersion for Perfect Absorption of Light

Abstract Besides the localized manipulation of phase and polarization, catenary optical fields can be used to realize perfect absorption of light. First, the catenary coupling occurring at structured holes or gaps may help to couple light into the subwavelength structures. Second, the localized resonance strongly increases the localized intensity as well as the absorption probability of incident photons. Third, the catenary fields may change the dispersion of electromagnetic modes, thus broadband absorption becomes possible. We also noted that the counter-propagating waves in a thin lossy slab would form catenary-shaped intensity profile, which means that the catenary is a universal characteristic for the absorption of light in structured materials.

7.1 Critical Coupling Associated with Catenary Optical Fields

Homogeneous electromagnetic absorbers attenuate incident waves along the direction of the refraction direction. In order to reduce the Fresnel reflection, the refractive index should be close to the surrounding material (~1 for air). As a result, the attenuation factor is often small and the required thickness for near-total absorption is much larger than the wavelength. In order to reduce the thickness, one natural idea is to gradually increase the attenuation factor along the propagation direction. The other is to convert longitudinal propagating waves to the horizontal direction. Since the dimension of the material surface is often much larger than the thickness in many practical applications, the incident wave could be absorbed along a much longer distance. This approach has been demonstrated to be useful for both electromagnetic absorbers and solar cells [1, 2].

To the best of our knowledge, the first surface absorber dates back to 1902 when Robert Wood discovered the exotic diffraction effect of metallic grating, which was then called Wood's anomaly [3, 4]. Wood noticed a surprising phenomenon: "*I was astounded to find that under certain conditions, the drop from maximum illumination to minimum, a drop certainly of from 10 to 1, occurred within a range of wavelengths not greater than the distance between the sodium lines*". The drop of the reflection

© Springer Nature Singapore Pte Ltd. 2019

X. Luo, *Catenary Optics*, https://doi.org/10.1007/978-981-13-4818-1_7

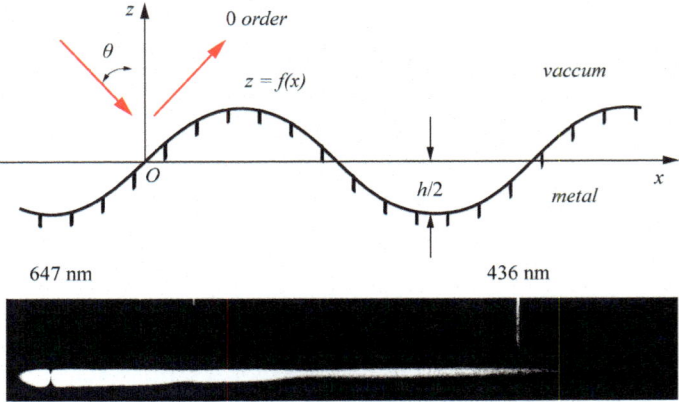

Fig. 7.1 Sinusoidal metal grating for total absorption of light. The bottom shows the spectrum of a collimated beam of white light after reflection from a gold grating where $h = 37$ nm, $\theta = 6.6°$. Adapted from [5] with permission. Copyright 1976, Published by Elsevier B.V

intensity is actually associated with the absorption induced by the excitation of surface plasmon polaritons (SPPs). In 1976, Maystre presented a rigorous theoretical and experimental description of a similar phenomenon in which the pitch of the grating is so fine that no diffracted orders are able to propagate [5] and called this "total absorption". As shown in Fig. 7.1, a sinusoidal grating with a pitch of 555 nm (~1800 grooves/mm) was illuminated with a collimated beam of white light from a tungsten lamp and the reflected beam was analyzed with a spectrograph. At the krypton laser wavelength (647 nm), there is an obvious drop of intensity. Since the wavelength is larger than the grating pitch, this reduction is not caused by normal diffraction.

Figure 7.2 shows the calculated and measured angle-resolved efficiency at 647 nm for a variety of geometric parameters. As the groove thickness increases, the absorption increases first, and then drops with an extended bandwidth after reaching the optimal point. The minimum recorded reflectance (0.3%) in the experiment was at an angle of incidence of 6.6° and a groove depth of 37 nm. For simulations, this occurs at an angle of 6.6° and a groove depth of 40 nm.

Although this gold-coated sinusoidal interference grating bears simultaneously low profile and high absorption, the fabrication accuracy cannot be easily guaranteed. In fact, due to a mishap during exposure, the groove depth of the fabricated grating varied from 13 nm on one side to 75 nm on the other. So, the measured results in Fig. 7.2 were obtained from a single sample at different places.

To make the fabrication more robust and repetitive, a two-dimensional planar grating shown in Fig. 7.3 is proposed to realize perfect absorption of light [6]. Note that, the top grating can also be treated as subwavelength hole array, which has drawn much attention in the early years of this century [7, 8]. By using the higher order surface plasmon modes in the metal-insulator-metal (MIM) waveguide, two distinct absorption peaks could be obtained with a single structure [6]. By further

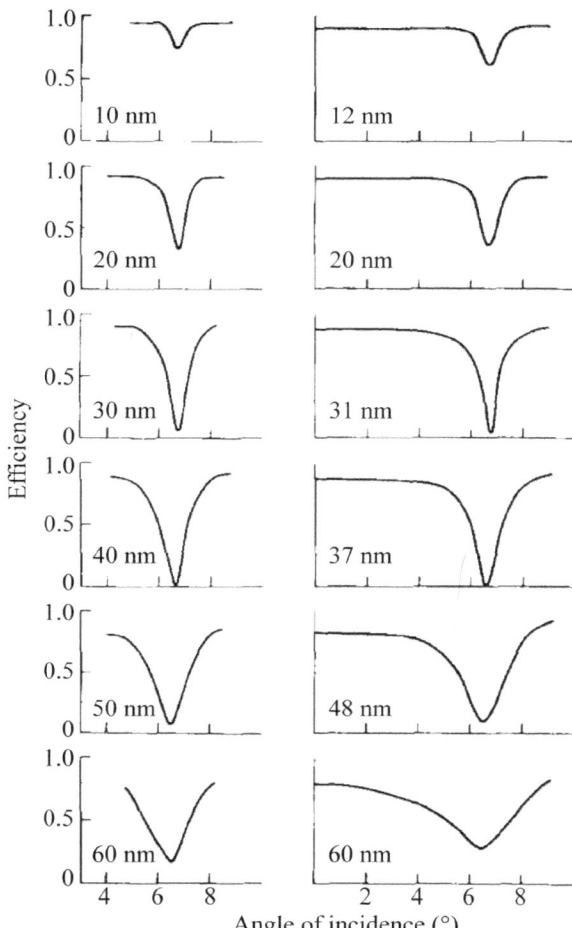

Fig. 7.2 Theoretical (left) and experimental (right) efficiency curves for gratings of various groove depths. Adapted from [5] with permission. Copyright 1976, Published by Elsevier B.V

perturbation of the geometric parameters to induce multi-absorption peaks at adjacent wavelengths, broader absorption bandwidth was also realized [9].

Figure 7.4a shows the instantaneous z-component of the electric field distribution in the dielectric spacer at a wavelength of 655 nm ($f = 458$ THz), which resembles the antisymmetric SPP mode (catenary plasmon) in a MIM waveguide. This can be further validated by the vectorial electric fields in the xz-plane (Fig. 7.4b). Similar to the reflective-type plasmonic lithographic experiment [10, 11], the holes act as couplers between the freely propagating waves and guided waves. Since the propagation constant of SPP is very large compared with the free space wavenumber, the period of the grating should be much smaller than the wavelength. Consequently, there is no higher order diffraction in free space.

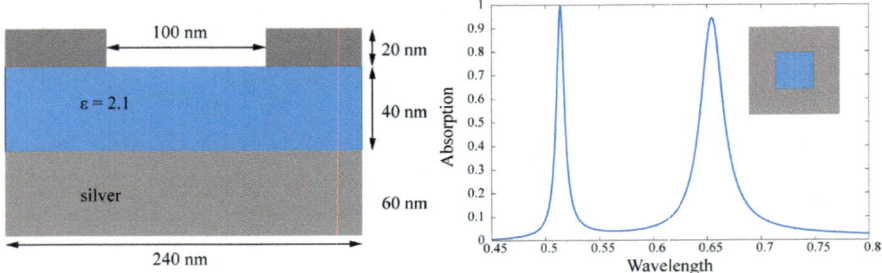

Fig. 7.3 Schematic and absorption for a planar structure comprised of a hole array, dielectric spacer, and metallic reflective layer. All the metals are chosen as silver. The permittivity of dielectric spacer is set to be 2.1. The absorption spectrum is different from that in [6] because of the chosen of material parameters

The physical mechanism of the perfect absorber based on propagating SPPs can be understood using the matching of boundary condition. In principle, when absorption is induced by the excitation of propagation waveguide modes, the following wavevector matching condition should be matched:

$$\beta = k_0 \sin \theta + mq, \tag{7.1.1}$$

where m is an integer, β is the propagation constant of waveguide mode, and $q = 2\pi/p$ is the lowest reciprocal vector, p is the period of the grating. Obviously, when the incident angle changes, the propagation constant and thus the absorption peak frequency would shift correspondingly.

To realize angle-independent absorption, the role of grating diffraction should be suppressed and localized effect need to be exploited. Such absorbers have been well studied in recent years. One excellent example is the metallic spherical voids array proposed by Teperik et al. [12]. Owing to the spherical symmetry, the absorption can be maintained for a large incidence angle up to 80°. Similar to this 3D design, another wide-angle yet flat structure was proposed recently in both microwave and optical regimes [13, 14]. Such structures also share a MIM configuration, but the holes array is replaced by isolated patches array. The gap between the top patch and the bottom metal film allows the formation of magnetic resonance with quasi-spherical symmetry, thus wide-angle absorption can be anticipated.

As illustrated in Fig. 7.5, circular patches arrayed in the hexagonal lattice are widely chosen to realize wide-angle absorption [14], owing to its high azimuthal symmetry and compatibility to large-area fabrication with microsphere self-assembly [15]. The radius of the cylindrical patch is r and the center-to-center distance between adjacent patches is a. The thicknesses of the patch and dielectric spacer are t and d, respectively. The thickness of the thick metal layer h is far larger (100 nm) than the penetration depth to suppress transmission. All metallic parts are chosen as gold and experimental values of permittivity were used [16]. Al_2O_3 is used as dielectric

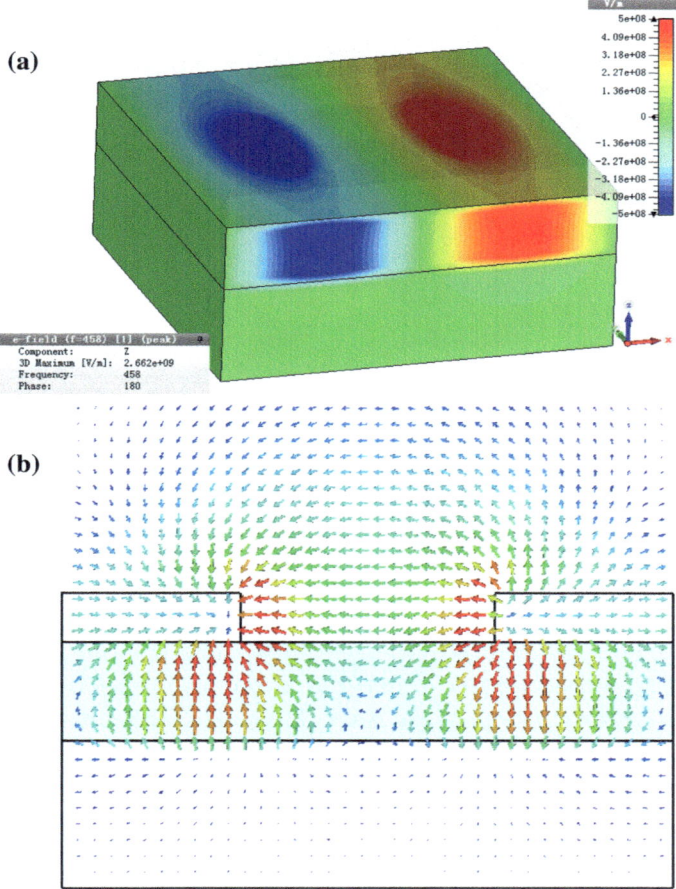

Fig. 7.4 **a** The distribution of E_z in the dielectric spacer at a wavelength of 655 nm, proving the existence of two counter-propagating catenary plasmons along the x-direction. The top layer is shown as transparent. **b** Electric field vectors in the xz-plane, showing two MIM gaps supporting coupled plasmons

spacer and the dielectric constant is 2.92 at wavelengths around 3 μm (Al_2O_3 is almost lossless in this frequency range).

The unit cell shown in Fig. 7.5a with dashed line was used in numerical simulations to mimic infinite cell array. The absorption is calculated as $A = 1 - S_{11}^2$ since there is no transmittance. For TE (TM) polarization, the E (H) field direction is along y-axis. At normal incidence, there are two absorption peaks located at 100 and 280 THz. Two spherically symmetrical resonances shown in Fig. 7.6b, c are responsible for them.

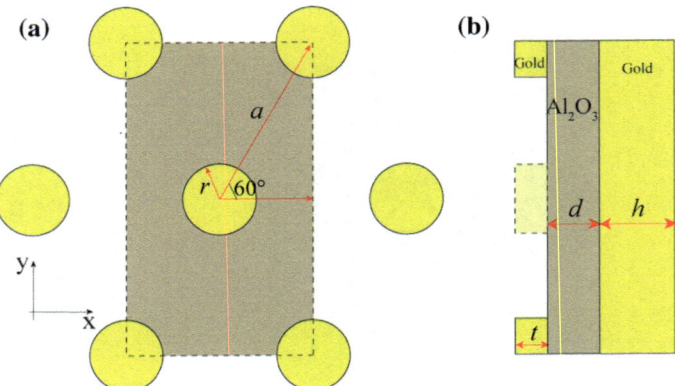

Fig. 7.5 **a** Front view and **b** side view of the absorber. The rectangular region in (**a**) is the unit cell used in numerical simulations. Reproduced from [14] with permission. Copyright 2011, Optical Society of America

The angle dependences of the absorption of the dual-band absorber are shown in Fig. 7.7. The absorption for TM (TE) polarization at 100 THz is larger than 90% at incidence angle of 70° (55°), while the absorption is still larger than 80% at incidence angle of 30° (45°) at 280 THz. Note that there is an angle dependent absorption curve, which is related to SPPs, as shown at the top right corner of Fig. 7.7b. The propagating SPPs cannot be excited when the period of the structure is in deep-subwavelength scale. However, this effect will become much more obvious when the period is comparable with the working wavelength. At large incidence angles, other absorption peaks also occur at frequencies around 180 THz for TM polarization due to excitation of horizontal resonance modes.

Besides the symmetry analysis, the above angle independence could be explained using an effective high-index material. The destructive interference condition for a typical Dallenbach absorber with a refractive index of n and thickness of d is written as [17]

$$nk_0 d\sqrt{1 - \sin^2\theta/n^2} = \pi/2. \qquad (7.1.2)$$

When $n \gg 1$, this condition is reduced to be $nk_0 d = \pi/2$, which is independent of the incident angle. Figure 7.8 shows a drawing of the absorbance of hypothetic Dallenbach absorber ($d = 1$ mm) as a function of the frequency and incident angle, where a striking similarity to that shown in Fig. 7.7 is found. Note that, this effective high-index material can be realized using metamaterial [18] with small magnetic resonance, thus there may be many different approaches to realize wide-angle perfect absorption.

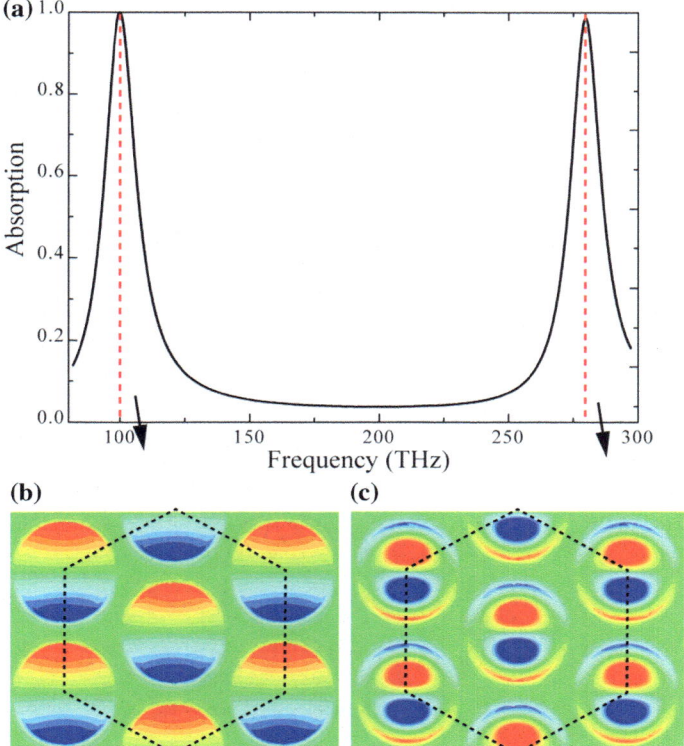

Fig. 7.6 a Absorption of the dual-band absorber for TM polarization wave. **b, c** Front view of the normal electric field and tangential magnetic field at the center of dielectric spacer. **b** $f = 100$ THz. **c** 280 THz. The dashed hexagons indicate the centers of the six patches surrounding the central patch. Copyright 2011, Optical Society of America

7.2   Broadband Absorption Based on Coupled Resonators

The absorbers proposed in the above section have relative narrow absorption bandwidth, which is detrimental for many practical applications. In this section, a planar structure with perturbed periodicity and sharp Fano resonance is presented. Generalized impedance theory and equivalent circuit theory are used to investigate the Fano-type resonance. Electric–magnetic coupling and periodicity perturbation induced magnetic–magnetic coupling are demonstrated unambiguously. It is shown that these resonances could greatly enhance the absorption at resonant frequencies. When combined together, these absorption peaks could form a continuous band with large absorption.

Fig. 7.7 Absorption of the dual-band absorber as a function of frequency and the angle of incidence for **a** TE and **b** TM polarizations. The angle-invariant absorption results from the localized plasmon, while the SPPs induced absorption is shown at the top right corner of (**b**). Reproduced from [14] with permission. Copyright 2011, Optical Society of America

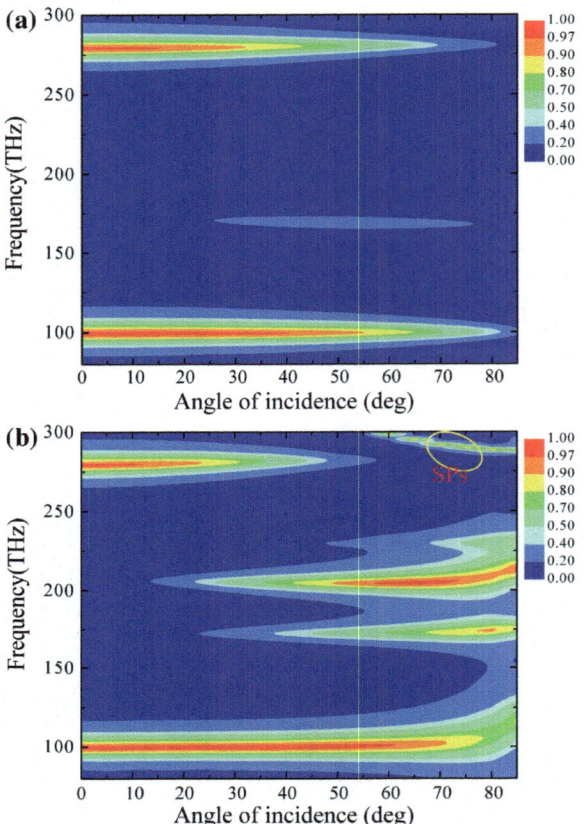

7.2.1 Structure and Generalized Impedance Theory

As illustrated in Fig. 7.9a, the structure being investigated is a periodic metallic wire pairs [19], which has been widely used to create artificial magnetic response and negative refractive index [20]. Different widths of wire pairs (depicted by gray, orange, and red colors) are used to perturb the periodicity and realize multiple resonances. In the simulations, these three wire pairs actually act as a new unit cell. The center-to-center distances between adjacent wire pairs are kept as $p = 5$ mm and the thickness of the dielectric spacer is set as $d = 0.75$ mm. Metal is assumed as perfect conductor while the permittivity of dielectric material is chosen as 12 with zero loss (the material loss will be considered later).

As is well known, the metallic wire pairs have strong magnetic response which can be described by effective magnetic dipoles. Giant magnetic response may be obtained with antiparallel surface currents when the electric field is perpendicular to the long axis of wire pairs. Sub-radiant modes, which couple weakly with the external field,

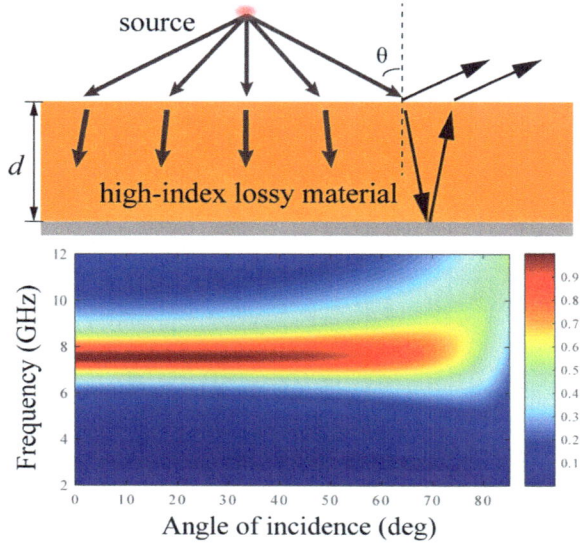

Fig. 7.8 Principle of the wide-angle absorbers based on multiple interferences. The bottom shows the simulated results for a Dallenbach absorber. The complex permittivity is chosen as 100 + 12i. Reproduced from [17] with permission. Copyright 2017, The Royal Society of Chemistry

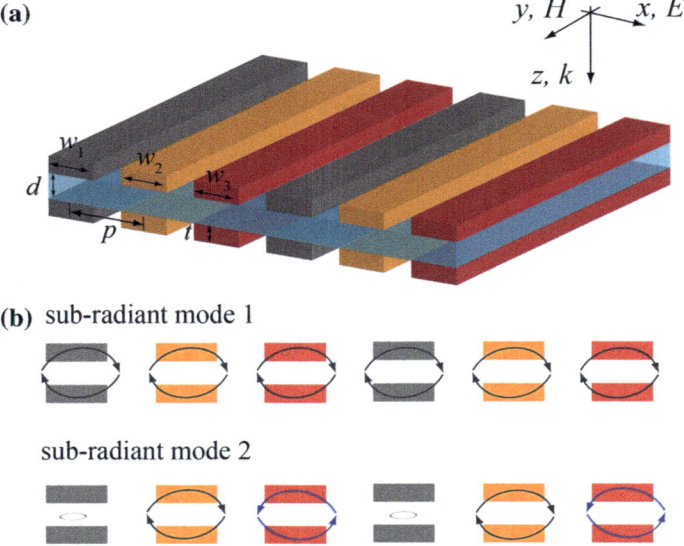

Fig. 7.9 a Schematic of the wire pairs with perturbed periodicity. The gray, orange, and red colors depict different widths. **b** Sketch map of the surface current distributions for the two sub-radiant modes. Reproduced from [19] with permission. Copyright 2013, Optical Society of America

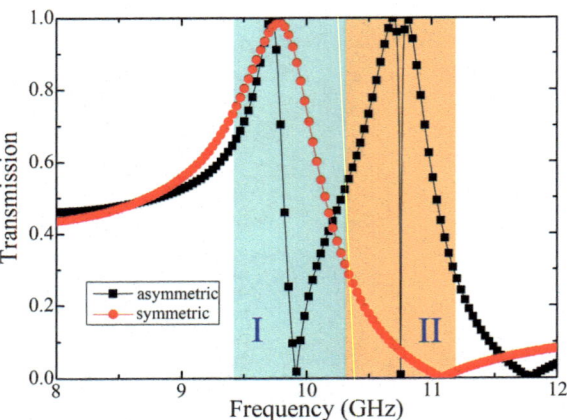

Fig. 7.10 Transmission coefficients for the symmetric and asymmetric slabs. Two regions are highlighted. Region I shows typical asymmetric line shape of Fano resonance. Region II demonstrates the ultra-sharp resonance peak induced by inter-element coupling. Copyright 2013, Optical Society of America

may exist at different resonance conditions. In the numerical simulation, Comsol Multiphysics, a commercial finite element method (FEM) solver is used with periodic boundary condition in x-direction and perfectly matched layer in $\pm z$-direction. The structure is assumed to be infinite long in y-direction while the electric and magnetic fields are along x -and y-directions, respectively.

In order to investigate the influence of structure asymmetry, the transmission coefficients of both symmetric and asymmetric metamaterials at normal incidence were calculated and shown in Fig. 7.10. For the symmetric structure ($w_1 = w_2 = w_3 = 4$ mm), the transmission peak at 9.75 GHz is induced by the magnetic resonance. Similar with Fano resonance, the transmission profile is a bit asymmetric and a transmission dip occurs at 11.1 GHz.

Figure 7.11 shows the electric fields distribution at 9.75 GHz in the xz-plane. The strong E_z components between the top and bottom metallic patches are related to the magnetic resonance, while the amplitude along the z-direction follows a catenary curve [21]. Besides the magnetic resonance, this structure also supports strong electric response. As shown in Fig. 7.11c, the amplitude of $|E|$ between two patches is also a catenary, which is often characterized by an effective capacitance.

For the asymmetric structure, the widths of three wires can have three or two different values. Without loss of generality, the following parameters were chosen as $w_1 = 4$ mm, $w_2 = 3.6$ mm and $w_3 = 3.56$ mm. Compared with the symmetric case, the line shape around 9.8 GHz is more asymmetric. Moreover, ultra-sharp transmission dip at the center of a transmission peak was observed at frequencies around 10.8 GHz with Q-factor up to 100. Interestingly, this is in contrary to the metamaterial analogy of electromagnetic induced transparency (EIT) [22], where a transmission sharp peak arises in a reflective frequency region.

The above exotic transmission properties can be understood by using generalized surface impedance theory, which is used to retrieve effective material parameters for thin sheets such as graphene and single-layer metasurfaces. As illustrated in

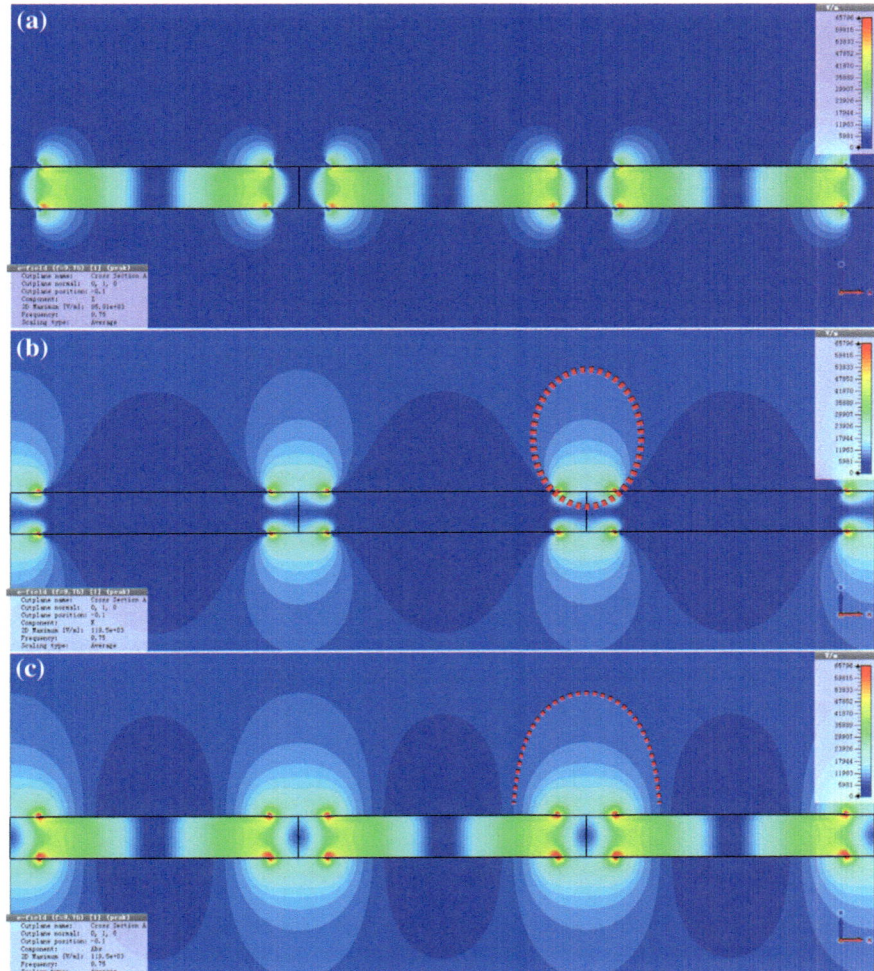

Fig. 7.11 Electric fields distribution at the transmission peak frequency of 9.75 GHz. **a** $|E_z|$. **b** $|E_x|$. **c** $|E|$. Note that, the equi-amplitude lines in (**b**) and (**c**) are approximately an ellipse and a catenary of equal strength, respectively

Fig. 7.12, a thin structured slab can be treated as a combination of electric impedance and magnetic impedance with the following boundary conditions [19, 23]:

$$\vec{E}_i + \vec{E}_r = \vec{E}_t + \hat{n} \times \vec{j}_m,$$
$$\vec{H}_i + \vec{H}_r = \vec{H}_t + \hat{n} \times \vec{j}_e. \tag{7.2.1}$$

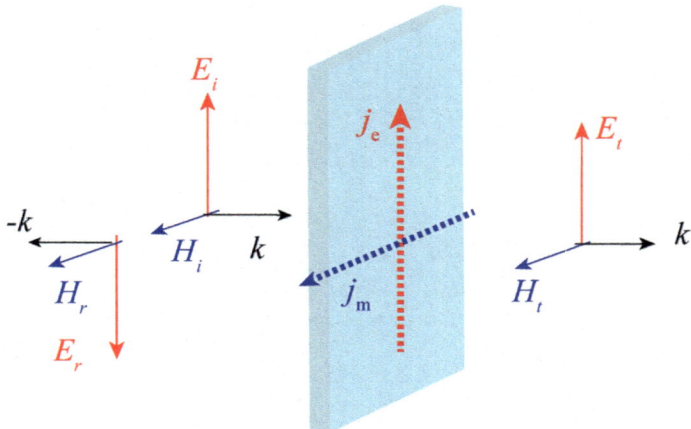

Fig. 7.12 Schematic of sheet impedance description for thin metamaterial. The magnetic and electric responses are described by equivalent magnetic and electric sheet currents. Reproduced from [19] with permission. Copyright 2013, Optical Society of America

By definition, there are $E_r = rE_i$, $H_r = rH_i$, $E_t = tE_i$, and $H_t = tH_i$, where $E_i = Z_0 H_i$ is the incident electric field, $Z_0 = 1/Y_0 = 377\,\Omega$ is the vacuum impedance, r and t are the reflection and transmission coefficients. The electric sheet current j_e and magnetic sheet current j_m are related to the average electric and magnetic field through the corresponding electric admittance Y_e and magnetic impedance Z_m:

$$\vec{j}_e = \frac{1}{Z_e}\vec{E}_{\text{average}} = Y_e\vec{E}_{\text{average}} = Y_e(1 + r + t)\vec{E}_i/2,$$
$$\vec{j}_m = Z_m\vec{H}_{\text{average}} = Z_m(1 - r + t)\vec{H}_i/2. \tag{7.2.2}$$

Combining Eqs. (7.2.1) and (7.2.2), one can obtain the following expressions:

$$Y_e = 2Y_0\frac{1 - r - t}{1 + r + t},$$
$$Z_m = 2Z_0\frac{1 + r - t}{1 - r + t}, \tag{7.2.3}$$
$$r = \frac{1}{2}\left(\frac{2Y_0 - Y_e}{2Y_0 + Y_e} + \frac{Z_m - 2Z_0}{Z_m + 2Z_0}\right),$$
$$t = \frac{1}{2}\left(\frac{2Y_0 - Y_e}{2Y_0 + Y_e} - \frac{Z_m - 2Z_0}{Z_m + 2Z_0}\right), \tag{7.2.4}$$

and

$$j_e = \frac{2Y_0 Y_e}{2Y_0 + Y_e}E_i,$$

$$j_m = \frac{2Z_0 Z_m}{Z_m + 2Z_0} H_i. \tag{7.2.5}$$

Equations (7.2.3)–(7.2.5) are the central results of the generalized sheet impedance theory. Using Eq. (7.2.3), the effective magnetic and electric impedances can be retrieved from simulated or measured S parameters. In the following, the generalized impedance theory is used to interpret the formation of Fano resonances in the structure shown in Fig. 7.9.

7.2.2 Fano Resonance Induced by Coupled Modes

For the structure without asymmetry ($w_1 = w_2 = w_3 = 4$ mm), the line shape shown in Fig. 7.10 is asymmetric. However, for the structure with perturbed periodicity, the adjacent magnetic resonators will couple with each other. As shown in Fig. 7.13, the electric and magnetic impedances as well as the corresponding sheet currents are calculated using Eqs. (7.2.3) and (7.2.5) [19]. Similar with the symmetric structure, the electric admittance is still rather flat. The magnetic impedance, however, shows complex resonant characteristics. Since there are three resonant peaks, the magnetic impedance shown in Fig. 7.13c can be written as a combination of individual magnetic resonances, as described by equivalent LC circuit model:

$$Z_m = \sum_{n=1}^{3} Z_{mn} = \sum_{n=1}^{3} \frac{i\omega L_n}{1 - \omega^2 L_n C_n}. \tag{7.2.6}$$

Here, L_1, L_2, L_3, C_1, C_2, and C_3 are the corresponding inductances and capacitances. The resonant frequencies are $\omega_1 = (L_1 C_1)^{-1/2}$, $\omega_2 = (L_2 C_2)^{-1/2}$, and $\omega_3 = (L_3 C_3)^{-1/2}$, respectively. By fitting Eq. (7.2.6) with the retrieved magnetic sheet impedance, the LC parameters can be obtained as $L_1 = 0.295$ nH, $L_2 = 0.273$ nH, $L_3 = 0.27$ nH, $C_1 = 0.896$ pF, $C_2 = 0.8$ pF, and $C_3 = 0.794$ pF. The corresponding resonant frequencies are 9.79, 10.77 and 10.87 GHz.

Similar with the magnetic sheet impedance, the magnetic sheet current is also a summation of all the individual sheet currents:

$$j_m = \sum_{n=1}^{3} j_{mn}, \tag{7.2.7}$$

where

$$j_{mn} = \frac{2Z_0 Z_{mn}}{Z_m + 2Z_0} H_i, \quad n = 1, 2, 3. \tag{7.2.8}$$

Using Eqs. (7.2.6) and (7.2.7), the sheet currents for different resonators can be easily calculated. As shown in Fig. 7.14a, the current of the first resonator dominates

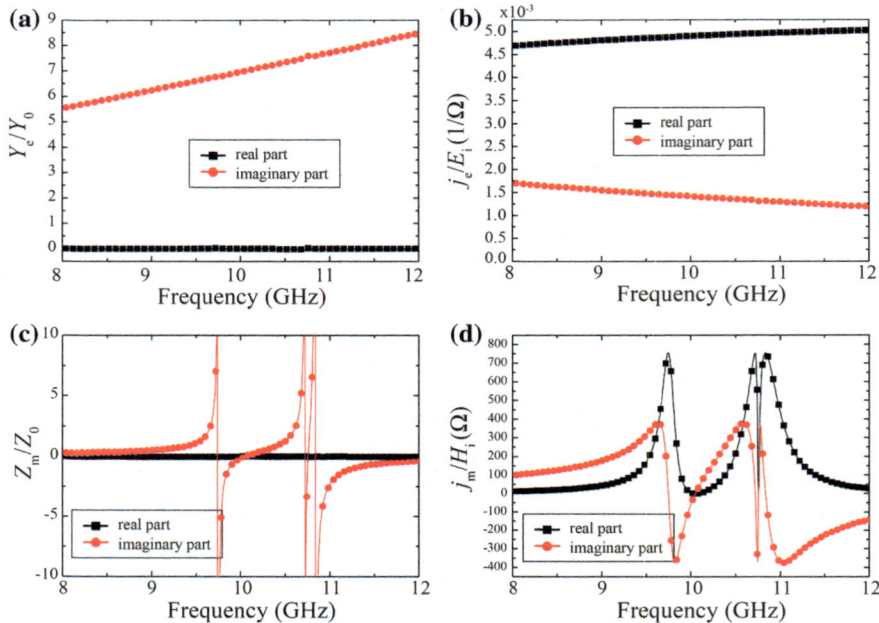

Fig. 7.13 **a, b** Electric admittance and electric sheet current for the asymmetric structure. **c** The corresponding magnetic impedance and **d** magnetic sheet current. Reproduced from [19] with permission. Copyright 2013, Optical Society of America

at frequencies around 9.79 GHz. On the contrary, the second and third resonators dominate the frequency region around 10.8 GHz. From Fig. 7.14b, one can find that the first resonator is out of phase with the other two resonators for 9.79 GHz $< f <$ 10.77 GHz. Also, for 10.77 GHz $< f <$ 10.87 GHz, the third one is out of phase with the others.

In general, the magnetic field inside the planar metamaterial is proportional to the sheet current. Recalling Eq. (7.2.1) and the Maxwell's equation:

$$\vec{E}_i + \vec{E}_r - \vec{E}_t = \hat{n} \times \vec{j}_{\mathrm{m}}$$
$$\oint_L \vec{E} \cdot \mathrm{d}\vec{l} = \frac{-\mathrm{d}}{\mathrm{d}t} \int_s \vec{B} \cdot \mathrm{d}\vec{S} \tag{7.2.9}$$

One can obtain that

$$\hat{n} \times \vec{j}_{\mathrm{m}} \cdot L = \frac{-\mathrm{d}}{\mathrm{d}t}(\vec{B} \cdot L \cdot \Delta), \tag{7.2.10}$$

where $L = p$ is the unit cell length and Δ is the effective thickness of the metamaterial layer. As a result, the magnetic field H_z can be calculated as

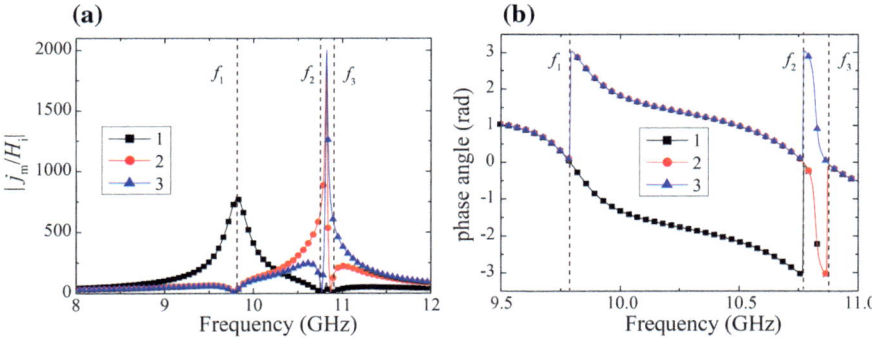

Fig. 7.14 **a** Amplitudes and **b** phases of the effective magnetic sheet current. The fitted parameters of inductances and capacitors are used (see text). The three frequencies where the phase changes occur are highlighted. At these frequencies, the corresponding amplitudes are zero. Reproduced from [19] with permission

$$\left| \vec{H}_z \right| = \frac{\left| \vec{j}_m \right|}{i\omega\mu_0\Delta}. \tag{7.2.11}$$

It should be noted that Eq. (7.2.11) is only an approximation under the condition $\Delta \ll \lambda$, for which the electric field integration along the thickness direction can be neglected.

In order to further prove the above discussion, the magnetic fields at the center of the three resonators for different frequencies are calculated using Comsol Multiphysics and illustrated in Fig. 7.15, which are very coincident with the magnetic sheet currents calculated from S parameters (Fig. 7.14a). At $f_1 = 8$ GHz, all the three wire pairs are out of resonance and the maximum magnetic field is only 3.6 H_i. At $f_2 = 9.79$ GHz, the first resonator is resonant and the maximum magnetic field increases to 69 H_i. At $f_3 = 10.77$ GHz and $f_5 = 10.87$ GHz, the second and the third wire pairs are resonant, respectively. Since f_3 and f_5 are spectrally close, a new resonant mode takes place between them. At $f_4 = 10.8$ GHz, the second and the third wire pairs are on resonance with opposite phase. The maximum magnetic field becomes as high as 140 H_i.

In order to further understand the influence of the periodicity perturbation on the Fano resonance, the effect of wire width is investigated. Two wires have the same width of 4 mm while the third wire width is changed from 3.8, 3.96 to 4 mm. As illustrated in Fig. 7.16, as the asymmetry decreases, two isolated resonances become very close and a third sharp resonance occurs. When the three wire pairs all have the same widths, the transmission spectrum will degenerate to a single magnetic resonance.

It should be noted that the Fano resonance presented here is a coherent process thus practical performance is highly dependent on the unit cell numbers of finite arrays [24]. Furthermore, the fabrication error may also limit the achievable bandwidth,

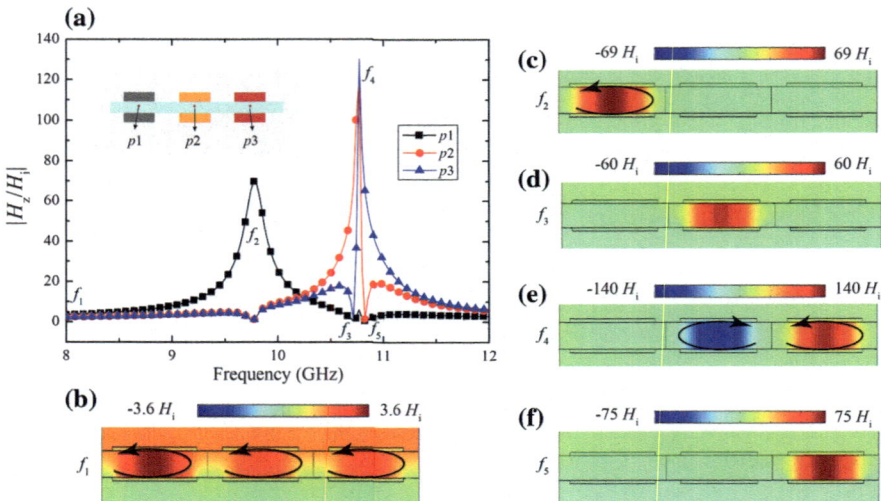

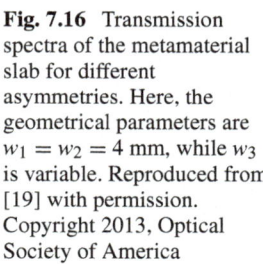

Fig. 7.15 **a** Normalized magnetic fields (H_z) at the center of these resonators. The inset shows the unit cell and the points where the magnetic fields are extracted. **b–f** Magnetic field distributions at frequencies of $f_1 = 8$ GHz, $f_2 = 9.79$ GHz, $f_3 = 10.77$ GHz, $f_4 = 10.82$ GHz and $f_5 = 10.87$ GHz. Reproduced from [19] with permission. Copyright 2013, Optical Society of America

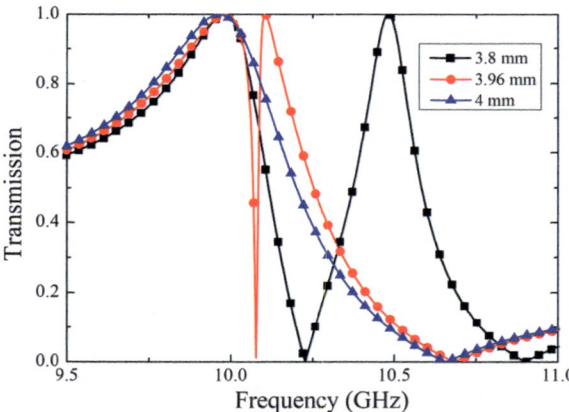

Fig. 7.16 Transmission spectra of the metamaterial slab for different asymmetries. Here, the geometrical parameters are $w_1 = w_2 = 4$ mm, while w_3 is variable. Reproduced from [19] with permission. Copyright 2013, Optical Society of America

especially if the structure is scaled to higher frequencies, where the intrinsic loss in metal is another restriction for high Q-factor resonance.

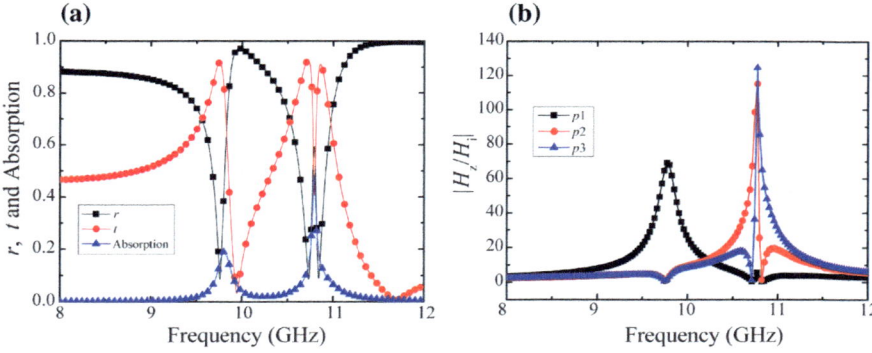

Fig. 7.17 **a** Reflection (r), transmission (t), and absorption of the lossy metamaterial slab. **b** Normalized magnetic fields extracted from Comsol Multiphysics. Reproduced from [19] with permission. Copyright 2013, Optical Society of America

7.2.3 Design of Broadband Absorbers

In the above discussion, the material loss is neglected since the dielectric material is lossless and metal is set as perfect electric conductor. In general, the loss of material has a great influence on the performance. In order to take it into account, a typical loss tangent of 0.003 was added in the dielectric material and metal is used as copper with conductivity of 5.7×10^7 S/m. As shown in Fig. 7.17a, obvious absorption was observed at the resonant frequencies 9.8 and 10.8 GHz. Nevertheless, the magnetic field enhancement factor only decreases a little, as shown in Fig. 7.17b.

As mentioned in the above discussion, the loss mechanism in the asymmetric structure can be utilized to realize wideband wide-angle absorber, which has been intensively studied in recent years [25–27]. The typical structure is shown in the inset of Fig. 7.18, where a metallic ground plane is added in the center of wire pair structure. The geometric parameters are optimized as $p = 5$ mm, $d = 0.75$ mm, $w_1 = 4$ mm, $w_2 = 4.12$ mm and $w_3 = 3.9$ mm. The permittivity is set as 12 with a higher loss tangent of $\tau = 0.015$. The achieved bandwidth for 90% absorption is about 0.5 GHz, which is much larger than that of metamaterial absorber without asymmetry (0.2 GHz) at the same thickness ($p = 5$ mm, $w = 4$ mm, and $\tau = 0.025$).

7.3 Dispersion Engineering for Broadband Metasurface Absorber

As discussed in the above section, stacking different sized structures is an efficient method to broaden the absorption bandwidth of narrow absorbers. In the following, an alternative approach is proposed by considering the frequency dispersion of the

Fig. 7.18 Absorption of the wideband absorber for different angles of incidence. The absorption for the symmetric structure is also shown for comparison. Inset is the schematic of the absorber composed of asymmetric wires and a metallic ground plane. Reproduced from [19] with permission. Copyright 2013, Optical Society of America

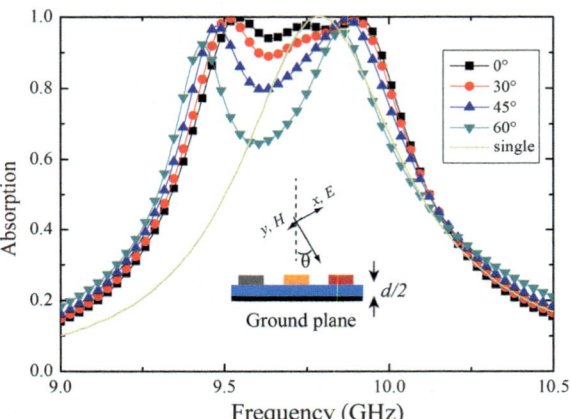

structured metasurface and the M-waves therein. A broadband yet thin absorber is proposed based on a metasurface and a metallic ground layer separated by a thin dielectric spacer [28]. The ideal impedance for the metasurface is theoretically deduced to absorb all electromagnetic radiations. Infrared dual-band absorption is numerically demonstrated and the conditions of perfect absorption are presented by utilizing equivalent circuit models.

As shown in Fig. 7.19, a metasurface is placed in front of a thick metal layer (ground plane) with a distance d. The metasurface can be thought as an interface characterized by its impedance. By matching the impedance of ground plane to that of vacuum, perfect absorption can be realized and the corresponding sheet is called as perfectly impedance-matched sheet (PIMS) [14].

The impedance of PIMS is obtained by setting reflection coefficient S_{11} to be zero. Assuming the dielectric spacer is nonmagnetic ($\mu = 1$), the boundary conditions for the sheet at normal incidence can be obtained (L denotes the left side of the impedance sheet and R denotes the right side) by

$$E^L = E^R : a + aS_{11} = \exp(-i\sqrt{\varepsilon_1}k_0 d) + r_m \exp(i\sqrt{\varepsilon_1}k_0 d)$$
$$H^L = H^R : Y_0(a - aS_{11}) = Y_1(\exp(-i\sqrt{\varepsilon_1}k_0 d) - r_m \exp(i\sqrt{\varepsilon_1}k_0 d)) + J,$$
$$J = Y_s E^{L(R)} = Y_s(a + aS_{11})$$

$$(7.3.1)$$

where Y_0, Y_1, and Y_m are the intrinsic admittance of vacuum, dielectric spacer, and metal (dependent on the incidence angle and polarization) [29], ε_1 and ε_m are permittivities of dielectric, and metal. k_0 is the wavevector in the vacuum and dielectric spacer. r_m is the reflectivity of the thick metal layer. J is the current flowing in the sheet while a is the amplitude of incident E-field. By eliminating a in Eq. (7.3.1), the admittance of PIMS is obtained as

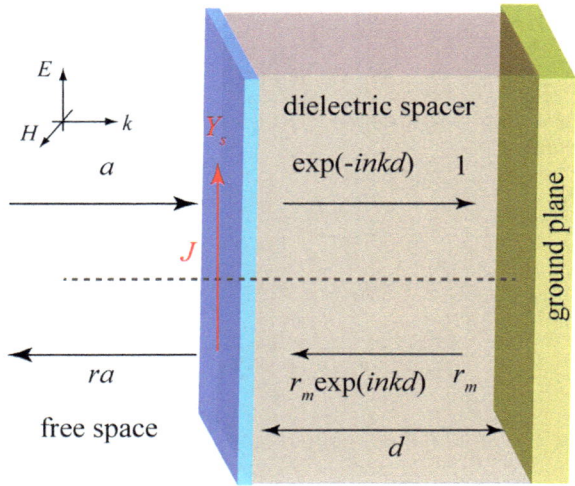

Fig. 7.19 Schematic of the absorber. It composes of a metasurface, dielectric spacer, and metallic ground plane. In the dielectric spacer and surrounding space, total fields consist of both the forward and backward going waves. The reflection of the thick metal layer and the absorber is denoted as r_m and S_{11}. The amplitude of E-field of forward going wave at the ground plane is chosen to be 1 in order to utilize the transfer matrix formulation while the amplitude of incident E-field is denoted as a. Adapted from [14] with permission. Copyright 2011, Optical Society of America

$$Y_s = 1/Z_s = Y_0 - Y_1 \frac{\exp(-i\sqrt{\varepsilon_1}k_0 d) - r_m \exp(i\sqrt{\varepsilon_1}k_0 d)}{\exp(-i\sqrt{\varepsilon_1}k_0 d) + r_m \exp(i\sqrt{\varepsilon_1}k_0 d)}. \tag{7.3.2}$$

The ideal impedance ($Z_s = R - iX$, where R and X are resistance and reactance) for PIMS at normal incidence is independent of polarization. As depicted in Fig. 7.20, the impedance should be pure resistive ($R = 377\ \Omega$, $X = 0$) corresponding to the traditional Salisbury screen when d is near $\lambda/4n$ (n is the refractive index of the dielectric spacer). In Zone I, the required resistance should be very small and the corresponding reactance should be capacitive ($R > 0, X < 0$). In contrast, the required reactance is inductive ($R > 0, X > 0$) for Zone II.

In an alternative approach, the metasurface can be interpreted as a homogeneous slab with effective permittivity ε_{eff}, which is directly related to the sheet impedance Z_{eff}:

$$\varepsilon_{\text{eff}} = 1 + \frac{i\sigma_{\text{eff}}}{\varepsilon_0 \omega} = 1 + \frac{i}{\varepsilon_0 \omega t_{\text{eff}} Z_{\text{eff}}}, \tag{7.3.3}$$

where $\sigma_{\text{eff}} = 1/(t_{\text{eff}} Z_{\text{eff}})$ is effective complex conductivity, while t_{eff} is the effective thickness. By inserting the impedance for PIMS into Eq. (7.3.3), the ideal permittivity could be obtained as

Fig. 7.20 Ideal impedances of PIMS versus the effective thickness of dielectric layer ($2nd/\lambda$). n is chosen to be 2 and r_m is -1 corresponding to the reflection coefficient of PEC. Reproduced from [14] with permission. Copyright 2011, Optical Society of America

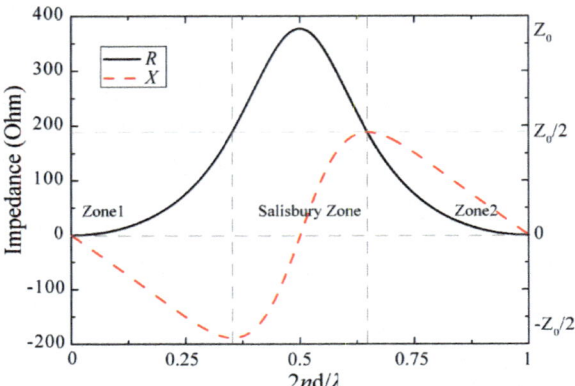

$$\varepsilon_{ideal} = 1 + \frac{i}{\varepsilon_0 \omega t_{eff}}\left(Y_0 - Y_1 \frac{\exp(-i\sqrt{\varepsilon_1}k_0 d) - r_m \exp(i\sqrt{\varepsilon_1}k_0 d)}{\exp(-i\sqrt{\varepsilon_1}k_0 d) + r_m \exp(i\sqrt{\varepsilon_1}k_0 d)}\right). \quad (7.3.4)$$

Obviously, such dispersion is very complex and cannot be found in naturally occurring material. To realize high-efficiency broadband absorption, metasurface should be designed properly to mimic the dispersion of ideal absorbing sheet.

Figure 7.21a shows a practical design for infrared broadband absorption [28]. A single unit cell of the absorber consists of metallic (for example, gold) substrate, dielectric spacer, and a cross resonator made of nichrome. The dielectric spacer is assumed as transparent and has a permittivity of 2.25 which is assumed to be constant in the range of frequencies considered here. The lossy material is chosen as nichrome due to its larger resistivity compared with noble metals. It is worth noting that the permittivity of nichrome is dependent on film thickness and can be well described by the Drude model. The plasma frequency ω_p and collision frequency $\Gamma = 1/\tau$ are chosen as 2.9×10^{15} rad/s and 1.65×10^{14} Hz for a 15 nm thick nichrome layer by fitting the experimental results [30].

FEM was used in the numerical calculations with periodic boundary conditions in x- and y-directions. As shown in Fig. 7.21b, broadband absorption at normal incidence is realized at frequencies between 21 and 44 THz with absorption greater than 97%. The corresponding geometrical parameters are given in the caption. For comparison, the absorption curves for several Salisbury-type absorbers are also depicted, where the metasurface is replaced by a homogeneous layer of nichrome with a thickness of 10 nm. Obviously, the bandwidth of the proposed absorber is two times larger than that of Salisbury absorber without increase of overall thickness. In order to investigate the role of dispersion on the absorption bandwidth, the impedances retrieved from the calculation as well as that of the ideal absorbing sheet are illustrated in Fig. 7.22. As expected, both the real and imaginary parts of effective impedance ($Z = R - jX$) approach the ideal curves in a rather wide frequency range, thus, a broadband absorber is realized.

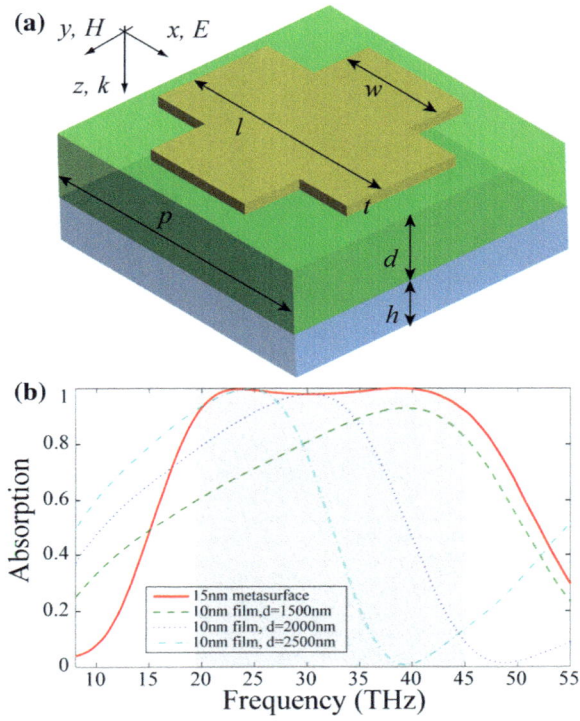

Fig. 7.21 **a** Unit cell of the broadband absorber and its optimized dimensions: $p = 1.5$, $d = 1.5$, $t = 0.015$, $l = 1.25$, and $w = 0.8$ in microns. **b** Absorption at normal incidence as a function of frequency. The absorption curves for Salisbury-type absorbers are shown to illustrate the bandwidth enhancement effect. The thicknesses of the spacers are 1.5, 2, and 2.5 in microns, respectively. Reproduced from [28] with permission. Copyright 2012, Optical Society of America

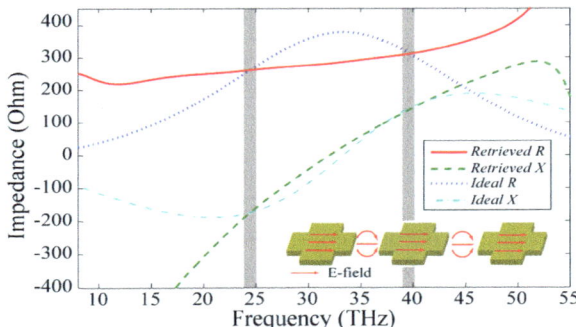

Fig. 7.22 Retrieved sheet impedance and the corresponding impedance of an ideal absorbing sheet. The retrieved impedances overlap with the ideal one at 24.5 and 39.5 THz. Inset is the schematic of the E-field distribution in the metasurface. Reproduced from [28] with permission. Copyright 2012, Optical Society of America

(a)

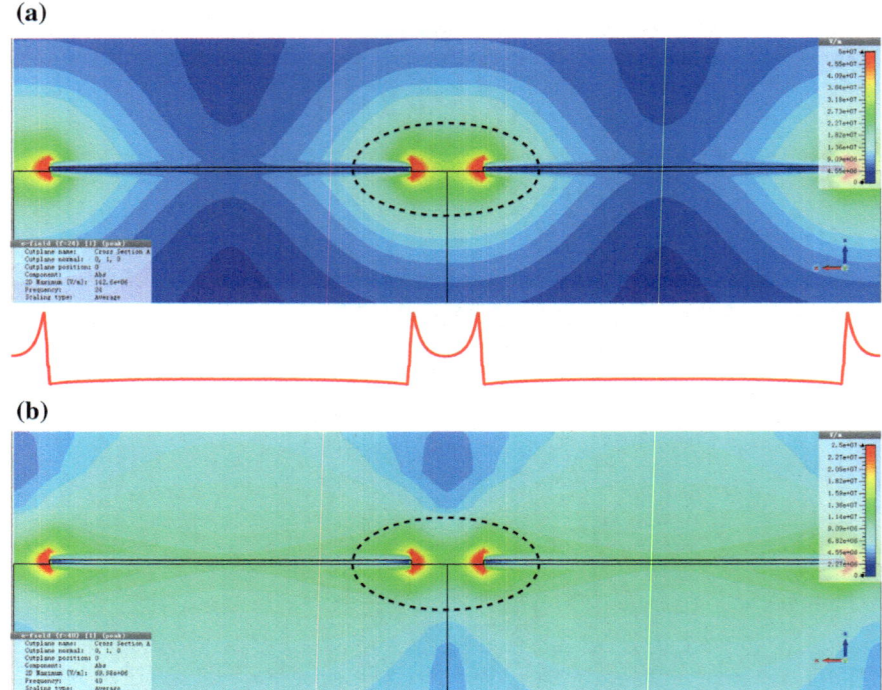

(b)

Fig. 7.23 Absolute value of the electric fields distribution for two unit cells at $f = 24$ THz (**a**) and 40 THz (**b**). The dashed circles highlight the region where catenary optical fields exist

To shed on the physical origin of the specific dispersion, the electric field (E-field) distributions at normal incidence are shown in the inset of Figs. 7.22 and 7.23. Both the fields at 24 THz and 40 THz show obvious catenary feature in the gap. As highlighted by Figs. 7.23 and 7.24, the electric field in the gap is well fitted by the catenary function $E = a \exp(bx) + a \exp(-bx) + c$ with $a = 0.01278$, $b = 34.86$, $c = 0.3814$ and a R-square of 0.9999.

In the equivalent circuit theory, this catenary fields can be represented by an equivalent capacitor, while the cross resonator behaves as a lossy plasmonic element described by an inductor and resistor. These circuit elements are connected in series thus, the effective sheet impedance can be written as $Z_{\text{eff}}(\omega) = R - j\omega L + j/(\omega C)$, where R, L, and C can be treated as constant and fitted from the retrieved impedance with $R \approx 285$ Ω, $L = 1.45$ pH, and $C = 0.017$ fF. Although this equivalent circuit model provides some insight to the frequency dispersion of the metasurface, it is helpful to interpret the circuit model using effective permittivity. Using Eq. (7.3.3), the effective permittivity can be expressed in a Lorentz form:

$$\varepsilon_{\text{eff}}(\omega) = 1 + \frac{\omega_1^2}{\omega_0^2 - \omega^2 - i\omega\gamma}, \tag{7.3.5}$$

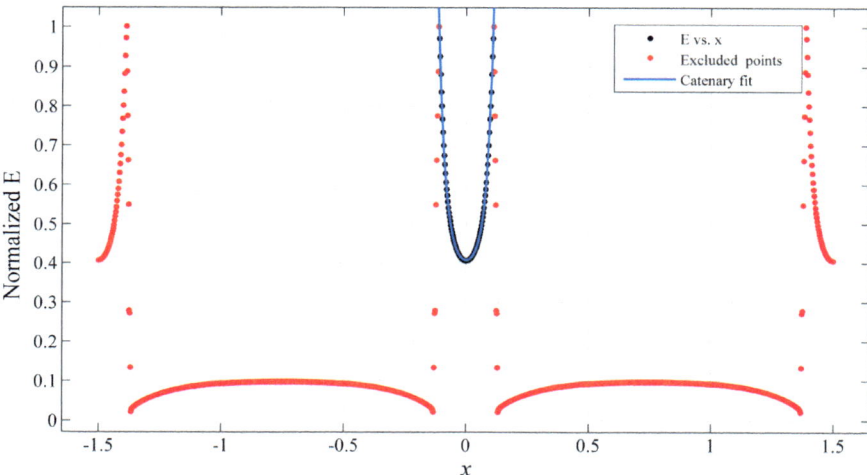

Fig. 7.24 Catenary fit of the normalized electric field in the gap. The points |x|>0.175 is excluded

Fig. 7.25 Effective permittivity of the metasurface retrieved from reflection coefficient and fitted by Lorentz model. The permittivity described by Drude model is also shown. Reproduced from [28] with permission. Copyright 2012, Optical Society of America

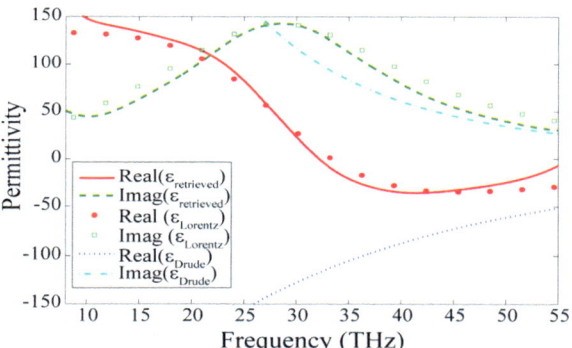

where $\omega_0 = 2\pi f_0 = (LC)^{-1/2}$ denotes the oscillation frequency of the bound electron in the metasurface, while $\omega_1 = 2\pi f_1 = (\varepsilon_0 Lt)^{-1/2}$ and $\gamma = R/L$ are related to the density and damping of the electrons. According to the above fitting results, f_0, f_1, and γ are 32.05, 362.71, and 196.5 THz, respectively. The effective permittivity retrieved from reflection coefficients as well as that fitted by Lorentz model is illustrated in Fig. 7.25. Clearly, the oscillation frequency f_0 is just the middle point of the absorption region. Also, the thickness of dielectric spacer $d = 1.5$ μm is nearly $\lambda/4$ at this frequency. At frequencies below f_0, the metasurface behaves as dielectric with high dielectric loss. At frequencies above f_0, however, it behaves as metal with negative permittivity. It is this specific dispersion which broadens the absorption bandwidth.

The transformation of the Drude model for metallic film to Lorentz model for metasurface is the key of the bandwidth enhancement in the proposed absorber.

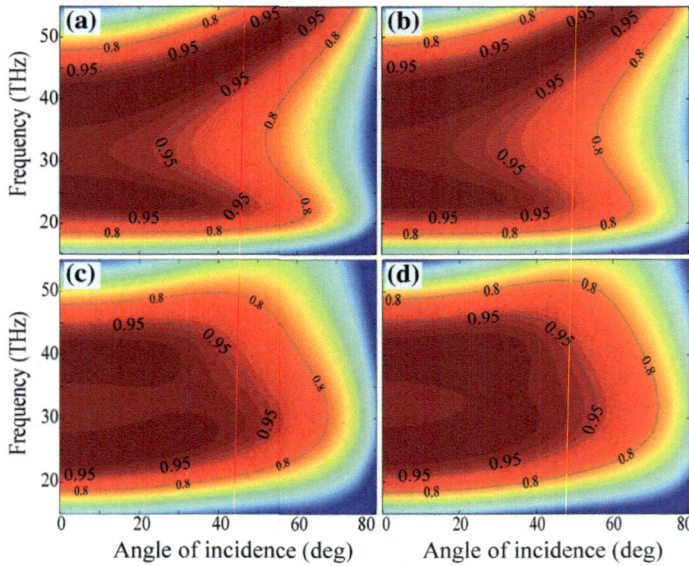

Fig. 7.26 Absorption versus incidence angle for **a, b** TE and **c, d** TM polarizations. The absorption for the equivalent absorber are shown in (**b**) and (**d**) for comparison. Reproduced from [28] with permission. Copyright 2012, Optical Society of America

Intuitively, it can be understood by an analog of interband transition in noble metals, where enhanced absorption occurs when electrons from the filled band below the Fermi surface are excited to higher bands. For the metasurface, similarly, the electrons are bounded in the metallic cross and form a bound state, thus enhanced absorption in this frequency region is achieved.

For many applications, absorption at large angle of incidence is required for both TE and TM polarizations. The numerical results for oblique incidence are shown in Fig. 7.26, where absorption bandwidth for both polarizations keeps almost the same for angle up to 50°. For TE polarization, the second absorption peak shift to higher frequency and is still larger than 97% at angle of 60°. For TM polarization, the frequency shift is very small and both absorption peaks drop linearly with increasing angle. More interestingly, the absorption of an equivalent absorber with metasurface described by Lorentz model agrees well with that of the proposed absorber at almost all incident angles. This further proves the accuracy of the equivalent material description used above.

7.4 Catenary Dispersion Model for Broadband Absorption

As shown in Chap. 6 and Appendix D, under static approximation, a metallic grating can be described by its sheet impedance:

$$Z_{\text{eff}} = \frac{1}{2i\,F},\tag{7.4.1}$$

where F is a catenary function that can be expressed as [31]

$$F(p, w) = \frac{p}{\lambda}\ln(\csc(\frac{\pi w}{2p})),\tag{7.4.2}$$

where λ and p are the operation wavelength and period of the structure, respectively. Equations (7.4.1) and (7.4.2) give a direct linkage between the geometry of the slit and its corresponding optical/electromagnetic response.

To extend this concept to a 2D version, a similar analysis has been made for closed-ring resonator (CRR) at microwave frequency. The catenary field distributions in CRRs can be divided into two parts: One is between the two arms of a resonator and the other is between adjacent resonators. The retrieved field profiles for both cases are shown in Fig. 7.27a, which indicate that at a fixed period p, there are two variables (g and w) that can alter the distribution of the overall optical field.

The experiential models for these structures have been proposed in the equivalent circuit theory in an exact catenary form as [33]

$$X_L = \omega L = \frac{p - g}{p}F(p, 2w),\tag{7.4.3}$$

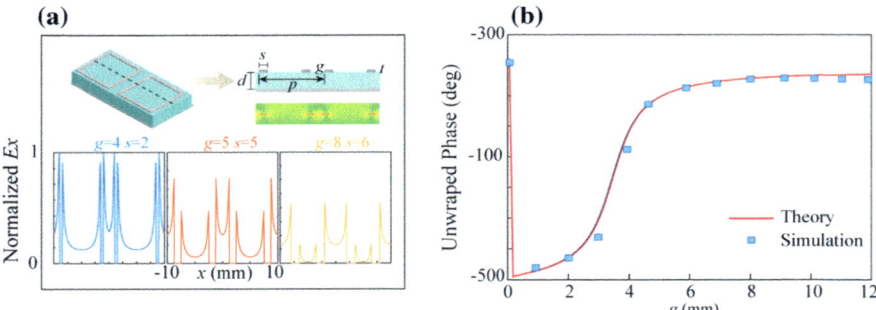

Fig. 7.27 The catenary model of metallic rings array. **a** The extracted electric field distribution in two adjacent CRRs with various g and w. The upper inset is the schematic diagram of the structure with $p=20$ mm, $d=5$ mm, and $t=0.01$ mm. **b** The comparison of unwrapped phase between simulation and theory with different g. The working frequency is $f=4$ GHz and w is fixed at 1 mm. Both results show great agreement even g approaches zero. Reproduced from [32] with permission. Copyright 2018, The Authors

$$B_C = \omega C = 4\frac{p - g}{p}F(p, g), \tag{7.4.4}$$

$$R = R_s\frac{p^2}{2lw}, \tag{7.4.5}$$

$$Z_{\text{eff}} = R + i(X_L - \frac{1}{B_C})Z_0, \tag{7.4.6}$$

where X_L is the reactance of the inductance L and B_C is the susceptance of the capacitance C. $\omega = 2\pi c/\lambda$ is the angular frequency in the operation band, where c is the vacuum light velocity. R is the resistance of the effective sheet and $R_s = 1/\sigma t$ (σ is the conductivity of the metal) is the sheet resistance for a thin film with thickness t.

To further validate this catenary model, the simulated and calculated phase responses with various g at 4 GHz is depicted in Fig. 7.27b. The simulated results matched their calculated counterparts perfectly even when g approaches zero.

7.4.1 Microwave Absorber

For a long time, the equivalent circuit theory was thought as fitted models thus the underlying physics is obscure. However, the catenary optical field distributions, as well as the catenary dispersion, make the circuit model more meaningful and useful. The aforementioned analysis also suggests that by properly controlling this catenary model, almost arbitrary responses can be achieved with such metasurfaces.

In the following, the catenary model is employed to design a broadband yet lightweight absorber operating at frequencies as low as P band [32]. The schematic of the proposed broadband absorber is shown in Fig. 7.28a. The geometry of the absorber consists of tri-layered CRR metasurfaces and a reflective ground plane that separated from each other by three dielectric layers (air) with different thicknesses. By employing the transmission line theory, the reflection coefficient r can be easily calculated and the absorption is obtained by $A = 1 - |r|^2$ because the reflective ground plane can prevent transmission.

Traditionally, the optimization of this kind of structure is mainly based on parameter scanning and mountain climbing. Such design process is innately suffered from human guided errors and the solution is often locally optimized. However, as the direct linkage between the structure and its performance can be well described by the catenary models, the automatic design process can be employed. Since the effective impedance of the metasurface can be expressed in a mathematical form, it can be well treated as an effective sheet and thus, the structure can be seen as a generalized multilayer [34]. With the help of the genetic algorithm (GA), the related parameters can be identified automatically after several iterations to approach the desired absorption spectrum. The optimized parameters, as well as the involved materials, are given in the figure caption of Fig. 7.28a and the calculated absorption performance under normal incidence are depicted in Fig. 7.28b. It should be noted that

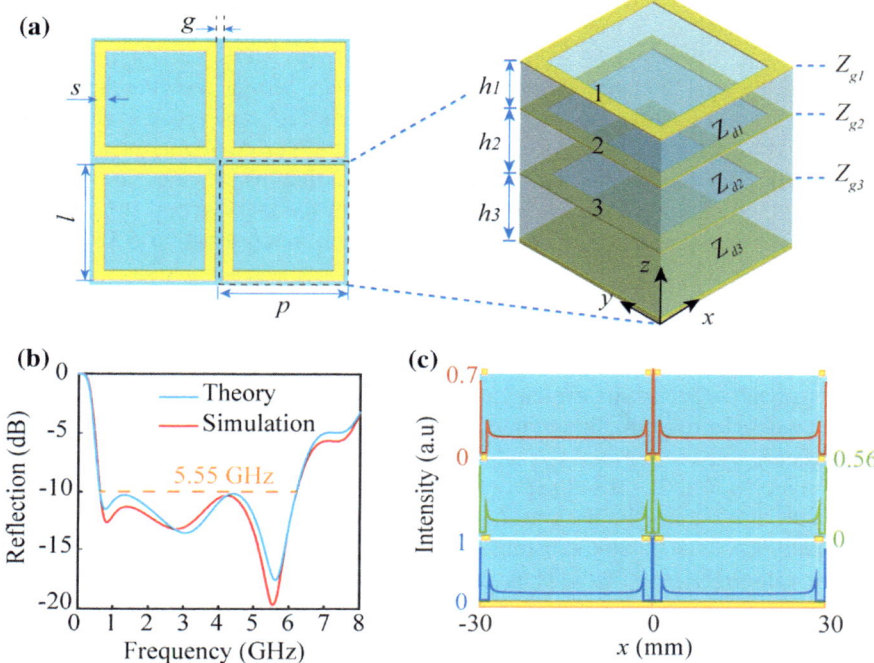

Fig. 7.28 The design of the broadband absorber. **a** Geometric configuration. The optimized parameters are $p = 30$ mm, $g = 0.2$ mm, $w_1 = w_2 = 1$ mm, $w_3 = 1.5$ mm. $h_1 = 13$ mm, and $h_2 = h_3 = 18$ mm. $R_1 = 30$ Ω, $R_2 = R_3 = 22$ Ω. The dielectric is set as air with $\varepsilon = 1$. Z_{gi}, Z_{di} ($i = 1, 2$, and 3) are the effective impedance of metasurface and the impedance of dielectric. **b** Simulated and theoretical reflection from 0 to 8 GHz. The -10 dB bandwidth is 5.55 GHz with nearly three-octave bandwidth. **c** The extracted electric field distribution along the tri-layered metasurface at 5.56 GHz. As the distance between adjacent resonators is much smaller than that between two arms of a resonator, the intensity of electric field is much stronger in the former case. Reproduced from [32] with permission. Copyright 2018, The Authors

in order to make this structure in subwavelength scale and reduce the complexity in optimization, the period of the metasurfaces is fixed to be $p = 30$ mm. In order to confirm the validity of the theoretical results, full wave simulation is carried out with the commercial software CST Microwave Studio. The boundary conditions are unit cell in the xy-plane and open in z-axis. As shown in Fig. 7.28b, both the theoretical and simulated results matched each other quite well and indicate that the -10 dB polarization-independent reflection is realized from 0.65 to 6.2 GHz by this tri-layered metasurface.

To evaluate the performance of the device comparatively, the theoretical limit under normal incidence, i.e., the Rozanov limit is calculated by [35]

$$d_l = \left| \int_0^\infty \ln|R(\lambda)|\mathrm{d}\lambda \right| / 2\pi^2, \tag{7.4.7}$$

where d_l is the total thickness for an ideal device. In the current case, the theoretical value of d_l with the same performance in Fig. 7.28b is 43.8 mm which is very close to the thickness of the designed device ($d_t = 49$ mm). The deviation ratio, defined as $(d_t - d_l)/d_t$, is about 10%, which is much smaller than its previous counterparts based on other design principles. To further confirm the existence of the catenary field, the extracted electric field distributions at each metasurface at 5.56 GHz is given in Fig. 7.28c.

As the metasurfaces in the simulation are set as surfaces with no physical thicknesses, to meet the requirement in fabrication, traditional printed circuit board (PCB) process is no longer suitable because the thicknesses of boards are in millimeter scale. In order to realize deep-subwavelength thickness for the structure, the CRRs must be printed on a dielectric film thinner than tens of micrometers. Here, 50 μm polyimide film was employed to serve as the substrate for the metasurface and the thicknesses of nichrome ($\sigma = 2.2 \times 10^5$ S/m) for the tri-layered metasurface are varied from 100 to 200 nm to meet the desired R_s. In fact, the precise fabrication for subwavelength patterns on the flexible substrate is quite challenging, especially for large-scale devices. Although some methods have been proposed such as laser processing and film printing, these techniques are not suitable for this case owing to their low fabrication precisions. To overcome this issue, a new process based on moving exposure technique was proposed to enable large-scale fabrication with high precision.

The schematic diagram of the fabrication process is shown in Fig. 7.29. First, a clean polyimide film is stretched tightly by a 900-mm-inner-diameter carbon fiber holder. With a home-made stretching machine, the flatness and uniformity of the film can be guaranteed for further processing. The deposition of nichrome is realized by ion beam assisted deposition at the speed of 0.3 nm/s and room temperature. As this process is performed at a relatively low temperature, it can avoid the thermal effect to the film. After that, a set of equipment was developed to realize large-scale spin coating, prebaking, exposure, developing, and wet etching. It should be noted that as the sample is in meter scale, its exposure cannot be achieved directly. Instead, moving exposure techniques were developed to realize uniform step-by-step exposure. The maximum diameter for the sample can be as large as 1.2 m.

After the metasurfaces are fabricated, for ease of characterization, polymethacrylimide (PMI) foam was employed in the experiment instead of air to serve as the dielectric layer. As the permittivity of PMI is only 1.05, such replacement has little influence on the response. The experiment was carried out in an anechoic chamber with a vector network analyzer. The image of the sample is depicted in Fig. 7.30a and the measured results are shown in Fig. 7.30b accompanied with the simulated results. It can be seen from Fig. 7.30a that the subwavelength structures are well patterned on the film with excellent uniformity. The measured results matched the simulated ones quite well with nearly the same -10 dB bandwidth. As there is no obvious discontinuity occurred in the measured region, the consistency of the results further confirmed the validity of the measurement. The slight discrepancy between

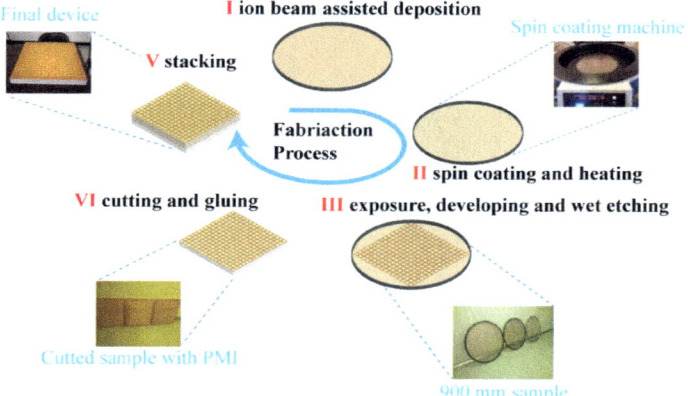

Fig. 7.29 Fabrication process of the broadband absorber. The proposed technique could enable large-scale fabrication with high precision on flexible substrates which is promising in industrial applications. Reproduced from [32] with permission. Copyright 2018, The Authors

measurement and simulation can be attributed to two reasons. First, the fabrication error and the roughness of the film cannot be fully removed, which unavoidably leads to some deviations. Second, epoxy glue for adhering the film and PMI in the experiment is not considered in the simulation, which can also modify the response spectrum. To further confirm the lightweight property of the fabricated device, a 600 mm × 600 mm × 49 mm sample was measured with a total mass of 1.09 kg. The corresponding density is only 0.06 g/cm^3, which is much lighter than traditional PCB boards. In fact, the density of this device can be further decreased if the PMI layers are removed. In that case, the absorber will be a truly flexible structure that can be employed for conformal applications.

The bandwidth of this kind of absorber can be further broadened if absorbing dielectric layers are employed. To demonstrate this, the bottom PMI was replaced with commercial honeycomb-structured absorbing material. In this case, the structural parameters for the tri-layered metasurface and the thicknesses of the dielectric layers are the same as the aforementioned device. The fabricated sample and corresponding measured results from 0.3 to 18 GHz are shown in Fig. 7.30c, d. The operation bandwidth of this device shows a slightly blue shift compared with the former one that the −10 dB reflection bandwidth ranges from 0.9 to 18 GHz except for two narrow bands near 4.9 GHz (−9.5 dB) and 9 GHz (−8.5 dB). This can be attributed to the impedance mismatch of the whole structure with air. With some optimization on the simulation and fabrication, a better absorption performance can be expected.

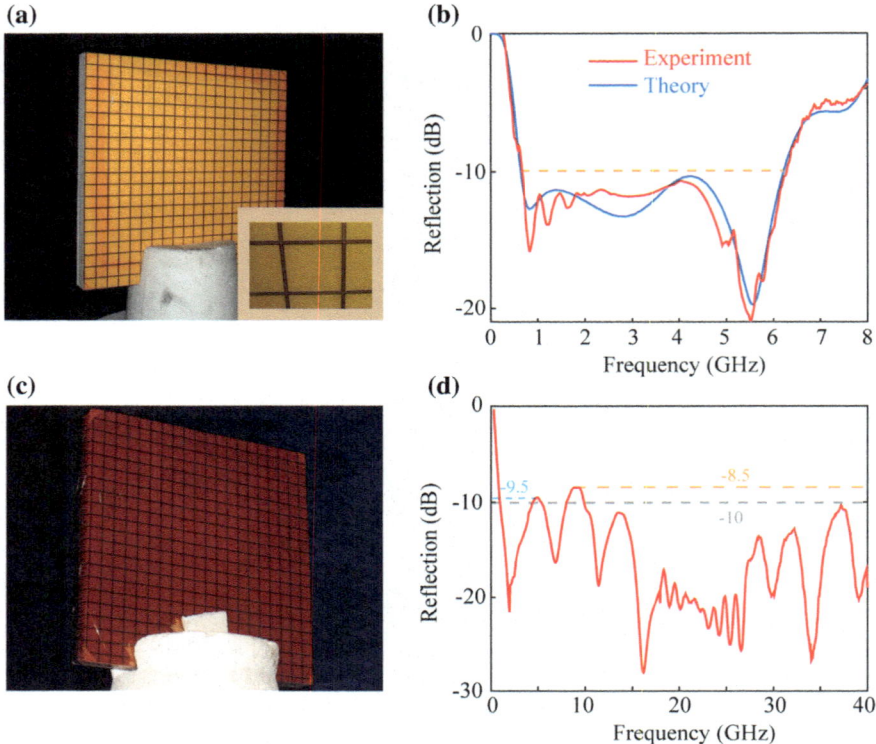

Fig. 7.30 The fabricated samples and their measured results. **a, b** The image of the foam-based absorber and its performance from 0 to 8 GHz. **c, d** The photos of the absorbing honeycomb-structured-based absorber and its measured reflection from 0.3 to 40 GHz. Reproduced from [32] with permission. Copyright 2018, The Authors

7.4.2   Terahertz Absorber

The above design principle can be also applied in other frequency such as terahertz band. Figure 7.31a–d shows the geometric configuration and equivalent circuit model for a terahertz broadband absorber comprising two metasurfaces, three dielectric spacers and one ground plane [36]. As illustrated in Fig. 7.31e, the results of the circuit model agree well with the full-wave simulation, and the absorption is larger than 90% at frequencies ranging from 0.52 to 4.4 THz.

According to the equivalent circuit theory, the frequency-dependent impedances shown in Fig. 7.31a, b can be expressed as

$$Z_{s1} = R_1 + i\omega L_1 + \frac{1}{i\omega C_1}, \tag{7.4.8}$$

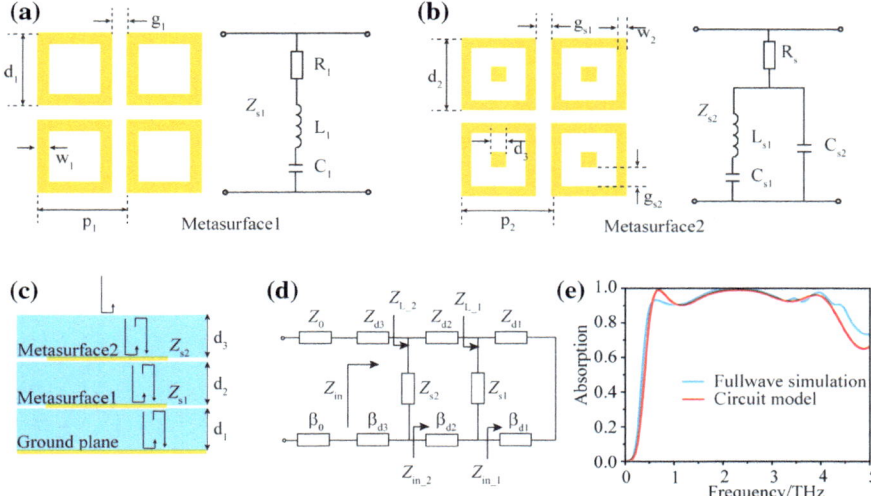

Fig. 7.31 Theoretical simulations of the broadband THz absorber. **a, b** Geometric parameters and corresponding equivalent circuit model for metasurface 1 and metasurface 2. **c** Schematic of the transmission line model for broadband THz absorption. **d** Equivalent circuit model of the structure with dual-metasurface based on meta mirror. **e** A comparison of the calculated absorption obtained with full-wave simulation and circuit model. Reproduced from [36] with permission. Copyright 2018, Honglin Yu et al

$$Z_{s2} = R_s + \frac{Z_{g1} Z_{g2}}{Z_{g1} + Z_{g2}}. \qquad (7.4.9)$$

In Eq. (7.4.9), Z_{g1} and Z_{g2} are the equivalent frequency-dependent impedance of two resonant circuits in metasurface 2, which can be expressed as

$$Z_{g1} = i\omega L_{s1} + \frac{1}{i\omega C_{s1}}, \qquad (7.4.10)$$

$$Z_{g2} = \frac{1}{i\omega C_{s2}}. \qquad (7.4.11)$$

Similar to the discussion in the previous section, the equations for calculating the resistance, capacitance, and inductance in the equivalent circuit may be expressed as

$$R_s = R_0 \frac{p_2^2}{2d_2 w_2 + d_3^2}, \qquad (7.4.12)$$

$$L_{s1} = \frac{F(p_1, 2w_2, \lambda)}{\omega}, \qquad (7.4.13)$$

$$C_{s1} = \frac{0.75 \times 4d_2 \varepsilon_{\text{eff}}}{\omega p_2} F(p_2, g_{s1}, \lambda), \qquad (7.4.14)$$

$$C_{s2} = \frac{d_3}{\omega p_1}\left(\frac{C'C''}{C' + C''}\right), \qquad (7.4.15)$$

where $R_0 = 1/(\sigma_{Cr}t)$ is the resistance of Cr film (σ_{Cr} and t are the conductivity and thickness of chromium). ε_{eff} is the permittivity of SU-8 photoresist. C' and C'' are induced by the outer and inner gap, which can be expressed as

$$C' = \frac{4\varepsilon_{eff} F(p_2, g_{s1}, \lambda)}{\omega} \qquad (7.4.16)$$

and

$$C'' = \frac{4\varepsilon_{eff} F(p_2, g_{s2}, \lambda)}{\omega}. \qquad (7.4.17)$$

Here, the catenary function F is expressed as

$$F(p, s, \lambda) = \frac{p}{\lambda}\cos\theta \ln\left[\csc\left(\frac{\pi s}{2p}\right)\right], \qquad (7.4.18)$$

where θ is the angle of incidence and λ is the wavelength. In the same way, the resistance, inductance and capacitance in the equivalent circuit model of metasurface 1 can be written as

$$R_1 = R_0 \frac{p_1^2}{2d_1 w_1}, \qquad (7.4.19)$$

$$L_1 = \frac{d_1}{\omega p_1} F(p_1, 2w_1, \lambda), \qquad (7.4.20)$$

$$C_1 = \frac{4d_1\varepsilon_{eff}}{\omega p_1} F(p_1, g_1, \lambda). \qquad (7.4.21)$$

Assuming that a plane wave normally impinges on the structure, due to reflections occurring at the metasurface and background plane, both the forward and backward scattering waves exist in the dielectric spacer and surrounding space, as depicted in Fig. 7.31c. According to the transmission line theory, the equivalent circuit model can be obtained, as shown in Fig. 7.31d. Z_{s1} and Z_{s2} are the equivalent impedances of the two cascaded metasurfaces. Z_{d1}, Z_{d2}, and Z_{d3} are the characteristic impedance of three SU-8 colloids based spacer layers, respectively. β_0 and β_d are the propagation constant of the free space and medium. Z_0 and Z_{in} are the characteristic impedance of free space and input impedance of the whole circuit model. Different from the previous section, here, the input impedance is used to calculate the transfer matrix ($m = 1, 2$) as

$$Z_{in\_(m+1)} = Z_{dm} \frac{Z_{L\_m} + i Z_{d(m+1)} \tan(\beta_d d_{(m+1)})}{Z_{d(m+1)} + i Z_{L\_m} \tan(\beta_d d_{(m+1)})}, \tag{7.4.22}$$

$$Z_{L\_m} = \frac{Z_{sm} Z_{in\_m}}{Z_{sm} + Z_{in\_m}}, \tag{7.4.23}$$

$$\beta_d = \frac{\omega \sqrt{\varepsilon_d(1 - i \tan \delta)}}{c}, \tag{7.4.24}$$

$$Z_d = Z_0 \frac{1}{\sqrt{\varepsilon_d(1 - i \tan \delta)}}. \tag{7.4.25}$$

Here, ε_d and $\tan \delta$ are the relative permittivity and dielectric dissipation factor of SU-8 colloids. Utilizing iterative algorithm, the input impedance of the entire equivalent circuit model can be derived from the bottom layer recursively. The reflection coefficient is calculated by the following equation:

$$\Gamma = \frac{Z_{in} - Z_0}{Z_{in} + Z_0}. \tag{7.4.26}$$

The theoretical results obtained by transmission line model are shown in Fig. 7.31e, which show good agreements with the numerical results obtained by full-wave simulation.

To further illustrate the relationship between the effective impedance and the geometry of the metasurface, the amplitude of electric field for dual-metasurface in the xz-plane at resonant frequencies (0.6 and 2.5 THz) is plotted in the Fig. 7.32. The instantaneous electric field distributions under normal incidence at these resonant frequencies are also illustrated. Apparently, the strong resonance occurred in the gap of the adjacent unit cell are responsible for the absorption at the lower frequency (0.6 THz). For the higher frequency (2.5 THz), the resonance occurred in the gap of the single unit cell, and additionally, there appears a hybrid mode between two metasurfaces. To take a closer look at this issue, the E-field amplitude is extracted at these resonances. Interestingly, all profiles in the gaps can be well described by the hyperbolic cosine catenary curves.

When changing the geometry of metasurface, the catenary field distributions would be altered as well, leading to the modulation of the effective impedance at corresponding metasurface. By properly optimizing the profile of the catenary field and coupling the different catenary fields in structure, the desired effective impedance of metasurfaces can be obtained. The discrepancy between the extracted field and fitting catenary in Fig. 7.32d is caused by the coupling of closely adjacent catenary field in metasurface 1, which implies a hybrid mode. Similar to the dispersion management (equivalent circuit theory and transmission line model) [37, 38], it is believed that this catenary E-field model offers a new interpretation to understand the physics behind strongly coupled metasurfaces.

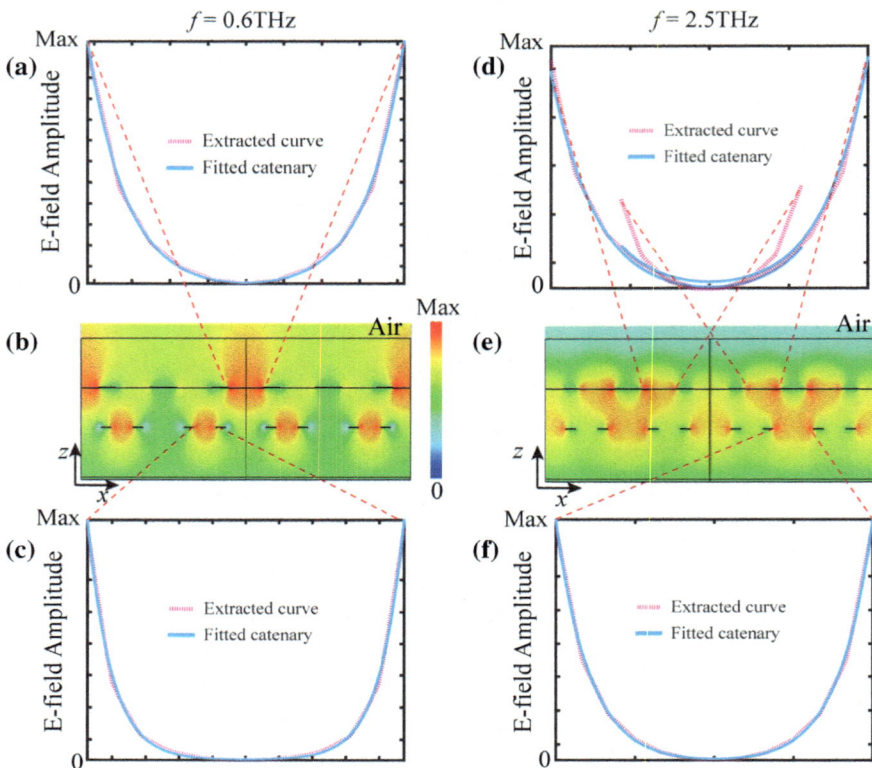

Fig. 7.32 The extracted electric field amplitude (red dotted line) and fitting catenary curve (blue solid line) between two arms of the resonator and adjacent resonators for dual-metasurface at 0.6 THz (**a** and **c**) and 2.5 THz (**d** and **f**). The slight discrepancy is because that the extracted field is the superposition of catenary field (evanescent field) and incident field. **b**, **e** Electric field distributions in *xz*-plane at two resonant frequencies. Reproduced from [36] with permission. Copyright 2018, Honglin Yu et al

7.5 Coherent Perfect Absorption in Metallic Thin Films

In a common absorber, the incident energy is delivered to the systems via a single channel, for instance by a plane wave illuminated on one side of the absorber. However, it was found that the incident energy could be perfectly absorbed under incidence on opposite sides of an absorber. This interference-assisted absorption is known as "coherent perfect absorption", and was experimentally demonstrated in a silicon slab under coherent monochromatic illumination [39, 40]. In a coherent perfect absorber (CPA), two counter-propagating input beams of identical amplitudes and phases interfere destructively and result in perfect absorption. Such a mechanism provides a new way for the control of electromagnetic absorption [41, 42]. Interestingly, owing to the symmetry of the configuration, the two counter-propagating waves

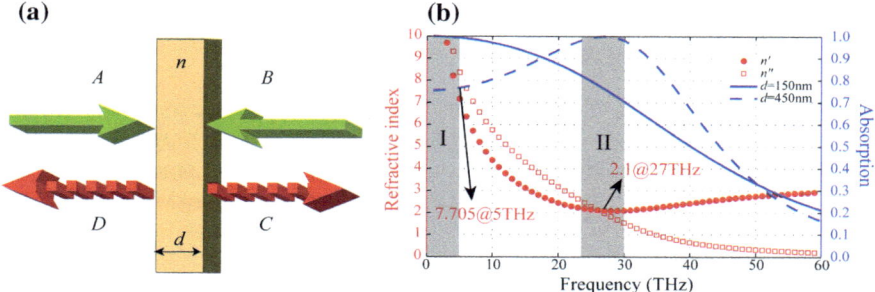

Fig. 7.33 Concept of coherent perfect absorption based on a thin film. **a** Sketch of a coherent perfect absorber. The output beams consist of reflected and transmitted components. When they are properly designed, the output waves can destructively interfere on each side, resulting in perfect absorption. **b** Refractive index of doped silicon described by the Drude model and the absorption curves for different thicknesses. The two absorption regions are highlighted as Zone I and Zone II. Reproduced from [44] with permission. Copyright 2012, Optical Society of America

inside highly lossy absorber would form a catenary function, which is critical for the perfect absorption.

Figure 7.33a shows a typical planar CPA structure. Two coherent beams normally illuminate from two opposite sides. An input beam (A_1 or A_2) of this two-port system is partially transmitted and reflected, while an output beam (B_1 or B_2) consists of reflected and transmitted components. To analyze this system, one can solve the Maxwell's equations by using scattering matrix (S matrix) [43]. In general, the relationship between input and output can be described by

$$\begin{bmatrix} B_1 \\ B_2 \end{bmatrix} = S \begin{bmatrix} A_1 \\ A_2 \end{bmatrix} = \begin{bmatrix} r_{11} & t_{12} \\ t_{21} & r_{22} \end{bmatrix} \begin{bmatrix} A_1 \\ A_2 \end{bmatrix}, \tag{7.5.1}$$

where r_{ij} and t_{ij} are the reflection and transmission coefficients of an input beam (A_1 or A_2), respectively. The scattering matrix S depends on operating wavelength, geometry of the structure and material properties. For simplicity, a symmetric two-port structure is considered, where $r_{12} = r_{21} = r$. If the system is in steady-state, linear regime and exclusive of magneto-optically gyrotropic medium, S matrix will be constrained by optical reciprocity, which suggests that S is symmetric. For the two-port cases, this constraint implies that $t_{12} = t_{21} = t$. On the basis of optical reciprocity, the output beams can be written as a function of the incidence as

$$\begin{bmatrix} B_1 \\ B_2 \end{bmatrix} = S \begin{bmatrix} A_1 \\ A_2 \end{bmatrix} = \begin{bmatrix} r & t \\ t & r \end{bmatrix} \begin{bmatrix} A_1 \\ A_2 \end{bmatrix}. \tag{7.5.2}$$

The reflection and transmission coefficients can be described by

$$r = \frac{(n^2 - 1)(-1 + e^{i2nkd})}{(n + 1)^2 - (n - 1)^2 e^{i2nkd}}, \tag{7.5.3}$$

$$t = \frac{4ne^{inkd}}{(n+1)^2 - (n-1)^2 e^{i2nkd}}, \tag{7.5.4}$$

where d is the thickness of the slab and $n = n' + in''$ is the complex refractive index. Coherent perfect absorption occurs when output wave components vanish ($B_1 = B_2 = 0$) and corresponding inputs are so-called CPA eigenmodes. Due to the mirror symmetry of the system, coherent perfect absorption can only be achieved for symmetrical inputs ($A_1 = A_2, r + t = 0$) or antisymmetrical inputs ($A_1 = -A_2, r - t = 0$). In both cases, the magnitude of reflection and transmission are equal, indicating that a CPA can act as a beam splitter when illuminated by a single beam. Using Eqs. (7.5.3) and (7.5.4), the CPA condition for normal incidence can be obtained as [39]

$$\exp(inkd) = \pm\frac{n-1}{n+1}. \tag{7.5.5}$$

Note that, n should be replaced by the impedance Z for materials with magnetic response [45]. The $\pm$ sign corresponds to the symmetrical and antisymmetrical inputs. As shown in [40], nearly infinite discrete solutions of Eq. (7.5.5) have been obtained for $kd \gg 1$. Another important case of the two-port system is a film structure much thinner than the operating wavelength ($d \ll \lambda$, $|nkd| \ll 1$), such as a dielectric or a metal film. In this case, the left and right side of Eq. (7.5.5) can be approximated as $1 + inkd$ and $\pm(1 - 2/n)$. Since $|nkd| \ll 1$, only plus sign term in the right side should be selected. The real and imaginary parts of the refractive index are approximated as [44]

$$n' \approx n'' \approx \frac{1}{\sqrt{kd}} = \sqrt{\frac{c}{\omega d}}. \tag{7.5.6}$$

Compared with the general CPA condition, the CPA condition for the ultrathin film is clearer and easier to meet with the ultrathin film made of metal or heavily doped semiconductor. Over a broad frequency range, the complex dielectric function can be explained by the Drude model

$$n^2 = \varepsilon_1 + i\varepsilon_2 = \varepsilon_\infty - \frac{\omega_p^2}{\omega(\omega + i\Gamma)}, \tag{7.5.7}$$

$$\begin{aligned}\varepsilon_1 &= \varepsilon_\infty - \frac{\omega_p^2 \tau^2}{1 + \omega^2 \tau^2}, \\ \varepsilon_2 &= \frac{\omega_p^2 \tau^2}{\omega(1 + \omega^2 \tau^2)}\end{aligned} \tag{7.5.8}$$

where ε_∞ is the dielectric constant, $\Gamma = 1/\tau$ is collision frequency (τ is the scattering rate), and ω_p is the plasma frequency. In the very low-frequency range, where $\omega \ll 1/\tau$, the real and imaginary parts of the refractive index are of comparable magnitude with

$$n' \approx n'' \approx \sqrt{\frac{\varepsilon_2}{2}} = \sqrt{\frac{\tau \omega_p^2}{2\omega}}. \tag{7.5.9}$$

Inserting Eq. (7.5.9) into Eq. (7.5.6), the thickness for CPA at this frequency range can be written as

$$d_w \approx \frac{2c}{\omega_p^2 \tau}. \tag{7.5.10}$$

This characteristic length is so-called Woltersdorff thickness [46]. It quantifies the thickness of a metallic film with maximal absorption for incoherent input in the low-frequency range.

When the working frequency increases, the absorption at Woltersdorff thickness decreases. However, the absorption can still be nearly perfect by adjusting the thickness. By setting ε_1 in Eq. (7.5.8) to be zero, the real and imaginary parts of the refractive index become

$$n' = n'' = \sqrt{\frac{\varepsilon_2}{2}} = \sqrt{\frac{\varepsilon_\infty}{2\omega\tau}}. \tag{7.5.11}$$

By combining Eq. (7.5.11) with Eq. (7.5.6), a second characteristic length called Plasmon thickness can be obtained by

$$d_p \approx \frac{2c\tau}{\varepsilon_\infty}. \tag{7.5.12}$$

It should be noted that Eq. (7.5.11) is fully met at a single frequency, thus the absorption bandwidth is narrow. When $\omega_p^2 \tau^2 = \varepsilon_\infty$, the Plasmon thickness is equal to the Woltersdorff thickness, which ensures a larger absorption bandwidth. It is seen that both the Plasmon thickness and Woltersdorff thickness requires the real and imaginary parts of the refractive index to be equal. Figure 7.33b shows that the calculated Woltersdorff thickness and Plasmon thickness (150 and 450 nm) are in good agreement with the theoretical values (151 and 416 nm). Generally, the two characteristic lengths depart from each other due to their different dependences on scattering time. Both of them are in subwavelength scale according to the approximation made in Eq. (7.5.6).

Under coherent illumination, the counter-propagating waves inside the slab have equal amplitudes. The total electric fields can be written as

$$E \sim \exp(in'kx)\exp(-n''kx) \pm \exp(-in'kx)\exp(n''kx). \tag{7.5.13}$$

The real part can be written as

$$\mathrm{Re}(E) \sim \cos(n'kx)\left[\exp(-n''kx) \pm \exp(n''kx)\right], \tag{7.5.14}$$

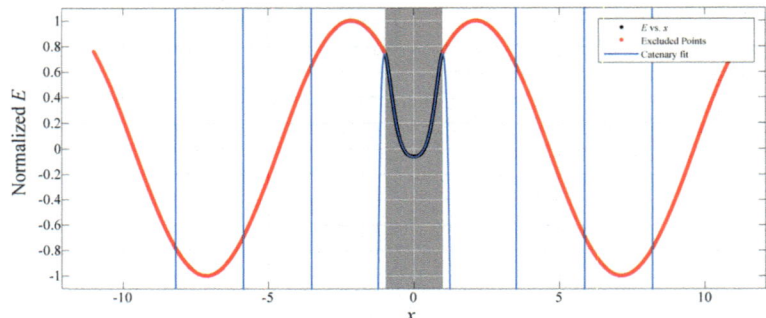

Fig. 7.34 Normalized electric field (real part) in lossy slab under symmetric coherent incidence. The fitting function is $a*\cos(b*x)*(\exp(c*x) + \exp(-c*x)) + d$. The fitted parameters are $a = 0.01569$, $b = 1.34$, $c = 5.454$, $d = 0.09735$, and R-square = 0.9998. The small error is owing to the numerical error in FDTD simulations

which resembles the catenary function. Figure 7.34 shows the normalize electric fields (real part) under symmetric coherent illumination with a wavelength of 10 μm. The lossy metallic slab has a thickness of 1 μm and a conductivity of 10^5 S/m.

The coherent perfect absorption can also be explained as a time-reversed process of a laser. According to semiclassical lasing theory, the first lasing mode is an eigenvector of S matrix with an infinite eigenvalue. This means that lasing occurs when poles (points at which an eigenvalue of S diverges) of the S matrix move upward to the real axis by adding gains. In contrast, a time-reversed process can arise when a specific degree of dissipation is added to the resonator. In this case, the positive imaginary part of the refractive index equals in absolute value to that at the lasing threshold. When the system is illuminated coherently by the time reverse of the output of a laser, the coherent perfect absorption occurs [39]. Since lasers are often narrowband, it is often thought that the CPA should be narrowband. The broadband absorption in the regime $|nkd| \ll 1$, however, breaks down this conclusion. In fact, it was proposed that the broadband absorption can be treated as a time-reversed electromagnetic radiation [44]. In the impedance theory, the thin film CPA may be approximated as a resistive sheet with $Z = 1/(d_w\sigma_0) = Z_0/2$ as the thickness of the slab is much smaller than the skin depth. Here, σ_0 is the static conductivity and Z_0 is the impedance of vacuum. Then, consider the radiation property of an infinite oscillating current sheet in the xy-plane. Assuming that the current is $\vec{J} = K \sin(\omega t)\vec{x}$, the electric field at $z = 0$ can be written as $\vec{E} = -0.5\mu_0 cK \sin(\omega t)\vec{x}$. The effective sheet impedance, defined as E/J, is $-Z_0/2$, which is just in opposite to the thin film CPA condition. Such a radiation may be thought as the time-reversed process of the broadband CPA, although the infinite oscillating current sheet may be not applicable in practical conditions.

As illustrated in Fig. 7.35a, a thin film CPA was realized using a heavily doped silicon film [44]. Mach–Zehnder geometry was used to obtain the relative phase of input beams. The phase difference between the two arms in air can be denoted as $\Delta\varphi$

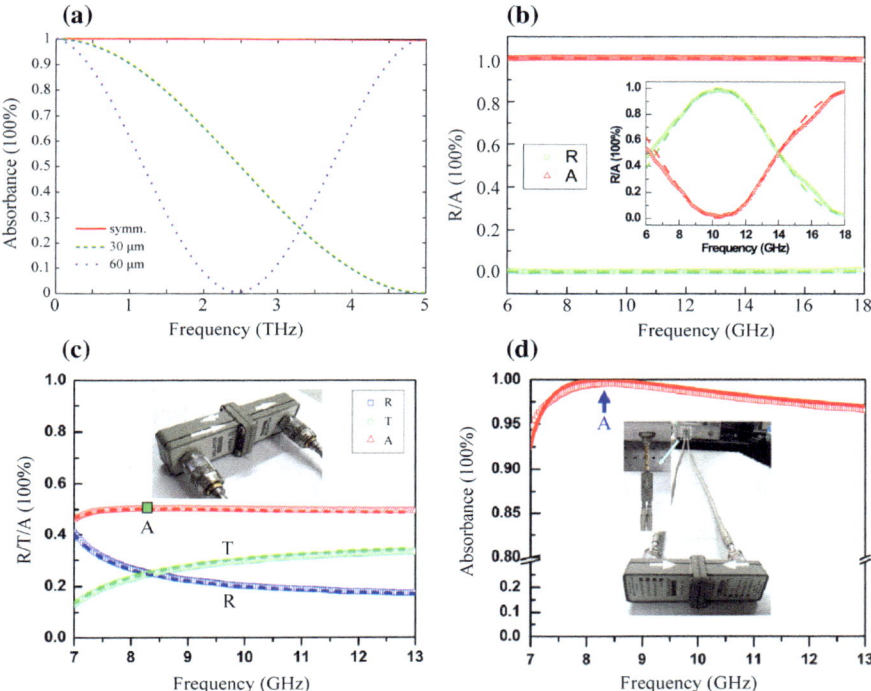

Fig. 7.35 Coherent perfect absorption in thin films. **a** Coherent absorption of 150 nm thick doped silicon. The path difference between the two arms in Mach-Zehnder geometry varies from 0, 30 to 60 μm. Reproduced with permission from [44]. Copyright 2012, Optical Society of America. **b** Coherent absorption of an ultrathin conductive film. The measured (open symbols) and calculated (dash lines) results in free space for the cases of $L = 0$ and $L = 14.5$ mm (inset), where L represents the difference in the distance of the film to the two horn antennas. The sheet resistance is $R_s = 180$ Ω. Reproduced with permission from [47]. Copyright 2015, American Physical Society. **c** Measured (symbols) and calculated (dashed lines) reflectance, transmittance, and absorbance of the graphene/PET sample with $R_s = 310$ Ω. The inset is the photo of the experimental setup, where the thick arrows denote the incidence, reflection, and transmission. **d** The measured (symbols) and calculated (line) coherent absorption of the graphene monolayer with $R_s = 310$ Ω. The inset is the photo of the experimental setup, where the two thick white arrows illustrate the coherent illumination (**c**), (**d**). Reproduced with permission from [48]. Copyright 2015, AIP Publishing LLC

$= k \Delta l$, where Δl is the path difference between the two arms. The absorbance of doped silicon film can be adjusted from unit to near zero at 2.5 THz when Δl increases from 0 to 60 μm. Moreover, it was demonstrated that a CPA operated at microwave frequencies can be realized using a single-layer ultrathin conductive film [47]. The CPA experiment was implemented in free space and the complete absorption is nearly frequency-independent with a relative bandwidth of 100%, as shown in Fig. 7.35b. The absorption approaches zero when the relative phase difference of the two incident beams is π.

A monolayer graphene can be also exploited to realize CPA with the assistance of suitable phase modulation between two incident beams at the quasi-CPA frequencies [48]. Such a graphene-based CPA holds broadband angular selectivity and the absorbance can be tuned substantially by varying the carrier concentration through chemical doping. As demonstrated in Fig. 7.35c, d, it was experimentally demonstrated that a graphene-based CPA can operate over the microwave X-band (7–13 GHz). The absorbance of the unpatterned graphene monolayer is observed to be greater than 94% over the working band.

Note that the results in Figs. 7.35c, d are measured using waveguide, which is easier to implement especially in the low frequency. Based on this approach, nearly perfect absorption of electromagnetic waves was experimentally demonstrated using polyimide films in a waveguide system within the microwave UHF band (0.3–0.6 GHz). As illustrated in Fig. 7.36, two pairs of coax-to-waveguide adapters were used to measure the absorbance so that the combined working bands cover 0.3–0.6 GHz.

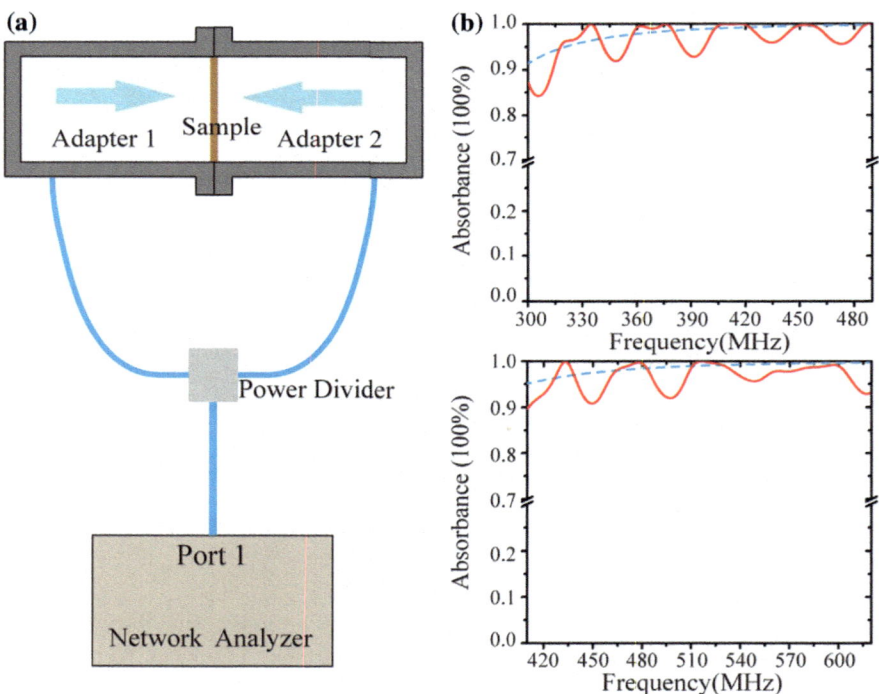

Fig. 7.36 Coherent perfect absorption in waveguides. **a** Schematic of the experimental setup realizing coherent perfect absorption in a waveguide system. **b** Measured (red solid lines) and calculated (blue dashed lines) absorbance of the polyimide film with sheet resistance 200 Ω. Reproduced from [42] with permission. Copyright 2018, Published by Elsevier Ltd

7.6 Catenary Plasmons for Solar Cell Enhancement

7.6.1 Localized Field Enhancement in Plasmonic Grating

It is well known that metallic gratings can be used as transparent conducting materials (TCOs), which are very important for photovoltaic devices. Owing to the coupled catenary plasmons in the metallic gratings, strong localized field enhancement may be anticipated and exploited to further increase the energy efficiency. In the following, a representative example is shown to illustrate this fact [49].

As shown in Fig. 7.37, this work exploits a MIM structure comprised of organic semiconductors sandwiched by silver metallic layers. One metal layer is continuous and acts as cathode, and the other is a periodic grating acting as semi-transparent anode. The grating structure can efficiently couple the incident plane wave to surface plasmons. Figure 7.37b illustrates SEM image of the fabricated sample. The deposited organic materials have the sinusoidal shape due to the height profile of the Ag nanowire.

As shown in Fig. 7.38a, the transmittance of the silver grating for TM polarized incidence is a bit smaller than the ITO. However, owing to strongly localized and coupled plasmons (Fig. 7.38b), the photocurrent density and external quantum efficiency (EQE) could be greatly increased. The short circuit current J_{sc} of the nanowire device, ~5.21 mA/cm^2, is about 40% higher than that of ITO device ~3.72 mA/cm^2. Therefore, the overall power conversion efficiency (PCE) of the nanowire device (~1.32%) is enhanced about 35% than that of the ITO device (~0.96%). Simultaneously, a 2.5-fold EQE enhancement around the wavelength of 560 nm was clearly observed. The measured current density and EQE enhancement are shown in Fig. 7.38c, d.

(a)   **(b)**

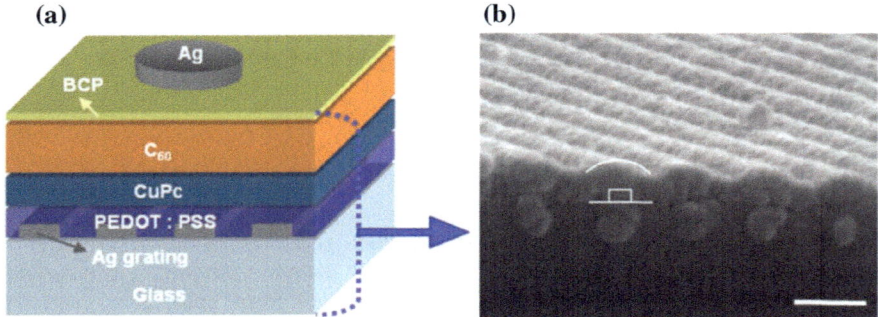

Fig. 7.37 **a** Schematic of the small molecular weight organic solar cell. The device has layered structure of Ag nanowire anode, PEDOT:PSS, CuPc, C60, BCP, and cathode (thick Ag film) from bottom to top. **b** Cross-sectional view of one of the fabricated devices without 70 nm thick Ag cathode. Scale bar: 200 nm. Organic layers are depicted with solid lines and the square shows the position of the Ag nanowire on glass substrate. Reproduced from [49] with permission. Copyright 2010, WILEY-VCH Verlag GmbH & Co. KGaA, Weinheim

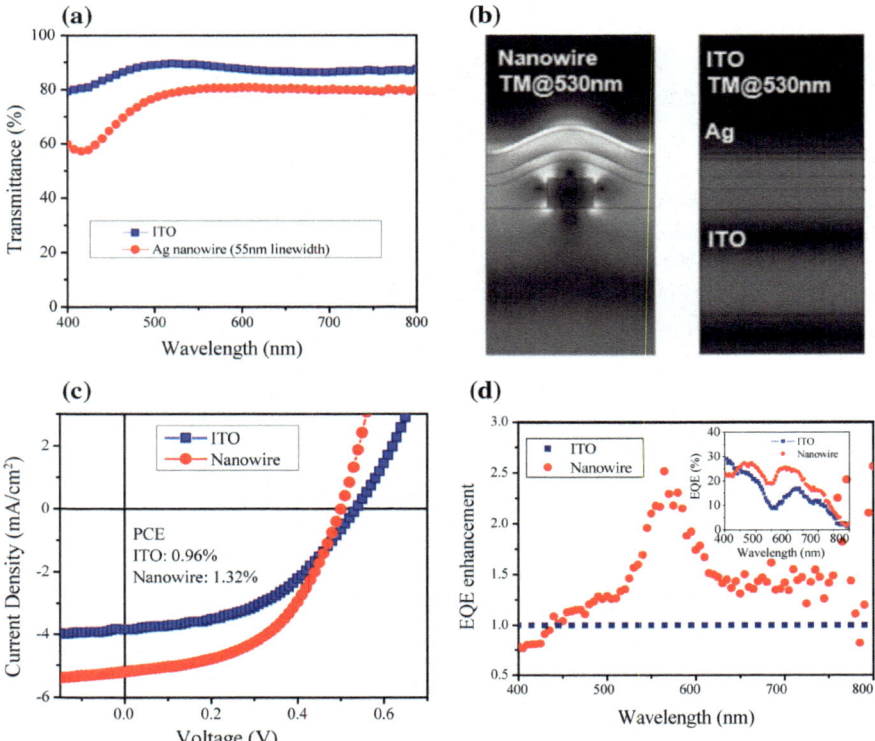

Fig. 7.38 **a** Transmittance spectra of ITO and Ag grating electrodes. **b** Simulated electric field profiles of nanowire and ITO devices at the wavelength 530 nm for TM illumination. Catenary plasmons are observed owing to the coupling of metallic wires and silver reflective layer (top cathode). **c** *J–V* curve of the nanowire and control ITO devices. **d** EQE enhancement of nanowire device with reference to ITO device. The inset gives the measured EQE of the nanowire and ITO devices. Reproduced from [49] with permission. Copyright 2010, WILEY-VCH Verlag GmbH & Co. KGaA, Weinheim

7.6.1.1 Localized Field Enhancement with Coupled Nanoparticles

As discussed in Chap. 3, the plasmonic field may be greatly changed by coupling multilayered metal–dielectric films. Similarly, the catenary coupling of plasmons in nanoparticles may lead to further enhancement of local fields. For instance, Kawawaki et al., examined the plasmonic enhancement of dye-sensitized photocurrents by Au nanoparticle (NP) ensembles with different particle densities to study the effects of plasmon coupling on the photocurrent enhancement [50].

Figure 7.39 illustrates the operating principle. When light illuminates the nanoparticles, the fields in the gap would greatly increase along with the decrease of gap distance. The enhanced electric fields would enhance the photocurrents by excit-

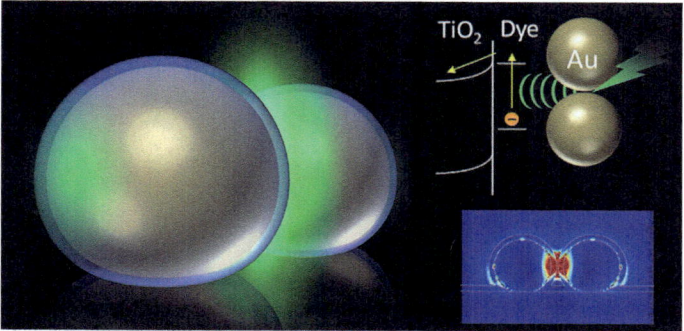

Fig. 7.39 Operating principle of dye-sensitized photocurrents enhancement by gold nanoparticles. Reproduced from [50] with permission. Copyright 2013, American Chemical Society

ing electrons of dye molecules more efficiently than the incident far-field light. The electrons will then transfer to the conduction band of TiO_2 and produce photocurrent.

Figure 7.40 shows the calculated localized electric fields around the spherical Au NPs of 15, 40, 100, and 150 nm diameters by the FDTD method. Three typical structures, i.e., the isolated NP, coupled NPs with positive and negative distances, are compared. In the case of isolated NPs, 40 nm NPs exhibit the most intense localized electric fields. However, in the case of dense NP ensembles, plasmon coupling effects contribute to the photocurrent enhancement and 100 nm NPs exhibit the highest electric field intensity and the greatest enhancement factor. The coupled fields follow the catenary function and can be called catenary plasmons.

In general, the localized field enhancement is dependent on both the NP size and gap distance. As shown in Fig. 7.41a, d, the field enhancement factor for both 40-nm-sized and 100-nm-sized NPs increases as gap becomes smaller. The absorption spectra (extinction-specular reflection forward and backward scattering) were measured by using an integrating sphere, while the photocurrent action spectra (photocurrent ratio of the electrode with Au NPs to that without NPs) were measured by a potentiostat (SI1280B, Solartron). In the experiment, the gap width was controlled by varying the particle occupancy A_{NP}/A_E, where A_{NP} is projected area of Au NPs and A_E is that of the electrode surface. Obviously, both the two spectra show strong enhancement for smaller gap width, and the 100-nm-sized NPs have better responsivity at longer wavelength.

7.6.2 Catenary Model for the Localized Field Enhancement

In this section, a detailed analysis of the catenary plasmons in coupled metallic spherical particles is given [51]. First of all, 3D full-wave simulations were carried to obtain the electric field distribution within the gap of spherical nanoparticle pair,

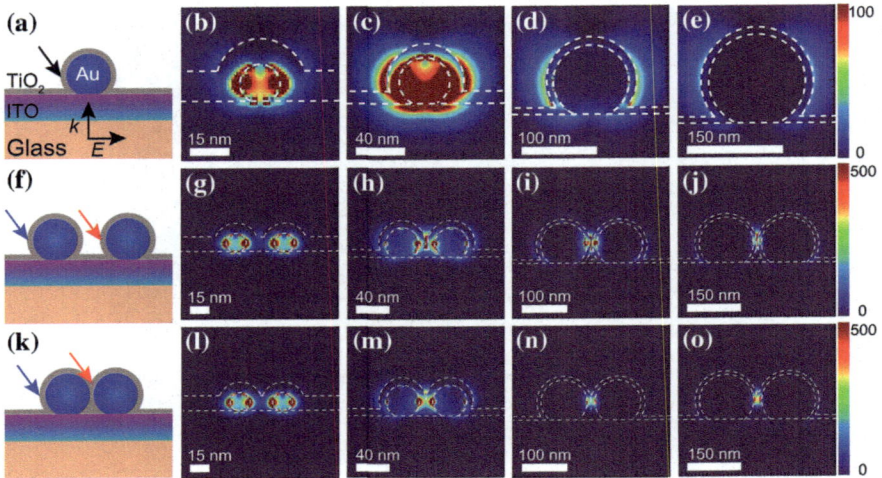

Fig. 7.40 Models of **a** an isolated Au NP, **f** coupled NPs with 2 nm gap, and **k** coupled NPs with 2 nm gap coated with TiO₂ (10 nm thick). **b–e, g–j**, and **l–o** The corresponding electric field distributions calculated for NPs of 15, 40, 100, and 150 nm diameter at the wavelength that gives the maximum electric field intensity at the arrows. Reproduced from [50] with permission. Copyright 2013, American Chemical Society

where the Au–Au nanosphere pair with a diameter D of 50 nm is modeled. As shown in Fig. 7.42a, S represents the interparticle edge-to-edge separation gap and the surrounding medium of the dimer is air. Considering the fact that the coupling behavior of the nanoparticles is closely related to the polarization of the incident light, here we focus on the polarization along the axis of the dimer. Since the gap between spherical nanoparticle pair is very small, the mesh size surrounding the whole dimer is setting as $\Delta x = \Delta y = \Delta z = 0.5$ nm.

Figure 7.42b shows the normalized scattering spectrum of the metallic sphere pair when the separation gap is $S = 30$ nm, which has a resonant peak at 523 nm. Electric field distribution of the spherical nanoparticle pair along the cross section of the dimer at the peak wavelength is shown in Fig. 7.42c. Obviously, the field distribution is symmetric and local field enhancement is obtained due to the evanescent coupling between them. The maximal electric field appears at where the edge-to-edge distance becomes minimum. In order to further check the field distribution within the gap, the electric field distribution on the white curve was retrieved. The hyperbolic cosine catenary function was utilized to fit the electric field profile with $R^2 = 0.9987$.

The coupling coefficient of the nanosphere pair strongly dependents on the separation distance between them. In order to investigate how the coupling properties change with the separation gap, full-wave simulations were carried at separation distances ranging from 2 to 120 nm. Several typical field distributions are shown in Fig. 7.43a–c. The relationship between the maximum field enhancement and the normalized separation distance, i.e., S/D, is plotted in Fig. 7.43d (dotted line). Obviously,

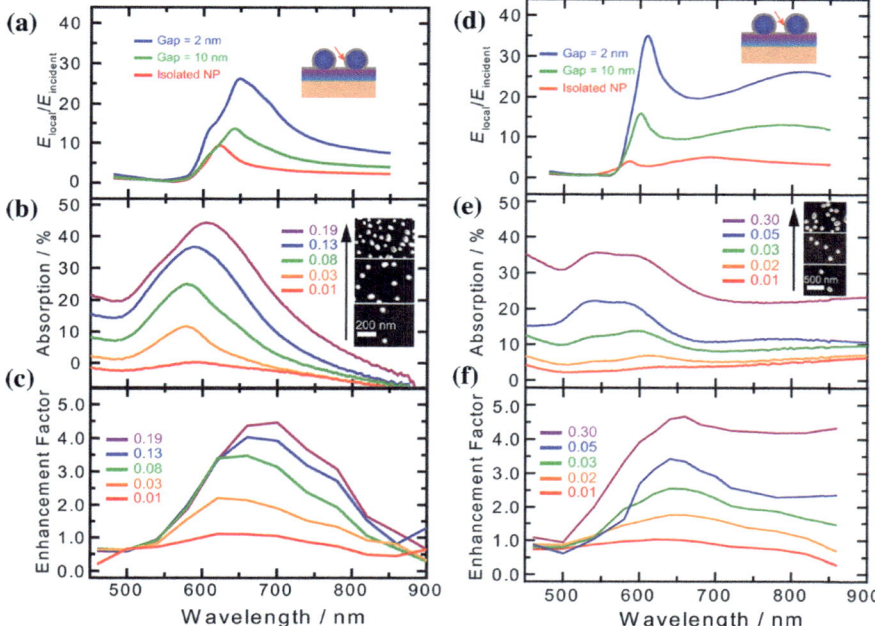

Fig. 7.41 **a** Spectra of the electric field intensity at outside and inside of coupled TiO$_2$-coated 40 nm Au NPs (simulated). The data for isolated NPs is also shown. **b, c** Absorption spectra and photocurrent action spectra of photocurrent enhancement factor for the TiO$_2$-coated 40 nm Au NP electrodes with different particle occupancies (A$_{NP}$/A$_E$ = 0.01–0.19) (experimental). **d–f** Similar to **a–c** but with 100 nm Au NPs. Reproduced from [50] with permission. Copyright 2013, American Chemical Society

the local field enhancement decay with the increasing separation distance. Generally, it is thought that the local field enhancement follows exponential attenuation expressed as

$$f(x) = a \exp(-bx). \tag{7.6.1}$$

However, inspired by low-dimensional Coulomb's law of static electric fields, here a refined model defined as

$$f(x) = a/x^b \tag{7.6.2}$$

is utilized to model the relationship between them. From the fitting results, one can find that although both of the two functions can approach the simulation results well, the modified model has much higher fitting precision. Note that when the gap width is too small, quantum effects such as Landau damping must be considered [52], which set a limit for the local field enhancement.

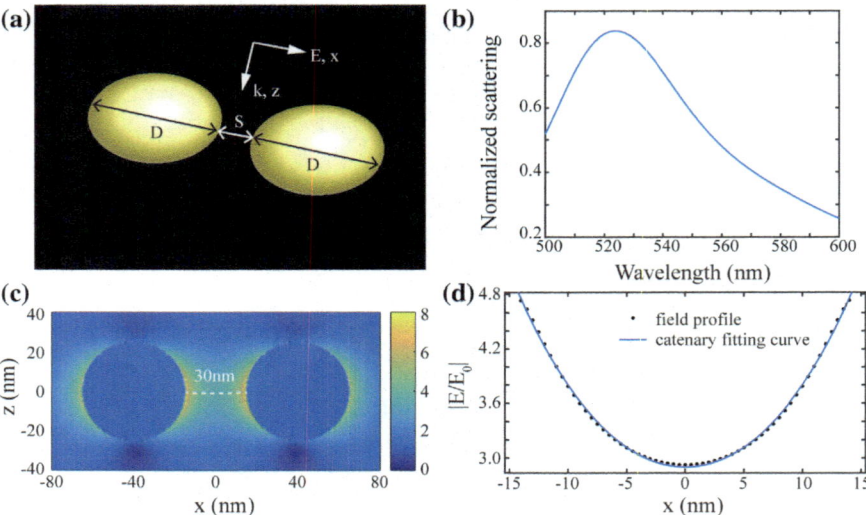

Fig. 7.42 **a** Schematic of the Au–Au dimer. **b** Scattering spectrum of the coupled Au–Au dimer with separation distance $S = 30$ nm. **c** Simulated electric field distribution of the spherical nanoparticle pair along the cross section of the dimer at the peak wavelength of 523 nm. **d** Retrieved field profile on the white curve in (**c**), which is well fitted by catenary function with $a = 1.45$, $1/b = 0.07708$. Reproduced from [51] with permission. Copyright 2019, Springer Science + Business Media

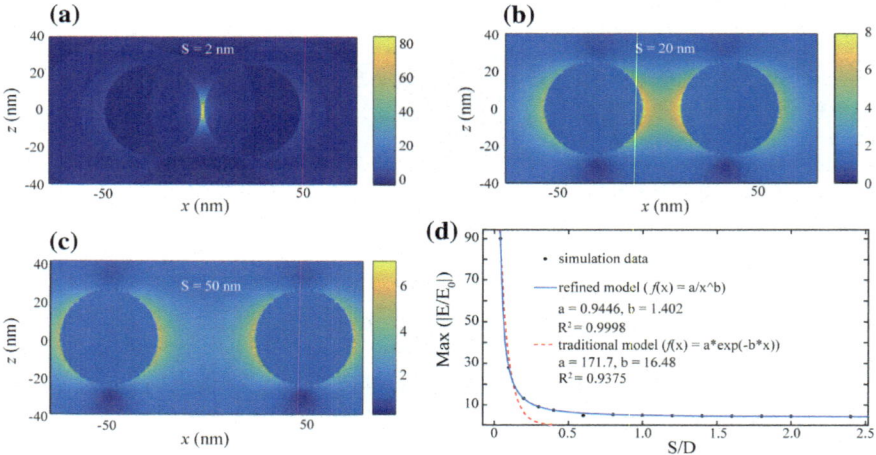

Fig. 7.43 Simulated electric field distribution of the spherical nanoparticle pair along the cross section of the dimer at the corresponding peak wavelength when the separation distance S is equal to **a** 2 nm, **b** 20 nm, and **c** 50 nm. **d** Retrieved field profile on the white curve in (**c**). Different fitting curves were utilized to model the near-field coupling property. Reproduced from [51] with permission. Copyright 2019, Springer Science + Business Media

The above catenary model may provide a quick guidance for the design of the coupled plasmonic system. Besides solar cells, biochemical sensors and nonlinear optical devices could also benefit from it [53, 54]. As a final remark, it is worthy to mention that the absorption and radiation are two deeply related physical processes, and many interesting phenomena have been found in the optical spectrum, such as the Fano resonance in the Rydberg spectral atomic lines [55] and the electromagnetically induced transparency [56]. Inspired by the fact that molecular and atomic resonance are responsible for many light absorption processes, the complex resonant modes in subwavelength structures provide a new platform for the investigation of optical absorbing materials.

References

1. H.A. Atwater, A. Polman, Plasmonics for improved photovoltaic devices. Nat. Mater. **9**, 205–213 (2010)
2. X. Luo, Subwavelength artificial structures: opening a new era for engineering optics. Adv. Mater. 1804680 (2018)
3. R.W. Wood, On a remarkable case of uneven distribution of light in a diffraction grating spectrum. Proc. R. Soc. Lond. **18**, 269 (1902)
4. D. Maystre, *Theory of wood's anomalies*, ed. by S. Enoch, N. Bonod. Plasmonics (Springer Series in Optical Sciences No. 167, Springer, 2012)
5. M.C. Hutley, D. Maystre, The total absorption of light by a diffraction grating. Opt. Commun. **19**, 431–436 (1976)
6. C. Hu, Z. Zhao, X. Chen, X. Luo, Realizing near-perfect absorption at visible frequencies. Opt. Express **17**, 11039–11044 (2009)
7. T.W. Ebbesen, H.J. Lezec, H.F. Ghaemi, T. Thio, P.A. Wolff, Extraordinary optical transmission through sub-wavelength hole arrays. Nature **391**, 667–669 (1998)
8. H. Liu, P. Lalanne, Microscopic theory of the extraordinary optical transmission. Nature **452**, 728–731 (2008)
9. C. Hu, L. Liu, Z. Zhao, X. Chen, X. Luo, Mixed plasmons coupling for expanding the bandwidth of near-perfect absorption at visible frequencies. Opt. Express **17**, 16745–16749 (2009)
10. C. Wang, P. Gao, Z. Zhao, N. Yao, Y. Wang, L. Liu, K. Liu, X. Luo, Deep sub-wavelength imaging lithography by a reflective plasmonic slab. Opt. Express **21**, 20683–20691 (2013)
11. P. Gao, N. Yao, C. Wang, Z. Zhao, Y. Luo, Y. Wang, G. Gao, K. Liu, C. Zhao, X. Luo, Enhancing aspect profile of half-pitch 32 nm and 22 nm lithography with plasmonic cavity lens. Appl. Phys. Lett. **106**, 093110 (2015)
12. T.V. Teperik, F.J. García de Abajo, A.G. Borisov, M. Abdelsalam, P.N. Bartlett, Y. Sugawara, J.J. Baumberg, Omnidirectional absorption in nanostructured metal surfaces. Nat. Photonics **2**, 299–301 (2008)
13. S.A. Tretyakov, S.I. Maslovski, Thin absorbing structure for all incidence angles based on the use of a high-impedance surface. Microw. Opt. Technol. Lett. **38**, 175–178 (2003)
14. M. Pu, C. Hu, M. Wang, C. Huang, Z. Zhao, C. Wang, Q. Feng, X. Luo, Design principles for infrared wide-angle perfect absorber based on plasmonic structure. Opt. Express **19**, 17413–17420 (2011)
15. T.D. Dao, K. Chen, S. Ishii, A. Ohi, T. Nabatame, M. Kitajima, T. Nagao, Infrared perfect absorbers fabricated by colloidal mask etching of Al–Al$_2$O$_3$–Al trilayers. ACS Photonics **2**, 964–970 (2015)
16. E.D. Palik, *Handbook of Optical Constants of Solids* (Academic press, 1985)

17. M. Pu, X. Ma, X. Li, Y. Guo, X. Luo, Merging plasmonics and metamaterials by two-dimensional subwavelength structures. J. Mater. Chem. C **5**, 4361 (2017)
18. M. Choi, S.H. Lee, Y. Kim, S.B. Kang, J. Shin, M.H. Kwak, K.-Y. Kang, Y.-H. Lee, N. Park, B. Min, A terahertz metamaterial with unnaturally high refractive index. Nature **470**, 369–373 (2011)
19. M. Pu, C. Hu, C. Huang, C. Wang, Z. Zhao, Y. Wang, X. Luo, Investigation of Fano resonance in planar metamaterial with perturbed periodicity. Opt. Express **21**, 992–1001 (2013)
20. G. Dolling, C. Enkrich, M. Wegener, J.F. Zhou, C.M. Soukoulis, S. Linden, Cut-wire pairs and plate pairs as magnetic atoms for optical metamaterials. Opt. Lett. **30**, 3198–3200 (2005)
21. M. Pu, Y. Guo, X. Li, X. Ma, X. Luo, Revisitation of extraordinary Young's interference: from catenary optical fields to spin-orbit interaction in metasurfaces. ACS Photonics **5**, 3198–3204 (2018)
22. N. Liu, L. Langguth, T. Weiss, J. Kastel, M. Fleischhauer, T. Pfau, H. Giessen, Plasmonic analogue of electromagnetically induced transparency at the Drude damping limit. Nat. Mater. **8**, 758–762 (2009)
23. A. Epstein, G.V. Eleftheriades, Huygens' metasurfaces via the equivalence principle: design and applications. J. Opt. Soc. Am. B **33**, A31–A50 (2016)
24. V.A. Fedotov, N. Papasimakis, E. Plum, A. Bitzer, M. Walther, P. Kuo, D.P. Tsai, N.I. Zheludev, Spectral collapse in ensembles of metamolecules. Phys. Rev. Lett. **104**, 223901 (2010)
25. J. Grant, Y. Ma, S. Saha, A. Khalid, D.R.S. Cumming, Polarization insensitive, broadband terahertz metamaterial absorber. Opt. Lett. **36**, 3476–3478 (2011)
26. C. Long, S. Yin, W. Wang, W. Li, J. Zhu, J. Guan, Broadening the absorption bandwidth of metamaterial absorbers by transverse magnetic harmonics of 210 mode. Sci. Rep. **6**, 21431 (2016)
27. Y. Guo, L. Yan, W. Pan, B. Luo, X. Luo, Ultra-broadband terahertz absorbers based on 4 × 4 cascaded metal-dielectric pairs. Plasmonics **9**, 951–957 (2014)
28. Q. Feng, M. Pu, C. Hu, X. Luo, Engineering the dispersion of metamaterial surface for broadband infrared absorption. Opt. Lett. **37**, 2133–2135 (2012)
29. X. Luo, Principles of electromagnetic waves in metasurfaces. Sci. China-Phys. Mech. Astron. **58**, 594201 (2015)
30. G. Biener, A. Niv, V. Kleiner, E. Hasman, Metallic subwavelength structures for a broadband infrared absorption control. Opt. Lett. **32**, 994–996 (2007)
31. M. Pu, X. Ma, Y. Guo, X. Li, X. Luo, Theory of microscopic meta-surface waves based on catenary optical fields and dispersion. Opt. Express **26**, 19555–19562 (2018)
32. Y. Huang, J. Luo, M. Pu, Y. Guo, Z. Zhao, X. Ma, X. Li, X. Luo, Catenary electromagnetics for ultrabroadband lightweight absorbers and large-scale flat antennas. Adv. Sci. 1801691 (2019)
33. R.J. Langley, E.A. Parker, Equivalent circuit model for arrays of square loops. Electron. Lett. **18**, 294–296 (1982)
34. Y. Wang, X. Ma, X. Li, M. Pu, X. Luo. Perfect electromagnetic and sound absorption via subwavelength holes array. Opto-Electron. Adv. **1**, 180013 (2018)
35. K.N. Rozanov, Ultimate thickness to bandwidth ratio of radar absorbers. IEEE Trans. Antennas Propag. **48**, 1230–1234 (2000)
36. M. Zhang, F. Zhang, Y. Ou, J. Cai, H. Yu, Broadband terahertz absorber based on dispersion-engineered catenary coupling in dual metasurface. Nanophotonics **8**, 117–125 (2019)
37. M. Pu, P. Chen, Y. Wang, Z. Zhao, C. Huang, C. Wang, X. Ma, X. Luo, Anisotropic meta-mirror for achromatic electromagnetic polarization manipulation. Appl. Phys. Lett. **102**, 131906 (2013)
38. Y. Guo, Y. Wang, M. Pu, Z. Zhao, X. Wu, X. Ma, C. Wang, L. Yan, X. Luo, Dispersion management of anisotropic metamirror for super-octave bandwidth polarization conversion. Sci. Rep. **5**, 8434 (2015)
39. Y.D. Chong, L. Ge, H. Cao, A.D. Stone, Coherent perfect absorbers: time-reversed lasers. Phys. Rev. Lett. **105**, 053901 (2010)
40. W. Wan, Y. Chong, L. Ge, H. Noh, A.D. Stone, H. Cao, Time-reversed lasing and interferometric control of absorption. Science **331**, 889–892 (2011)

41. D.G. Baranov, A.E. Krasnok, T. Shegai, A. Alù, Y.D. Chong, Coherent perfect absorbers: linear control of light with light. Nat. Rev. Mater. **2**, 17064 (2017)
42. C. Yan, M. Pu, J. Luo, Y. Huang, X. Li, X. Ma, X. Luo, Coherent perfect absorption of electromagnetic wave in subwavelength structures. Opt. Laser Technol. **101**, 499–506 (2018)
43. B.E.A. Saleh, M.C. Teich, *Fundamentals of Photonics*, 2 edn. (Wiley, 2007)
44. M. Pu, Q. Feng, M. Wang, C. Hu, C. Huang, X. Ma, Z. Zhao, C. Wang, X. Luo, Ultrathin broadband nearly perfect absorber with symmetrical coherent illumination. Opt. Express **20**, 2246–2254 (2012)
45. M. Pu, Q. Feng, C. Hu, X. Luo, Perfect absorption of light by coherently induced plasmon hybridization in ultrathin metamaterial film. Plasmonics **7**, 733–738 (2012)
46. W. Woltersdorff, Über die optischen Konstanten dünner Metallschichten im langwelligen Ultrarot. Z. Für Phys. Hadrons Nucl. **91**, 230–252 (1934)
47. S. Li, J. Luo, S. Anwar, S. Li, W. Lu, Z.H. Hang, Y. Lai, B. Hou, M. Shen, C. Wang, Broadband perfect absorption of ultrathin conductive films with coherent illumination: superabsorption of microwave radiation. Phys. Rev. B **91**, 220301(R) (2015)
48. S. Li, Q. Duan, S. Li, Q. Yin, W. Lu, L. Li, B. Gu, B. Hou, W. Wen, Perfect electromagnetic absorption at one-atom-thick scale. Appl. Phys. Lett. **107**, 181112 (2015)
49. M.-G. Kang, T. Xu, H.J. Park, X. Luo, L.J. Guo, Efficiency enhancement of organic solar cells using transparent plasmonic Ag nanowire electrodes. Adv. Mater. **22**, 4378 (2010)
50. T. Kawawaki, Y. Takahashi, T. Tatsuma, Enhancement of dye-sensitized photocurrents by gold nanoparticles: effects of plasmon coupling. J. Phys. Chem. C **117**, 5901–5907 (2013)
51. X. Ma, Y. Guo, M. Pu, X. Li, X. Luo, Refined model for plasmon ruler based on catenary shaped optical fields. Plasmonics (2019)
52. J. Khurgin, W.-Y. Tsai, D.P. Tsai, G. Sun, Landau damping and limit to field confinement and enhancement in plasmonic dimers. ACS Photonics **4**, 2871–2880 (2017)
53. S.S. Aćimović, M.P. Kreuzer, M.U. González, R. Quidant, Plasmon near-field coupling in metal dimers as a step toward single-molecule sensing. ACS Nano **3**, 1231–1237 (2009)
54. H. Aouani, M. Rahmani, M. Navarro-Cia, S.A. Maier, Third-harmonic-upconversion enhancement from a single semiconductor nanoparticle coupled to a plasmonic antenna. Nat. Nanotechnol. **9**, 290–294 (2014)
55. A.E. Miroshnichenko, S. Flach, Y.S. Kivshar, Fano resonances in nanoscale structures. Rev. Mod. Phys. **82**, 2257–2298 (2010)
56. M. Fleischhauer, A. Imamoglu, J.P. Marangos, Electromagnetically induced transparency: optics in coherent media. Rev. Mod. Phys. **77**, 633–673 (2005)

Chapter 8
Catenary Optical Fields for Thermal Emission Engineering

Abstract In this chapter, the basic concepts and examples of thermal emission engineering with catenary optical fields are discussed. First, it is shown that the near-field coupling featured by catenary function can be used to break the far-field limit on thermal radiation. Second, by leveraging the complex catenary optical fields in strongly coupled subwavelength structures, many of the radiation properties such as coherence, spectral and polarization selectivity could be readily controlled.

Keywords Catenary electromagnetics · Thermal emission · Metamaterials

8.1 Introduction

Thermal radiation is one important origin of electromagnetic wave. It is well-known that every physical body with nonzero temperature would spontaneously emit and absorb electromagnetic energy. In order to investigate the law of thermal radiation regardless of the body's shape or composition, an idealized physical body, i.e., the blackbody was proposed by Gustav Kirchhoff in 1806s. Ideally, a black body is a standard object that can absorb all incident EM radiation, regardless of frequency or angle of incidence. In Kirchhoff's original definition, black body is described as [1]:

> …the supposition that bodies can be imagined which, for infinitely small thicknesses, completely absorb all incident rays, and neither reflect nor transmit any. I shall call such bodies *perfectly black*, or, more briefly, *black bodies*.

Note that Kirchhoff treated the blackbody as infinitely thin, which is not physical owing to the fact that anybody would have enough thickness to absorb incident light, as pointed out by Planck [2]. This requirement was recently rigorously demonstrated on the basis of electromagnetic theory and causality relation by Rozanov [3]. Consequently, in the modern textbook, the term "infinitely small thicknesses" has been dropped.

No matter how the blackbodies are constructed, they would radiate energy in the form of electromagnetic wave. Planck's Radiation Law gives the spectral emittance of the blackbody as a function of wavelength and temperature [4]:

© Springer Nature Singapore Pte Ltd. 2019
X. Luo, *Catenary Optics*, https://doi.org/10.1007/978-981-13-4818-1_8

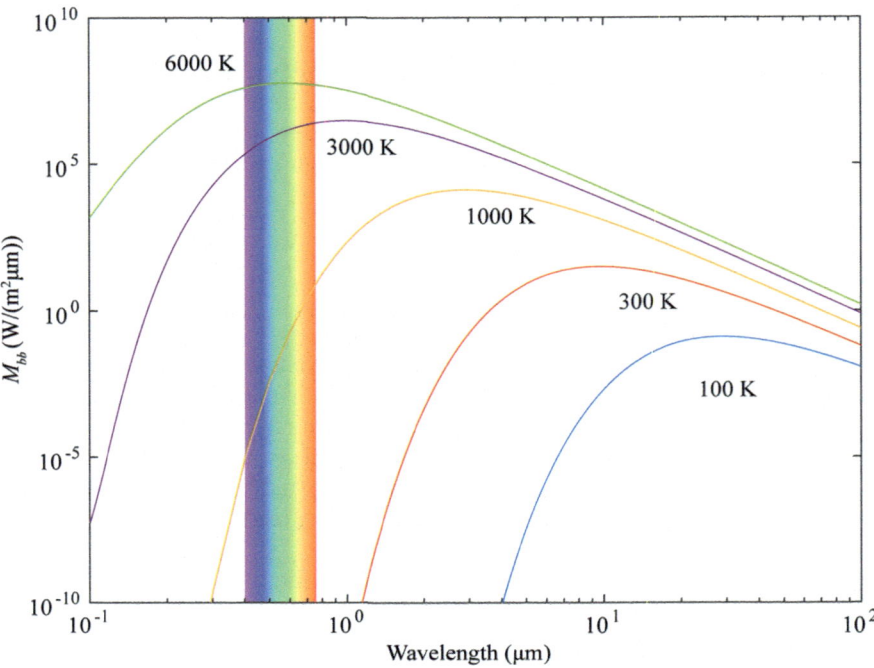

Fig. 8.1 The radiation spectrum of blackbody under different temperature. The color bar represents the visible spectral regime

$$M_{bb}(\lambda, T) = \frac{c_1}{\lambda^5 \left(e^{c_2/\lambda T} - 1 \right)}, \qquad (8.1.1)$$

where c_1 is the first radiation constant (3.7415×10^{-16} W m^4), c_2 the second radiation constant (1.4388×10^{-2} m K), T the absolute temperature (K), λ is the wavelength (m). For a Lambertian source which emits or reflects a radiance that is independent of angle, M is related to the spectral radiance by

$$M = \int_0^{2\pi} d\phi \int_0^{\pi/2} L \, \cos\theta \, \sin\theta d\theta = \pi L. \qquad (8.1.2)$$

Obviously, Planck's radiation law sets a fundamental limit on the emittance of materials. Traditional approaches to increase the emittance rely on the increase of temperature. Once temperature is fixed, the maximal radiation is also limited. As shown in Fig. 8.1, when the temperature is increased from 300 to 3000 K, the peak radiance increased by more than 10^5 times.

The emissivity of blackbody sets a maximal limit for any real radiators. For flat materials, the emissivity can be obtained from the absorption coefficient via the

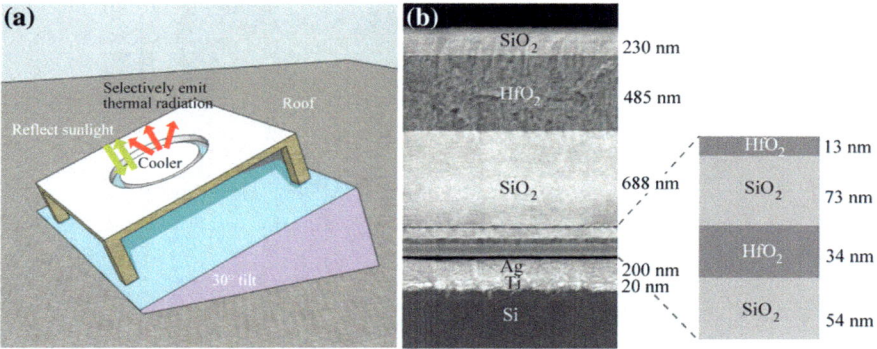

Fig. 8.2 Rooftop apparatus and photonic radiative cooler. **a** Three-dimensional schematic of the apparatus and radiative cooler, showing the general mode of operation of the radiative cooler. The apparatus is designed to minimize conductive and convective heat exchange to the cooler. **b** SEM image of the photonic radiative cooler that is designed, implemented and tested in the experiments. It consists of seven layers of HfO_2 and SiO_2 on top of 200 nm of Ag, a 20-nm-thick Ti adhesion layer, and a 750-mm-thick, 200-mm-diameter Si wafer substrate. Reproduced from [5] with permission. Copyright 2014, Springer Nature

Fresnel's formula. Here, we present some common characteristics of the thermal emission. First, the manipulation of infrared radiation is of particular importance for thermal control, which may be used for passive cooling of buildings and satellites [5–7]. As shown in Fig. 8.2, Raman et al., introduced a simple design based on one-dimensional photonic films that is amenable to large-scale fabrication [5]. When exposed to direct sunlight exceeding 850 W/m^2 on a rooftop, the photonic radiative cooler cools to 4.9 °C below ambient air temperature and has a cooling power of 40.1 W/m^2 at ambient air temperature. Another example of this application is based on SiO_2 microspheres and polymethylpentene. Both the materials are highly transparent in the visible band and absorptive in the thermal infrared regime [8]. Experiments show that the hybrid structure has a noon-time radiative cooling power of 93 W/m^2 under direct sunshine, which is a bit larger than the value of the multilayer design [5].

Note that thermal radiation in the far infrared band is not the only aspect for thermal control. Under sunlight illumination, the near-infrared absorption could be used to control the temperature. In 2007, Synnefa et al., developed 10 prototype cool-colored coatings using near-infrared reflective color pigments in comparison to color matched, conventionally pigmented coatings [7]. An infrared camera was used to depict differences in the thermal performance of the coatings. Figure 8.3 shows visible and infrared images of the cool and standard black coatings. Although the color is the same, the infrared image demonstrates that the temperature of standard black coating is much higher owing to the additional absorption in the near-infrared band.

Another important property of thermal radiation is that its polarization is dependent on material and observation direction, through which the objects can be dis-

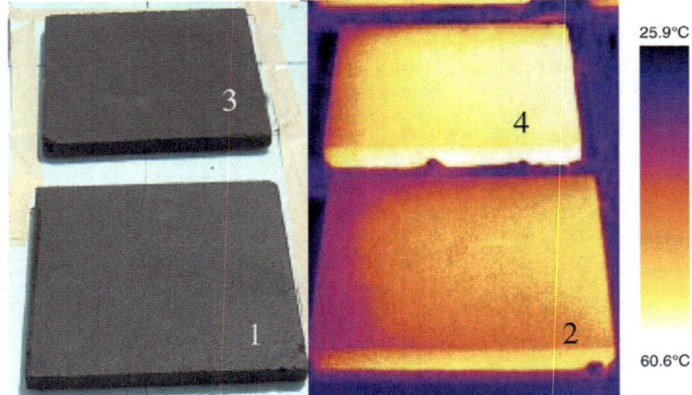

Fig. 8.3 Visible and infrared images of cool (1, 2) and standard (3, 4) black coatings. Reproduced from [7] with permission. Copyright 2006 Elsevier Ltd

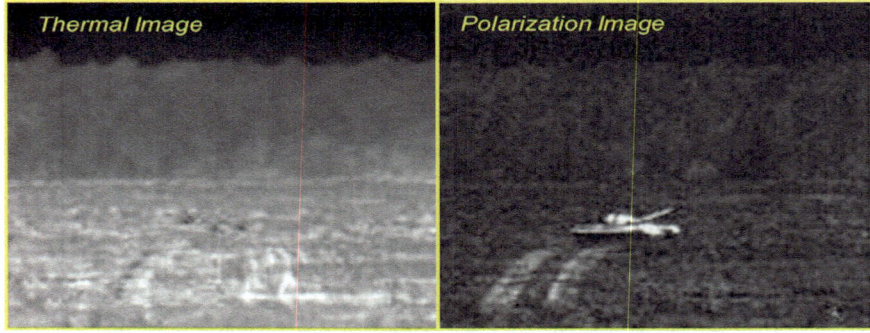

Fig. 8.4 Comparison of thermal and polarization image taken using a long wave infrared (LWIR) imaging polarimeter. Reproduced from [10] with permission. Copyright 2014, SPIE

criminated from backgrounds with even equal thermal radiation intensity. As early as 1926, Worthing has demonstrated that the polarization of thermal radiation of metal is highly dependent on the radiation angle, which was called a deviation from Lambert's law [9]. Figure 8.4 shows two example images to demonstrate this from a military application. These images are taken late at night after the vehicle has realized thermal equilibrium with its background. Since the temperature of the tank equals the temperature of the background and their emissivity is similar, the thermal image shows that the thermal contrast is essentially zero. In the polarization image, however, the tank is strikingly visible because the tank reflects and emits polarization states different from the ground [10].

Note that the wavelength of infrared radiation is much longer than the visible light, so a much rough surface may reflect infrared emission well. For example, varnished wood has very smooth surfaces, thus specular reflections would occur, in particular for large angle of incidence. According to Fresnel's equations, the reflection is highly polarized, thus could be eliminated with a polarizer.

8.2 Beyond Planck's Thermal Radiation Law

Planck's law describes the maximum heat radiation between two surfaces in vacuum. Apparently, the Planck's law sets an upper limit for any radiator at thermal equilibrium, whatever its chemical composition or macroscopic surface structure. While the validity of this law has been argued since its proposition and even Planck himself recognized that the law may not valid when the characteristic length scales are comparable to the wavelength of thermal radiation. This can be seen in Plank's book *The Theory of Heat Radiation* [2]:

> it will be assumed that the linear dimensions of all parts of space considered, as well as the radii of curvature of all surfaces under consideration, are large compared with the wave lengths of the rays considered. With this assumption we may, without appreciable error, entirely neglect the influence of diffraction caused by the bounding surfaces, and everywhere apply the ordinary laws of reflection and refraction of light.

Obviously, Planck has omitted the diffraction effect, which dominates the light—matter interaction when structures' characteristic dimension is comparable to the wavelength. With vectorial diffraction, surface modes such as surface plasmon polaritons, surface-phonon polaritons, or adsorbate vibrational modes, may exert non-negligible effects. From the point of near-field view, it is acknowledged that surface waves can be excited thermally and can generate resonant near-field effects. As the distance is smaller than the characteristic thermal wavelength, the effect of near-field radiation (i.e., the tunneling effect assisted by evanescent waves) may become important [11]. Since the characteristic thermal wavelength is large for lower temperature, the near-field radiation may become much stronger. For instance, the wavelength corresponding to a temperature of 273 K is 8.4 μm, which is rather large for microelectromechanical systems (MEMS) and Micro-optical-electromechanical Systems (MOEMS). We note that Planck's law can also be exceeded in the meso-field, where diffraction rather than evanescent wave dominates the physical process [12].

The theoretical foundation of near-field radiation is the fluctuational electrodynamics theory [13]. In 1971, Polder et al., employed this method to study near-field radiation between closely spaced bodies [14]. In 1999, John Pendry proposed a macroscopic theory to elucidate that it is the evanescent wave which enhances radiative heat transfer between two surfaces separated by a gap [15]. It is assumed that the mechanism of radiative heat exchange between surfaces is due to freely propagating modes and evanescent photon tunneling modes as shown in Fig. 8.5. By using geometric laws of reflection and refraction, Planck's theory only considers the propagating waves and ignores the contribution of evanescent waves. By employing rigorous calculations, Pendry pointed out that the heat transport across the vacuum is not independent of the distance between two parallel surfaces when the distance is

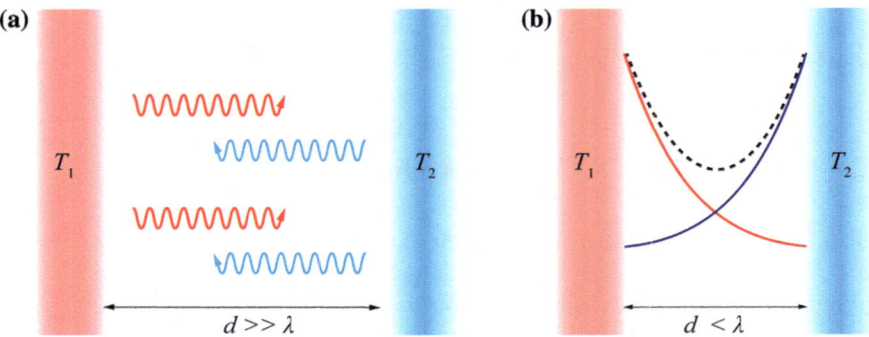

Fig. 8.5 **a** Conventional radiative transfer based on propagating waves. **b** Photon tunneling via evanescent states. The latter mechanism dominates at short distances. The field distributions for evanescent waves fulfil typical hyperbolic cosine catenary functions. Adapted from [15] with permission. Copyright 1999, IOP Publishing Ltd

smaller than the characteristic thermal wavelength. Owing to the photon tunneling induced by evanescent waves, the heat flow to one surface can be hugely enhanced compared with that available from blackbody radiation at the same temperature. Practically, this near-field radiation enhancement can be employed for many applications such as lithography and surface modification [15].

Here, we present the calculating methods given by Pendry [15]. Consider two parallel flat surfaces separated by a distance d, which have reflection coefficients $R_1(k, \omega)$ and $R_2(k, \omega)$. Suppose that surface 1 radiates a transverse magnetically (TM or p)-polarized electromagnetic wave, the electric field may be written as:

$$\mathbf{E}_p = E_{0p}\hat{\mathbf{K}}_p^+ \exp(i\mathbf{k} \cdot \mathbf{r}_\parallel - \alpha z), \tag{8.2.1}$$

where

$$\alpha = +\sqrt{k^2 - \omega^2 c_0^{-2}} \tag{8.2.2}$$

is the attenuation factor, while

$$\hat{\mathbf{K}}_P^\pm = \frac{i c_0 \alpha}{\omega k}\left[k_x, \; k_y, \; \pm i k^2 \alpha^{-1} \right] \tag{8.2.3}$$

is the polarization vector normalized to unity. In the evanescent regime, the reflection from the second surface, and further contributions from multiple reflections between the two surfaces modify the wave field:

$$E'_p = E_{0p}\left[\hat{\mathbf{K}}_p^+ \exp(i\mathbf{k} \cdot \mathbf{r}_\parallel - \alpha z) + R_{2p}(k, \omega)\hat{\mathbf{K}}_p^- \exp(i\mathbf{k} \cdot \mathbf{r}_\parallel + \alpha z - 2\alpha d) \right]$$
$$\times \left[1 - R_{1p}(k, \omega)R_{2p}(k, \omega) \right]^{-1}. \tag{8.2.4}$$

Calculating the Poynting vector for this field gives the flow of energy:

$$\dot{q}(k, \omega) = \frac{2|E_{0p}|^2}{\omega\mu_0}\alpha e^{-2\alpha d}\frac{\mathrm{Im}R_{2p}(k, \omega)}{\left|1 - R_{1p}(k, \omega)R_{2p}(k, \omega)e^{-2\alpha d}\right|^2}. \tag{8.2.5}$$

By using electromagnetic Green function to calculate the density of states outside the first surface, and populating each state with energy

$$\hbar|\omega|\left[\frac{1}{2} + \frac{1}{\exp(\beta\hbar|\omega|) - 1}\right], \tag{8.2.6}$$

one could obtain

$$|E_{0p}|^2 = \hbar|\omega|\left[\frac{1}{2} + \frac{1}{\exp(\beta\hbar|\omega|) - 1}\right]\frac{\omega\mathrm{Im}R_{1p}(k, \omega)}{\varepsilon_0 c_0^2 \alpha}\frac{2}{\pi}. \tag{8.2.7}$$

Hence, there is

$$\begin{aligned}\dot{q}_p(k, \omega) = {} & \frac{8\hbar|\omega|}{\pi}\left[\frac{1}{2} + \frac{1}{\exp(\beta\hbar|\omega|) - 1}\right] \\ & \times e^{-2\alpha d}\frac{\mathrm{Im}R_{1p}(k, \omega)\mathrm{Im}R_{2p}(k, \omega)}{\left|1 - R_{1p}(k, \omega)R_{2p}(k, \omega)e^{-2\alpha d}\right|^2}.\end{aligned} \tag{8.2.8}$$

The expression for tunneling through the transverse electric (TE or s-) polarized channel is exactly the same except that R_p should be replaced by R_s. Nevertheless, for metallic structure, R_p may dominate for large k. The complete evanescent heat transfer from surface 1–2 can be calculated by

$$\dot{Q}_{EV}(T, d) = \sum_k \int_0^\infty \left[\dot{q}_p(k, \omega) + \dot{q}_s(k, \omega)\right]\mathrm{d}\omega. \tag{8.2.9}$$

8.2.1 Near-Field Thermal Radiation with Flat Surfaces

To quantitatively investigate the thermal radiation between two surfaces in a high-vacuum environment, Green's dyadic and the fluctuation dissipation theorem are used to directly calculate the heat transfer coefficient. Figure 8.6 shows that the transfer coefficient could approach 30,000 W/m^2 K for a distance of 10 nm, which is larger than the blackbody limit by a factor of 4 orders. As a comparison, the thermal conduction effect is also investigated. If the gap is filled with air [thermal conductivity is about 0.023 W/(mK)], the equivalent transfer coefficient would be 2.3 × 10^6 W/m^2 K. Although the heat transfer induced by the near-field radiation is still small, it cannot be neglected for high-precision application.

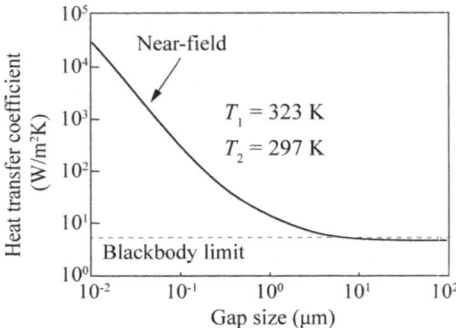

Fig. 8.6 Gap-size dependent heat transfer coefficient calculated by Green's dyadic and fluctuation–dissipation theorem. Reproduced from [16] with permission. Copyright 2008, American Institute of Physics

Chen's group also reported experimental studies on radiative heat flux between two parallel glass surfaces [16, 17]. As shown in Fig. 8.7a, two identical glass optical flat slabs are used as emitter and receiver, respectively. To maintain a gap between the emitter and the receiver, polystyrene microspheres are placed between the two surfaces as spacers. As the distance between the two surfaces is close enough, the radiation enhancement can be attributed to surface-phonon polaritons at the interface between glass and vacuum, which is consistent with Pendry's theory. The measured heat flux data shows that the far-field heat flux stays below the upper limit predicted by Planck's law, while the near-field heat flux clearly exceeds the blackbody upper limit by more than 35% in the entire temperature range (Fig. 8.7b). This experiment is a convincing demonstration that the catenary-shaped distributions of evanescent waves can lead to breakdown of the Planck's blackbody radiation law in the near field.

8.2.2 Near-Field Thermal Radiation with Structured Surfaces

The key of near-field thermal emission enhancement is the contribution of evanescent waves. Consequently, by patterning surfaces with subwavelength structures and exciting evanescent waves more efficiently, the radiative heat flux can be further improved by more than one order of magnitude in certain range of thickness. Figure 8.8 gives the radiative heat flux for two metasurfaces with varying volume filling ratios f, which is defined as W/P and W^2/P^2 for one-dimensional (1D) and two-dimensional (2D) cases (Here, W and P are the width and periodicity of the grating). The unit cell of 2D metasurface contains a square nanopillar, while that of 1D metasurface contains a stripe extending to the infinity in the y direction. Doped

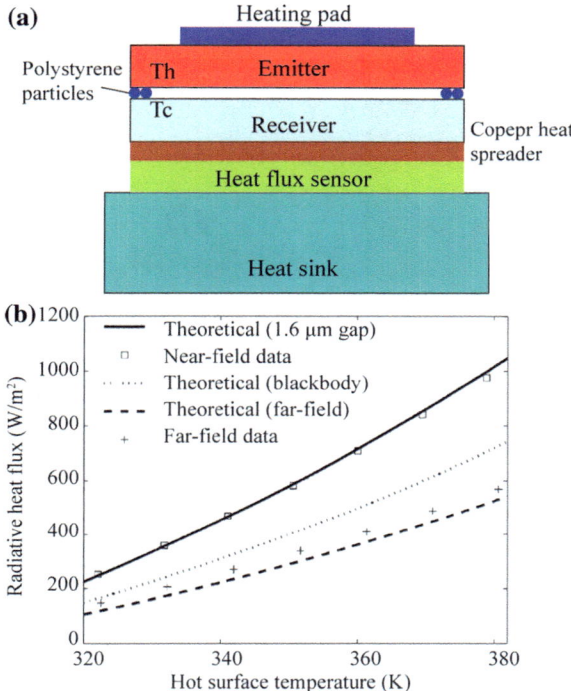

Fig. 8.7 **a** Schematic of the experiment setup. By carefully choosing the number of particles, the gap between the emitter and the receiver can be maintained in micron scale. **b** Measured radiative heat flux. The far-field heat flux stays below the upper limit predicted by Planck's law, while the near-field heat flux clearly exceeds the blackbody upper limit. Reproduced from [16] with permission. Copyright 2008, American Institute of Physics

silicon is used as the base material since SPPs can be thermally excited and the resonance frequency could be tuned by doping. The doping level is set to be 10^{20} cm^{-3}, and the dielectric function is described by Drude model.

The radiative heat flux between thin films (i.e., $f = 1$) of the same thickness is denoted by the dash-dotted line and is 1.86 kW/m^2. This value exceeds that between bulk doped silicon of 1.629 kW/m^2 and is more than 15 times that predicted by Planck's formula. The underlying physical mechanism for the enhancement is attributed to the coupling of SPPs inside the thin film. Patterning the film into 1D metasurface can enhance thermal radiation for all practical volume filling ratios. Interestingly, while the 2D metasurface yields a radiative heat flux higher than that of thin films at moderate filling ratios, it does not support a heat flux as high as that of the 1D metasurface. Beyond $f = 0.36$, 2D patterning will deteriorate the radiative transfer because the effective radiating area is reduced.

The effective medium corresponding to both 1D and 2D metasurfaces is uniaxial. The effective dielectric functions for ordinary ε_O (electric field is perpendicular

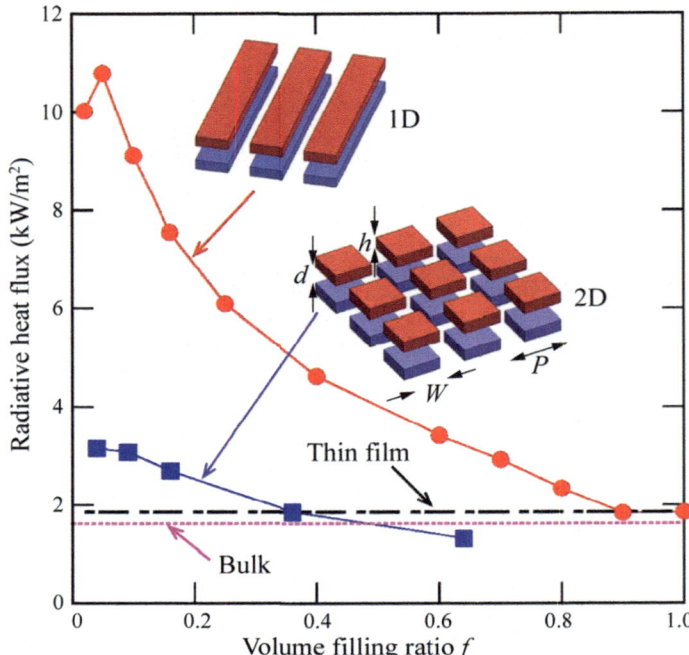

Fig. 8.8 Radiative heat flux as a function of the volume filling ratio at $P = 100$ nm, $d = 100$ nm, and the thickness of film $h = 400$ nm. For all calculations, $T_1 = 310$ K and $T_2 = 290$ K. Inset shows the 1D and 2D metasurfaces. Adapted from [18] with permission. Copyright 2015, American Chemical Society

to the optical axis) and extraordinary waves (electric field is parallel to the optical axis) ε_E can be calculated. The dielectric tensor of 1D and 2D effective medium is diag(ε_E, ε_O, ε_O) and diag(ε_O, ε_O, ε_E), respectively. The calculated values for the real parts of ε_O and ε_E as functions of the angular frequency are plotted in Fig. 8.9 for $f = 0.16$ and at a temperature of 300 K. It is assumed that the slight perturbation of temperature for the two metamaterials does not affect their dielectric functions. At angular frequencies below 2.46×10^{14} rad/s, the dielectric functions of orthogonal directions have opposite signs, implying that the effective material has a hyperbolic dispersion. Therefore, both 1D and 2D metasurfaces are hyperbolic metamaterials. For p-polarized electromagnetic waves with high-k, evanescent waves in conventional elliptic materials become propagating. In other words, the coupling of evanescent waves would induce one particular catenary optical fields which could propagate inside such materials. The evanescent waves in the vacuum are coupled with the propagating high-k modes, leading to broadband high local density of states (LDOS). This may be the basic mechanism for the efficient radiative transfer supported by 1D and 2D metasurfaces. It should be noted that effective medium theory may overpredict the heat flux since the period of the subwavelength structures are

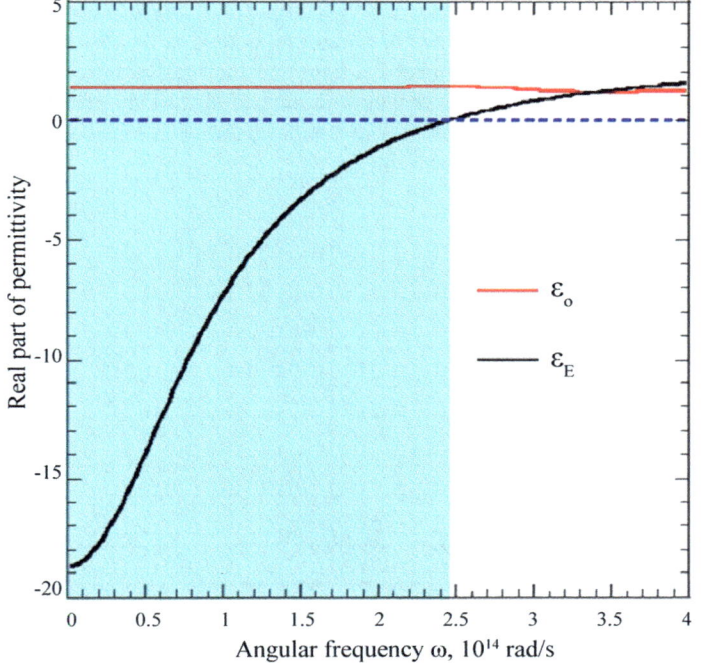

Fig. 8.9 Real part of the effective dielectric function for orthogonal directions for $f = 0.16$, the shaded region supports hyperbolic dispersion. Adapted from [18] with permission. Copyright 2015, American Chemical Society

not infinitely small. Consequently, rigorous analyses are still needed to give more accurate results.

8.2.3 Far-Field Super-Planckian Thermal Radiation

In 2016, Biehs and Ben-Abdallah investigated the super-Planckian thermal emission using the Landauer formalism to deal with radiative heat exchanges between 2 or N objects in the far-field regime [19]. They proved that the flux radiated by indefinite planar media is bounded by the blackbody emission, but there is in principle no upper limit demonstrating such possibility for a super-Planckian thermal emission with finite size systems. In a more recent work, it was shown that the radiative coupling could also lead to super-Planckian thermal emission [12].

For instance, it is well-known that the scattering and absorption cross sections of a small object can be much larger than the geometric cross section in Mie's scattering theory, which means that such object could scatter or absorb much more light than its geometric area can collect [20, 21]. According to Kirchhoff's theory, at thermal equilibrium condition, the emittance of a material equals to its absorbance (defined by the intensity of light), therefore the amount of thermal radiation can be much larger than that defined with the object's geometric size. This effect can also be illustrated by the Poynting vector in a resonant absorber comprised of a metallic patch, dielectric spacer and reflective layer.

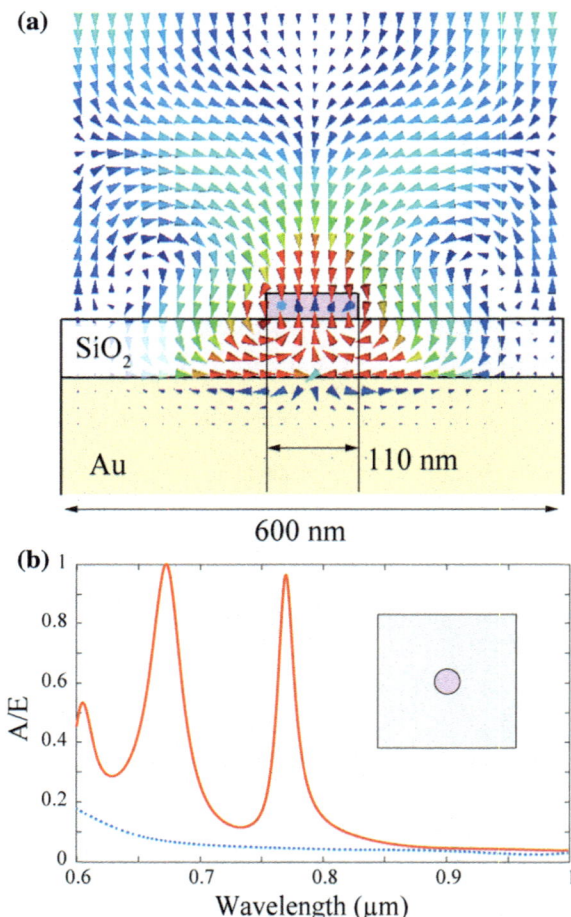

Fig. 8.10 a Illustration of the power flow for a structured absorber comprised of a metallic patch, 70-nm-thick SiO_2 spacer and a gold mirror. The diameter, thickness and period of the patch are 110, 30 and 600 nm. **b** Absorbance/emittance for structures with (solid red curve) and without (dashed blue curve) metallic patches. The inset shows the front view of the unit cell. When normalized to ideal blackbody with area equal to that of patches, the maximal enhancement factor could be as large as 38 at $\lambda = 673$ nm. Adapted from [21] with permission

As shown in Fig. 8.10, although the top patch is very small compared with the unit cell, almost all incident light energy at $\lambda = 673$ nm can be concentrated into the cavity formed between the patch and reflective layer. When the effective absorption/emission area is defined by the patch size, the Planck's law can be violated in the far field. Of course, once the size of radiators becomes much larger than the wavelength, the diffraction effect would be much smaller, thus there would be negligible enhancement over the classic law. Moreover, it should be noted that the integral power emitted at all wavelengths may remain sub-Planckian for any body formed by passive and causal components [22]. Based on the above considerations, in many practical applications such as radiative cooling, the focus of interest is moved from breaking the Planck's limit to other concepts including coherent and spectral-selective thermal emission. As discussed in the following sections, the catenary optical fields are also very useful to control the coherence and spectral-selectivity.

8.3 Coherent Thermal Radiation

A thermal emitting source, such as a black body or the incandescent filament of a light bulb, is often treated as a typical example of an incoherent source and is in sharp contrast to a laser. While a laser is highly monochromatic and directional, a thermal source has a broad spectrum and is usually quasi-isotropic, thus called as a Lambertian radiator. Traditionally, thermal radiators suffered from lack of directivity as the radiation direction cannot be manipulated and the emitted waves are incoherent. By incoherent, it means that different points of a thermal source emit uncorrelated fields that do not interfere. Nevertheless, this may be no longer true when the material supports surface waves, thus the thermally induced random fluctuating currents become correlated. In this section, we show that both surface plasmons and quasi-guided waves in graphene multilayers could be exploited to realize directional thermal emission.

8.3.1 Coherent Thermal Radiation Based on Surface Waves

In 2002, Greffet et al. experimentally validated the effect of surface waves on radiative properties and discussed the physical origin of coherent thermal emission [23]. They demonstrated that by introducing a periodic microstructure into a polar material (SiC), a thermal infrared source can be fabricated that was coherent over large distances compared with the wavelength and radiated in well-defined directions. Narrow angular emission lobes similar to antenna lobes were observed and the emission spectra of the source was depending on the observation angle. The origin of the coherent emission was attributed to the diffraction of surface-phonon polaritons by the grating.

(a)

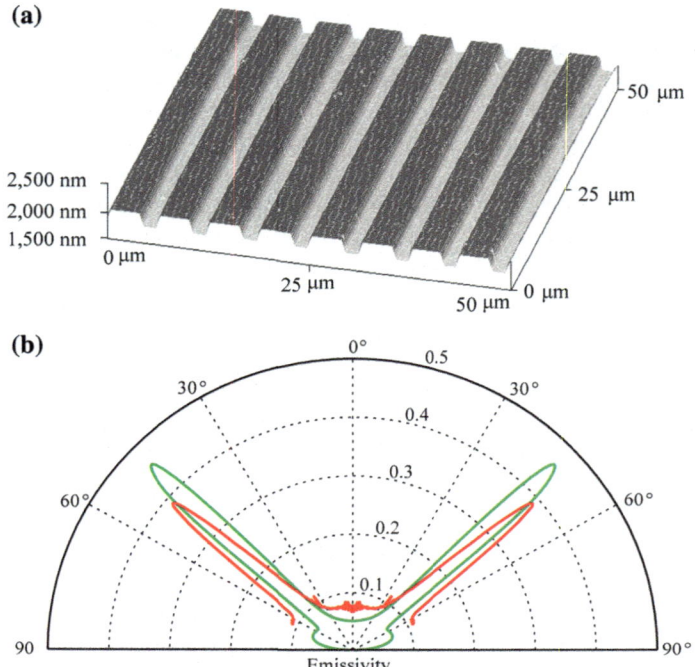

(b)

Fig. 8.11 **a** Image of the grating obtained by atomic force microscopy. **b** Measured (red) and simulated (green) polar plot of the emissivity of the grating at $\lambda = 11.36$ μm for p-polarization. Reproduced from [23] with permission. Copyright 2002, Springer Nature

Figure 8.11a shows the atomic force microscope (AFM) image of the fabricated grating which is optimized to produce a strong peak of emissivity at around $\lambda = 11.36$ μm. The measured and simulated results of the thermal emission in a plane perpendicular to the grating lines are shown in Fig. 8.11b. Clearly, the emission is highly directional and looks similar to the angular pattern of an antenna. It should be noted that the small angular width of the emission pattern is a signature of the local spatial coherence of the source. In order to prove experimentally the role of the surface wave, spectral measurements of the emissivity have been done for both the s- and p-polarizations. The peaks can never be observed for s-polarization nor for p-polarization in the spectral region where surface waves cannot exist.

The angle-selective thermal emission can be understood using the Kirchhoff's theorem. Under thermal equilibrium condition, the emission and the absorption coefficient should equal for any polarization, wavelength and angle. Consequently, angle-selective absorption discussed in Chap. 7 can be directly utilized for this application. Figure 8.12 shows the principle for angle-selective absorption and emission. To match the boundary condition, the following condition should be met

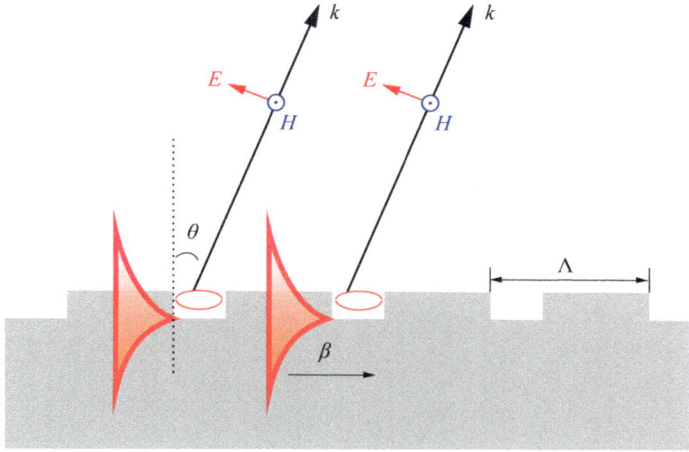

Fig. 8.12 Schematic of surface plasmon-coupled with far-field radiation. The propagation constant should satisfy the wavevector matching condition. The electric field lines of SPPs follow the equal-strength catenary functions. The intensity of gap SPP modes follows ordinary catenary function

$$\beta = k_0 \sin \theta + m \frac{2\pi}{\Lambda}, \qquad (8.3.1)$$

where β is the propagation constant of surface wave, Λ is the period of the grating, m is an integer. After some mathematical manipulation, this angle could be rewritten as

$$\sin \theta = \frac{\beta}{k_0} - m \frac{\lambda}{\Lambda}. \qquad (8.3.2)$$

To further validate the above analysis, the angle-dependent absorption of the infrared SiC grating was measured as shown in Fig. 8.13a. Using Kirchhoff's law, the emission spectra for $\lambda = 11.04\,\mu m$, $\lambda = 11.36\,\mu m$ and $\lambda = 11.86\,\mu m$ are plotted in Fig. 8.13b, which closely resemble that shown in Fig. 8.11. The discrepancy is because of the experimental resolution limit of the direct emissivity measurement.

8.3.2 Coherent Thermal Radiation in Graphene

Graphene is a two-dimensional material with exotic electronic, optical and thermal properties. The optical absorption in monolayer graphene is limited by the fine structure constant $\alpha = e^2/\hbar c \approx 1/137.036$ [24]. Strong enhancement of light absorption and thermal radiation in homogeneous graphene was demonstrated. Interestingly, it was shown that the thermal radiation can be highly directive.

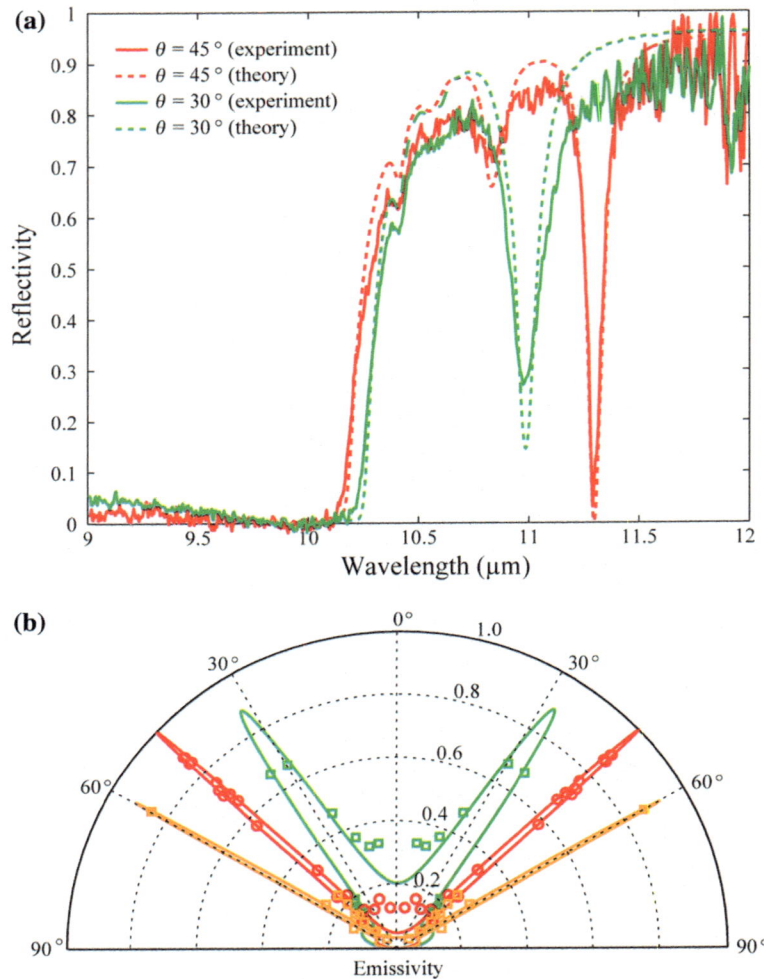

Fig. 8.13 **a** Comparison between measured and calculated spectral reflectivities of a SiC grating at room temperature. **b** Emissivity of a SiC grating in *p*-polarization at different wavelengths. Green, $\lambda = 11.04\,\mu m$; red, $\lambda = 11.36\,\mu m$; yellow, $\lambda = 11.86\,\mu m$. Reproduced from [23] with permission. Copyright 2002, Springer Nature

The surface (two-dimensional) electric conductivity of graphene is highly dependent on the working frequency and chemical potential (or Fermi energy). When there is no external magnetic field, the local conductivity is isotropic, i.e., there is no Hall conductivity. In this case, the conductivity can be approximated for $k_B T \ll \mu_c$, $\hbar\omega$ as [25]:

$$\sigma_{2D}(\omega) \approx \frac{ie^2}{4\pi\hbar} \ln\left[\frac{2|\mu_c| - (\omega + i2\Gamma)\hbar}{2|\mu_c| + (\omega + i2\Gamma)\hbar}\right]$$
$$+ \frac{ie^2 k_B T}{\pi\hbar^2(\omega + i2\Gamma)}\left[\frac{\mu_c}{k_B T} + 2\ln(e^{-\mu_c/k_B T} + 1)\right], \qquad (8.3.3)$$

where $k_B T$ is the thermal energy, μ_c is the chemical potential and Γ is the scattering rate, e, k_B, and $\hbar$ are electron charge, Boltzmann constant, and reduced Plank constant (Dirac constant), respectively. The first term in Eq. (8.3.3) is due to the contribution of inter-band transition and the second term results from intra-band transition. The frequency of transition between these two regimes is dependent on the chemical potential. The scattering rate is assumed to be $\Gamma = 0.43$ meV. The sheet impedance of graphene can be calculated as

$$Z_s(\omega) = \frac{1}{\sigma_{2D}(\omega)}, \qquad (8.3.4)$$

which can also be transformed into traditional form by writing the 3D conductivity as $\sigma_{3D}(\omega) = \sigma_{2D}(\omega)/t$, where $t \approx 0.5$ nm is the thickness of graphene [26, 27]. When light propagates forward and backward inside this lossy material, the interference would form a catenary-like intensity distribution. However, since the thickness is much smaller than the wavelength, the catenary is actually flattened and often not observed.

For a perfect absorber comprised of a graphene layer, a metallic reflection layer, a dielectric spacer with thickness d and permittivity ε, the required sheet impedance for graphene can be written as [28]

$$Y_s = \frac{Z_0}{Z_s} = \cos\theta - i\sqrt{\varepsilon - \sin^2\theta}\cot\left(\sqrt{\varepsilon - \sin^2\theta}kd\right) \qquad (8.3.5)$$

for TE polarization and

$$\frac{Z_0}{Z_s} = \frac{1}{\cos\theta} - i\frac{\varepsilon}{\sqrt{\varepsilon - \sin^2\theta}}\cot\left(\sqrt{\varepsilon - \sin^2\theta}kd\right) \qquad (8.3.6)$$

for TM polarization.

At normal incidence, the above two required impedances may be reduced to be

$$Z_s = \frac{Z_0}{1 - in\cot(nkd)}, \qquad (8.3.7)$$

where n is the refractive index of the dielectric spacer. If there is $Z_s = Z_0 = 377\ \Omega$, the maximal absorption condition becomes $nkd = \pi/2$, corresponding to traditional Salisbury screen (the lossy material can be either carbon or nichrome) [29].

In order to understand the different absorption mechanism at large incidence angle for TE and TM polarizations, the characteristics of Eqs. (8.3.5) and (8.3.6) are analyzed. As the impedance of graphene is larger than Z_0 for frequencies larger than 5 THz and chemical potential less than 200 meV, the prefect absorption condition can only be achieved when

$$kd \approx \frac{1}{\sqrt{\varepsilon - \sin^2 \theta}} \left(\frac{\pi}{2} + m\pi \right). \quad m = 0, 1, 2, 3 \dots \qquad (8.3.8)$$

The corresponding angles are

$$\theta = \arccos \left(\frac{Z_0}{Z_s} \right) = \arccos(377\sigma_{2D}) \qquad (8.3.9)$$

for TE polarization, and

$$\theta = \arccos \left(\frac{Z_s}{Z_0} \right) \qquad (8.3.10)$$

for TM polarization. Obviously, there is no solution for TM polarization because $Z_s > Z_0$. Note that the above discussions are only valid for the graphene–dielectric–metal sandwiched structure. For multilayers, the absorption condition would change correspondingly.

Since graphene possesses exotic optical and thermal properties, it provides an ideal platform for the engineering of thermal radiation. At near-infrared and visible frequencies, the conductivity of monolayer graphene ($\mu_c = 0$) becomes a universal constant equal to $e^2/4\hbar$, which can be obtained from the asymptotic behavior of graphene conductivity. The optical sheet resistance can be calculated as 16.4 kΩ. In numerical simulations, the emission of metal-backed graphene is calculated by the product of emissivity (absorbance) and the blackbody radiation. As an example, the temperature is set as 1500 K and the corresponding radiation center frequency is 155.44 THz ($\lambda = 1.93 \, \mu$m). The radiation angle is then calculated to be $\theta = 88.5°$. Since the absorption at $\theta = 0°, 30°$, and $60°$ is only $0.09, 0.1, 0.17$ for TE polarization (the case for TM polarization is even smaller), the emissivity is highly dependent on the angle, implying a good spatial coherence.

Due to the strong angle dependence of the thermal radiation, it can be utilized as highly directive thermal source. The thickness of the dielectric layer can be changed to achieve different kinds of radiation property such as narrow band single peak radiation (small d) or multifrequency comb-like radiation (large d). Figure 8.14 shows the thermal emission of the structure for different dielectric thicknesses. When the thickness of dielectric layer is set as 50, 5, and 0.5 μm, the corresponding frequency interval between adjacent peaks becomes 2.86, 28.6 and 286 THz, respectively. Furthermore, we note that the thermal radiation property can be adjusted by the chemical potential. Along with the increase of μ_c, the inter-band transition shifts to higher frequency. Due to the abrupt change of conductivity, the emission spectrum will also experience a rapid alteration.

The above discussion is limited for monolayer graphene. In fact, the radiation angle can be changed by altering the layer number of multilayer graphene. Since

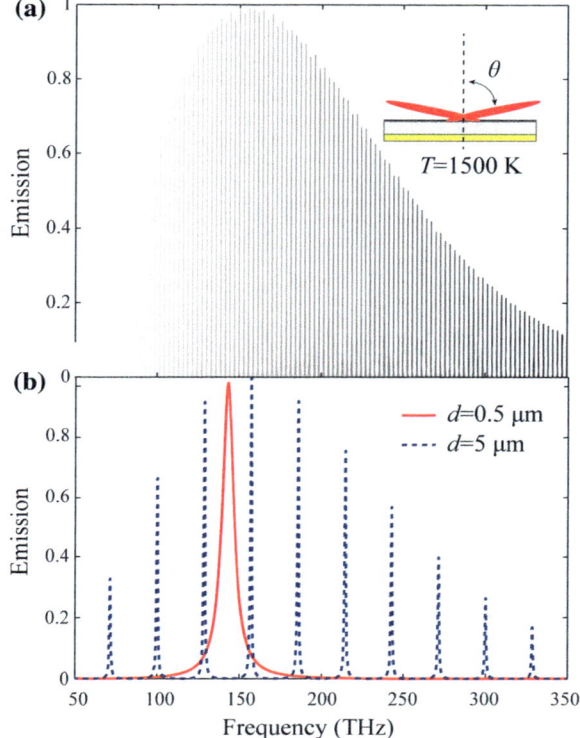

Fig. 8.14 **a** TE-polarized thermal emission of the graphene structure at $\theta = 88.5°$ and $T = 1500$ K for $d = 50$ μm. Inset is the schematic of the radiation pattern. **b** TE-polarized emission spectra for $d = 0.5$ and 5 μm. Reproduced from [28] with permission. Copyright 2013, Optical Society of America

the optical conductivity is proportional to the layer number, the conductivity of bilayer and trilayer graphene would be twice and triple of $e^2/4\hbar$. In Fig. 8.15, the radiation patterns at $f = 155$ THz ($d = 0.46$ μm) for 1, 5, and 10 layers of graphene are calculated using transfer matrix method for different radiation angles. As the increase of the layer number, the radiation angle would decrease. Meanwhile, the radiation at small angle (such as normal direction) will increase and the beam width would be larger. Thus, for practical applications, a trade-off between radiation angle and beam width is required.

The thermal radiation of graphene can hardly be achieved using traditional metal film. For example, the conductivity of nichrome at 150 THz is about 10^5 S/m, thus a thickness of 0.6 nm would be required for the same sheet resistance (16.4 kΩ). Although material with smaller conductivity seems to be possible solution, the requirement for thermal property may not be fulfilled as the temperature is too high.

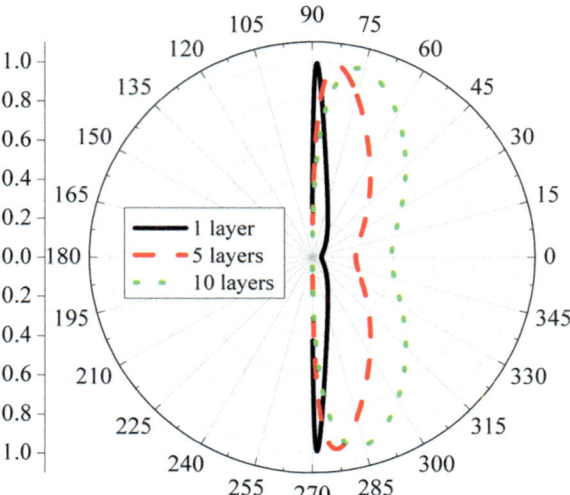

Fig. 8.15 Polar plot of the radiation pattern of graphene with 1, 5, and 10 layers. The sample is placed vertically (in the direction along 90°–270°). Reproduced from [28] with permission. Copyright 2013, Optical Society of America

As a final remark to the graphene absorbers and thermal radiators, it is interesting to investigate the relation between fine constant α, Euler's number ($e = 2.71828$), and catenary function. The Euler's number, when multiplied with the cosine of $1/\alpha$, nearly equals to 1 by a very high accuracy. Maybe this is a coincidence, but we believe there are still many secrets behind these phenomena.

8.4 Perfect Thermal Radiation

Based on Kirchhoff's law, it is well-known that the emissivity of an object can be designed with subwavelength structures in the same way as the perfect absorbers [30–34]. Like high-temperature absorbers, the materials' parameters at specific temperature should be given to predict the correct emissivity [35]. In the following, the role of catenary optical fields in thermal radiation spectra modulation is discussed.

8.4.1 Metamaterials

Metamaterials have been proved to be very flexible in the control of thermal emission spectrum. To reveal the basic physical mechanism, structures operating at single or several discrete wavelengths should be investigated first. Figure 8.16

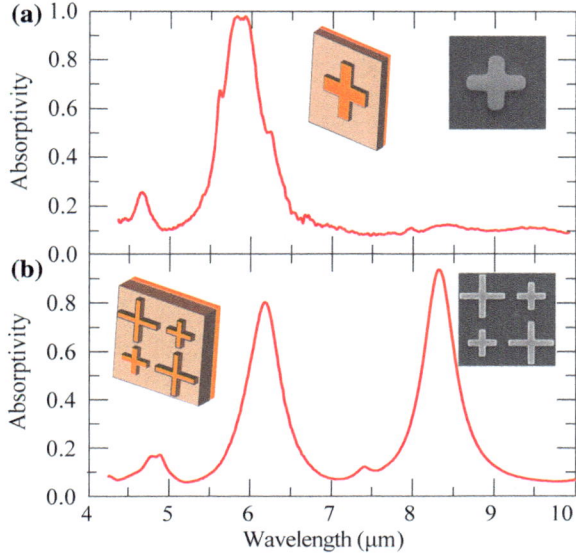

Fig. 8.16 **a** Experimental absorptivity of the single band metamaterial absorber. The period is 3.2 μm, the length and width of the metallic bars are 1.7 and 0.5 μm. **b** Experimental absorptivity of the dual-band metamaterial absorber. The period is 7 μm. The width of all the metallic bars are 0.4 μm. The length of the two different bars are 3.2 and 2 μm. Inset shows 3D views and SEM images of one unit cell for the fabricated single and dual-band absorbers. Reproduced from [31] with permission. Copyright 2011, American Physical Society

shows the typical unit cells of a single-band and double-band infrared metamaterial absorber/emitter, which consists of a cross-shaped resonator, ground plane, and a dielectric layer between them. The thickness of both metal layer is 100 nm and thickness of the Al_2O_3 dielectric spacer is 200 nm. This trilayered metamaterial couples to both the electric and magnetic components of incident electromagnetic waves, thus allowing for minimization of the reflectance at one and two resonant frequencies, respectively.

A Fourier transform infrared microscope was used to characterize the absorptivity. The experimental absorptivity of the two absorbers is shown in Fig. 8.16. The single band absorber has an absorption peak at 5.8 μm with 97% absorption and the dual-band absorber has two absorption bands located at 6.18 and 8.32 μm with 80 and 93.5% absorption, respectively. Unlike a single resonator supporting two modes [36], the two absorption peaks can be separately modulated by tuning the geometric parameters of the metallic crosses. This kind of device can be utilized to match the characteristic spectra of atoms and molecules [37].

The measured emittance of the single and dual-band metamaterial samples is shown in Fig. 8.17. The absolute value emissivity can be characterized through the temperature-dependent emittance of the black carbon, which is used as the blackbody reference emittance at 300 °C. It can be observed that the metamaterial emissivities

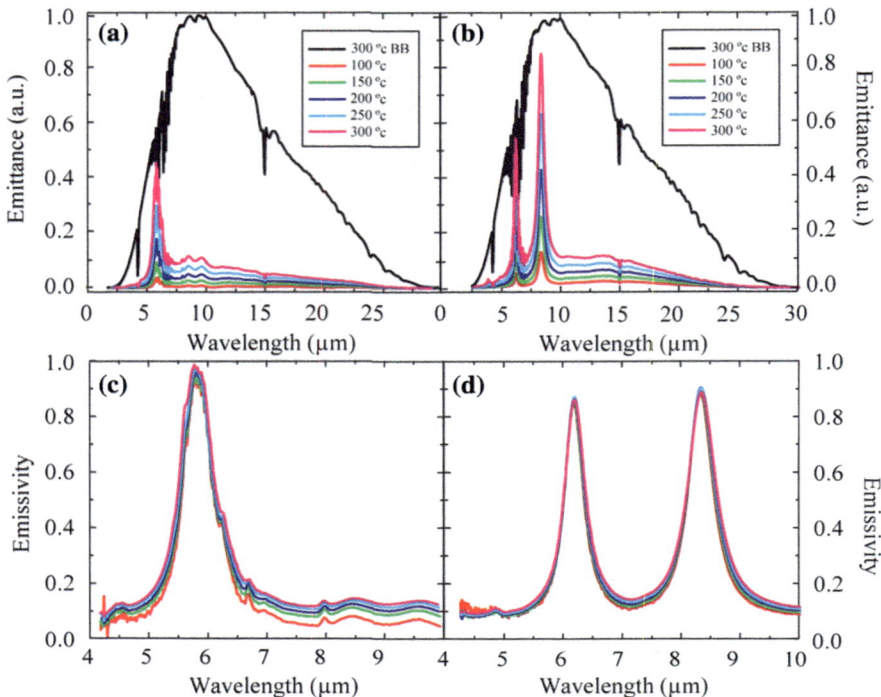

Fig. 8.17 Experimental emittance and normalized emissivity. **a** Emittance of the single band meta-material emitter at five different temperatures and emittance of the blackbody reference at 300 °C. **b** Emittance of the dual-band metamaterial emitter at five different temperatures and emittance of the blackbody reference at 300 °C. **c** Normalized emissivity of single band metamaterial emitter. **d** Normalized emissivity of dual-band metamaterial emitter. Reproduced from [31] with permission. Copyright 2011, American Physical Society

are nearly equal to that of the blackbody at the absorption peaks. In other bands, the emitted energy of the blackbody is significantly greater than that of the metamaterial emitters at the same temperature. In some special cases, this could lead to much higher emittance larger than the Planck's limit. For instance, if both the blackbody and the metamaterial are heated with the same power, the equilibrium temperature for the metamaterial may be much higher than the blackbody, thus higher emittance could be obtained [38].

The emissivity of metamaterial could be calculated by division with the emittance of a perfect blackbody at the same temperature, as shown in Fig. 8.17c, d. The temperature dependence of the emissivity may be attributed to the temperature-dependent material parameters and thermal expansion of the dielectric spacer layer.

Narrowband thermal emitters are also important in solar thermo-photovoltaic (STPV) applications. Figure 8.18 shows an operational principle combining a broad-band absorber with a perfect selective emitter in the opposite side [35]. Since the

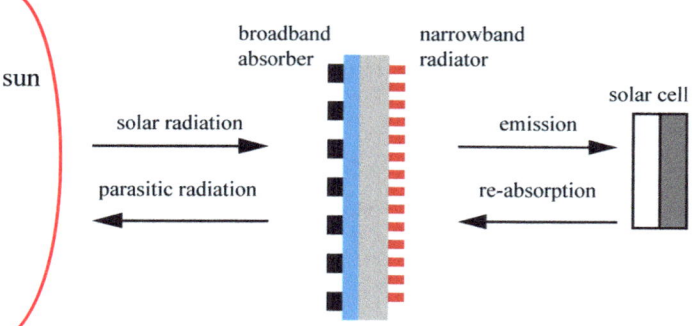

Fig. 8.18 The principle geometry of the harvesting/re-radiating system. Reproduced from [35] with permission. Copyright 2013, Optical Society of America

energy gap of the STPV cell is typically larger than 0.3 eV, the operating temperature of STPV systems should be at least 750 K according to Planck's law. As the temperature is so high, selecting a kind of thermal resistive material to avoid material melting is of great importance. Since tungsten could withstand high temperature (as high as 3650 K), it has been widely used in high-temperature applications. The absorber is heated via absorbing solar radiation and the radiation emitted by the emitter converts to electrical energy through a solar cell. The temperature at equilibrium state is set to be 966 K, corresponding to solar cell with an energy gap of 0.4 eV. Thus the emission peak for narrowband absorber is at 3.1 μm.

As illustrated in Chap. 7, narrowband absorber/emitter can be realized by a metal–insulator–metal (MIM) triple layered metamaterial [36]. Figure 8.19a depicted one design where the top layer is a tungsten circular plate separated from the bottom tungsten substrate by a fused silica interlayer [35]. The optimized absorption spectrum is plotted as the black dotted curve as shown in Fig. 8.19g. The corresponding electromagnetic fields distribution is shown in Fig. 8.19e, characterized by the catenary optical fields and antiparallel currents excited on the two separated tungsten layers.

Although the MIM structure performs good at normal temperature, it cannot form a sharp absorption peak since the electron scattering rate of tungsten at 966 K is quite large. To overcome this, the top layer is flattened to decrease surface scattering. Meanwhile, the resonant cavity is preserved as shown in Fig. 8.19b–d. The geometrical parameters are optimized for $T = 966$ K: $h = 0.9$ μm, $h_d = 0.72$ μm, $h_m = 0.27$ μm, $p = 1.17$ μm, $r_s = 0.495$ μm, $r_i = 0.225$ μm. Figure 8.19f describes the distribution of the electric fields, showing stronger horizontal coupling than vertical coupling. Since the electron scattering at the upper surface becomes much weaker, the quality-factor becomes larger as indicated in Fig. 8.19g. Consequently, the emission bandwidth is greatly reduced compared with the common MIM structure.

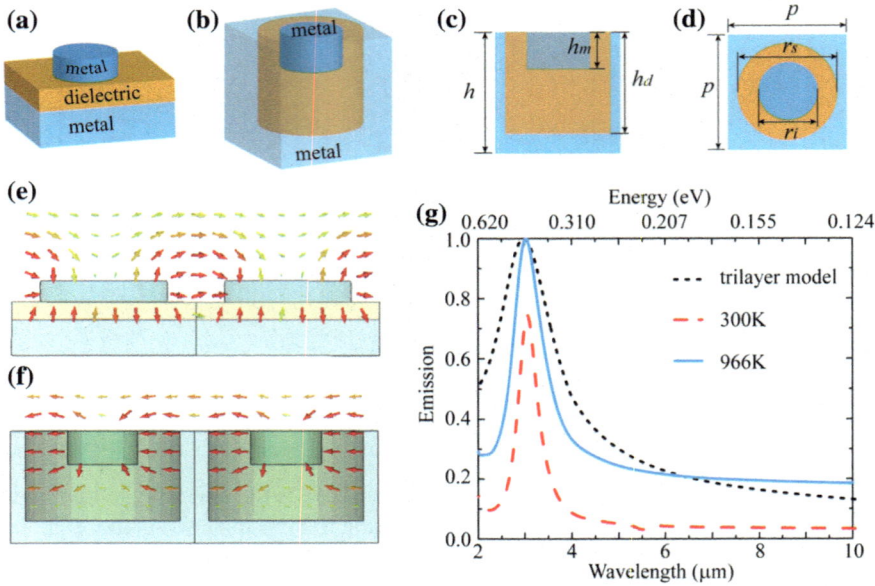

Fig. 8.19 Design of the high-temperature narrow band emitter. **a** Perspective view of the traditional trilayer structure unit cell. **b** Perspective view of the flat selective emitter unit cell. **c** Front view of the emitter unit cell. **d** Top view of the emitter unit cell. **e** The electric field distribution of the traditional trilayer structure. **f** The electric field distribution of the flat selective structure. **g** The emittance of the traditional MIM structure and the flat structure at different temperatures. Reproduced from [35] with permission. Copyright 2013, Optical Society of America

8.4.2 Metal–Dielectric Multilayers

Although metamaterials are promising candidates to realize optimal absorption of electromagnetic waves and thermal radiation, the subwavelength structures limited the fabrication cost and area. In this section, it is shown that multilayered structures could also result in good spectral-selectivity when proper materials are chosen. For instance, infrared absorption/radiation and visible transparency is demonstrated for potential applications in radiative cooling of optical windows. Once again, the interference of light fields in the highly lossy material layer would take the form of catenaries, which helps to realize impedance match and perfect absorption/radiation.

As depicted in Fig. 8.20a, the designed structure is composed of four alternating ITO and photoresist layers on top of a silica wafer [39]. As all these employed materials are intrinsically transparent at visible range, the structure is highly optically transparent. An 80-cm-diameter sample is fabricated by film evaporation and spin coating. The wavelength dependence of transmittance between 400 and 800 nm is illustrated in Fig. 8.20b with the measured average transmittance above 80%. To give a visual demonstration of the transparency, a photo of the sample is taken and shown in Fig. 8.20c. The logo SKLOTNM (State Key Laboratory of Optical Technolo-

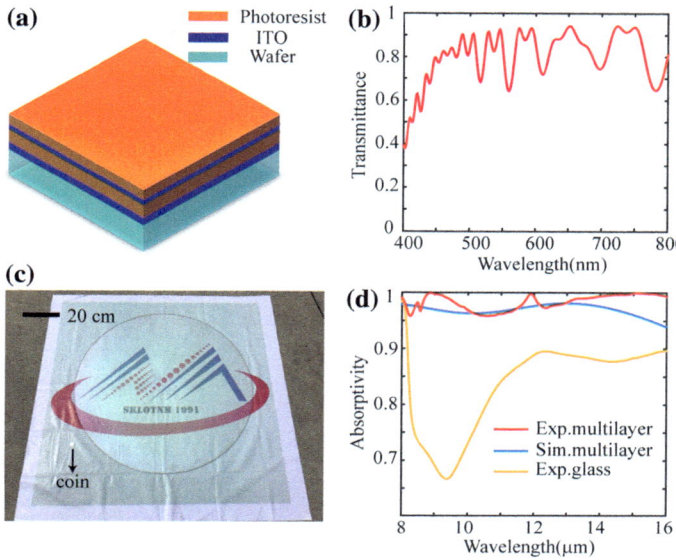

Fig. 8.20 **a** Schematic of the multilayer absorber with alternating ITO and photoresist layers. The thicknesses of the layers from top to bottom are 1.5, 0.1, 1.8, and 0.3 μm. **b** Transmittance of the metamaterial at visible wavelengths (400–800 nm). The structure has an average transmittance of more than 80% in this region. **c** Photo of an 80-cm-diameter metamaterial. The patterns covered by the sample are clearly visible. A coin is set beside the sample for comparison. **d** Measured (red) and simulated (blue) absorptivity of the metamaterial at $\theta = 15°$. The yellow line is the measured absorptivity of the glass wafer employed in the fabrication, which indicates that the broadband absorption is not induced by the wafer. Reproduced from [39] with permission. Copyright 2017, The Japan Society of Applied Physics

gies on Nanofabrication and Micro-engineering) underneath is clearly visible in the photo. The ultra-broadband infrared absorption is achieved by the combination of the inter-band transition of charge carriers of the two materials and the structural effect induced by the alternating alignment of ITO and photoresist layers. The measured and simulated wavelength dependences of absorptivity in far infrared are depicted in Fig. 8.20d, which are both much larger than 90%. The absorptivity of the bare silica wafer employed in the fabrication is also shown for comparison, which indicates that the absorption is not induced by the wafer.

In the ITO layer, the intensity of electric fields shows a catenary shape as a result of the interference of counter-propagating waves. By setting the interface between the bottom ITO and spacer as $z = 0$, the range of z for the top ITO is $z = [2.1, 2.2]$. As shown in Fig. 8.21, at different wavelengths, the symmetry of the catenary curves is varying. Since the catenary curve is an indicator of the propagation loss inside the film, it should be precisely manipulated to realize perfect absorption.

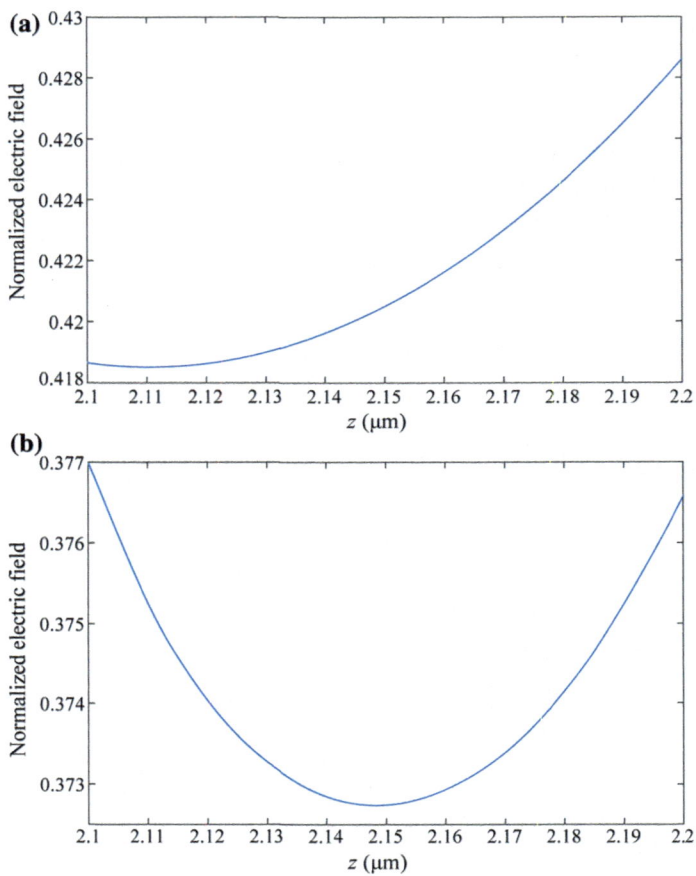

Fig. 8.21 Normalized electric field intensity distribution in the top ITO layer. **a** $f = 24$ THz. **b** $f = 30$ THz

8.5   Reduction of Thermal Radiation in Pseudo-Brewster Angle

As one of the most fundamental rules in classical optics, the Fresnel equations, accompanied with the Snell's law, determine the reflection and transmission of light incident on an interface of two media with different refractive indices. One important consequence of the Fresnel equations is the Brewster effect, where the reflectivity for the wave polarized in the plane of incidence (TM or p-polarization) vanishes at a particular incidence angle (Brewster angle). In the ideal case, the reflectivity of p-polarized wave at the Brewster angle would be zero. However, in general, most materials used in optics and electromagnetics are lossy to some extent, thus the reflectivity minimum is not exactly zero. For strongly lossy materials such as metals, the minimal reflectivity is often larger than 10%, thus the corresponding angle is termed as pseudo-Brewster angle [40, 41]. Note that metal is often treated as a good reflector, thus the reflection minimal seems to be an "abnormal" effect. The energy transmitted into metal would be totally absorbed since there is no way to let them transmit if the metal thickness is larger than the skin depth. This absorption is indeed very strong at the infrared and visible wavelengths, where metal could not be treated as perfect conductors.

The control of the polarization-dependent absorption at pseudo-Brewster angle has great significance in infrared applications such as functional devices and infrared polarimetric detection. According to the Kirchhoff's law [31], the thermal emission due to blackbody radiation would also be highly dependent on angle and polarization since the thermal emissivity at infrared region equals to the absorptivity. This also explains why the thermal radiation differs from that described by Lambert's cosine's law [9]. As a result, polarimetric imaging could provide a new route for infrared surveillance with the ability to reveal the hidden metallic objects [10].

In this section, the design of structured surfaces with inhibited pseudo-Brewster effect is presented. By eliminating the energy loss of surface electromagnetic wave propagating along a structured metallic surface, it is shown that both the polarimetric thermal emission and laser's reflection loss can be significantly reduced. Consequently, the proposed metasurfaces would serve as an efficient way to modify the polarization states, for either thermal infrared radiation or coherent laser beam [42].

To characterize the physical principle of the design, one should consider the Fresnel's reflection at a lossy metal surface under variant angles of incidence. Since metal is thick enough to absorb the transmitted light, the angle and polarization-dependent absorptivity should be:

$$A_s = 1 - \left(\frac{\sqrt{\varepsilon_m - \sin^2 \theta} - \cos \theta}{\sqrt{\varepsilon_m - \sin^2 \theta} + \cos \theta} \right)^2 \tag{8.5.1}$$

and

$$A_p = 1 - \left(\frac{\varepsilon_m \cos \theta - \sqrt{\varepsilon_m - \sin^2 \theta}}{\varepsilon_m \cos \theta + \sqrt{\varepsilon_m - \sin^2 \theta}} \right)^2, \tag{8.5.2}$$

where A_s and A_p are the absorptivity for s- and p-polarizations, ε_m and θ are the permittivity of the metal as well as the incidence angle. When $|\varepsilon_m|$ is much larger than unity and possesses negligible loss, A_s is almost zero for all angle of incidence.

In a phenomenological view, the increased absorption at the pseudo-Brewster angle is associated with the large vertical electric fields E_z above the metal surface, which makes the horizontal wave impedance matched to that of metal. To eliminate these vertical fields, periodic metallic posts have been introduced on the surface. As can be seen in Fig. 8.22b, the posts have intrinsic plasmonic modes propagating vertically in the posts array [43]. Since the electric fields are perpendicular to the

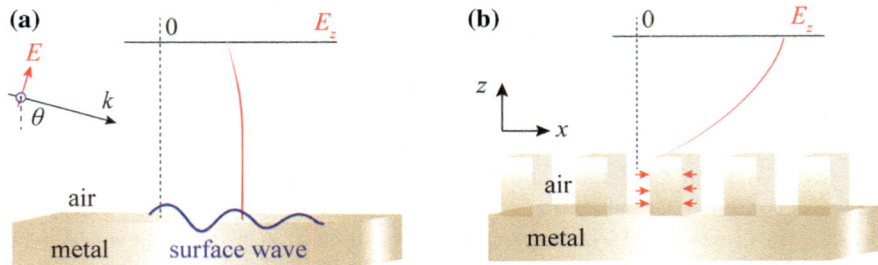

Fig. 8.22 Elimination of the surface wave. **a** Electrodynamic reflection of a flat surface at the pseudo-Brewster angle. Note that the reflection minimum is related to the surface wave that propagates along the interface between air and a lossy smooth metal. **b** Electric filed distributions at the pseudo-Brewster angle for a posts array featured by a catenary-shaped intensity. Reproduced from [42] with permission. Copyright 2019, Institute of Optics and Electronics, Chinese Academy of Sciences

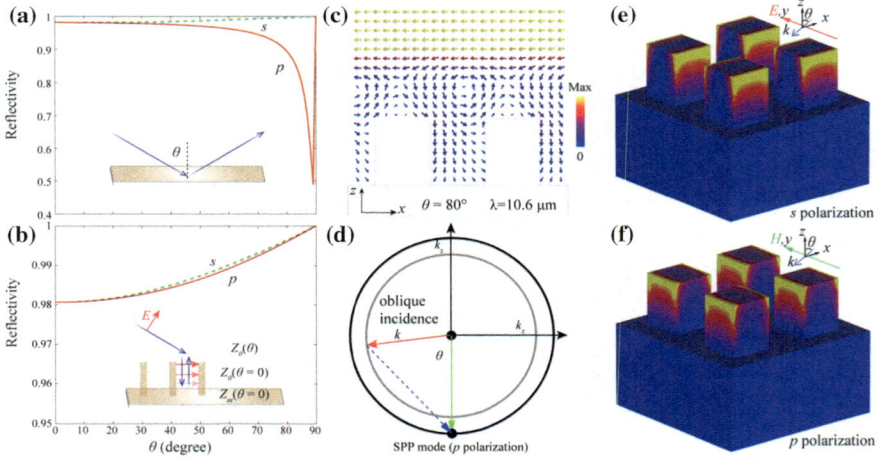

Fig. 8.23 Theoretical analysis of the metallic grating. **a** Reflectance of s- and p-polarized light for a smooth gold plate. **b** Reflectance of s- and p-polarized light for a gold plate decorated with subwavelength posts array. **c** Energy flow in the xz-plane when p-polarized light is incident with an angle of 80°. **d** Schematic of the wavevector mismatching at the air-posts interface. **e, f** Amplitudes of the electric fields for s- and p-polarization at $\theta = 80°$. Reproduced from [42] with permission. Copyright 2019, Institute of Optics and Electronics, Chinese Academy of Sciences

vertical axis, E_z is forced to vanish at the surface, which helps to reduce the absorption as well as the thermal emission.

To prove the above theory, a full-wave finite-element method (FEM) is adopted to solve Maxwell's Equations in the structured surface rigorously [44]. Figure 8.23a, b show the angle-resolved reflectivity at $\lambda = 10.6\,\mu$m for both the s- and p-polarizations at a smooth and a structured gold surfaces, respectively. The pseudo-Brewster angle for the smooth gold surface is about 86°, coinciding well with the large-index approximation. The p-polarized reflectivity at this angle is only about 50%, implying a strong ohmic loss in this case. For a periodic rectangular posts array with a period of 3 μm, a height of 1.9 μm and a width of 1.25 μm, the absorption peaks become nearly completely eliminated, while the reflectivity for the two polarizations is the same for almost all angles of incidence.

Figure 8.23c depicts the energy flow of the p-polarized light with an incident angle of 80°, which could provide a visual understanding of the new boundary layer. Two signatures can be observed: First, there is clear vortex flows just above the posts array; second, only very small part of energy is directed into the gaps between the posts. The vortex is similar to the fluid dynamics when the surface roughness is larger than the molecules of fluid [45]. The small energy penetration can be interpreted by considering the horizontal wave vector of the electromagnetic wave. As illustrated in Fig. 8.23d, the wavevector in the posts array is perpendicular to the surface for p-polarization and presents a huge mismatch with that of the incidence wave, leading to a dramatic reflection at the first interface between air and the posts array. More precisely, the optical performance of the posts array can be analyzed using a simple impedance model and transfer matrix method.

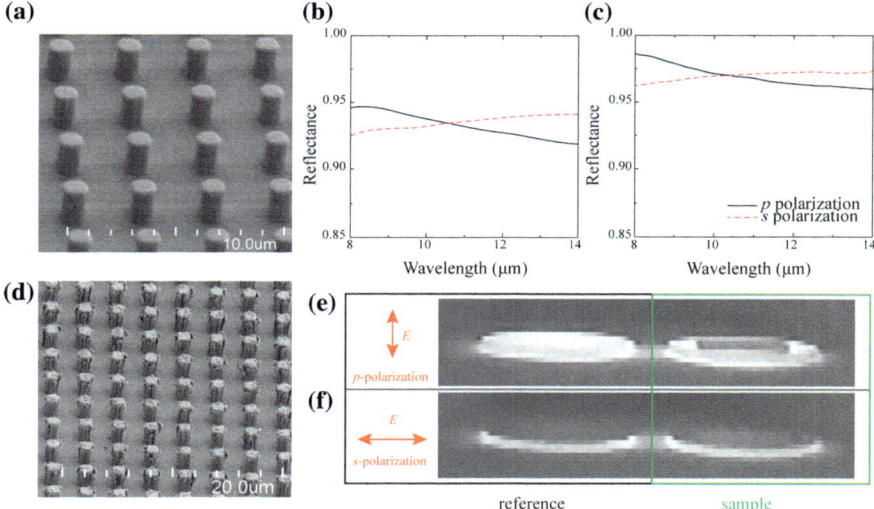

Fig. 8.24 Broadband reflectance and polarimetric imaging. **a** Perspective view of the gold-sample. **b, c** FTIR spectra of the gold-sample for incidence angles of 70°–80°. The p- and s-polarizations are indicated in each panel. **d** Perspective view of the chromium-sample. **e, f** Polarized infrared images of the chromium-sample and a reference sample for p- (**e**) and s- (**f**) polarizations. The strong emission at the circumference of the sample is from the silicon dioxide substrate (diameter: 2.5 cm, the structure was fabricated in a square region with a width of 1.5 cm). Reproduced from [42] with permission. Copyright 2019, Institute of Optics and Electronics, Chinese Academy of Sciences

In the experiments, the posts array was fabricated by triple resist technology followed by laser direct writing and electron beam deposition. The SEM image of a sample coated with gold is shown in Fig. 8.24a. The reflectance for different incident angles and polarizations measured by Fourier transform infrared spectroscopy (FTIR) are shown in Fig. 8.24b, c, showing good agreement with the theoretical results. The slightly increased absorption is attributed to the finite thickness and surface roughness of the metallic coating layer. To demonstrate the suppressed unusual thermal radiation, the infrared images of a posts array (Fig. 8.24d) and a reference plate coated with 200-nm-thick chromium are captured by an infrared polarimetric imaging system. The temperature is set to be 50 °C, corresponding to black radiation centered at about $\lambda = 9 \, \mu\text{m}$. The measured thermal emission angle with respect to the normal axis is near 70°. As shown in Fig. 8.24e, f, the emission intensity for the structured sample are almost identical for the two polarizations. Unlike the structured sample, the flat surface has much stronger intensity for p-polarization, which is a significant signal for infrared target recognition.

References

1. Black body, https://en.wikipedia.org/wiki/Black_body
2. M. Planck, *The Theory of Heat Radiation* (P. Blakiston's Son & Co., Philadelphia, 1914)
3. K.N. Rozanov, Ultimate thickness to bandwidth ratio of radar absorbers. IEEE Trans. Antennas Propag. **48**, 1230–1234 (2000)
4. M. Vollmer, K.-P. Mollmann, *Infrared Thermal Imaging: Fundamentals, Research and Applications* (Wiley-VCH Verlag GmbH & Co. KGaA, Germany, 2010)
5. A.P. Raman, M.A. Anoma, L. Zhu, E. Rephaeli, S. Fan, Passive radiative cooling below ambient air temperature under direct sunlight. Nature **515**, 540–544 (2014)
6. M.M. Hossain, M. Gu, Radiative cooling: principles, progress, and potentials. Adv. Sci. **2016**, 1500360 (2016)
7. A. Synnefa, M. Santamouris, K. Apostolakis, On the development, optical properties and thermal performance of cool colored coatings for the urban environment. Sol. Energy **81**, 488–497 (2007)
8. Y. Zhai, Y. Ma, S.N. David, D. Zhao, R. Lou, G. Tang, R. Yang, X. Yin, Scalable-manufactured randomized glass-polymer hybrid metamaterial for daytime radiative cooling. Science **355**, 1062–1066 (2017)
9. A.G. Worthing, Deviation from Lambert's law and polarization of light emitted by incandescent tungsten, tantalum and molybdenum and changes in the optical constants of tungsten with temperature. J. Opt. Soc. Am. **13**, 635–649 (1926)
10. J.L. Pezzaniti, D. Chenault, K. Gurton, M. Felton, Detection of obscured targets with IR polarimetric imaging, in *Proceedings of SPIE* (2014), vol. 9072, p. 90721D
11. E. Rousseau, A. Siria, G. Jourdan, S. Volz, F. Comin, J. Chevrier, J.-J. Greffet, Radiative heat transfer at the nanoscale. Nat. Photonics **3**, 514–517 (2009)
12. D. Thompson, L. Zhu, R. Mittapally, S. Sadat, Z. Xing, P. McArdle, M.M. Qazilbash, P. Reddy, E. Meyhofer, Hundred-fold enhancement in far-field radiative heat transfer over the blackbody limit. Nature **561**, 216–221 (2018)
13. S.M. Rytov, *Theory of Electric Fluctuations and Thermal Radiation* (Air Force Cambridge Research Center, USA, 1959)
14. D. Polder, M. Van Hove, Theory of radiative heat transfer between closely spaced bodies. Phys. Rev. B **4**, 3303–3314 (1971)

15. J.B. Pendry, Radiative exchange of heat between nanostructures. J. Phys. Condens. Matter **11**, 6621 (1999)
16. L. Hu, A. Narayanaswamy, X. Chen, G. Chen, Near-field thermal radiation between two closely spaced glass plates exceeding Planck's blackbody radiation law. Appl. Phys. Lett. **92**, 133106 (2008)
17. S. Shen, A. Narayanaswamy, G. Chen, Surface phonon polaritons mediated energy transfer between nanoscale gaps. Nano Lett. **9**, 2909–2913 (2009)
18. X. Liu, Z. Zhang, Near-field thermal radiation between metasurfaces. ACS Photonics **2**, 1320–1326 (2015)
19. S.-A. Biehs, P. Ben-Abdallah, Revisiting super-Planckian thermal emission in the far-field regime. Phys. Rev. B **93**, 165405 (2016)
20. J. Ng, H. Chen, C.T. Chan, Metamaterial frequency-selective superabsorber. Opt. Lett. **34**, 644–646 (2009)
21. X. Luo, Subwavelength artificial structures: opening a new era for engineering optics. Adv. Mater. 1804680 (2018)
22. S.I. Maslovski, C.R. Simovski, S.A. Tretyakov, Overcoming black body radiation limit in free space: metamaterial superemitter. New J. Phys. **18**, 013034 (2016)
23. J.-J. Greffet, R. Carminati, K. Joulain, J.-P. Mulet, S. Mainguy, Coherent emission of light by thermal sources. Nature **416**, 61–64 (2002)
24. R.R. Nair, P. Blake, A.N. Grigorenko, K.S. Novoselov, T.J. Booth, T. Stauber, N.M.R. Peres, A.K. Geim, Fine structure constant defines visual transparency of graphene. Science **320**, 1308–1308 (2008)
25. R. Alaee, M. Farhat, C. Rockstuhl, F. Lederer, A perfect absorber made of a graphene micro-ribbon metamaterial. Opt. Express **20**, 28017–28024 (2012)
26. A. Vakil, N. Engheta, Transformation optics using graphene. Science **332**, 1291–1294 (2011)
27. Q. Feng, M. Pu, C. Hu, X. Luo, Engineering the dispersion of metamaterial surface for broadband infrared absorption. Opt. Lett. **37**, 2133–2135 (2012)
28. M. Pu, P. Chen, Y. Wang, Z. Zhao, C. Wang, C. Huang, C. Hu, X. Luo, Strong enhancement of light absorption and highly directive thermal emission in graphene. Opt. Express **21**, 11618–11627 (2013)
29. W.W. Salisbury, Absorbent body for electromagnetic waves. U.S. Patent 2599944 (1952)
30. M. Diem, T. Koschny, C.M. Soukoulis, Wide-angle perfect absorber/thermal emitter in the terahertz regime. Phys. Rev. B **79** (2009)
31. X. Liu, T. Tyler, T. Starr, A.F. Starr, N.M. Jokerst, W.J. Padilla, Taming the blackbody with infrared metamaterials as selective thermal emitters. Phys. Rev. Lett. **107**, 045901 (2011)
32. N. Mattiucci, G.D. Aguanno, A. Alu, C. Argyropoulos, J.V. Foreman, M.J. Bloemer, Taming the thermal emissivity of metals: a metamaterial approach. Appl. Phys. Lett. **100**, 201109 (2012)
33. M.A. Kats, R. Blanchard, S. Zhang, P. Genevet, C. Ko, S. Ramanathan, F. Capasso, Vanadium dioxide as a natural disordered metamaterial: perfect thermal emission and large broadband negative differential thermal emittance. Phys. Rev. X **3**, 041004 (2013)
34. T. Inoue, M. De Zoysa, T. Asano, S. Noda, Realization of narrowband thermal emission with optical nanostructures. Optica **2**, 27–35 (2015)
35. M. Song, H. Yu, C. Hu, M. Pu, Z. Zhang, J. Luo, X. Luo, Conversion of broadband energy to narrowband emission through double-sided metamaterials. Opt. Express **21**, 32207–32216 (2013)
36. M. Pu, C. Hu, M. Wang, C. Huang, Z. Zhao, C. Wang, Q. Feng, X. Luo, Design principles for infrared wide-angle perfect absorber based on plasmonic structure. Opt. Express **19**, 17413–17420 (2011)
37. H.T. Miyazaki, T. Kasaya, M. Iwanaga, B. Choi, Y. Sugimoto, K. Sakoda, Dual-band infrared metasurface thermal emitter for CO_2 sensing. Appl. Phys. Lett. **105**, 121107 (2014)
38. M.D. Zoysa, T. Asano, K. Mochizuki, A. Oskooi, T. Inoue, S. Noda, Conversion of broadband to narrowband thermal emission through energy recycling. Nat. Photonics **6**, 535–539 (2012)

39. Y. Huang, M. Pu, P. Gao, Z. Zhao, X. Li, X. Ma, X. Luo, Ultra-broadband large-scale infrared perfect absorber with optical transparency. Appl. Phys. Express **10**, 112601 (2017)
40. G. Ohman, The pseudo-Brewster angle. IEEE Trans. Antennas Propag. **25**, 903–904 (1977)
41. R.M.A. Azzam, Complex reflection coefficients of p- and s-polarized light at the pseudo-Brewster angle of a dielectric–conductor interface. J. Opt. Soc. Am. A **30**, 1975–1979 (2013)
42. X. Ma, M. Pu, X. Li, Y. Guo, X. Luo, All-metallic wide-angle metasurfaces for multifunctional polarization manipulation. Opto-Electron. Adv. **2**, 180023 (2019)
43. K. Wang, D.M. Mittleman, Metal wires for terahertz wave guiding. Nature **432**, 376–379 (2004)
44. *CST Microwave Studios* (CST-Computer Simulation Technology AG, 2013)
45. S. Granick, Y. Zhu, H. Lee, Slippery questions about complex fluids flowing past solids. Nat. Mater. **2**, 221–227 (2003)

Chapter 9
From Catenary Optics to Engineering Optics 2.0

Abstract In this chapter, we summarize the applications of catenary optics in optical engineering. Based on the novel properties of catenary optical fields and catenary structures, it is shown that traditional optical laws and theories could be extended and generalized, which opens a door towards the next-generation engineering optics.

Keywords Optical engineering · Diffraction limit · Generalized snell's law · Subwavelength electromagnetics

9.1 Basic Laws of Traditional Engineering Optics

Engineering optics is a discipline that seeks to apply optical theories to practical applications in areas including but not limited to imaging, micro/nano-fabrication, remote sensing and communications to illumination, display, and energy harvest [1]. From a historical view, classic engineering optics has its root in the basic laws developed in the last four hundred years, such as Snell's law, Fermat's principle, Fresnel's equations, Huygens' principle, and Kirchhoff's diffraction theory (Fig. 9.1)). As highlighted by recent articles and reviews [2–6], traditional optical theories have brought serious limitations and restrictions on the performances of optical materials, devices, and systems. In what follows, we briefly list some representative challenges:

Diffraction limit. Unlike the prediction of geometric optics, the resolving ability of optical imaging system is not only limited by the aberration, but also the diffraction limit. For microscopic applications, the maximal spatial resolution is about half the wavelength. For telescope applications, the angular resolution is limited by the aperture, leading to numerous efforts to create large-aperture lenses and mirrors.

Curved optical devices. Base on Snell's law and the law of reflection, flat surfaces usually do not change the propagation direction of light, thus are not able to form magnified or demagnified images. In imaging applications, the shapes of lenses and mirrors must be curved to bend light correctly. For small devices, the curved shapes may not be a serious problem. However, it would be extremely difficult to control the surface quality over a large area.

© Springer Nature Singapore Pte Ltd. 2019 355
X. Luo, *Catenary Optics*, https://doi.org/10.1007/978-981-13-4818-1_9

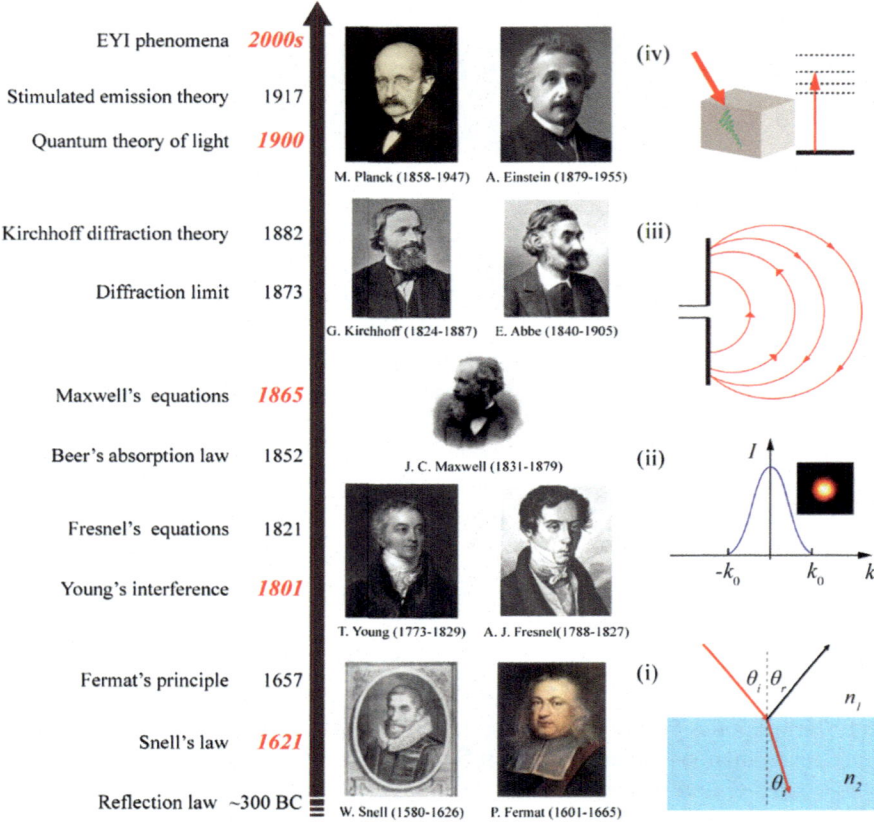

Fig. 9.1 Milestones in classic engineering optics. **i–iv** Schematics of the laws for reflection/refraction, diffraction, radiation and absorption. Reproduced with permission from [5]. Copyright 2018, American Chemical Society. All portraits are reproduced from Wikipedia [8–16]

Thick optical materials. The absorption and conversion of electromagnetic waves take place in volume space. It has been shown that there is a fundamental limit on the thickness of the materials to absorb or change the polarizations in a given bandwidth [7]. For thin film solar cells, however, the thickness should be as thin as possible to minimize the recombination of electrons and holes.

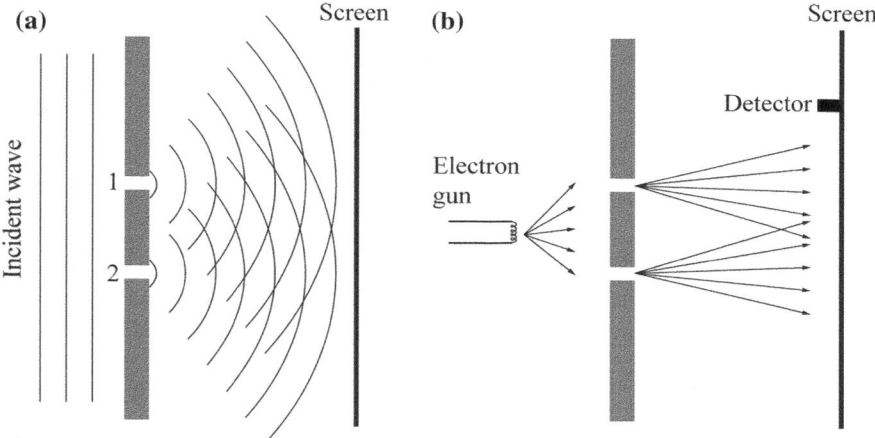

Fig. 9.2 Young' double slits interference for photons and electrons. From single photon and single electron experiment, it is known that the one photon or electron must transmit two slits simultaneously and interference with itself

9.2 Young's Interferences of Photons, Electrons and Coupled Plasmons

According to Abbe' imaging theory, the imaging process is actually the interference of various spectral components generated by the objects and lens systems. Thus, interference lies at the most basic level for optical imaging systems. Indeed, the diffraction and interference are the essence of wave optics, and it seems impossible to say which is more fundamental. Richard Feynman has written in his lectures on physics (Chapter 30, Volume I) [17]:

"No one has ever been able to define the difference between interference and diffraction satisfactorily. It is just a question of usage, and there is no specific, important physical difference between them. The best we can do, roughly speaking, is to say that when there are only a few sources, say two, interfering, then the result is usually called interference, but if there is a large number of them, it seems that the word diffraction is more often used."

In modern physics, probably the most important experiment is the Young's double-slits interference (Fig. 9.2), which was the key to demonstrate the wave characteristic of both the light and electron. According to a poll of Physics World readers, the most beautiful experiment in physics is thought to be the interference of single electrons in a Young's double slit, while the Young's original experiment was also one of the most beautiful 10 experiments [18]. Richard Feynman once stated that this phenomenon (Young's double-slit experiment applied to the interference of single electrons) is in "the heart of quantum mechanics", and "*absolutely* impossible, to explain in any classical way" [17].

Now that interference is the basis of wave optics, and Young's double slits experiment is the simplest and most powerful method to demonstrate the interference effect, any approach to break the diffraction limit should generate abnormal effects in the double slits experiment. Fortunately, such effects have been observed during the investigation of Young' experiment in the subwavelength scale, i.e., the width of the metallic slits and the distance between the slits are smaller than the wavelength [19–23]. It has been realized that the coupled surface plasmons are critical to understand these phenomena. Since photons, electrons, and plasmons are three basic elements in modern physics, it is, therefore, suitable to treat Young's interferences of photons, electrons, and plasmons as important platforms to study their physical properties. Like the photons and electrons, the coupled plasmons would have far-reaching influences and applications in future physics and optics. Consequently, it has been listed as one milestone in the history of optics and nanophotonics, as shown in Fig. 9.1.

For the convenience of discussion, the abnormal interference effects are called extraordinary Young's interference (EYI) [5, 24]. In the following, we shall review several kinds of EYI effects discovered with subwavelength slits perforated in metallic film [23].

9.2.1 Shrunk Interference Patterns in EYI

With electronic computers and advanced electromagnetic algorithm, the subwavelength Young's interference can be numerically simulated with high accuracy. In Fig. 9.3, a very interesting phenomenon is seen: at a wavelength of 365 nm, the peak-to-peak distance of optical field is close to 100 nm, and no interference pattern is observed; however, at a wavelength of 380 nm, the peak-to-peak distance is decreased to about 25 nm, which is only ~$\lambda/15$. The electric field distribution implies that surface plasmon resonance has been excited. When the localized fields at $\lambda = 365$ nm are taken as an interference pattern, a more counter-intuitive conclusion may be drawn: the increase of wavelength has led to a much smaller interference pattern. This EYI effect means that one need not reduce the wavelength to achieve higher resolution, thus the classical diffraction limit could be overcome. Note that the above phenomena can only be observed for transverse magnetic (TM) but not transverse electric (TE) polarization.

Owing to the small feature size below the diffraction limit, subwavelength interference is difficult to record with traditional instruments such as charge coupled devices (CCD) and complementary metal oxide semiconductor (CMOS) detectors. By placing photoresist (PR) layer below the perforated silver film, a simple method was proposed to record the subwavelength interference fringes. This configuration also functions as a novel scheme for sub-diffraction-limited nano-fabrication. As illustrated in Fig. 9.4, with a light source operating at 436 nm, the width of 1D interference fringe was reduced to be smaller than 50 nm, which is almost only one-ninth of the vacuum wavelength. According to numerical simulations and dispersion dia-

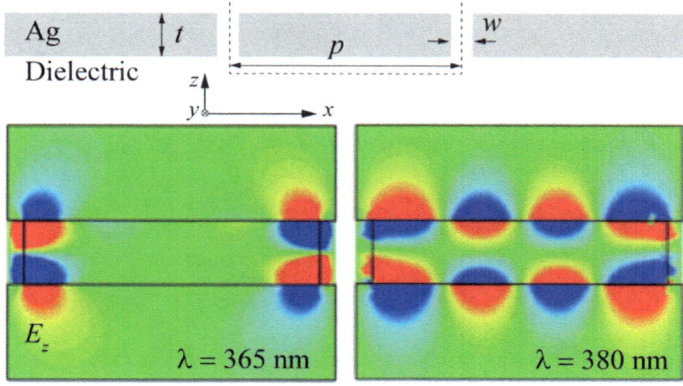

Fig. 9.3 EYI at the silver–dielectric interface. The top panel shows the geometric configuration. The thickness of the silver film and the width of the slits are chosen to be 20 and 10 nm. The bottom shows the z-component of the electric fields with the same magnitude scale. Reproduced with permission from [24]. Copyright 2018, American Chemical Society

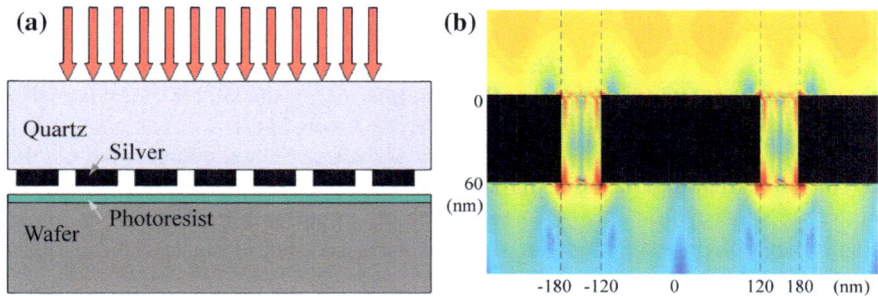

Fig. 9.4 Interference lithography based on EYI effect. **a** Schematic of the experimental configuration. **b** Simulated intensity distribution for the electric fields. Reproduced from [19] with permission. Copyright 2004, American Institute of Physics

gram of SPP, this approach may provide a promising alternative to traditional bulky and costly photolithographic systems.

9.2.2 Shifted Interference Patterns in EYI

The shrunk interference patterns described in above section reveal the fact that the effective wavelength of coupled plasmon (catenary plasmon) is dependent on the thickness of metallic film [20]. Similarly, the catenary plasmon in a metal–insulator–metal (MIM) configuration also has a propagation constant determined by the thickness of the dielectric core. This dependence is also responsible for EYI effect

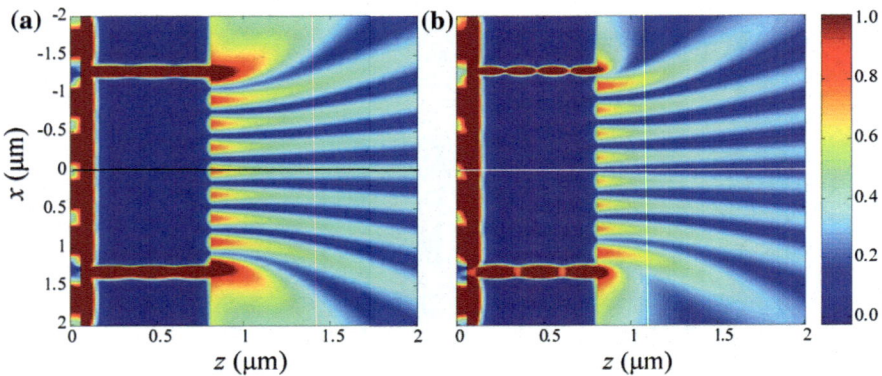

Fig. 9.5 Near-field|Hy| distribution for the symmetric and asymmetric configurations. The thickness of the metal film is $h = 700$ nm and the distance between the two slits is $d = 2.6$ μm. **a** $w_1 = w_2 = 100$ nm. **b** $w_1 = 100$ nm and $w_2 = 25$ nm. Reprinted from [25] with permission. Copyright 2007, Optical Society of America

with unequal slit width [25]. As shown in Fig. 9.5, when light passes through two slits with the same thickness and different widths (w_1 and w_2), light in the two channels may acquire a phase shift of π, thus the intensity at the center of the two slits becomes zero. This is in contrast to the bright line formed in traditional interference experiment, where the slit width only influences the transmitted intensity but not the phase shift. Based on this effect, constructive and destructive interferences were used to direct the energy flow of SPP to one particular direction [26]. Meanwhile, more slits or wider slits filled with dielectrics have been adopted to increase the exciting efficiency [27].

9.2.3   Modulated Transmission in EYI

To render the subwavelength interference effect into practical imaging or lithographic devices, the energy efficiency must be optimized. In general, the separation distance and width of the slits, together with the film thickness will influence the efficiency at a given wavelength. As a result, these geometric parameters must be optimized in simulations and experiments [19, 20]. In a follow-up work, Young's interference at the metal surface was revisited with a detailed discussion of the influence of the incident wavelength. It was found that the total intensity of the far-field diffraction pattern is reduced or enhanced as a function of the wavelength of the incident light beam [21], as shown in Fig. 9.6.

The above phenomenon was attributed to the constructive and destructive interference of SPP propagating along the surfaces. When the incident field is TE-polarized, the transmission of the double slits is small and weakly modulated as a function of

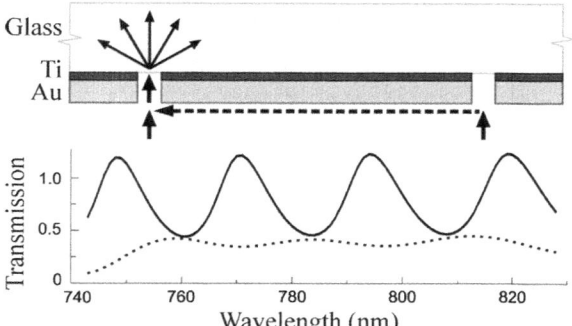

Fig. 9.6 Mutual coupling in Young's double slits experiments. The normalized transmission oscillates as the change of wavelength. The solid and dashed lines indicate the transmission for TM and TE polarization, respectively. Adapted with permission from [21]. Copyright 2005, The American Physical Society

wavelength. This experiment is also an additional physical explanation of the famous extraordinary optical transmission (EOT) effect [28]. From the perspective of electromagnetic theory, this unusual interference effect is related to the mutual coupling of antennas, which has been widely known in antenna theory [29].

9.2.4 Extraordinary Vertical Fabry–Perot Interference

In Young's double slits experiment, the subwavelength interference takes place in the plane perpendicular to the original propagation direction of incident light. In contrary, F-P interferometers are characterized by interference along the propagation direction. By adding subwavelength surface structures on the two sides of the classical F-P cavity (a single dielectric slab), the interference phenomena would change dramatically. Figure 9.7a shows a simple configuration based on metallic slits separated by a dielectric slab of $\varepsilon = 2.25$. Without these subwavelength slits, a 10-mm-thick dielectric slab would have a transmission peak at $f = 10$ GHz ($\lambda = 20$ mm). However, the introduction of subwavelength slits makes the peak shift to 5.13 GHz owing to the additional phase shift induced by the catenary optical fields [30], as shown in Fig. 9.7b. Since the thickness is only about one-sixth of the wavelength, this subwavelength interference provides an efficient approach to reduce the thickness of similar devices.

The above EYI effects imply that the catenary plasmons and catenary optical fields in subwavelength structures may have a much smaller effect wavelength, tunable phase shift, and amplitude. In a more general sense, these catenary fields are also called metasurface-waves (M-wave) to stress their linkage with surface structures and surface waves [30, 31]. Consequently, classic optical theories such as the diffraction limit, the laws of reflection and refraction, and many others could be extended to a new level. In the next section, we give a concise description of these theoretical advancements.

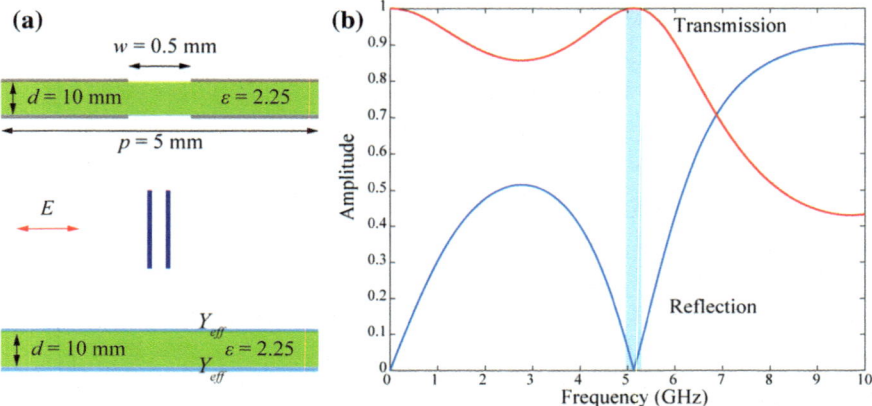

Fig. 9.7 Subwavelength F-P interference. **a** Definition of the geometric parameters. The bottom shows an effective medium model based on effective admittance. **b** Calculated transmission and reflection coefficients for TM polarization. The transmission peak locates at $f = 5.13$ GHz. Adapted with permission from [23]. Copyright 2018, the Optical Society of America

9.3   Generalized Optical Theories and Their Applications

As described in previous chapters, catenary optics has been demonstrated to be a promising way to revolutionize traditional optics in many manners. On the first hand, catenary structures could produce linear and continuous geometric phase, thus form a basic building block for phase-gradient metasurface lenses and other functional devices. On the other hand, the hyperbolic cosine catenary function is an intrinsic characteristic of evanescent coupling, which is a new solution to Maxwell's equations in the subwavelength scale. By properly tuning the catenary optical fields with metallic and dielectric subwavelength structures, the high-order evanescent waves can be harnessed to realize super-resolution imaging beyond the diffraction limit, localized phase modulation, and broadband perfect absorption/radiation, among others. These breakthroughs are important bases of the next-generation engineering optics, i.e., the Engineering Optics 2.0 (EO 2.0). Table 9.1 shows a brief comparison of Engineering Optics 1.0 and 2.0, which highlights the fundamental differences in the basic optical laws and theories with and without subwavelength structures.

9.3.1   Generalized Diffraction Theory Based on Catenary Optical Fields

In classic optical theory, the information carried by evanescent waves are not collected, thus optical performances such as imaging resolution, divergence of laser beams are restricted by the well-known diffraction limit. This fact is easy to be

Table 9.1 Comparison of engineering optics 1.0 and 2.0. [5]

Properties	Status of EO 1.0	Promise of EO 2.0
Refraction and Reflection	• Propagation direction is determined by refractive index and shape • Bulky and heavy lenses and mirrors	• Tunable propagation direction without changing the geometric shape • Thin and lightweight lenses, mirrors and other functional optical elements
Diffraction	• Resolution is limited by wavelength and aperture • High cost and long time are required for large optical systems	• Increased resolution beyond the diffraction limit • Reduced time and cost bearable for common users
Radiation	• Electromagnetic radiation is limited by the size of antenna and laser • Thermal radiation is limited by the temperature	• Enhanced radiation for small antenna and laser • Boosted thermal radiation at given temperature
Absorption	• Absorption takes place inside volume materials • Larger thickness is required for wider bandwidth	• Absorption takes place along the surface • Broadband absorption realized with vanishing thickness

understood in the spatial frequency domain. With Fourier transform [32], a given electromagnetic field could be decomposed into the propagating and evanescent components:

$$
\begin{aligned}
\mathbf{E}(\mathbf{r}, t) &= \int_{-\infty}^{\infty} \int_{-\infty}^{\infty} \mathbf{A}(k_x, k_y) \exp(i k_x x + i k_y y) dk_x dk_y \\
&= \int_{k_x^2 + k_y^2 < k_0^2} \mathbf{A}(k_x, k_y) \exp(i k_x x + i k_y y) dk_x dk_y \quad, \\
&\quad + \int_{k_x^2 + k_y^2 \geq k_0^2} \mathbf{A}(k_x, k_y) \exp(i k_x x + i k_y y) dk_x dk_y
\end{aligned}
\tag{9.3.1}
$$

where k_0 is the wavenumber of light in the surrounding space, the first and second items indicate the propagating and evanescent components. In optical materials with positive dielectric constant, the evanescent components have an imaginary wavenumber in the propagating direction, thus cannot be received by far-field detectors. Although traditional near-field techniques could break the far-field Abbe–Rayleigh diffraction limit, the resolution is still limited by the distance between the imaging and source plane, and this limit is called the near-field diffraction limit [31, 33], which can be defined according to the working distance d at which the highest spatial component decays to its $1/(e)$ in amplitude:

$$\delta \geq \frac{\lambda}{2} \frac{1}{\sqrt{1 + \left(\frac{\lambda}{2\pi d}\right)^2}}. \tag{9.3.2}$$

One possible way to overcome this limit is to recover and amplify high-order evanescent waves by near-field coupling. Taking the plasmonic nanolithography system as an example [19, 34–37], the resolution has been pushed down to sub-22 nm via the combination of superlens and reflective lens [34], while the system size and weight have been greatly reduced, providing a promising alternative to traditional complex and costly projection lithography [38–40]. By contrast, the projection lens of Zeiss Starlith 1900i has a height of 129 cm and a weight of 1080 kg [41], while the height and weight of plasmonic lens are only about 20 cm and 2.5 kg [40].

The coupled SPPs, i.e., the catenary plasmons on the metal–dielectric interfaces are critical to achieving super-resolution imaging. Figure 9.8a–i illustrate the coupling effects in single-layered metal film and metal–dielectric multilayers. With a simple Young's double slits interference experiment [19, 20], it has been demonstrated that a thinner thickness promises a higher resolution.

Figure 9.8c, d show a silver-film superlens and a cavity lens comprised of a superlens and a reflective layer [34, 42]. The reflective layer has two important roles. First, similar to the tapetum lucidum of cat's eyes [47], the light intensity in the photosensitive region will be increased by reflection to enable weak light exposure. Second, the catenary plasmon modes formed between the two metallic layers ensure a larger depth of field [24, 48, 49]. As a result, the cavity lens configuration has been widely adopted in many recent plasmonic imaging experiments [33, 34, 50].

Besides superlens and cavity lens, metal–dielectric multilayer-based hyperbolic lens, i.e., the so-called hyperlens has also gained much interest of researchers because of its versatility in sub-diffraction light manipulation via plasmonic coupling (Fig. 9.8e) [51, 52]. Such multilayers could let light propagate in a ray-like way [53, 54], which makes magnifying or demagnifying imaging beyond the diffraction limit become possible. Figure 9.8f, g represents the experimental demonstrations of spherical hyperlenses for microscopy and cylindrical hyperlens for demagnifying lithography, respectively [43, 44, 55].

Compared with curved hyperlens, planar hyperlenses are not only easier to fabricate but also bear many other novel properties such as negative refraction and spatial frequency filtering, as shown in Fig. 9.8h, i. It is shown that the interference of light spectra generated by a concentric grating combined with planar hyperlens will lead to circularly polarized Bessel beam [46]. As a sub-diffraction form of the so-called diffractionless solution of the Helmholtz equation [56], the near-field Bessel beam could be utilized as a virtual scanning tip for scanning imaging or nanolithography.

With the invention of various sub-diffraction-limited metalenses, it is possible to construct novel optical systems to realize super-resolution imaging for practical applications. Figure 9.8j illustrates a sample machine for plasmonic lithography, where classic optical elements are combined with subwavelength structures. Among

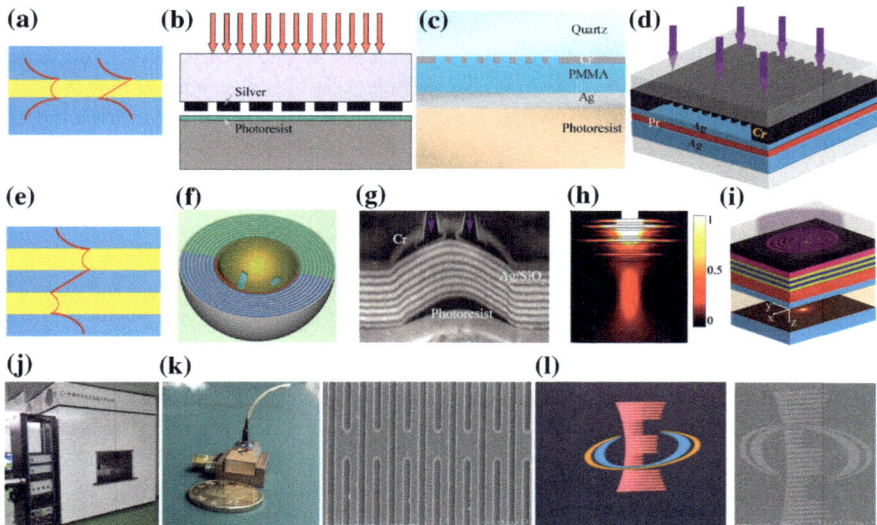

Fig. 9.8 Coupled surface plasmons for sub-diffraction-limited imaging and nanolithography. **a** Schematic of the even and odd modes in a thin metallic layer. **b** Schematic for the EYI and super-resolution imaging. Reproduced from [19] with permission. Copyright 2004, American Institute of Physics. **c** Sketch map of the superlens lithography. Reproduced from [42] with permission. Copyright 2005, The Authors. **d** Configuration of the cavity lens. **e** Schematic of the plasmonic coupling in metal–dielectric multilayers. **f** Spherical hyperlens for magnifying super-resolution imaging. Reproduced from [43] with permission. Copyright 2010, Springer Nature. **g** Cylindrical hyperlens for demagnifying lithography. Reproduced from [44] with permission. Copyright 2016, The Royal Society of Chemistry. **h** Near-field imaging based on negative refraction. Reproduced from [45] with permission. Copyright 2013, Springer Nature. **i** Realization of near-field Bessel beam via the combination of hyperlens and concentric rings. Reproduced from [46] with permission. Copyright 2017, The Royal Society of Chemistry. **j** Photography of the plasmonic lithography machine. **k** Photography and SEM image of the fabricated chip for single photon detection. **l** Photograph and SEM image of the device for color display. **j–l** Reproduced from [40] with permission. Copyright 2018, The Authors. The whole figure is adapted from [6] with permission. Copyright 2018, WILEY-VCH Verlag GmbH & Co. KGaA, Weinheim

various functional devices fabricated using this approach, the integrated chip for single photon detection and the device for full-color displaying shown in Fig. 9.8k, l represent two initial demonstration of the super-resolution fabrication capability. More complex structures such as the geometric metasurfaces based on nanoslits array have also been demonstrated using reflective and cavity-enhanced metalenses [50, 57].

In the abovementioned approach, the recovery of evanescent wave is the key to realize super-resolution imaging, thus their operation modes are limited to work in the near-field. Since far-field imaging applications such as cameras and telescopes are more frequently involved in our daily life and scientific researches, it is important to answer whether sub-diffraction approach can be applied for far-field applications

[58]. In optical telescopes, both the objects and images are located at the far-field of lens, so it seems impossible to make use of the evanescent waves excited by subwavelength structures on the lens. Nevertheless, it has been shown that evanescent waves can have significant impacts even in this condition. Since subwavelength structures may introduce broadband phase shift required to modulate the pupil function [59], which results in a superoscillatory function in the focal plane, the classic Abbe–Rayleigh's diffraction limit can be surpassed [60].

9.3.2 Generalized Laws of Reflection and Refraction

The laws of reflection and refraction result in the technology convention that uses curved geometry to control the propagation direction of light. To meet the ever-growing requirement for higher resolution, the aperture size of both lenses and mirrors must be greatly increased to be larger than several meters. Consequently, the thickness or height should grow correspondingly, leading to a significant increase of weight and cost [61]. In such a condition, flat optical devices such as diffractive optical elements (DOEs) including Fresnel lens and Fresnel zone plates, have been intensively investigated [62, 63]. Nevertheless, since the element size in traditional DOEs are much larger than the operational wavelength, the numerical aperture must be very small. Furthermore, the multilevel structures in DOEs are not compatible with micro-electronic fabrication technologies [4].

In an effort to replace bulky and heavy optical elements with thinner and more compact ones, nanoslits array with gradient widths (Fig. 9.9a) are used to introduce localized gradient phase by tuning the coupling strength of catenary optical fields [27, 64]. As a variation of the thickness-dependent plasmon dispersion on a single metallic film [20], the propagation constant as well as the phase shift of gap plasmons in MIM slits can be tuned via the width [27]. Since the phase shift is controlled at the deep subwavelength scale, almost arbitrary phase profile could be produced, which eliminates the spherical aberration easily. After the experimental demonstration of the nanoslits lens in the visible regime (Fig. 9.9b) [65], both numerical and experimental efforts have been devoted to extend such a design to be polarization independent with square or circular shaped holes (Fig. 9.9c) [66, 67]. Figure 9.9d illustrates a catenary metalens based on geometric phase [68]. Unlike conventional diffractive optical elements with wavelength-dependent diffraction efficiency, almost all the diffracted light could be directed to predefined directions in an ultra-wideband frequency range owing to the judicious combination of geometric phase and continuous structures [68, 69]. This high-efficiency and achromatic phase shift can be understood from the generalized Fresnel's equations [31] and geometric phases.

The above gradient subwavelength structures are critical to the "generalized laws of reflection and refraction" [31, 72]:

$$n_1 k_0 \sin \theta_i + \nabla \Phi_r(\tau) = n_1 k_0 \sin \theta_r,$$

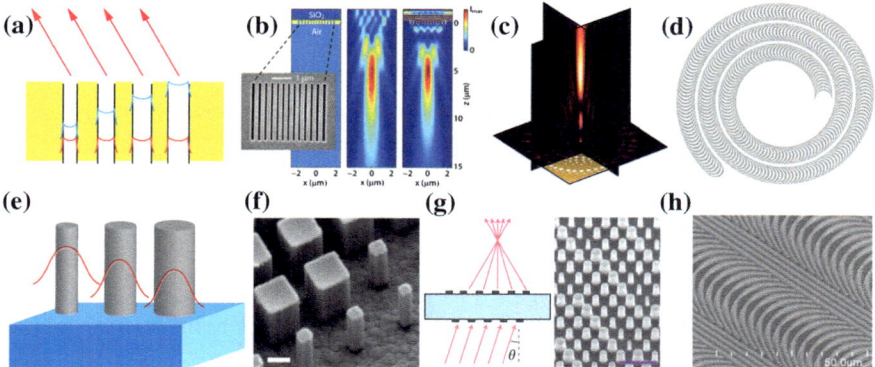

Fig. 9.9 Plasmonic and dielectric metasurfaces for flat optics. **a** Schematic of the plasmonic flat optical devices based on coupled surface plasmons. **b** SEM image, simulated and experimental field distributions for a plasmonic slits lens. Reproduced from [65] with permission. Copyright 2009, American Chemical Society. **c** Plasmonic flat lens based on circular holes. Reproduced from [66] with permission. Copyright 2013, American Chemical Society. **d** Schematic of the catenary metalens. Reproduced from [68] with permission, Copyright 2015, The Authors. **e** Sketch map of the coupled dielectric rods for wavefront manipulation. **f** SEM image of flat lens based on rectangular dielectric pillars. Scale bar: 200 nm. Reproduced from [70] with permission. Copyright 2017, American Chemical Society. **g** Schematic and SEM images of the doublet wide-angle flat lens. Reproduced from [71] with permission. Copyright 2016, The Authors. **h** SEM image of the dielectric catenary structures. The whole figure is adapted from [6] with permission. Copyright 2018, WILEY-VCH Verlag GmbH & Co. KGaA, Weinheim

$$n_1 k_0 \sin \theta_i + \nabla \Phi_t(\tau) = n_2 k_0 \sin \theta_t, \tag{9.3.3}$$

where $\nabla \Phi$ is the phase gradient in the metasurface plane and can be understood as a horizontal wavevector. n_1 and n_2 the refractive indices of media at the incident and transmit sides, θ_i, θ_r and θ_t the angles for incident, reflected and refracted light, τ denotes the time dependence. Since $\nabla \Phi$ is a vector, the reflective and refractive angles can not only be tuned in amplitudes, but also inside or outside the incident plane. Via external stimuli, $\nabla \Phi(\tau)$ may vary with time, providing dynamic tunability of the reflection and refraction.

Owing to significantly reduced loss compared with metallic structures, dielectric structures have been intensively investigated to increase the efficiency of phase modulation. However, compared with the metallic counterparts, the dielectric resonators are less isolated thus unwanted energy cross-talking may occur especially in the case of oblique incidence. To overcome this problem, super cells consisting of several identical single resonators are often adopted to achieve the designed phase and amplitude response [73]. In some cases, as noted by Lalanne et al. [74], by increasing the refractive index and carefully choosing the geometric parameters, the unwanted coupling can be minimized (Fig. 9.9e) to realize isolated and local phase response similar to the metallic waveguides. Figure 9.9f, g show a dielectric metalens and a double-layered wide-angle lens based on dynamic propagation phase with high-index dielectrics such as titanium dioxide (TiO_2) and silicon [70, 71]. Furthermore, by integrating dielectric resonance in high-aspect-ratio gratings [73] with the catenary structures shown in Fig. 9.9h, the overall efficiency can be further improved to be above 90% in a continuous wideband and wide-angle of incidences.

Although flat metalens has greatly reduced the thickness compared with refractive lens and has much higher efficiency than diffractive zone plate, this technology itself is still not perfect. One big drawback of these lenses, no matter composed of metallic or dielectric resonators, is that they have chromatic dispersion similar to DOEs [75]. For an achromatic lens with a focal length of f, the ideal phase profile is inversely proportional to the wavelength. For a metalens designed based on geometric phase, the phase shift is independent of the wavelength. In other words, the actual focal length varies across a wavelength range. As a result, the geometric phase is intrinsically related to chromatic dispersion and a wavelength-dependent phase shift is required to compensate these dispersions. Although the phase shift induced by metallic and dielectric resonance can be utilized to realize the required phase compensation [76, 77], such methods are only suitable for small-aperture lens, which has relatively small chromatic dispersion. Based on these considerations, we expect the combination of traditional optics and flat optics will be a promising route to reshape the engineering imaging optics [59, 62].

9.3.3 Generalized Theory for Absorption and Radiation

Engineering optics not only deals with propagation and transformation of light fields but also seeks to manage the absorption and emission processes of photons for applications such as electromagnetic detectors, solar cells, lasers, and radiative cooling. In this section, we focus on how subwavelength structures could greatly improve the performance of optical absorbers in terms of efficiency, bandwidth, thickness, operational temperatures, etc.

From a fundamental point of view, there is an ultimate limitation on the bandwidth and thickness of optical and electromagnetic absorbing materials. As early as in the 1900s, Max Planck revealed that the thickness of electromagnetic absorber should be large enough to achieve complete absorption [7, 78]. This judgment was rigorously demonstrated in 2000 by Konstantin Rozanov using causality and Kramers–Kronig relation [79]. For non-magnetic layered absorbers, the minimal thickness can be written as:

$$h \geq \frac{(\lambda_{max} - \lambda_{min})\Gamma}{172} = \frac{\Delta\lambda\Gamma}{172}, \tag{9.3.4}$$

where Γ is the reflectance in dB. This formula sets an ultimate object for classic optics, but the thicknesses of classic absorbing materials such as doped silicon and carbon-filled composites are far from this goal. For example, it was reported that

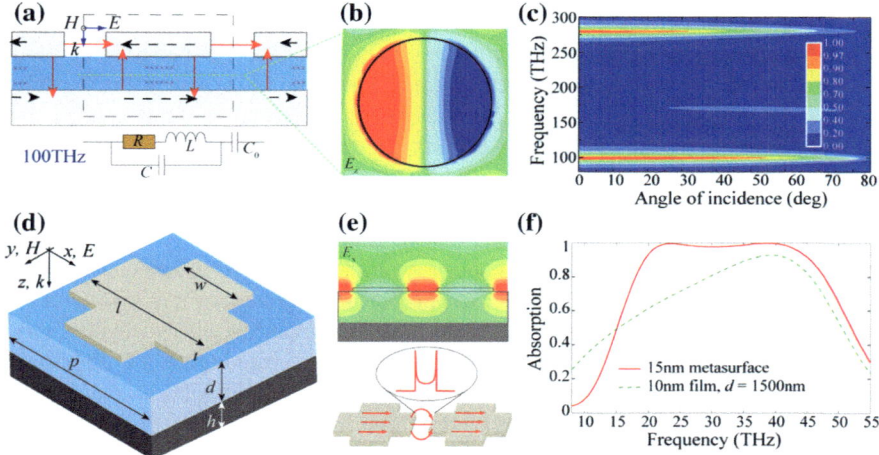

Fig. 9.10 Coupled subwavelength structures for broadband absorbers. **a** Schematic of the electric fields distribution of a perfect absorber. **b** Simulated vertical electric fields in the dielectric spacer. **c** Dependence of the dual-band absorption over the incidence angle and frequency. **a–c** Adapted from [83] with permission. Copyright 2011, Optical Society of America. **d** Schematic of a broadband absorber based on cross-shaped surface resonators. **e** Electric fields distribution illustrating a series connected equivalent circuit. The magnified region illustrates the catenary optical fields. **f** Comparison of the absorbance for the structured absorber and a Salisbury-type absorber with the same thickness. **d–f** Reproduced from [85] with permission. Copyright 2012, Optical Society of America. The whole figure is adapted from [6] with permission. Copyright 2018, WILEY-VCH Verlag GmbH & Co. KGaA, Weinheim

a 2-μm-tall carbon nanotube forest would have an emittance/absorbance of 0.97 in 0.2–12 μm, which is much thicker than the theoretical limit of 1 μm [80].

When subwavelength structures are used as electromagnetic absorbers, four obvious advantages over traditional absorbers have been revealed in recent studies. First of all, by converting the propagating direction of incident wave from vertical axis to horizontal axis parallel to the surface [81, 82], the thickness of absorbers can be significantly reduced. Figure 9.10a is a schematic of the most commonly used unit cell configuration that can be found in various literature [83, 84], where a metallic patch is placed above a dielectric spacer and a metallic reflective layer. The scattering induced by the patch array could convert freely propagating waves into localized or horizontally propagating light inside the dielectric layer. Figure 9.10b shows the localized vertical electric fields originating from the magnetic coupling, which enables wide-angle absorption depicted in Fig. 9.10c [83]. In principle, both the horizontal and vertical couplings in the form of catenary shapes are of critical importance to optimize the absorption efficiency [24]: while the vertical coupling provides the magnetic resonance, the horizontal coupling is responsible for the dispersion engineering of the metasurface.

Second, when the dispersion of structured materials cancels with the propagation dispersion, broadband destructive interference, and absorption may be realized within limited thickness [86, 87]. As can be deduced from a general transfer matrix method, a structured sheet material with designed impedance dispersion may double the absorption bandwidth [85] of commonly layered absorbers with Salisbury or Jaumann configurations [88]. It is well-known that destructive interference is the main absorption mechanism for Salisbury-type absorbers, therefore the bandwidth enlarging phenomenon is attributed to the dispersion compensation of phase shift in structured surfaces. Figure 9.10d, e illustrate a cross-shaped surface structure and the simulated catenary-shaped electric fields distribution [85]. Figure 9.10f plots the absorption curves comparing the Salisbury-type and structured absorbers, showing a two-fold enhancement of bandwidth without increasing the thickness. To obtain a larger absorption bandwidth, multilayer designs [89], together with horizontal mixture of resonances [90, 91] have been intensively investigated since the dispersion property of a single-layered metasurface is physically limited.

Third, by combining the concepts of coherent perfect absorption and dispersion engineering of metallic thin film, one can construct absorbers with unparalleled bandwidth and thin-profile advantages beyond the Planck–Rozanov limit, which means achieving ultrabroadband absorption with an ultrathin thickness [92, 93]. As predicted by Pu et al., this broadband absorption can be seen as a time-reverse process of oscillating electric current sheet [93], which is intrinsically broadband, unlike the resonant nature of lasing and anti-lasing process [94]. It is striking that even a sub-nanometric thick resistive metal or graphene can absorb almost all incoming microwaves under coherent conditions [95]. Therefore, the broadband coherent perfect absorption provides a general guidance on how to design broadband absorbers with significantly reduced thickness and increased bandwidth.

At last, classic absorbers are often not tunable once fabricated. Nevertheless, the electromagnetic spectrum of subwavelength absorbers may be dynamically tuned via external electric and optical stimuli, since tunable materials and elements, including semiconductors, graphene, phase-change materials and Micro-Electromechanical System (MEMS) devices, could be easily integrated with metallic structures [96–98]. When the absorption spectra are merged to form an effective broadband absorption curve, the fundamental limit on bandwidth and thickness may be surpassed to some extent.

The above results indicate that the super-absorbers based on subwavelength structures are very promising as an alternative to traditional absorbers, which may find extensive applications in both the optical and microwave regimes. Compared with traditional absorbing materials and devices, the subwavelength structured absorbers have more degrees of design freedom. However, it is still hard to realize multispectral absorption design, e.g. perfect absorption in the infrared atmosphere window while transparency in the visible range, with simple metal, dielectrics and semiconductors. Consequently, it may be essential to combine subwavelength structures and various traditional materials with natural resonances to realize multispectral optical manipulation. These on-demand absorption spectra may have widespread applications in daytime radiative cooling [99, 100], detectors and radiators [101, 102].

As an inverse process of absorption, electromagnetic radiation has many different sources, ranging from the oscillation of electrons to spontaneous and stimulated emission of atomic systems, which are described by classic electrodynamics and quantum theories. In either case, single radiators or emitters usually have limited efficiency due to either impedance mismatch or limited density of state. For engineering optics, we are more concerned about quantum light sources such as thermal radiators and lasers. First of all, although Planck's formula presents a universal description of the maximal thermal radiation of ideal blackbody, recent results have shown that the thermal emission of structured materials can exceed this limit [103, 104]. There are two physical mechanisms: On the first hand, the near-field radiation is not included in Planck's original formula [105]. Owing to the evanescent coupling described by a catenary function, the radiative heat transfer coefficient can be larger than the far-field value by several orders of magnitude [106]. On the other hand, since Planck's law is defined upon infinite large absorber, it is possible to break this limit with finite-size absorbers. For instance, it is well-known that the scattering and absorption cross sections of a small object can be much larger than the geometric cross section in Mie's scattering theory, which means that such object could scatter or absorb much more light than its geometric area can collect [107]. According to Kirchhoff's theory, at thermal equilibrium condition, the emittance of a material equals to its absorbance (defined by the intensity of light), therefore the amount of thermal radiation can be much larger than that defined with the object's geometric size.

9.4 Conclusions and Outlooks

In this chapter, the role of catenary optics in EO 2.0 is highlighted. It is shown that catenary optical fields and catenary structures are indispensable in the subwavelength optical materials and devices to revolutionize classic engineering optics. Based on catenary optics, many classic optical theories have been revised and generalized, opening a door for many fascinating applications. Since a detailed description of EO 2.0 is planned in another monograph, here we only give a concise discussion.

As a final remark, it should be noted that engineering optics is only one important application of catenary optical fields and catenary subwavelength structures. In the foreseeable future, catenary optics would bring a series of revolutions in optical materials, devices, and systems, in a wide range of areas such as optical imaging, nanolithography, wireless communications, sensors/detectors, display, solar cells, thermal management as well as stealth technology. In the fundamental aspects, catenary optics may boost related researches including nanophotonics, acoustics, quantum tunneling, and superstring theory.

References

1. K. Iizuka, *Engineering Optics*, 3rd edn. (Springer, 2008)
2. P. Lalanne, P. Chavel, Metalenses at visible wavelengths: past, present, perspectives. Laser Photonics Rev. **11**, 1600295 (2017)
3. X. Luo, Subwavelength optical engineering with metasurface waves. Adv. Opt. Mater. **6**, 1701201 (2018)
4. F. Capasso, The future and promise of flat optics: a personal perspective. Nanophotonics **7**, 953 (2018)
5. X. Luo, Engineering optics 2.0: a revolution in optical materials, devices, and systems. ACS Photonics **5**, 4724-4738 (2018)
6. X. Luo, Subwavelength artificial structures: opening a new era for engineering optics. Adv. Mater. 1804680 (2018)
7. X. Luo, M. Pu, X. Ma, X. Li, Taming the electromagnetic boundaries via metasurfaces: from theory and fabrication to functional devices. Int. J. Antennas Propag. **2015**, 204127 (2015)
8. Willebrord Snellius, https://commons.wikimedia.org/wiki/File:Willebrord_Snellius.jpg
9. Pierre de Fermat, https://commons.wikimedia.org/wiki/File:Pierre_de_Fermat.jpg
10. Thomas Young,https://commons.wikimedia.org/wiki/File:LifeOfThomasYoung1855PeacockG. jpg
11. Augustin Fresnel, https://commons.wikimedia.org/wiki/File:Augustin_Fresnel.jpg
12. James Clerk Maxwell, https://commons.wikimedia.org/wiki/File:James_Clerk_Maxwell.png
13. Gustav   Robert   Kirchhoff,   https://commons.wikimedia.org/wiki/File:Gustav_Robert_Kirchhoff.jpg
14. Ernst Abbe, https://commons.wikimedia.org/wiki/File:Ernst_Abbe.jpg
15. Max Planck,https://commons.wikimedia.org/wiki/File:Max_Planck_1933.jpg
16. Albert Einstein, https://commons.wikimedia.org/wiki/File:Albert_Einstein_(Nobel).png
17. R.P. Feynman, R.B. Leighton, M. Sands, *The Feynman Lectures on Physics* (Basic Books, 1963)
18. R.P. Crease, The most beautiful experiment. Phys. World **15**, 19 (2002)
19. X. Luo, T. Ishihara, Surface plasmon resonant interference nanolithography technique. Appl. Phys. Lett. **84**, 4780–4782 (2004)
20. X. Luo, T. Ishihara, Subwavelength photolithography based on surface-plasmon polariton resonance. Opt. Express **12**, 3055–3065 (2004)
21. H.F. Schouten, N. Kuzmin, G. Dubois, T.D. Visser, G. Gbur, P.F.A. Alkemade, H. Blok, G. W.'t Hooft, D. Lenstra, E.R. Eliel, Plasmon-assisted two-slit transmission: Young's experiment revisited. Phys. Rev. Lett. **94**, 053901 (2005)
22. R. Zia, M.L. Brongersma, Surface plasmon polariton analogue to Young's double-slit experiment. Nat. Nanotechnol. **2**, 426 (2007)
23. X. Luo, D. Tsai, M. Gu, M. Hong, Subwavelength interference of light on structured surfaces. Adv. Opt. Photonics **10**, 757–842 (2018)
24. M. Pu, Y. Guo, X. Li, X. Ma, X. Luo, Revisitation of extraordinary Young's interference: from catenary optical fields to spin-orbit interaction in metasurfaces. ACS Photonics **5**, 3198–3204 (2018)
25. H. Shi, X. Luo, C. Du, Young's interference of double metallic nanoslit with different widths. Opt. Express **15**, 11321–11327 (2007)
26. T. Xu, Y. Zhao, D. Gan, C. Wang, C. Du, X. Luo, Directional excitation of surface plasmons with subwavelength slits. Appl. Phys. Lett. **92**, 101501 (2008)
27. T. Xu, C. Wang, C. Du, X. Luo, Plasmonic beam deflector. Opt. Express **16**, 4753–4759 (2008)
28. T.W. Ebbesen, H.J. Lezec, H.F. Ghaemi, T. Thio, P.A. Wolff, Extraordinary optical transmission through sub-wavelength hole arrays. Nature **391**, 667–669 (1998)
29. R. Welti, Light transmission through two slits: the Young experiment revisited. J. Opt. Pure Appl. Opt. **8**, 606 (2006)

30. M. Pu, X. Ma, Y. Guo, X. Li, X. Luo, Theory of microscopic meta-surface waves based on catenary optical fields and dispersion. Opt. Express **26**, 19555–19562 (2018)
31. X. Luo, Principles of electromagnetic waves in metasurfaces. Sci. China-Phys. Mech. Astron. **58**, 594201 (2015)
32. J.B. Pendry, Negative refraction makes a perfect lens. Phys. Rev. Lett. **85**, 3966–3969 (2000)
33. Z. Zhao, Y. Luo, W. Zhang, C. Wang, P. Gao, Y. Wang, M. Pu, N. Yao, C. Zhao, X. Luo, Going far beyond the near-field diffraction limit via plasmonic cavity lens with high spatial frequency spectrum off-axis illumination. Sci. Rep. **5**, 15320 (2015)
34. P. Gao, N. Yao, C. Wang, Z. Zhao, Y. Luo, Y. Wang, G. Gao, K. Liu, C. Zhao, X. Luo, Enhancing aspect profile of half-pitch 32 nm and 22 nm lithography with plasmonic cavity lens. Appl. Phys. Lett. **106**, 093110 (2015)
35. D.O.S. Melville, R.J. Blaikie, Super-resolution imaging through a planar silver layer. Opt. Express **13**, 2127–2134 (2005)
36. N. Fang, H. Lee, C. Sun, X. Zhang, Sub-diffraction-limited optical imaging with a silver superlens. Science **308**, 534–537 (2005)
37. L. Pan, Y. Park, Y. Xiong, E. Ulin-Avila, Y. Wang, L. Zeng, S. Xiong, J. Rho, C. Sun, D.B. Bogy, X. Zhang, Maskless plasmonic lithography at 22 nm resolution. Sci. Rep. **1**, 175 (2011)
38. F.J. Garcia-Vidal, L. Martin-Moreno, T.W. Ebbesen, L. Kuipers, Light passing through sub-wavelength apertures. Rev. Mod. Phys. **82**, 729–787 (2010)
39. E. Ozbay, Plasmonics: merging photonics and electronics at nanoscale dimensions. Science **311**, 189–193 (2006)
40. X. Luo, Plasmonic metalens for nanofabrication. Natl. Sci. Rev. **5**, 137–138 (2018)
41. T. Laufer, Thermal Fluid-Structure analysis of an optical device including radiation and conduction, in *Star European Conference* (2011)
42. H. Lee, Y. Xiong, N. Fang, W. Srituravanich, S. Durant, M. Ambati, C. Sun, X. Zhang, Realization of optical superlens imaging below the diffraction limit. New J. Phys. **7**, 255 (2005)
43. J. Rho, Z. Ye, Y. Xiong, X. Yin, Z. Liu, H. Choi, G. Bartal, X. Zhang, Spherical hyperlens for two-dimensional sub-diffractional imaging at visible frequencies. Nat. Commun. **1**, 143 (2010)
44. L. Liu, K. Liu, Z. Zhao, C. Wang, P. Gao, X. Luo, Sub-diffraction demagnification imaging lithography by hyperlens with plasmonic reflector layer. RSC Adv. **6**, 95973–95978 (2016)
45. T. Xu, A. Agrawal, M. Abashin, K.J. Chau, H.J. Lezec, All-angle negative refraction and active flat lensing of ultraviolet light. Nature **497**, 470–474 (2013)
46. L. Liu, P. Gao, K. Liu, W. Kong, Z. Zhao, M. Pu, C. Wang, X. Luo, Nanofocusing of circularly polarized Bessel-type plasmon polaritons with hyperbolic metamaterials. Mater. Horiz. **4**, 290–296 (2017)
47. J.A. Coles, Some reflective properties of the tapetum lucidum of the cat's eye. J. Physiol. **212**, 393–409 (1971)
48. T. Xu, L. Fang, J. Ma, B. Zeng, Y. Liu, J. Cui, C. Wang, Q. Feng, X. Luo, Localizing surface plasmons with a metal-cladding superlens for projecting deep-subwavelength patterns. Appl. Phys. B **97**, 175–179 (2009)
49. L. Bourke, R.J. Blaikie, Herpin effective media resonant underlayers and resonant overlayer designs for ultra-high NA interference lithography. J. Opt. Soc. Am. A **34**, 2243–2249 (2017)
50. L. Liu, X. Zhang, Z. Zhao, M. Pu, P. Gao, Y. Luo, J. Jin, C. Wang, X. Luo, Batch fabrication of metasurface holograms enabled by plasmonic cavity lithography. Adv. Opt. Mater. **5**, 1700429 (2017)
51. A. Poddubny, I. Iorsh, P. Belov, Y. Kivshar, Hyperbolic metamaterials. Nat. Photonics **7**, 948–957 (2013)
52. A.A. Orlov, S.V. Zhukovsky, I.V. Iorsh, P.A. Belov, Controlling light with plasmonic multilayers. Photonics Nanostruct. - Fundam. Appl. **14**, 213–230 (2014)
53. W. Wang, H. Xing, L. Fang, Y. Liu, J. Ma, L. Lin, C. Wang, X. Luo, Far-field imaging device: planar hyperlens with magnification using multi-layer metamaterial. Opt. Express **16**, 21142–21148 (2008)

54. S. Han, Y. Xiong, D. Genov, Z. Liu, G. Bartal, X. Zhang, Ray optics at a deep-subwavelength scale: a transformation optics approach. Nano Lett. **8**, 4243–4247 (2008)
55. J. Sun, T. Xu, N.M. Litchinitser, Experimental demonstration of demagnifying hyperlens. Nano Lett. **16**, 7905–7909 (2016)
56. A. Dudley, M.P.J. Lavery, M.J. Padgett, A. Forbes, Unraveling Bessel beams. Opt. Photonics News **22**, 24–29 (2013)
57. J. Luo, B. Zeng, C. Wang, P. Gao, K. Liu, M. Pu, J. Jin, Z. Zhao, X. Li, H. Yu, X. Luo, Fabrication of anisotropically arrayed nano-slots metasurfaces using reflective plasmonic lithography. Nanoscale **7**, 18805–18812 (2015)
58. F. Qin, M. Hong, Breaking the diffraction limit in far field by planar metalens. Sci. China Phys. Mech. Astron. **60**, 044231 (2017)
59. Z. Li, T. Zhang, Y. Wang, W. Kong, J. Zhang, Y. Huang, C. Wang, X. Li, M. Pu, X. Luo, Achromatic broadband super-resolution imaging by super-oscillatory metasurface. Laser Photonics Rev. **12**, 1800064 (2018)
60. C. Wang, D. Tang, Y. Wang, Z. Zhao, J. Wang, M. Pu, Y. Zhang, W. Yan, P. Gao, X. Luo, Super-resolution optical telescopes with local light diffraction shrinkage. Sci. Rep. **5**, 18485 (2015)
61. H.P. Stahl, Survey of cost models for space telescopes. Opt. Eng. **49**, 053005 (2010)
62. R.A. Hyde, Eyeglass. 1. Very large aperture diffractive telescopes. Appl. Opt. **38**, 4198–4212 (1999)
63. P.D. Atcheson, C. Stewart, J. Domber, K. Whiteaker, J. Cole, P. Spuhler, A. Seltzer, J.A. Britten, S.N. Dixit, B. Farmer, L. Smith, MOIRE: initial demonstration of a transmissive diffractive membrane optic for large lightweight optical telescopes, in (2012), Vol. 8442, pp. 844221-8442–14
64. Y. Li, X. Li, M. Pu, Z. Zhao, X. Ma, Y. Wang, X. Luo, Achromatic flat optical components via compensation between structure and material dispersions. Sci. Rep. **6**, 19885 (2016)
65. L. Verslegers, P.B. Catrysse, Z. Yu, J.S. White, E.S. Barnard, M.L. Brongersma, S. Fan, Planar lenses based on nanoscale slit arrays in a metallic film. Nano Lett. **9**, 235–238 (2009)
66. S. Ishii, V.M. Shalaev, A.V. Kildishev, Holey-metal lenses: sieving single modes with proper phases. Nano Lett. **13**, 159–163 (2013)
67. L. Lin, X.M. Goh, L.P. McGuinness, A. Roberts, Plasmonic lenses formed by two-dimensional nanometric cross-shaped aperture arrays for Fresnel-region focusing. Nano Lett. **10**, 1936 (2010)
68. M. Pu, X. Li, X. Ma, Y. Wang, Z. Zhao, C. Wang, C. Hu, P. Gao, C. Huang, H. Ren, X. Li, F. Qin, J. Yang, M. Gu, M. Hong, X. Luo, Catenary optics for achromatic generation of perfect optical angular momentum. Sci. Adv. **1**, e1500396 (2015)
69. X. Li, M. Pu, Z. Zhao, X. Ma, J. Jin, Y. Wang, P. Gao, X. Luo, Catenary nanostructures as highly efficient and compact Bessel beam generators. Sci. Rep. **6**, 20524 (2016)
70. M. Khorasaninejad, Z. Shi, A.Y. Zhu, W.T. Chen, V. Sanjeev, A. Zaidi, F. Capasso, Achromatic metalens over 60 nm bandwidth in the visible and metalens with reverse chromatic dispersion. Nano Lett. **17**, 1819–1824 (2017)
71. A. Arbabi, E. Arbabi, S.M. Kamali, Y. Horie, S. Han, A. Faraon, Miniature optical planar camera based on a wide-angle metasurface doublet corrected for monochromatic aberrations. Nat. Commun. **7**, 13682 (2016)
72. Y. Xu, Y. Fu, H. Chen, Planar gradient metamaterials. Nat. Rev. Mater. **1**, 16067 (2016)
73. D. Lin, P. Fan, E. Hasman, M.L. Brongersma, Dielectric gradient metasurface optical elements. Science **345**, 298–302 (2014)
74. P. Lalanne, P. Chavel, Metalenses at visible wavelengths: an historical fresco. Proc SPIE **10113**, 101130F (2017)
75. A.A. Fathnan, D.A. Powell, Bandwidth and size limits of achromatic printed-circuit metasurfaces. Opt. Express **26**, 29440–29450 (2018)
76. W.T. Chen, A.Y. Zhu, V. Sanjeev, M. Khorasaninejad, Z. Shi, E. Lee, F. Capasso, A broadband achromatic metalens for focusing and imaging in the visible. Nat. Nanotechnol. **13**, 220–226 (2018)

77. S. Wang, P.C. Wu, V.-C. Su, Y.-C. Lai, M.-K. Chen, H.Y. Kuo, B.H. Chen, Y.H. Chen, T.-T. Huang, J.-H. Wang, R.-M. Lin, C.-H. Kuan, T. Li, Z. Wang, S. Zhu, D.P. Tsai, A broadband achromatic metalens in the visible. Nat. Nanotechnol. **13**, 227–232 (2018)
78. M. Planck, *The Theory of Heat Radiation* (P. Blakiston's Son & Co., 1914)
79. K.N. Rozanov, Ultimate thickness to bandwidth ratio of radar absorbers. IEEE Trans. Antennas Propag. **48**, 1230–1234 (2000)
80. K. Mizuno, J. Ishii, H. Kishida, Y. Hayamizu, S. Yasuda, D.N. Futaba, M. Yumura, K. Hata, A black body absorber from vertically aligned single-walled carbon nanotubes. Proc. Natl. Acad. Sci. U. S. A. **106**, 6044–6047 (2009)
81. C. Hu, Z. Zhao, X. Chen, X. Luo, Realizing near-perfect absorption at visible frequencies. Opt. Express **17**, 11039–11044 (2009)
82. A. Moreau, C. Ciraci, J.J. Mock, R.T. Hill, Q. Wang, B.J. Wiley, A. Chilkoti, D.R. Smith, Controlled-reflectance surfaces with film-coupled colloidal nanoantennas. Nature **492**, 86–89 (2012)
83. M. Pu, C. Hu, M. Wang, C. Huang, Z. Zhao, C. Wang, Q. Feng, X. Luo, Design principles for infrared wide-angle perfect absorber based on plasmonic structure. Opt. Express **19**, 17413–17420 (2011)
84. T.D. Dao, K. Chen, S. Ishii, A. Ohi, T. Nabatame, M. Kitajima, T. Nagao, Infrared perfect absorbers fabricated by colloidal mask etching of Al–Al2O3–Al trilayers. ACS Photonics **2**, 964–970 (2015)
85. Q. Feng, M. Pu, C. Hu, X. Luo, Engineering the dispersion of metamaterial surface for broadband infrared absorption. Opt. Lett. **37**, 2133–2135 (2012)
86. D. Ye, Z. Wang, K. Xu, H. Li, J. Huangfu, Z. Wang, L. Ran, Ultrawideband dispersion control of a metamaterial surface for perfectly-matched-layer-like absorption. Phys. Rev. Lett. **111**, 187402 (2013)
87. M.R. Singh, K. Davieau, J.J.L. Carson, Effect of quantum interference on absorption of light in metamaterial hybrids. J. Phys. Appl. Phys. **49**, 445103 (2016)
88. W.W. Salisbury, Absorbent body for electromagnetic waves, U.S. patent 2599944 (1952)
89. A. Naqavi, S.P. Loke, M.D. Kelzenberg, D.M. Callahan, T. Tiwald, E.C. Warmann, P. Espinet-González, N. Vaidya, T.A. Roy, J.-S. Huang, T.G. Vinogradova, H.A. Atwater, Extremely broadband ultralight thermally-emissive optical coatings. Opt. Express **26**, 18545–18562 (2018)
90. C. Hu, L. Liu, Z. Zhao, X. Chen, X. Luo, Mixed plasmons coupling for expanding the bandwidth of near-perfect absorption at visible frequencies. Opt. Express **17**, 16745–16749 (2009)
91. C. Wu, G. Shvets, Design of metamaterial surfaces with broadband absorbance. Opt. Lett. **37**, 308–310 (2012)
92. S. Li, J. Luo, S. Anwar, S. Li, W. Lu, Z.H. Hang, Y. Lai, B. Hou, M. Shen, C. Wang, Broadband perfect absorption of ultrathin conductive films with coherent illumination: superabsorption of microwave radiation. Phys. Rev. B **91**, 220301(R) (2015)
93. M. Pu, Q. Feng, M. Wang, C. Hu, C. Huang, X. Ma, Z. Zhao, C. Wang, X. Luo, Ultrathin broadband nearly perfect absorber with symmetrical coherent illumination. Opt. Express **20**, 2246–2254 (2012)
94. W. Wan, Y. Chong, L. Ge, H. Noh, A.D. Stone, H. Cao, Time-reversed lasing and interferometric control of absorption. Science **331**, 889–892 (2011)
95. S. Li, Q. Duan, S. Li, Q. Yin, W. Lu, L. Li, B. Gu, B. Hou, W. Wen, Perfect electromagnetic absorption at one-atom-thick scale. Appl. Phys. Lett. **107**, 181112 (2015)
96. M.A. Kats, D. Sharma, J. Lin, P. Genevet, R. Blanchard, Z. Yang, M.M. Qazilbash, D.N. Basov, S. Ramanathan, F. Capasso, Ultra-thin perfect absorber employing a tunable phase change material. Appl. Phys. Lett. **101**, 221101 (2012)
97. P.-Y. Chen, C. Argyropoulos, A. Alù, Broadening the cloaking bandwidth with non-Foster metasurfaces. Phys. Rev. Lett. **111**, 233001 (2013)
98. X. Wu, C. Hu, Y. Wang, M. Pu, C. Huang, C. Wang, X. Luo, Active microwave absorber with the dual-ability of dividable modulation in absorbing intensity and frequency. AIP Adv. **3**, 022114 (2013)

  99. A.P. Raman, M.A. Anoma, L. Zhu, E. Rephaeli, S. Fan, Passive radiative cooling below ambient air temperature under direct sunlight. Nature **515**, 540–544 (2014)
 100. Y. Huang, M. Pu, P. Gao, Z. Zhao, X. Li, X. Ma, X. Luo, Ultra-broadband large-scale infrared perfect absorber with optical transparency. Appl. Phys. Express **10**, 112601 (2017)
 101. H.T. Miyazaki, T. Kasaya, M. Iwanaga, B. Choi, Y. Sugimoto, K. Sakoda, Dual-band infrared metasurface thermal emitter for CO2 sensing. Appl. Phys. Lett. **105**, 121107 (2014)
 102. A. Kohiyama, M. Shimizu, H. Yugami, Unidirectional radiative heat transfer with a spectrally selective planar absorber/emitter for high-efficiency solar thermophotovoltaic systems. Appl. Phys. Express **9**, 112302 (2016)
 103. Y. Guo, C.L. Cortes, S. Molesky, Z. Jacob, Broadband super-Planckian thermal emission from hyperbolic metamaterials. Appl. Phys. Lett. **101**, 131106 (2012)
 104. S.I. Maslovski, C.R. Simovski, S.A. Tretyakov, Overcoming black body radiation limit in free space: metamaterial superemitter. New J. Phys. **18**, 013034 (2016)
 105. J.B. Pendry, Radiative exchange of heat between nanostructures. J. Phys.: Condens. Matter **11**, 6621 (1999)
 106. L. Hu, A. Narayanaswamy, X. Chen, G. Chen, Near-field thermal radiation between two closely spaced glass plates exceeding Planck's blackbody radiation law. Appl. Phys. Lett. **92**, 133106 (2008)
 107. J. Ng, H. Chen, C.T. Chan, Metamaterial frequency-selective superabsorber. Opt. Lett. **34**, 644–646 (2009)

Appendix A
Catenary Function for a Freely Hanging Chain

The catenary function can be deduced using differential equation [1]. As shown in Fig. A.1, the forces acting on a segment include the tension T_0 at A (the lowest point on the chain), the tension T at B, and the weight of the chain $(0, -\rho g s)$, where ρ is the mass per unit length, g is the acceleration of gravity, and s is the length of chain between A and B.

The tension at A is horizontal, and it pulls the section to the left so it may be written as $(-T_0, 0)$ where T_0 is the magnitude of the force. The tension at B is parallel to the curve and pulls the section to the right, so it may be written as $(T\cos\varphi, T\sin\varphi)$, where T is the magnitude of the force and φ is the angle between the curve at B and the x-axis. Finally, the weight of the chain is represented by $(0, -\rho g s)$. The chain is in equilibrium so the sum of three forces is 0, therefore

$$T \cos \varphi = T_0 \tag{A.1}$$

and

$$T \sin \varphi = \rho g s, \tag{A.2}$$

and dividing these gives

$$\frac{dy}{dx} = \tan \varphi = \frac{\rho g s}{T_0}. \tag{A.3}$$

It is convenient to write

$$a = \frac{T_0}{\rho g}, \tag{A.4}$$

© Springer Nature Singapore Pte Ltd. 2019
X. Luo, *Catenary Optics*, https://doi.org/10.1007/978-981-13-4818-1

Fig. A.1 Schematic of the force acting on one segment of a catenary from A to B. Under equilibrium the sum of these forces must be zero

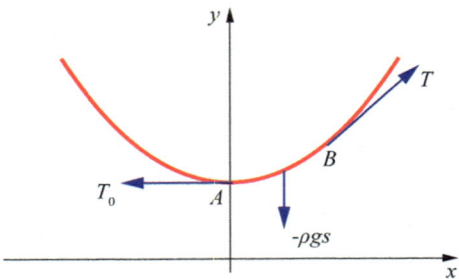

which is the length of chain whose weight is equal in magnitude to the tension. Then there is

$$\frac{dy}{dx} = \frac{s}{a}.$$ (A.5)

The formula for arc length is

$$\frac{ds}{dx} = \sqrt{1 + \left(\frac{dy}{dx}\right)^2} = \frac{\sqrt{a^2 + s^2}}{a}.$$ (A.6)

Then

$$\frac{dx}{ds} = \frac{1}{\frac{ds}{dx}} = \frac{a}{\sqrt{a^2 + s^2}}$$ (A.7)

and

$$\frac{dy}{ds} = \frac{\frac{dy}{dx}}{\frac{ds}{dx}} = \frac{s}{\sqrt{a^2 + s^2}}.$$ (A.8)

Equation (A.8) can be integrated to give

$$y = \sqrt{a^2 + s^2} + c_1,$$ (A.9)

and by shifting the position of the x-axis, c_1 can be taken to be 0. Then

$$y = \sqrt{a^2 + s^2}, \ y^2 = a^2 + s^2.$$ (A.10)

The x-axis thus chosen is called the directrix of the catenary. It follows that the magnitude of the tension at a point $T = \rho g y$ which is proportional to the distance between the point and the directrix. The integral of expression for dx/ds can be found using standard techniques giving

$$x = a \sinh^{-1}(s/a) + c_2 \tag{A.11}$$

and, again, by shifting the position of the y-axis, c_2 can be taken to be 0. Then

$$x = a \sinh^{-1}(s/a), \quad s = a\sinh\frac{x}{a}. \tag{A.12}$$

The y-axis thus chosen passes through the vertex. These results can be used to eliminate s giving

$$y = a \cosh\frac{x}{a}. \tag{A.13}$$

For catenary of equal strength, a similar approach can be used to obtain the function: In a catenary of equal strength, cable is strengthened according to the magnitude of the tension at each point, so its resistance to breaking is constant along its length. Assuming that the strength of the cable is proportional to its density per unit length, the weight, w, per unit length of the chain can be written T/c, where c is constant, and the analysis for nonuniform chains can be applied.

The equations for tension are

$$T \cos \varphi = T_0, \tag{A.14}$$

and

$$T \sin \varphi = \frac{1}{c} \int T ds. \tag{A.15}$$

By combining them there is

$$\tan \varphi = y' = \frac{1}{c} \int \frac{1}{\cos \varphi} ds. \tag{A.16}$$

And by differentiation with s, there are

$$\frac{dy'}{ds} = \frac{1}{c \cos \varphi}, \tag{A.17}$$

and

$$\frac{d}{dx}\frac{dx}{ds}y' = \frac{1}{c\cos\varphi}.$$ (A.18)

Since $\cos\varphi = dx/ds$, there are

$$\frac{1}{\frac{ds}{dx}}y'' = \frac{1}{c}\frac{ds}{dx},$$ (A.19)

$$y'' = \frac{1}{c}\left(\frac{ds}{dx}\right)^2,$$ (A.20)

and

$$y'' = \frac{1}{c}\left[1 + y'^2\right].$$ (A.21)

By separating variables, there are

$$\frac{dy'}{1 + y'^2} = \frac{1}{c}dx,$$ (A.22)

$$\int\frac{dy'}{1 + y'^2} = \frac{1}{c}\int dx,$$ (A.23)

and

$$\arctan(y') = \frac{1}{c}x + c_3.$$ (A.24)

By choosing c_3 to be zero, there is

$$y' = \tan\left(\frac{1}{c}x\right).$$ (A.25)

The solution to this is

$$y = c\ln\sec\frac{x}{c}.$$ (A.26)

Appendix B
Matlab Codes for GDSII File Generation

Note: All files are generated with the help of a Matlab toolbox provided by Ulf Griesmann, which can be downloaded from https://sites.google.com/site/ulfgri/numerical/gdsii-toolbox.

B.1 Matlab codes for a single catenary of equal strength.

```
clear all
clc
M=10;
N=50;
p=2;
w=0.2;
gdsii_units
gs = gds_structure('catenary');
x=linspace(-0.46*p,0.46*p,N);
for n=1:N
y1(n)=p/pi*log(abs(cos(pi*x(n)/p)));
end
y2=y1+w;
xx=[x,fliplr(x)];
yy=[y1,fliplr(y2)];
xy=[xx',yy'];
gs(end+1)=gds_element('boundary', 'xy',xy, 'layer',1);
glib = gds_library('array', 'uunit',1e-6, 'dbunit',1e-9, gs);
write_gds_library(glib, 'catenaryunit_2micron.gds')
```

B.2 Matlab codes for a 1D paraxial (wide-angle) catenary lens

In the following, the parameters are set as $f = 10~\mu m$ and $\lambda = 632.8$ nm.

```
clear all
clc
```

© Springer Nature Singapore Pte Ltd. 2019
X. Luo, *Catenary Optics*, https://doi.org/10.1007/978-981-13-4818-1

```matlab
M=6;
N=50;
p=2;
w=0.2;
gdsii_units
gs = gds_structure('catenary');
k=2*pi/0.632;
f=15;
for m=1:10
rt(m)=sqrt(4*f*(-pi/2+m*pi)/k);
end
fun=@(x) tan((k*x.^2./(4*f)));

Q=100;
for m=1:M
if m==1;
    rt0=0;
    x=linspace(0,rt(m)-0.01,Q);
else
    rt0=rt(m-1)+0.1;
    x=linspace(rt(m-1)+0.01,rt(m)-0.01,Q);
end
for qq=1:Q
    y(qq)=integral(fun,rt0,x(qq));
end
for n=1:N+1
    xt=x;
    ip=find(y==min(y));
    y1=y-y(ip);
    y11=y1;
    ymax=0.5*(max(x)-min(x));
    ip1=find(y1>ymax);

        y1(ip1)=ymax;

y2=y11-w;
ip2=find(y2>ymax);
y2(ip2)=ymax;

xt(ip2)=[];
y1(ip2)=[];
y2(ip2)=[];

xx=[xt,fliplr(xt)];
yy=[y1,fliplr(y2)]+2*n*w;
xy=[xx',yy'];
```

```
  gs(end+1)=gds_element('boundary', 'xy',xy, 'layer',1);
end
end

for m=1:M
if m==1;
    rt0=0;
    x=linspace(0,rt(m)-0.01,Q);
else
    rt0=rt(m-1)+0.1;
    x=linspace(rt(m-1)+0.01,rt(m)-0.01,Q);
end

for qq=1:Q

    y(qq)=integral(fun,rt0,x(qq));

end

for n=1:N+1
    xt=x;
    ip=find(y==min(y));
    y1=y-y(ip);
    y11=y1;
    ymax=0.5*(max(x)-min(x));
    ip1=find(y1>ymax);
    y1(ip1)=ymax;
    y2=y11-w;
    ip2=find(y2>ymax);
    y2(ip2)=ymax;
    xt(ip2)=[];
    y1(ip2)=[];
    y2(ip2)=[];
    xx=-[xt,fliplr(xt)];
    yy=-[y1,fliplr(y2)]+2*n*w+w;
    xy=[xx',yy'];
    gs(end+1)=gds_element('boundary', 'xy',xy, 'layer',1);
    end
end
glib = gds_library('array', 'uunit',1e-6, 'dbunit',1e-9, gs);
write_gds_library(glib, 'catenary1dfocus.gds')
```

B.3 Matlab codes for 2D catenary lens

The parameters are chosen as $|l| = 0$, $f = 10$ μm and $\lambda = 632.8$ nm.

```
clear all
clc
gdsii_units
gs = gds_structure('focuscatenary');
gs1 = gds_structure('focuscatenary');
k=2*pi/0.632;
N=1000;
k=2*pi/0.632;
f=10;
phi0=18*2*pi+pi;
l=-2;
t=0*pi;
for m=1:N
a00=(phi0-l*(t+pi))/k;
rt00=sqrt(a00.^2-f^2);
t=t+0.2/rt00;
a=(phi0-l*(t+1*pi))/k;
a1=(phi0-l*(t+2*pi))/k;
rt1=sqrt(a.^2-f^2);
rt2=sqrt(a1.^2-f^2);
fun = @(x) tan((k*sqrt(x.^2+f^2)+l*t)/2);
zz=50;
r=linspace(rt1+0.1,rt2-0.1,zz);
zz=size(r);
zz=zz(2);
for n=1:zz
    y(n)=integral(fun,rt1+0.1,r(n));
    dt=(atan2(y(n),r(n)));
    rr(n)=sqrt(r(n)^2+y(n)^2);
    x1(m,n)=r(n)*cos(t+dt);
    y1(m,n)=r(n)*sin(t+dt);
end
    x2(m)=x1(m,zz);
    y2(m)=y1(m,zz);
    x3(m)=x1(m,1);
    y3(m)=y1(m,1);
if m>3
    if rem(m,2)==0
    xx=[x1(m-1,:)';fliplr(x1(m,:))';x3(m-1)'];
```

```
            yy=[y1(m-1,:)';fliplr(y1(m,:))';y3(m-1)'];
            xy=[xx,yy];
            gs(end+1)=gds_element('boundary', 'xy',xy, 'layer',1);
            end
    end
end
t=0*pi;
for m=1:N
a00=(phi0-l*(t+0.5*pi))/k;
rt00=sqrt(a00.^2-f^2);
t=t+0.2/rt00;
a=(phi0-l*t)/k;
a1=(phi0-l*(t+1*pi))/k;
rt1=sqrt(a.^2-f^2);
rt2=sqrt(a1.^2-f^2);
fun = @(x) tan((k*sqrt(x.^2+f^2)+l*t)/2);
zz=50;
r=linspace(rt1+0.1,rt2-0.1,zz);
zz=size(r);
zz=zz(2);
for n=1:zz
    y(n)=integral(fun,rt1+0.1,r(n));
    dt=(atan2(y(n),r(n)));
    rr(n)=sqrt(r(n)^2+y(n)^2);
    x1(m,n)=r(n)*cos(t+dt);
    y1(m,n)=r(n)*sin(t+dt);
end
x2(m)=x1(m,zz);
y2(m)=y1(m,zz);
x3(m)=x1(m,1);
y3(m)=y1(m,1);
if m>3
    if rem(m,2)==0
    xx=[x1(m-1,:)';fliplr(x1(m,:))';x3(m-1)'];
    yy=[y1(m-1,:)';fliplr(y1(m,:))';y3(m-1)'];
    xy=[xx,yy];
    gs(end+1)=gds_element('boundary', 'xy',xy, 'layer',1);
    end
end
end
glib = gds_library('array', 'uunit',1e-6, 'dbunit',1e-9, gs);
write_gds_library(glib, 'focusl00.gds')
```

B.4 Matlab codes for chiral catenary lens

The parameters are chosen as $|l| = 2$, $f = 10$ μm and $\lambda = 632.8$ nm.

```
clear all
clc
gdsii_units
gs = gds_structure('focuscatenary');
gs1 = gds_structure('focuscatenary');
k=2*pi/0.632;
N=200;
k=2*pi/0.632;
f=10;
phase0=18*2*pi+pi;
l=0;
for p=1:5
a=(phase0+p*2*pi)/k;
a1=(phase0+2*pi+p*2*pi)/k;
rt1=sqrt(a.^2-f^2);
rt2=sqrt(a1.^2-f^2);
phi=0;
N=round(round(pi*(rt1+rt2)/0.2)/2)*2;
for m=1:N
    phi=phi+2*pi/(N-2);
    fun = @(x) tan((k*sqrt(x.^2+f^2)+l*phi)/2);
    zz=50;
    r=linspace(rt1+0.05,rt2-0.05,zz);
    zz=size(r);
    zz=zz(2);
for n=1:zz
    y(n)=-integral(fun,rt1+0.1,r(n));
    dt=(atan2(y(n),r(n)));
    rr(n)=sqrt(r(n)^2+y(n)^2);
    x1(m,n)=r(n)*cos(phi+dt);
    y1(m,n)=r(n)*sin(phi+dt);
end
    x2(m)=x1(m,zz);
    y2(m)=y1(m,zz);
    x3(m)=x1(m,1);
    y3(m)=y1(m,1);
if m>3
    if rem(m,2)==0
    xx=[x1(m-1,:)';fliplr(x1(m,:))';x3(m-1)'];
    yy=[y1(m-1,:)';fliplr(y1(m,:))';y3(m-1)'];
```

```
      xy=[xx,yy];
      gs(end+1)=gds_element('boundary', 'xy',xy, 'layer',1);
      end
end
end
end
glib = gds_library('array', 'uunit',1e-6, 'dbunit',1e-9, gs);
write_gds_library(glib, 'focusl2.gds')
```

B.5 Matlab codes for Bessel beam generation

The parameters are chosen as $l = \pm4$ and $p = 2$ μm.

```
clear all
clc
gdsii_units
gs = gds_structure('besselcatenary');
gs1 = gds_structure('besselcatenary');
phi0=5.0*pi;
r0=phi0/2;
N=300;
p=2;
rinner=3*pi;
router=20*pi/p;
for pp=1:6
N=360+(pp-1)*80;
tt=linspace(0,2*pi+0.012,N);
r0=3.5*pi+(pp-1)*pi;
for m=1:N
t=tt(m);
r=r0+0.3:0.05:r0+pi-0.23;
zz=size(r);
zz=zz(2);
y=real(log(cos(r)));
for n=1:zz

    dt=(atan2(y(n),r(n)));
    rr(n)=sqrt(r(n)^2+y(n)^2);
    x1(m,n)=r(n)*cos(t+dt);
    y1(m,n)=r(n)*sin(t+dt);

end
x2(m)=x1(m,zz);
y2(m)=y1(m,zz);
x3(m)=x1(m,1);
y3(m)=y1(m,1);
if m>3
```

```
    if rem(m,2)==0
    xx=[x1(m-1,:)';fliplr(x1(m,:))';x3(m-1)'];
    yy=[y1(m-1,:)';fliplr(y1(m,:))';y3(m-1)'];
    xy=[xx,yy]*p/pi;
    gs(end+1)=gds_element('boundary', 'xy',xy, 'layer',1);
    end
end
end
end

for pp=1:6
    r0=3.5*pi+(pp-1)*pi;
    r1=r0-0.05;
    r2=r0+0.15;
    phi=tt;
    x1=r1.*cos(phi);
    y1=r1.*sin(phi);
    x2=r2.*cos(phi);
    y2=r2.*sin(phi);
    xx=[x1';fliplr(x2)'];
    yy=[y1';fliplr(y2)'];
    xy=[xx,yy]*p/pi;
    gs(end+1)=gds_element('boundary', 'xy',xy, 'layer',2);
end
glib = gds_library('array', 'uunit',1e-6, 'dbunit',1e-9, gs);
write_gds_library(glib, 'bessel2.gds')
```

B.6 Matlab codes for dielectric catenary grating

```
clear all
clc
tic
S = gds_structure('array');
gs = gds_structure('cate');
un=load('unit_data.txt');
ps0=reshape(un(:,1),9,26)*1e3;
ws0=reshape(un(:,2),9,26)*1e3;
ph0=reshape(un(:,5),9,26);
am0=reshape(un(:,3),9,26);
sp0=116;
a=(ps0-ws0);
w_s=min(min(ws0)):20:max(max(ws0));
p_s=min(min(ps0)):20:max(max(ps0));
[ps,ws]=meshgrid(p_s,w_s);
```

```
ph=interp2(ps0,ws0,ph0,ps,ws,'cubic');
al=30;
N=100;
px=10.6/sind(al)*1e3;
px0=0.75*px;
py=3.3*1e3;
a1=140e3;a2=10e3;b1=0.8e3;b2=0.55e3;
w_min=(b1-b2)/(a1-a2)^2*(px0-a2)^2+b2;
x01=(px0+0.2e3)/2*cos(linspace(-pi,0,N));
y01=px/pi*log(abs(sec(pi*x01/px)));
theta=pi*x01/px;
m01=2.9e4*px^(-1.205)+0.007238;
m1=m01*(cos(theta-pi)+1)+1;
m02=-2681*px^(-0.9634)+0.9956;
m2=(1-m02)*cos(theta)+m02;
dd=py/2*cos(theta).*(m1+m2);
for i=1:N
    cc1=find(abs(dd(i)-p_s)==min(abs(dd(i)-p_s)));
    cc2=find(abs(ph(:,cc1(1))-sp0)==min(abs(ph(:,cc1(1))-sp0)));
    w(i)=ws(cc2(1),cc1(1));
end
d=dd-w;
dw=zeros(1,N);
dw(d<w_min)=w_min-d(d<w_min);
dx=(w-dw).*sin(theta)/2;
dy=-(w-dw).*cos(theta)/2;
x2=fliplr(x01+dx);
y2=fliplr(y01+dy);
x1=x01-dx;
y1=y01-dy;
x0=[x1 x2];
y0=[y1 y2];
y10=interp1(x1,y1,px0/2);
x0(x0>px0/2)=px0/2;x0(x0<-px0/2)=-px0/2;
y0(y0>y10)=y10;
M=1;
N=5;
xmax=px*M;
ymax=py*N;
xy0=cell(M,N);
h=max(y1)-min(y1)-1;
dr=px-px0;
if dr<=4e3&&dr>=2.4e3
```

```
        Mm=1;pr=dr-0.8e3;
elseif dr<2.4e3
        Mm=1;pr=1.5e3;
else
        M01=fix((dr-0.8e3)/1.6e3);M02=fix((dr-0.8e3)/3e3);
        Mm=round((M01+M02)/2);pr=(dr-0.8e3)/Mm;
end
cr1=find(abs(pr-p_s)==min(abs(pr-p_s)));
cr2=find(abs(ph(:,cr1(1))-sp0)==min(abs(ph(:,cr1(1))-sp0)));
wr=ws(cr2(1),cr1(1));
xr=[wr/2 wr/2 -wr/2 -wr/2];
yr=[N*py/2 -N*py/2 -N*py/2 N*py/2]+h;
for m=1:M
        X0=(m)*px;
        XX=X0-M*px/2-px/2;
        for j=1:Mm
            XX0=XX+px0/2+dr/2-(Mm-1)*pr/2+(j-1)*pr;
            xx0=xr+XX0;
            xy0=[xx0',yr'];
            gs(end+1)=gds_element('boundary', 'xy',xy0, 'layer',1);
            end
        for n=1:N
            Y0=(n)*py;
            YY=Y0-N*py/2-py/2;
            x=1*XX+x0;
            y=1*YY+y0;
            xy=[x',y'];
            gs(end+1)=gds_element('boundary', 'xy',xy, 'layer',1);
        end
end
glib = gds_library('array', 'uunit',1e-9, 'dbunit',1e-9, gs);
write_gds_library(glib, 'catenarygrating.gds');
toc
```

B.7 Matlab codes for dielectric catenary lens

```
clear all
clc
tic
close all
un=load('unit_data.txt');
ps0=reshape(un(:,1),9,26)*1e3;
ws0=reshape(un(:,2),9,26)*1e3;
```

```
ph0=reshape(un(:,5),9,26);
am0=reshape(un(:,3),9,26);
sp0=116;
a=(ps0-ws0);
w_s=min(min(ws0)):10:max(max(ws0));
p_s=min(min(ps0)):10:max(max(ps0));
[ps,ws]=meshgrid(p_s,w_s);
ph=interp2(ps0,ws0,ph0,ps,ws,'cubic');
f=10.24e4;
R=12e4;
k=2*pi/10600;
num=round(k*R^2/(2*f*2*pi)/2);
NUM=180;
dph=pi/4;
dp0=-pi/2;
for n=1:NUM
    phi0(n)=(n-1)*2*pi/NUM;
    for m=1:num
        rt1(m,n)=sqrt(4*f*(dp0+pi/2+dph/2+phi0(n)+(2*m-2)*pi)/k);
        rt2(m,n)=sqrt(4*f*(dp0+pi/2+pi-dph/2+phi0(n)+(2*m-2)*pi)/k);
        rt02(m,n)=sqrt(4*f*(dp0+pi/2+pi+phi0(n)+(2*m-2)*pi)/k);
        drt(m,n)=rt2(m,n)-rt1(m,n);
        if m==1
            drt0(m,n)=rt02(m,n)-sqrt(4*f*(dp0+pi/2+phi0(n)+(2*m-2)*pi)/k);
else
    drt0(m,n)=rt02(m,n)-rt02(m-1,n);
 end
 dr(m,n)=sqrt(4*f*(dp0+pi/2+dph/2+phi0(n)+(2*m-1)*pi)/k)-rt2(m,n);
 end
end
for m=1:num
    if m==num
        ll(m)=2*pi*rt1(m,1);
else
    ll(m)=2*pi*rt1(m,1)+pi*(rt1(m+1,1)-rt1(m,1));
 end
end
py=3.3e3;
gdsii_units
gs = gds_structure('catenary');
S = gds_structure('array');
```

```
dot=0;ot0=0;
xx1=0;xx2=0;yy1=0;yy2=0;id=ones(1,100);
for m=1:num
    N=round(4/3*ll(m)/py);
    Q=round(drt(m,1)/2500)+15;
    ot=ot0-dot;
    n0=0;
    for n=1:N
        if ot>=2*pi

            dot=2*pi-(ot-ot0);
            break
end
nn=ceil((ot+0.000001)/(2*pi/NUM));
a1=rt1(m,nn);
a2=rt2(m,nn);
if nn==NUM
    if m==num
        break;
    else
        dx0=mod(ot,(2*pi/NUM))/(2*pi/NUM)*(rt1(m+1,1)-rt1(m,nn));
    end
else
    dx0=mod(ot,(2*pi/NUM))/(2*pi/NUM)*(rt1(m,nn+1)-rt1(m,nn));
    end
    if dr(m,nn)<=4e3&&dr(m,nn)>=2.4e3
        M=1;pr=dr(m,nn)-0.8e3;
    elseif dr(m,nn)<2.4e3
        M=1;pr=1.5e3;
    else
        M01=fix((dr(m,nn)-0.8e3)/1.6e3);M02=fix((dr(m,nn)-0.8e3)/3e3);
        M=round((M01+M02)/2);pr=(dr(m,nn)-0.8e3)/M;
    end
    cr1=find(abs(pr-p_s)==min(abs(pr-p_s)));
    cr2=find(abs(ph(:,cr1(1))-sp0)==min(abs(ph(:,cr1(1))-sp0)));
    wr=ws(cr2(1),cr1(1));
    zr=1;
    for j=1:M
        r=dx0+a2+dr(m,nn)/2-(M-1)*pr/2+(j-1)*pr;
        if id(j)<1
xx1(j,id(j))=(r-wr/2)*cos(ot);
yy1(j,id(j))=(r-wr/2)*sin(ot);
```

```
xx2(j,id(j))=(r+wr/2)*cos(ot);
yy2(j,id(j))=(r+wr/2)*sin(ot);
id(j)=id(j)+1;
else
if mod(n,zr)==0

    xx1(j,id(j))=(r-wr/2)*cos(ot);
    yy1(j,id(j))=(r-wr/2)*sin(ot);
    xx2(j,id(j))=(r+wr/2)*cos(ot);
    yy2(j,id(j))=(r+wr/2)*sin(ot);
    id(j)=id(j)+1;
end
end
end
if nn>n0

    x01=(a2-a1+0.2e3)*(cos(linspace(-pi,0,Q))+1)/2+a1-0.1e3;
    fun=@(x)tan((k*x.^2./(4*f))-phi0(nn)-dp0);
    y01=zeros(1,Q);
    ot0=2*py/(a1+a2);
    for qq=1:Q
        y01(qq)=integral(fun,a1,x01(qq));
end
theta=k*x01.^2./(4*f)-(2*m-1)*pi-phi0(nn)-dp0;
m01=2.9e4*drt0(m,nn)^(-1.205)+0.007238;
m1=m01*(cos(theta-pi)+1)+1;
m02=-2681*drt0(m,nn)^(-0.9634)+0.9956;
m2=(1-m02)*cos(theta)+m02;
dd=0.5*tan(ot0)*x01.*cos(theta).*(m1+m2);
w=zeros(1,Q);
for i=1:Q
    cc1=find(abs(dd(i)-p_s)==min(abs(dd(i)-p_s)));
    cc2=find(abs(ph(:,cc1(1))-sp0)==min(abs(ph(:,cc1(1))-sp0)));
    w(i)=ws(cc2(1),cc1(1));
end
w_min=0;
d=dd-w;
dw=zeros(1,Q);
dw(d<w_min)=w_min-d(d<w_min);
dx=(w-dw).*sin(theta)/2;
dy=-(w-dw).*cos(theta)/2;
x2=fliplr(x01+dx);
```

```
    y2=fliplr(y01+dy);
    x1=x01-dx;
    y1=y01-dy;
    x0=[x1 x2];
    y0=[y1 y2];
    y10=interp1(x1,y1,a1);
    y20=interp1(x1,y1,a2);
    x0(x0>a2)=a2;x0(x0<a1)=a1;
    y0(y0>y10&x0<(a1+a2)/2)=y10; y0(y0>y20&x0>(a1+a2)/2)=y20;
    y0=y0-(max(y1)+min(y2))/2;
        end
        x01=dx0+x0;
        xx=x01*cos(ot)-y0*sin(ot);
        yy=y0*cos(ot)+x01*sin(ot);
        xy=[xx',yy'];
        gs(end+1)=gds_element('boundary', 'xy',xy, 'layer',1);
        ot=ot+ot0;
        n0=nn;
     end
    end
    n_l=length(xx1(:,1));
    for i=1:n_l
        n_p=length(xx1(1,:))-length(intersect(find(xx1(i,:)==0),find(yy1(i,:)==0)));
        nu=ceil(n_p/4000);
        for j=1:nu
            if j<nu
                pp=(1:1:4001)+(j-1)*4000;
                else
                pp=((nu-1)*4000+1):1:n_p;
                end
                x=[xx1(i,pp) fliplr(xx2(i,pp))];
                y=[yy1(i,pp) fliplr(yy2(i,pp))];
                xy0=[x',y'];
                gs(end+1)=gds_element('boundary', 'xy',xy0, 'layer',1);
                end
    end
    ads.angle=180;
    S = add_ref(S,gs,'xy',[0 0],'strans',ads);
    S = add_ref(S,gs);
    kk={};
    kk{end+1}=eval('gs');
    kk{end+1}=eval('S');
```

glib = gds_library('array', 'uunit',1e-9, 'dbunit',1e-9);
glib=add_struct(glib, kk);
write_gds_library(glib,'catenarylens.gds')

B.8 Data for the design of optimal catenary
Filename: "unit_data.txt".

1.5	0.7	0.886531	0.0409639	127.434
1.5	0.8	0.921376	0.0107319	108.146
1.5	0.9	0.947764	0.0274874	86.9907
1.5	1	0.948336	0.0725442	63.785
1.5	1.1	0.903904	0.16413	39.472
1.5	1.2	0.830039	0.273507	15.4336
1.5	1.3	0.752418	0.406055	−9.82625
1.5	1.4	0.601589	0.63958	−45.3195
1.5	1.5	0.00026378	0.840557	−5.85155
1.6	0.7	0.873663	0.0631576	133.862
1.6	0.8	0.907865	0.0190956	116.232
1.6	0.9	0.938539	0.0190379	97.0254
1.6	1	0.952377	0.0490386	76.0078
1.6	1.1	0.933606	0.119524	53.5096
1.6	1.2	0.877018	0.216684	30.8013
1.6	1.3	0.804701	0.318643	8.73124
1.6	1.4	0.730901	0.449805	−15.4285
1.6	1.5	0.570725	0.6702	−49.7945
1.7	0.7	0.863567	0.0822338	139.264
1.7	0.8	0.895371	0.0354284	122.958
1.7	0.9	0.927	0.0155335	105.324
1.7	1	0.948881	0.034915	86.0824
1.7	1.1	0.947603	0.0870003	65.5388
1.7	1.2	0.911082	0.172779	43.9313
1.7	1.3	0.847613	0.266118	23.0308
1.7	1.4	0.779536	0.363604	2.43725
1.7	1.5	0.706554	0.493698	−20.9105
1.8	0.7	0.85475	0.0990508	143.864
1.8	0.8	0.885717	0.0499741	128.801
1.8	0.9	0.916995	0.0208267	112.344
1.8	1	0.942059	0.0279181	94.5119
1.8	1.1	0.951814	0.0666907	75.4085
1.8	1.2	0.932695	0.13809	55.2221
1.8	1.3	0.883084	0.22483	35.2053
1.8	1.4	0.818462	0.313131	15.8916
1.8	1.5	0.755484	0.406954	−3.47203

(continued)

(continued)

1.9	0.7	0.848246	0.114694	147.86
1.9	0.8	0.876174	0.0642482	133.539
1.9	0.9	0.90736	0.0305542	118.254
1.9	1	0.933647	0.0246709	101.66
1.9	1.1	0.950701	0.0531456	83.858
1.9	1.2	0.944467	0.11191	64.9584
1.9	1.3	0.908601	0.193269	45.5746
1.9	1.4	0.850774	0.277669	27.0252
1.9	1.5	0.790005	0.35949	9.20828
2	0.7	0.843777	0.128755	151.43
2	0.8	0.868857	0.0765928	137.769
2	0.9	0.897407	0.038873	123.39
2	1	0.926055	0.0255898	107.759
2	1.1	0.946771	0.0443098	91.0017
2	1.2	0.950547	0.0944561	73.0803
2	1.3	0.925915	0.167263	54.6386
2	1.4	0.877612	0.248748	36.7121
2	1.5	0.819692	0.326179	19.6093
2.1	0.7	0.838975	0.140788	154.614
2.1	0.8	0.861935	0.0882095	141.447
2.1	0.9	0.889721	0.0481753	127.76
2.1	1	0.9177	0.0293733	113.009
2.1	1.1	0.941243	0.0386973	97.1454
2.1	1.2	0.951423	0.0780348	80.2842
2.1	1.3	0.938379	0.145295	62.5807
2.1	1.4	0.898887	0.224522	45.1084
2.1	1.5	0.843884	0.302552	28.3174
2.2	0.7	0.836734	0.151993	157.465
2.2	0.8	0.856085	0.0980641	144.666
2.2	0.9	0.882843	0.0563613	131.636
2.2	1	0.911203	0.0330857	117.673
2.2	1.1	0.935038	0.0346813	102.653
2.2	1.2	0.949702	0.0663304	86.4974
2.2	1.3	0.946072	0.127084	69.5313
2.2	1.4	0.914605	0.20508	52.2422
2.2	1.5	0.865529	0.280614	36.1117
2.3	0.7	0.834358	0.162345	160.076
2.3	0.8	0.853031	0.105783	147.685
2.3	0.9	0.877212	0.0624701	135.043
2.3	1	0.904189	0.0370045	121.562
2.3	1.1	0.929424	0.031783	107.369

(continued)

(continued)

2.3	1.2	0.947725	0.0579472	91.7949
2.3	1.3	0.949209	0.111482	75.5425
2.3	1.4	0.926852	0.185044	58.8259
2.3	1.5	0.881337	0.264621	42.5917
2.4	0.7	0.832281	0.172675	162.313
2.4	0.8	0.848978	0.112797	150.205
2.4	0.9	0.87259	0.0694248	137.96
2.4	1	0.898506	0.0394201	125.214
2.4	1.1	0.924499	0.0310203	111.426
2.4	1.2	0.9435	0.0502868	96.5215
2.4	1.3	0.95085	0.0974159	80.8715
2.4	1.4	0.934535	0.170548	64.4806
2.4	1.5	0.895086	0.248624	48.5235
2.5	0.7	0.831599	0.181209	164.405
2.5	0.8	0.846259	0.119268	152.577
2.5	0.9	0.865987	0.0710931	140.968
2.5	1	0.892456	0.0398438	128.48
2.5	1.1	0.919296	0.0301979	114.899
2.5	1.2	0.940968	0.0426801	100.791
2.5	1.3	0.950349	0.0840036	85.5286
2.5	1.4	0.941086	0.154001	69.5206
2.5	1.5	0.906997	0.232516	53.702
2.6	0.7	0.83059	0.188141	166.336
2.6	0.8	0.844452	0.124002	154.677
2.6	0.9	0.864424	0.075802	143.104
2.6	1	0.888948	0.0431996	130.946
2.6	1.1	0.915231	0.0279735	118.114
2.6	1.2	0.938111	0.0372361	104.253
2.6	1.3	0.95166	0.076114	89.4204
2.6	1.4	0.945227	0.138782	73.9617
2.6	1.5	0.917784	0.217519	58.3353
2.7	0.7	0.830099	0.195804	168.069
2.7	0.8	0.842535	0.128573	156.526
2.7	0.9	0.86302	0.0796965	145.064
2.7	1	0.884792	0.042365	133.428
2.7	1.1	0.91111	0.0250268	120.821
2.7	1.2	0.934058	0.0302169	107.397
2.7	1.3	0.950468	0.064299	92.9644
2.7	1.4	0.949828	0.122866	77.9206
2.7	1.5	0.924301	0.202402	62.4117
2.8	0.7	0.829812	0.202537	169.587

(continued)

(continued)

2.8	0.8	0.84165	0.132886	158.243
2.8	0.9	0.861906	0.0809987	146.987
2.8	1	0.882527	0.0437928	135.322
2.8	1.1	0.907909	0.0214381	123.146
2.8	1.2	0.931481	0.0241505	110.123
2.8	1.3	0.947538	0.0513055	96.1784
2.8	1.4	0.95212	0.108415	81.2663
2.8	1.5	0.93213	0.187903	65.9051
2.9	0.7	0.83081	0.207096	171.081
2.9	0.8	0.842036	0.137631	159.453
2.9	0.9	0.85876	0.0823477	148.663
2.9	1	0.88177	0.0423629	137.177
2.9	1.1	0.9065	0.0167538	125.26
2.9	1.2	0.929048	0.0158434	112.556
2.9	1.3	0.947364	0.0403958	98.7241
2.9	1.4	0.953113	0.0943324	84.1676
2.9	1.5	0.937119	0.171849	68.9952
3	0.7	0.830733	0.212637	172.295
3	0.8	0.842292	0.138964	160.751
3	0.9	0.858976	0.0811716	150.016
3	1	0.881546	0.0390815	138.774
3	1.1	0.905086	0.011735	126.917
3	1.2	0.928446	0.00983448	114.383
3	1.3	0.947872	0.0289095	100.922
3	1.4	0.954917	0.0781608	86.716
3	1.5	0.941849	0.155345	71.5418
3.1	0.7	0.832742	0.214829	173.363
3.1	0.8	0.844107	0.136321	161.937
3.1	0.9	0.860473	0.078406	150.991
3.1	1	0.882655	0.0330875	139.888
3.1	1.1	0.906094	0.00390418	128.1
3.1	1.2	0.92792	0.0129143	115.698
3.1	1.3	0.946626	0.0189388	102.373
3.1	1.4	0.955843	0.0615023	88.2648
3.1	1.5	0.945913	0.13679	73.4111
3.2	0.7	0.829468	0.214261	173.894
3.2	0.8	0.839356	0.128826	162.479
3.2	0.9	0.852802	0.0670342	151.322
3.2	1	0.869058	0.0207074	139.983
3.2	1.1	0.888628	0.0198708	128.346
3.2	1.2	0.905046	0.0335902	116.084

(continued)

(continued)

3.2	1.3	0.918838	0.0319908	103.031
3.2	1.4	0.926875	0.0535227	89.1061
3.2	1.5	0.917772	0.121351	74.3765
3.3	0.7	0.821951	0.215344	174.967
3.3	0.8	0.831344	0.131893	163.224
3.3	0.9	0.84189	0.0681929	152.347
3.3	1	0.854664	0.023443	141.33
3.3	1.1	0.869517	0.0280541	129.95
3.3	1.2	0.884266	0.0419522	117.944
3.3	1.3	0.898348	0.0412012	105.187
3.3	1.4	0.90707	0.0523966	91.5464
3.3	1.5	0.90151	0.11015	77.1152
3.4	0.7	0.819705	0.217043	176.015
3.4	0.8	0.826435	0.130373	164.314
3.4	0.9	0.832129	0.0679038	153.364
3.4	1	0.846278	0.0285279	142.679
3.4	1.1	0.857429	0.0348655	131.458
3.4	1.2	0.870469	0.0486488	119.699
3.4	1.3	0.882481	0.0489175	107.401
3.4	1.4	0.893398	0.0510243	93.777
3.4	1.5	0.889608	0.101999	79.4691
3.5	0.7	0.818391	0.220763	177.046
3.5	0.8	0.822348	0.131198	165.444
3.5	0.9	0.827109	0.0669355	154.628
3.5	1	0.835917	0.0305573	144.119
3.5	1.1	0.848313	0.0397344	132.991
3.5	1.2	0.861089	0.0570653	121.416
3.5	1.3	0.871976	0.0562021	109.112
3.5	1.4	0.881352	0.0509946	96.321
3.5	1.5	0.881081	0.0929234	81.5603
3.6	0.7	0.815394	0.222624	177.766
3.6	0.8	0.817227	0.132725	166.064
3.6	0.9	0.82383	0.0686983	155.499
3.6	1	0.829287	0.0346574	144.827
3.6	1.1	0.839631	0.0454899	134.044
3.6	1.2	0.851454	0.0617857	122.737
3.6	1.3	0.861941	0.0625698	110.656
3.6	1.4	0.872234	0.0549132	97.7306
3.6	1.5	0.873242	0.0848341	83.5117
3.7	0.7	0.811888	0.227692	178.446
3.7	0.8	0.813608	0.131843	166.974

(continued)

(continued)

3.7	0.9	0.817653	0.0693935	156.313
3.7	1	0.821652	0.0380721	145.888
3.7	1.1	0.832237	0.0525707	135.394
3.7	1.2	0.842354	0.0679953	124.097
3.7	1.3	0.854351	0.0702366	112.273
3.7	1.4	0.864219	0.0598955	99.3574
3.7	1.5	0.86653	0.0765994	85.6968
3.8	0.7	0.811087	0.226446	179.509
3.8	0.8	0.808836	0.136582	167.391
3.8	0.9	0.812831	0.0682814	157.346
3.8	1	0.816519	0.0407031	146.772
3.8	1.1	0.823311	0.0563715	136.512
3.8	1.2	0.834022	0.0748716	125.522
3.8	1.3	0.845868	0.0776033	113.692
3.8	1.4	0.855742	0.066005	101.073
3.8	1.5	0.860709	0.0742835	87.277
3.9	0.7	0.807641	0.229648	179.99
3.9	0.8	0.809745	0.134412	168.273
3.9	0.9	0.810515	0.0697742	158.136
3.9	1	0.813273	0.0447206	147.774
3.9	1.1	0.817298	0.0619912	137.536
3.9	1.2	0.827257	0.0792489	126.457
3.9	1.3	0.837862	0.0845356	114.998
3.9	1.4	0.849943	0.0732626	102.48
3.9	1.5	0.854815	0.0739364	88.5397
4	0.7	0.806306	0.232526	180.691
4	0.8	0.805471	0.133375	169.06
4	0.9	0.806068	0.0702294	158.582
4	1	0.809041	0.0490083	148.488
4	1.1	0.814275	0.0663045	138.397
4	1.2	0.821826	0.0863447	127.568
4	1.3	0.829823	0.0909976	116.268
4	1.4	0.842561	0.0790628	103.717
4	1.5	0.848853	0.072802	89.9051

Appendix C
Matlab Codes for Vectorial Diffraction

The parameters in the following code correspond to that shown in Fig. 4d in Ref. [2].

```
lambda=0.632;
N=1000;
k0=2*pi/lambda;
lmax=20;
x=linspace(-lmax/2,lmax/2,N);
y=linspace(-lmax/2,lmax/2,N);
kmax=pi*N/lmax;
[X,Y]=meshgrid(x,y);
phi=atan2(Y,X);
r=sqrt(X.^2+Y.^2);
p=3;
w=0.3;
g=zeros(N,N);
xindex_min=round(N/2-0.5*p*N/lmax)+2;
xindex_max=round(N/2+0.5*p*N/lmax)-2;
for ind=xindex_min:xindex_max
xx=x(ind);
if xx<=0
yy1=p/pi*log(abs(cos(pi*xx/p)));
yy2=yy1+w;
else
yy1=-p/pi*log(abs(cos(pi*xx/p)));
yy2=yy1+w;
end
gg=zeros(1,N);
ig=find(y<yy2 & y>yy1);
gg(ig)=1;
```

© Springer Nature Singapore Pte Ltd. 2019
X. Luo, *Catenary Optics*, https://doi.org/10.1007/978-981-13-4818-1

```
g(:,ind)=gg;
end
g=g.*exp(-r.^2/2^2);
sigma=1;
Exi=g.*(1*exp(-0.52i*(X)/p-1i*sigma*2*pi/p*X.*sign(X)));
Eyi=g.*(1i*exp(0.52i*(X)/p-1i*sigma*2*pi/p*X.*sign(X)));
%%% To analyze the numerical results exported from numerical packages such
as FDTD,
%%% the imported fields should be loaded here.
Ax=fftshift(fft2((Exi)));
Ay=fftshift(fft2((Eyi)));
mm=linspace(-kmax-pi/lmax,kmax-pi/lmax,N)/(2*pi);
nn=linspace(-kmax-pi/lmax,kmax-pi/lmax,N)/(2*pi);
[MM,NN]=meshgrid(mm,nn);
q=sqrt(1/lambda^2-MM.^2-NN.^2);
Az=-(MM.*Ax+NN.*Ay)./q;
zz=0.5;
qq=exp(1j*2*pi*q*zz);
Ex=ifft2(ifftshift(Ax.*qq));
Ey=ifft2(ifftshift(Ay.*qq));
Ez=ifft2(ifftshift(Az.*qq));
I=abs(Ey).^2+abs(Ex).^2+0*abs(Ez).^2;
I=imrotate(I,0);
h=figure(6);
imagesc(x,y,I)
axis equal
axis xy
axis([-3,3,-3,3])
colormap wavelight
%%%%%%%%The Matlab function for wavelight is as follows:
function h = wavelight(m)
if nargin < 1, m = size(get(gcf,'colormap'),1); end
n = fix(1/2*m);
n1=m-n;
% r = [(1:n)'/n; ones(m-n,1)];
% g = [(1:n)'/n; ones(n,1); flip(1:n1)'/n1];
% b = [ones(n,1); ones(n,1); flip(1:n1)'/n1];
r = [(1:n)'/n; ones(n1,1)];
g = [(1:n)'/n; flip(1:n1)'/n1];
b = [ones(n,1); flip(1:n1)'/n1];
h = [r g b];
```

Appendix D
Quasi-Stationary Catenary Optical Fields

Figure D.1a is the schematic of an array of metallic slits. A transverse electric plane wave is normally incident on this array. In order to deduce the explicit functions to describe this subwavelength structure, a modified treatment is applied as shown in Fig. D.1b that PEC sheets are inserted in the metallic films. Owing to the symmetry of the structure, the insertion of PEC sheets parallel to y-axis and perpendicular to the plane of the slits through the central lines of the films does not disturb the electric field. Therefore, the electric field of the whole array can be divided into numerous identical fields such as (I), (II), and (III) in Fig. D.1b. If one can obtain the models for any of the small unit, they can be applied to other units as well. Thus, to simplify the case, we choose unit (II) in Fig. D.1b as an example, where the field is constrained from $x = -p/2$ to $x = p/2$.

As the electric field vibrates perpendicularly to the edges of the slits, the structure behaves as a purely capacitive device. First, let us calculate the background capacity of the modified structure that we remove the metallic slits and maintain the PEC sheets as shown in Fig. D.1c. In this case, the plane wave propagates in the negative direction of y-axis between two parallel PEC sheets. If we further constrain the space in y-axis direction from $y = -l$ to $y = l$, the capacity of the structure in Fig. D.1c is

$$C'' = \frac{2\kappa l}{p}, \tag{D.1}$$

where $\kappa = p\varepsilon_0$ is the characteristic capacity. Then we return to the case as shown in Fig. D.1d. The existence of the slits will distort the original electric field distributions and the influence can be described by the theory of Schwarz's conformal transformations in terms of equipotentials V and line of force U at (x, y) plane:

$$\sin(V + iU) = \csc(\pi w/2p)\sin[(x + iy)\pi/p]. \tag{D.2}$$

© Springer Nature Singapore Pte Ltd. 2019
X. Luo, *Catenary Optics*, https://doi.org/10.1007/978-981-13-4818-1

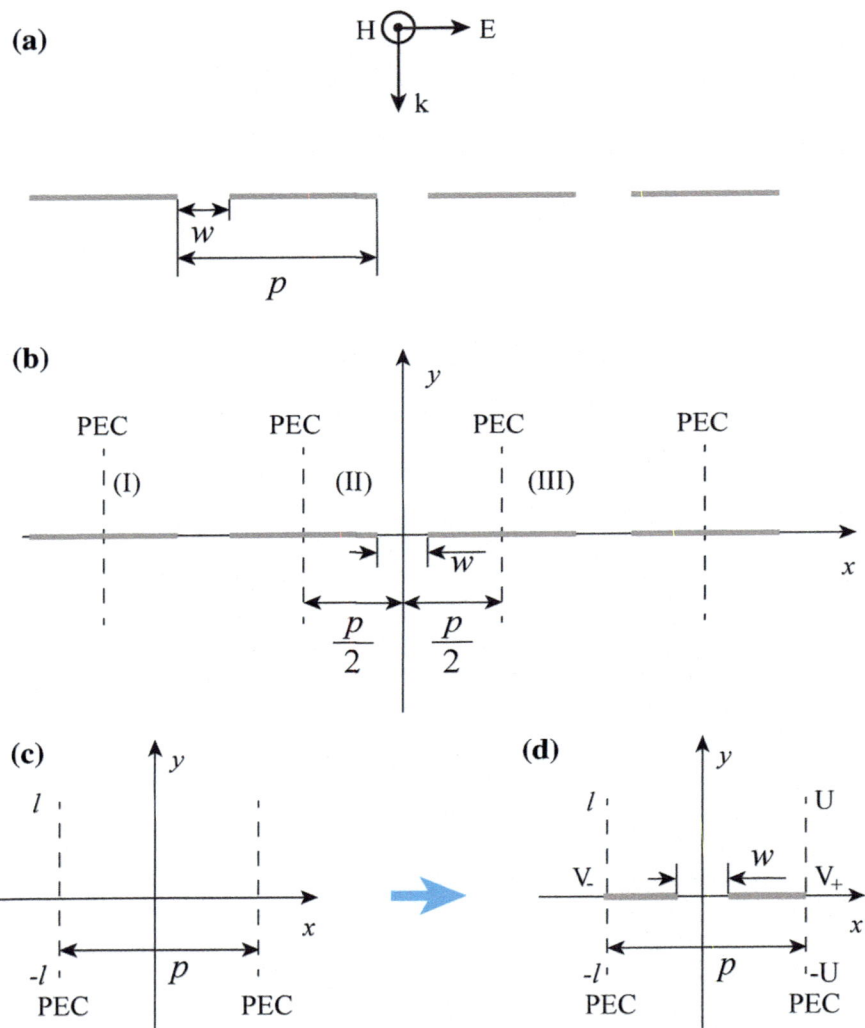

Fig. D.1 **a** Schematic of an array of metallic slits. **b** The modified treatment to the original structure that perfect electrical conducting (PEC) sheets are inserted in the central lines of the metallic films. **c** A single unit cell in (**b**) with metallic slits removed. **d** A single unit cell as shown in (**b**)

By employing this function, the field distributions in x-y plane can be calculated. As the potential difference between two identical conductors is $V = \pi$, we assume that the potentials at $x = \pm p/2$ are equal to $V_{\pm} = \pm\pi/2$. Then, at the boundary of $x = p/2$, Eq. (D.2) can be calculated as

$$\sin\left(\frac{\pi}{2} + iU\right) = \csc\left(\frac{\pi w}{2p}\right)\sin\left(\frac{\pi}{2} + i\frac{y\pi}{p}\right), \tag{D.3}$$

which can be simplified as

$$\cos(iU) = \csc\left(\frac{\pi w}{2p}\right)\cos\left(i\frac{y\pi}{p}\right). \tag{D.4}$$

As $\cos(ia) = \cosh(a)$, U can be expressed as

$$U(y) = \pm\cosh^{-1}\left[\csc\left(\frac{\pi w}{2p}\right)\cosh\left(\frac{y\pi}{p}\right)\right]. \tag{D.5}$$

Next, the total charge Q on the right PEC sheet can be calculated by the characteristic capacity $\kappa = p\varepsilon_0$ times U at the limits of the space from $y = -l$ to $y = l$.

$$Q = \kappa[U(l) - U(-l)] = 2\kappa U(l). \tag{D.6}$$

Thus, the capacity of the structure in Fig. D.1d can be calculated by

$$C' = \frac{Q}{V} = \frac{2\kappa}{\pi}\cosh^{-1}\left[\csc\left(\frac{\pi w}{2p}\right)\cosh\left(\frac{\pi l}{p}\right)\right]. \tag{D.7}$$

By using the equation

$$\cosh^{-1}x = \ln\left(x + \sqrt{x^2 - 1}\right), \tag{D.8}$$

the difference in capacity (due to the existence of the slits) can be deduced by

$$
\begin{aligned}
C = \lim_{l\to\infty}(C' - C'') &= \lim_{l\to\infty}\left[\frac{2\kappa}{\pi}\ln\left(2\csc\left(\frac{\pi w}{2p}\right)\cosh\left(\frac{\pi l}{p}\right)\right) - \frac{2\kappa l}{p}\right] \\
&= \lim_{l\to\infty}\left[\frac{2\kappa}{\pi}\ln\left(\csc\left(\frac{\pi w}{2p}\right)\right) + \frac{2\kappa}{\pi}\ln\left(2\cosh\left(\frac{\pi l}{p}\right)\right) - \frac{2\kappa l}{p}\right] \\
&= \frac{2\kappa}{\pi}\ln\left(\csc\left(\frac{\pi w}{2p}\right)\right) + \frac{2\kappa}{\pi}\ln\left(\exp\left(\frac{\pi l}{p}\right)\right) - \frac{2\kappa l}{p} \\
&= \frac{2\kappa}{\pi}\ln\left(\csc\left(\frac{\pi w}{2p}\right)\right) + \frac{2\kappa}{\pi}\frac{\pi l}{p} - \frac{2\kappa l}{p} \\
&= \frac{2\kappa}{\pi}\ln\left(\csc\left(\frac{\pi w}{2p}\right)\right).
\end{aligned} \tag{D.9}
$$

The corresponding impedance of the structure can be expressed by

$$Z_1 = 1/i\omega C = \frac{1}{i\frac{4p}{\lambda}\ln\csc\left(\frac{\pi w}{2p}\right)}, \tag{D.10}$$

where $\omega = 2\pi c/\lambda$ is the angular frequency in the operation band, c is the vacuum light velocity. After solving the case under transverse electric plane wave, the equivalent circuit model under the transverse magnetic wave can be easily obtained by applying the Babinet's principle that the product of the impedance for complementary slits equal to one-quarter the square of the vacuum impedance. Thus, the model in this case is

$$Z_2 = i\frac{p}{\lambda}\ln\csc\left(\frac{\pi w}{2p}\right). \tag{D.11}$$

When the metallic grating is surrounded by dielectrics, the capacitive impedance should be revised as

$$Z_1 = \frac{1}{i\frac{2p\left(n_1^2 + n_2^2\right)}{\lambda}\ln\csc\frac{\pi w}{2p}}. \tag{D.12}$$

Note that the impedance of inductive sheet is independent of the dielectric constant.

It should be noted that w is the length of a metallic film in this case instead of the width of the slit.

From the Schwarz's conformal transformation shown by Eq. (D.2), the electric field in the center ($y = 0$, $U = 0$) of the slit can also be obtained

$$|E| = \frac{\partial V}{\partial x}. \tag{D.13}$$

Considering the facts

$$\sin V = \csc(\pi w/2p)\sin(x\pi/p) \tag{D.14}$$

and

$$\cos V = \sqrt{1 - \csc^2(\pi w/2p)\sin^2(x\pi/p)}, \tag{D.15}$$

There is

$$
\begin{aligned}
|E| = \frac{\partial V}{\partial x} &= \frac{\csc(\pi w/2p)\cos(x\pi/p)\pi/p}{\cos V} \\
&= \frac{\cos(x\pi/p)\pi/p}{\sqrt{\sin^2(\pi w/2p) - \sin^2(x\pi/p)}} \\
&= \frac{\sqrt{2}\pi\cos(x\pi/p)}{p\sqrt{\cos(2\pi x/p) - \cos(w\pi/p)}}.
\end{aligned}
\tag{D.16}
$$

This equation can be further expanded to a polynomial form as

$$
E(x) = A + Bx^2 + Cx^4 + o(x^6),
\tag{D.17}
$$

where A, B, and C are the coefficients in terms of p and w that can be obtained by Taylor expansion, and o is the high-order term that can be omitted. Interestingly, the catenary function can be also expanded to the same mathematical form that $E(x)$ can be transformed into a typical catenary function

$$
E(x) = a\cosh(bx) + c,
\tag{D.18}
$$

where

$$
a = \frac{2\pi\sin\left(\frac{\pi w}{2p}\right)\cot^2\left(\frac{\pi w}{2p}\right)}{p\left[17 + \cos\left(\frac{\pi w}{p}\right)\right]},
\tag{D.19}
$$

$$
b = \frac{\pi\cot\left(\frac{\pi w}{2p}\right)}{P\sqrt{\frac{1 + \cos\left(\frac{\pi w}{p}\right)}{17 + \cos\left(\frac{\pi w}{p}\right)}}},
\tag{D.20}
$$

$$
c = \frac{\pi}{p\sin\left(\frac{\pi w}{2p}\right)} - \frac{2\pi\sin\left(\frac{\pi w}{2p}\right)\cot^2\left(\frac{\pi w}{2p}\right)}{p\left[17 + \cos\left(\frac{\pi w}{p}\right)\right]}.
\tag{D.21}
$$

The above catenary function is a little complicated and can be simplified to an experiential expression when $w \ll p$:

$$
|E| \approx \frac{1}{2\pi w}\cosh\left(\frac{7\cos(\pi w/(2p))}{w\cos(\pi w/(3p))}x\right) + \frac{\pi}{p\sin(\pi w/(2p))} - \frac{1}{2\pi w}.
\tag{D.22}
$$

Appendix E
Matlab Codes for the Transfer Matrix Method

Note: The parameters are corresponding to the curve shown in Fig. 4a (at an incidence angle of 60°) in Ref. [3].

```
clear all
close all
clc
f0=linspace(0,20,1000)*1e9;
tm=1;
sc=10;
theta=60;
theta=theta*pi/180;
ep=1;
Z0=377;
Y0=1/Z0;
mu=1;
g=5e-3;
a=0.5*0.5e-3;
d1=sc*1e-3;
d2=d1;
d3=d1;
d4=d1;
d5=d1;
d6=d1;
d7=d1;
d8=d1;
d9=d1;
for m=1:980
```

© Springer Nature Singapore Pte Ltd. 2019
X. Luo, *Catenary Optics*, https://doi.org/10.1007/978-981-13-4818-1

```
        f(m)=f0(m);
        lambda=3e8/f(m);
        omega=2*pi*f(m);
        k=omega/3e8;
        kk(m)=k;
        ky=k*sin(theta);
        kx0=sqrt(k^2-ky^2);
        kx1=sqrt(ep*mu*k^2);
   if tm==1
        Y00=Y0*k/kx0;
        Yi=Y0*ep*k/kx1;
   else
        Y00=Y0*kx0/k;
        Yi=Y0*kx1/k;
   end
        a1=0.031;
        b1=0.023;
        c1=0.918;
        abc=a1*cos(2*pi*2*a/g)+b1*sin(2*pi*2*a/g)+c1;
        Z1=abc*2/(1+ep)*1i*Z0*(4*g*log(csc(pi*a/g)))./lambda).^(-1);
        Z11=abc*2/(ep+ep)*1i*Z0*(4*g*log(csc(pi*a/g)))./lambda).^(-1);
        Y1=1/Z1;
        Y2=1/Z11;
        Y3=1/Z11;
        Y4=1/Z11;
        Y5=1/Z11;
        Yf=1/Z1;
        v=[1;0];
        v=0.5/Yi*[Yi+Y00+Y1 Yi-Y00+Y1;Yi-Y00-Y1 Yi+Y00-Y1]*v;
        v=[exp(-i*kx1*d1) 0;0 exp(i*kx1*d1)]*v;
        v=0.5*[2+Y2/Yi Y2/Yi;-Y2/Yi 2-Y2/Yi]*v;
        v=[exp(-i*kx1*d2) 0;0 exp(i*kx1*d2)]*v;
        v=0.5*[2+Y3/Yi Y3/Yi;-Y3/Yi 2-Y3/Yi]*v;
        v=[exp(-i*kx1*d3) 0;0 exp(i*kx1*d3)]*v;
        v=0.5*[2+Y4/Yi Y4/Yi;-Y4/Yi 2-Y4/Yi]*v;
        v=[exp(-i*kx1*d4) 0;0 exp(i*kx1*d4)]*v;
        v=0.5*[2+Y5/Yi Y5/Yi;-Y5/Yi 2-Y5/Yi]*v;
        v=[exp(-i*kx1*d5) 0;0 exp(i*kx1*d5)]*v;
        v=0.5/Y00*[Y00+Yi+Yf Y00-Yi+Yf;Y00-Yi-Yf Y00+Yi-Yf]*v;
        v1=v(1);
        v2=v(2);
        r(m)=(v2/v1);
        t(m)=(1/v1);
```

```
end
f=f*1e-9;
figure(1)
plot(f,abs(r).^2,f,abs(t).^2)
hold on; grid;
xlabel('Frequency [GHz]');ylabel('S Parameters')
```

Appendix F
Gallery of Pictures for Catenary

This section illustrates some interesting figures related to the catenary functions (Figs. F.1, F.2, F.3, F.4, F.5, F.6, F.7, F.8, F.9, F.10, F.11).

Fig. F.1 Cutaway of a nautilus shell showing the chambers arranged in an approximately logarithmic spiral. The dissepiments resemble asymmetric catenaries. Reproduced from Wikipedia [4]

© Springer Nature Singapore Pte Ltd. 2019   413
X. Luo, *Catenary Optics*, https://doi.org/10.1007/978-981-13-4818-1

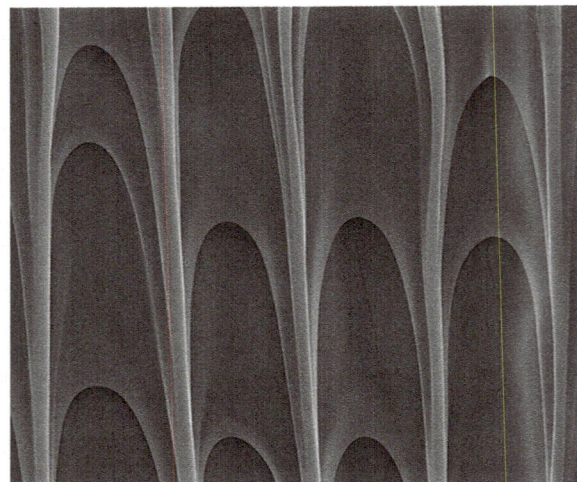

Fig. F.2 SEM image of the microgrooves of peristome surface of the tropical plant *Nepenthes alata*, each containing periodic duck-billed microcavities with arch-shaped edges. These arches are very similar to the catenary of equal strength. Reproduced from [5] with permission

Fig. F.3 Catenaries and their reflection images taken at the Century Hotel, Chengdu

Fig. F.4 The catenary curve formed by a cane. Note that there is also a straight line

Fig. F.5 U-shaped valley in Leh valley, Ladakh, NW Indian Himalaya. The glacier visible at the head of the valley is the last remnants of the formerly much more extensive valley glacier which carved this valley. Reproduced from Wikipedia [6]

Fig. F.6 Landscape Arch is the longest of the many natural rock arches located in the Arches National Park in Utah, United States. The arch is among many in the area known as Devil's Garden in the north area of the park. It was named by Frank Beckwith, leader of the Arches National Monument Scientific Expedition, who explored the area in the winter of 1933–1934. Reproduced from Wikipedia [7]

Fig. F.7 Sails from ancient Egypt featuring catenary curves. Sails from ancient Egypt date back to 3200 BCE, where reed boats sailed upstream against the River Nile's current. Reproduced from Wikipedia [8]

Fig. F.8 Main Terminal of Washington Dulles International Airport at dusk in Virginia. The roof is a suspended catenary curve. Reproduced from Wikipedia [9]

Fig. F.9 A photograph of Terminal 2 of Shuangliu Airport, Chengdu

Fig. F.10 The oldest operating McDonald's restaurant was the third one built, opening in 1953. It is located at 10207 Lakewood Blvd. at Florence Ave. in Downey, California. Reproduced from Wikipedia [10]

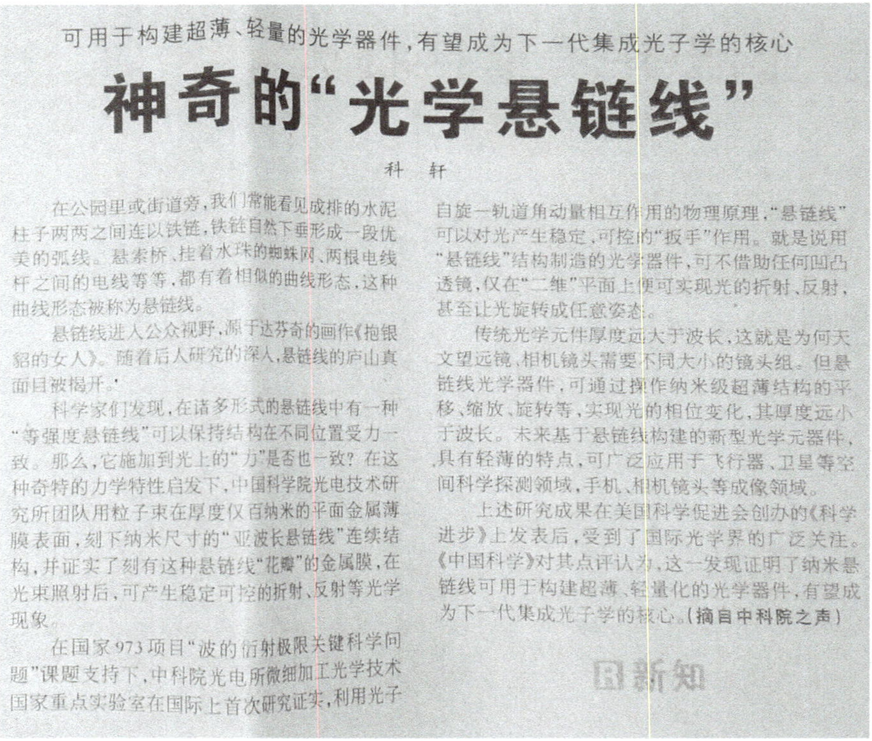

可用于构建超薄、轻量的光学器件，有望成为下一代集成光子学的核心

神奇的"光学悬链线"

科 轩

在公园里或街道旁，我们常能看见成排的水泥柱子两两之间连以铁链，铁链自然下垂形成一段优美的弧线。悬索桥、挂着水珠的蜘蛛网、两根电线杆之间的电线等等，都有着相似的曲线形态，这种曲线形态被称为悬链线。

悬链线进入公众视野，源于达芬奇的画作《抱银貂的女人》。随着后人研究的深入，悬链线的庐山真面目被揭开：

科学家们发现，在诸多形式的悬链线中有一种"等强度悬链线"可以保持结构在不同位置受力一致。那么，它施加到光上的"力"是否也一致？在这种奇特的力学特性启发下，中国科学院光电技术研究所团队用粒子束在厚度仅百纳米的平面金属薄膜表面，刻下纳米尺寸的"亚波长悬链线"连续结构，并证实了刻有这种悬链线"花瓣"的金属膜，在光束照射后，可产生稳定可控的折射、反射等光学现象。

在国家973项目"波的衍射极限关键科学问题"课题支持下，中科院光电所微细加工光学技术国家重点实验室在国际上首次研究证实，利用光子自旋一轨道角动量相互作用的物理原理，"悬链线"可以对光产生稳定，可控的"扳手"作用。就是说用"悬链线"结构制造的光学器件，可不借助任何凹凸透镜，仅在"二维"平面上便可实现光的折射、反射，甚至让光旋转成任意姿态。

传统光学元件厚度远大于波长，这就是为何天文望远镜，相机镜头需要不同大小的镜头组。但悬链线光学器件，可通过操作纳米级超薄结构的平移、缩放、旋转等，实现光的相位变化，其厚度远小于波长。未来基于悬链线构建的新型光学元件器件，具有轻薄的特点，可广泛应用于飞行器、卫星等空间科学探测领域，手机、相机镜头等成像领域。

上述研究成果在美国科学促进会创办的《科学进步》上发表后，受到了国际光学界的广泛关注。《中国科学》对其点评认为，这一发现证明了纳米悬链线可用于构建超薄、轻量化的光学器件，有望成为下一代集成光子学的核心。（摘自中科院之声）

Fig. F.11 A photograph of news reports in Renmin Ribao (People's Daily) at September 12, 2016. The English translation of the title is *Magical "Optical Catenaries"*

References

1. Catenary. http://www.en.wikipedia.org/wiki/Catenary
2. X. Luo, M. Pu, X. Li, X. Ma, Broadband spin Hall effect of light in single nanoapertures. Light Sci. Appl. **6**, e16276 (2017)
3. M. Pu, X. Ma, Y. Guo, X. Li, X. Luo, Theory of microscopic meta-surface waves based on catenary optical fields and dispersion. Opt. Express **26**, 19555–19562 (2018)
4. Nautilus. https://en.wikipedia.org/wiki/Nautilus
5. H. Chen, P. Zhang, L. Zhang, H. Liu, Y. Jiang, D. Zhang, Z. Han, L. Jiang, Continuous directional water transport on the peristome surface of Nepenthes alata. Nature **532**, 85–89 (2016)
6. U shaped valley. http://www.en.wikipedia.org/wiki/U_shaped_valley
7. Landscape Arch. https://en.wikipedia.org/wiki/Catenary/Landscape_Arch
8. Sail. https://en.wikipedia.org/wiki/Sail
9. Dulles Airport. https://en.wikipedia.org/wiki/Washington_Dulles_International_Airport
10. McDonald's. https://en.wikipedia.org/wiki/McDonald's

© Springer Nature Singapore Pte Ltd. 2019
X. Luo, *Catenary Optics*, https://doi.org/10.1007/978-981-13-4818-1

Printed by Printforce, the Netherlands